112630

WITHDRAWN

BEST MANAGEMENT PRACTICES

FOR AGRICULTURE AND SILVICULTURE

Proceedings of the 1978 Cornell Agricultural
Waste Management Conference

Edited By

RAYMOND C. LOEHR, Director
Environmental Studies Program

DOUGLAS A. HAITH, Associate Professor
Department of Agricultural Engineering

MICHAEL F. WALTER, Assistant Professor
Department of Agricultural Engineering

COLLEEN S. MARTIN, Administrative Aide
Environmental Studies Program

**College of Agriculture and Life Sciences
Cornell University, Ithaca, New York**

ANN ARBOR SCIENCE
PUBLISHERS INC
P.O. BOX 1425 • ANN ARBOR, MICH. 48106

TD
223
C69
1978

Copyright © 1979 by Ann Arbor Science Publishers, Inc.
230 Collingwood, P. O. Box 1425, Ann Arbor, Michigan 48106

Library of Congress Catalog Card No. 78-67494
ISBN 0-250-40271-8

Manufactured in the United States of America
All Rights Reserved

PREFACE

Historically, water pollution abatement policies have focused on control of municipal and industrial point sources of discharge. Interest in other potential sources of pollution has escalated as point sources have begun to come under control and as the nation has expanded its water pollution concerns to nutrients, persistent chemicals, and hazardous and toxic materials. The emphasis of national water pollution control policies is now in keeping to a minimum the amount of wastes entering the nation's waters from both point and nonpoint sources.

Agricultural and silvicultural nonpoint sources until recently have been considered as natural and generally uncontrollable. Such sources have included runoff from forestland, pastureland and cropland. Many of these nonpoint sources are difficult to control if emphasis is on the ability to establish and enforce effluent standards for a given runoff event. However, if control is approached by using appropriate management practices, it can be possible to reduce the pollutant load from nonpoint sources. Proper land and crop management methods such as erosion control, timing and integration of manure disposal practices on the land, timing of fertilizer applications to coincide closely with crop need, integrated pest management, and cropping and logging practices offer opportunities to minimize potential pollutants from agriculture and silviculture.

Thus the emphasis for control of nonpoint sources is being placed upon the use of "best management practices" rather than on the collection, treatment and effluent standards approach used for control of point sources. In general, a best management practice (BMP) is a practice which realizes water quality benefits while remaining consistent with sound agricultural and silvicultural practices. As used in New York State, a working definition of a BMP is

> Alternative combination of land use, conservation practices, and management techniques which, when applied to a unit of land, will result in the opportunity for a reasonable economic return within acceptable environmental standards.

The concept of BMPs is based on the assumption that environmental and traditional agriculture and silviculture management objectives are consistent, and that water quality goals can be met in a manner which is "practical" for the farmer and forest owner.

The federal regulations that have generated and enhanced this emphasis are PL 92-500, the Federal Water Pollution Control Act Amendments of 1972, and PL 95-217, the Clean Water Act of 1977. Other federal laws that strengthen the nation's role in water quality management are: the Forest and Rangeland Renewable Resources Planning Act of 1974, the Soil and Water Resources Conservation Act of 1977, and the Surface Mining Control and Reclamation Act of 1977. The legal and financial implications of these Acts are described in many papers, especially those of the conference keynote speakers. The presentations of these individuals are included in the section **Governmental Aspects.**

The planning process to identify water quality problems associated with agriculture and silviculture, and to identify the technical and institutional control of such problems, had been underway for over two years at the time of the conference. Information on the problems that have been identified and control approaches that have been considered are presented in several sections, especially **State and Watershed Approaches; Economic, Policy and Institutional Aspects; Approaches for BMP Selection;** and **Silviculture.**

Because nonpoint source control is a relatively new topic, technical information on the effects of specific BMPs is not abundant. Research and monitoring continue to be critical to refine existing information, gather new data on the effects of those practices that are implemented, and to review and revise practices and programs as new information becomes available. Research results representing the most up-to-date studies are included in these proceedings, especially in the sections entitled **Nutrient Management** and **Modeling Studies.**

As several of the keynote speakers noted, this conference and these proceedings may well be a benchmark, one that not only provides a comprehensive state-of-the-art knowledge of this subject, but also one that identifies an entry into an era of serious effort to control nonpoint sources from agriculture and silviculture.

The New York State College of Agriculture and Life Sciences, a Statutory College of the State University at Cornell University, was pleased to have the Environmental Research Laboratory, Environmental Protection Agency, Athens, Georgia, cosponsor the conference. This laboratory is the lead EPA facility for control of nonpoint sources of pollution. The program committee consisted of individuals from both organizations, and the participation of Dr. George Bailey, Dr. Douglas Haith, Mr. Lee Mulkey, Dr. Robert Swank, Mr. Thomas Waddell and Dr. Michael Walter is gratefully acknowledged.

The success of the conference was due in equal measure to the quality of the authors, their studies and their presentations, to the skill of the moderators, and to the interest and pertinent discussion of the participants. Attending the conference were 248 individuals from 35 states plus Washington, D.C., and Canada. Important assistance also was provided by Ms. Doreen Kirchgraber who played a major role in preparing for the conference and assisting at registration.

Any opinions, findings, conclusions or recommendations expressed in this publication are those of the authors and do not necessarily reflect the views of the College of Agriculture and Life Sciences or the Environmental Protection Agency.

                                      Raymond C. Loehr, Director
                                      Environmental Studies Program
                                      New York State College of Agriculture
                                          and Life Sciences
                                      Cornell University
                                      Conference Chairman

CONTENTS

Opening Remarks,   W. K. Kennedy xiii
Opening Remarks,   D. W. Duttweiler xv

SECTION I: GOVERNMENTAL ASPECTS KEYNOTE ADDRESSES 1

1. Nonpoint Source Pollution Control Strategy 3
   W. S. Groszyk
2. Improving Water Quality in Agriculture and Silviculture 11
   D. G. Unger
3. A State Perspective on Nonpoint Source Management 17
   P. A. A. Berle
4. Research Needs and Current Activities 25
   T. A. Murphy

SECTION II: APPROACHES FOR BMP SELECTION 31

5. Best Management Practices for Agriculture and Silviculture:
   An Integrated Overview 33
   G. W. Bailey and T. E. Waddell
6. Conservation District Involvement in
   208 Nonpoint Source Implementation 57
   R. E. Williams and J. E. Lake
7. The Role of Conservation Practices as
   Best Management Practices 69
   J. S. Johnson
8. Relationships Between Agricultural Land and Water Quality 79
   D. R. Coote, E. M. MacDonald and R. DeHaan
9. Integrating Water Quality and Best Management Practices 93
   A. F. Taverni and R. F. Dworsky

10. Environmental Implications of Trends in
    Agriculture and Silviculture 117
    S. G. Unger

11. Best Management Practices for Fertilizer Use 133
    W. C. White and H. Plate

SECTION III: NUTRIENT MANAGEMENT 143

12. Animal Manure Movement in Winter Runoff for
    Different Surface Conditions 145
    D. B. Thompson, T. L. Loudon and J. B. Gerrish

13. Estimation and Management of the Contribution by Manure
    from Livestock in the Ontario Great Lakes Basin
    to the Phosphorus Loading of the Great Lakes 159
    D. W. Draper, J. B. Robinson and D. R. Coote

14. Estimating Phosphorus Loading from Livestock Wastes:
    Some Wisconsin Results 175
    I. C. Moore, F. W. Madison and R. R. Schneider

15. Phosphorus—A Potential Nonpoint Source Pollution Problem
    in the Land Areas Receiving Long-Term Application of Wastes .. 193
    K. R. Reddy, R. Khaleel,
    M. R. Overcash and P. W. Westerman

16. Nutrient and Pesticide Movement from Field to Stream:
    A Field Study 213
    J. L. Baker, H. P. Johnson,
    M. A. Borcherding and W. R. Payne

17. The Fate of Nitrate in Small Streams
    and its Management Implications 247
    J. B. Robinson, H. R. Whiteley, W. Stammers,
    N. K. Kaushik and P. Sain

SECTION IV: SILVICULTURE 261

18. Best Management Practices for Silviculture 263
    W. C. Harper

19. An Approach to Water Resources Evaluation of
    Nonpoint Sources From Silvicultural Activities—
    A Procedural Handbook 271
    J. B. Currier, L. E. Siverts and R. C. Maloney

20. Estimating Impacts of Silvicultural Management Practices
on Forest Ecosystems 281
    F. R. Larson, P. F. Ffolliott, W. O. Rasmussen and D. R. Carder

21. Simulation of Stormwater Runoff and Sediment Yield for
Assessing the Impact of Silviculture Practices 295
    R. M. Li, K. G. Eggert and D. B. Simons

SECTION V: ECONOMIC, POLICY AND INSTITUTIONAL ASPECTS 309

22. Nonpoint Source Pollution from Agriculture: Some Sociological
Considerations for Implementing Policy 311
    J. C. van Es and L. C. Keasler

23. Social Costs and Effectiveness of Alternative Nonpoint
Pollution Control Practices 321
    K. F. Alt, J. A. Miranowski and E. O. Heady

24. Management and Financing of Agricultural BMPs 329
    J. M. Rice

25. The Economic Implications of Erosion and Sedimentation
Control Plans for Selected Pennsylvania Dairy Farms 341
    G. B. White and E. J. Partenheimer

26. Farm-Level Economic Evaluation of Erosion Control and
Reduced Chemical Use in Iowa 359
    J. M. McGrann and J. Meyer

27. Economic Impacts of Policies to Control Erosion and
Sedimentation in Illinois and Other Corn-Belt States 373
    W. D. Seitz, C. Osteen and M. C. Nelson

28. Procedure for Economic Evaluation of
Best Management Practices 383
    T. H. Dempster and J. H. Stierna

29. An Economic Analysis of Erosion Control Options in Texas 393
    D. R. Reneau and C. R. Taylor

30. The Policy Relevance of Alternative Institutional Approaches
to 208 Planning 419
    A. Hamilton and L. W. Libby

31. Economic, Institutional and Water Quality Considerations
in the Analysis of Sediment Control Alternatives:
A Case Study 429
    B. M. H. Sharp and S. J. Berkowitz

32. Institutional and Technical Aspects of the Development of Agricultural BMPs in a Five-County Rural/Urban Michigan Region 455
    J. P. Jones and J. C. Sutherland

SECTION VI: STATE AND WATERSHED APPROACHES 463

33. The Evaluation of Best Management Practices for the Reduction of Diffuse Pollutants in an Agricultural Watershed 465
    T. H. Cahill, R. W. Pierson, Jr. and B. Cohen

34. Sediment and Nutrient Contributions to the Maumee River from an Agricultural Watershed 491
    D. W. Nelson, E. J. Monke, A. D. Bottcher and L. E. Sommers

35. Water Quality Modeling in the Delaware Coastal Plain Region ... 507
    W. F. Ritter and P. A. Jensen

36. Estimation of Agricultural Nonpoint Loads to the Wakarusa River Basin Using the "Nonpoint Calculator" 525
    M. J. Davis and J. W. Nebgen

37. Approach for Analyzing and Managing Agricultural Nonpoint Sources in the State of Maryland 551
    R. F. Schoenhofer, W. A. Knight and C. V. Hancock

38. Development of a "208 Plan" for Agricultural Nonpoint Pollution Sources in Illinois 563
    D. H. Vanderholm, J. F. Frank and A. G. Taylor

39. Development of BMPs for Agriculture— New York State Strategy 581
    P. D. Robillard, M. F. Walter and R. Gilmour

SECTION VII: MODELING STUDIES 597

40. Evaluation of Controls for Agricultural Nonpoint Source Pollution 599
    J. J. Wineman, W. Walker, J. Kühner, D. V. Smith, P. Ginberg and S. J. Robinson

41. Mathematical Modeling of Water Quality Effects of Agricultural Best Management Practices 625
    C. Tang

42. Methodology for Determining the Optimal Mix of BMPs and Agricultural Production Modifications 649
    J. P. Heaney and D. C. Ammon
43. Modeling Nutrient Export in Rainfall and Snowmelt Runoff 665
    D. A. Haith and L. J. Tubbs
44. Modeling Soil and Water Conservation Practices 687
    D. C. Beyerlein and A. S. Donigian, Jr.
45. Interactive Effects of Pesticide Properties and Selected Conservation Practices on Runoff Losses: A Simulation Study 715
    J. D. Dean and L. A. Mulkey

Index 735

OPENING REMARKS

W. K. Kennedy

    Dean, College of Agriculture and Life Sciences
    Cornell University
    Ithaca, New York

It is a privilege to welcome participants to the 10th Annual Cornell University Conference "Best Management Practices for Agriculture and Silviculture." The New York State College of Agriculture and Life Sciences not only recognizes that Section 208 of the Federal Water Pollution Control Act Amendments of 1972 is the law, but it also strongly endorses the goal to have pollution-free waters by 1983. Society faces a tremendous challenge, but we are committed to use our research and educational resources in cooperation with other agencies to achieve this goal. As in other years we are delighted to have the U.S. Environmental Protection Agency (EPA) as a cosponsor of this conference. We also are pleased to have as keynote speakers our colleagues from EPA, the U.S. Department of Agriculture, and the New York State Department of Environmental Conservation.

The College of Agriculture and Life Sciences has no tolerance for individuals or industries that pollute our waters through carelessness or poor practices. It has been our experience that most farmers and food processing firms will readily adopt best management practices as we provide them with sound recommendations. The purpose of this conference is to exchange ideas and move a step closer in how we can make our fertilizer recommendations more precise, how we can handle animal manures without threat to water quality, and how we can further reduce erosion of our lands and sedimentation in our streams, lakes and reservoirs.

We know nonpoint sources of pollution can and are being reduced substantially by decreasing the level of nitrogen fertilizer applied to potatoes on Long Island. However, nonpoint sources of pollution do not always come from agricultural or forested lands. Another significant source of nitrates in the groundwater of Long Island is from the nitrogen fertilizer used on home lawns, golf courses and other turf areas. We do not have recommendations for turf grass which will maintain the color and growth desired by home owners and others and still not contribute nitrates to the groundwater. We are seeking such answers and will launch an extension program which may have to include information that the greenest lawn in the block is not a desirable goal. This is but one example of the type of nonpoint source questions that must be addressed and for which management solutions must be found and used.

Finally, a note of caution. We must seek ways of reducing nonpoint sources of pollution but also must realize that both the technological and economical limitations can be greater when controlling nonpoint sources than when controlling point sources. In our zeal to reduce nonpoint sources of pollution we should be ever vigilant to reduce point sources to the lowest feasible levels since the cost of point source reductions will be many fold less than for the nonpoint sources. If we are to be successful in obtaining the adoption of best management practices to reduce nonpoint sources of pollution, and this is a goal we must achieve, our recommendations must not impose high-cost practices on individuals or industries while there are point sources of pollution which can and should be controlled at a lower cost.

OPENING REMARKS

D. W. Duttweiler
    Director, Environmental Research Laboratory
    U.S. Environmental Protection Agency
    Athens, Georgia

On behalf of the U.S. Environmental Protection Agency and the Athens Environmental Research Laboratory I extend a cordial welcome to all of you who have gathered for this 10th Annual Cornell University Conference. It is gratifying to see the interest in this conference as evidenced by the large number of participants, and I am honored to help open the conference with Dean Kennedy.

Each year the Cornell Conference provides an important forum for the agricultural community concerned with environmental pollution control to exchange ideas with the environmental pollution control community concerned with agriculture. The Athens Laboratory is proud to join in sponsoring this year's conference, which focuses on a promising approach to a difficult water pollution control problem, because pollution by runoff from agricultural lands has been one of our major research interests since the laboratory's beginning in 1964.

Each year the Cornell Conference attracts a broad spectrum of participants, and this year is no exception. Both communities are represented by technical experts, consultants, government and university researchers, and public officials—almost every aspect of society's concern for this problem from policy to implementation. This is further evidence of the cooperation that is essential to continued high agricultural productivity in the U.S. in harmony with the national resolve to have a healthy environment.

The program in store for us is a comprehensive one. Not only are the technical aspects of the problem thoroughly covered, but also the economic, social and institutional considerations necessary to selecting and implementing best management practices will be discussed. With you, my appetite has been whetted to learn more about developing public policy, experiences in implementing best management practices, and research results that can contribute to our solutions of these problems. As a research manager, during these next few days I will be looking for gaps in knowledge that research should fill, since obviously much remains to be learned about practices to manage pollution from agriculture and silviculture. For example, we need to know how effective best management practices will be in protecting the environment and maintaining agricultural productivity, and what they cost in dollars and Btu.

This conference may well be a benchmark meeting—one that signals our entry into an era of serious effort to control agricultural nonpoint sources of water pollution. The conference will not give us all the answers, but it will be a springboard to develop sensible, economical, efficient strategies for managing agricultural pollution, based on the best agricultural and environmental science and technology.

SECTION I

GOVERNMENTAL ASPECTS
KEYNOTE ADDRESSES

NONPOINT SOURCE POLLUTION CONTROL STRATEGY

W. S. Groszyk
> Deputy Director
> Water Planning Division
> U.S. Environmental Protection Agency
> Washington, D.C.

I would like to discuss the two major elements of the Environmental Protection Agency's nonpoint source (NPS) pollution control strategy, and how two amendments in the Clean Water Act of 1977, signed last December, provide you who are working with Section 208 agencies and EPA with additional ways to implement NPS control programs, and thus solve some of our critical water quality problems. These two major elements of EPA's nonpoint source control program are: (1) to accelerate the implementation of NPS control programs and the application of Best Management Practices (BMPs) and, (2) to focus our available resources on solving the most severe problems first.

Before getting into details of these elements, let me first address two shibboleths about the NPS program. The first is that we do not know enough yet to do anything; the second is that we are demanding regulatory programs everywhere for every problem.

In the nonpoint source control area, EPA is frequently thought of as demanding abatement actions before all the answers are known. This is partly true, because we do often proceed, as I'm fond of saying, on a little data and a lot of faith. But let us review the situation for nonpoint source pollution control. We know that much of the nation's water pollution is from nonpoint source pollutants. We know that controls and methods exist which will reduce the nutrient, pesticide, sediment and coliform levels in the receiving streams. However, what we do not always know is the precise relationship between the application of one or more of these controls on an individual source and the resultant improvement in water quality. But, I submit that we should not and cannot wait until such clear demonstrations can be made in all cases. Why?—because the longer we defer introduction of NPS controls, the greater the likelihood we will be requiring and paying for higher levels of controls on point sources, when a cost-effective trade-off might promote a greater balance in control levels between point sources and nonpoint sources. There are statutory dates for point sources set out in the law which must be achieved and we cannot delay meeting these dates.

Having conceded that not all the answers to all the questions are or will be known in the next several years as the initial 208 plans come in, we can, through the continuing planning process and the update of initial 208 plans, fine-tune the methods and improve our understanding as progress is made in the state-of-the-art of correlating NPS controls with water quality effects.

On the second point, let me make it clear that EPA does not mandate regulatory programs be developed to control all nonpoint source problems. We will accept and approve other than regulatory or so-called voluntary programs. However, these voluntary efforts cannot be a reiteration of the "status quo." The status quo or business-as-usual approach is not acceptable.

Returning to the two major elements, the first was that accelerated implementation of 208 programs and the application of known controls for pollutants causing water quality problems is a major emphasis of EPA's water programs. One reason for this is to allow owners and operators to become eligible for cost-sharing money, and the second is that EPA and OMB have agreed that no further planning money is to be awarded to an agency beginning in Fiscal Year 1980 if implementation is not occurring.

Let me mention that one particular concern is the control of nonpoint source toxic pollutants. EPA is using the combined authorities of the Clean Water Act, the Resource Conservation and Recovery Act and the Safe Drinking Water Act to remove, contain and control toxic pollution from our ground and surface waters. Our water and effluent monitoring activities will give greater emphasis to monitoring of toxics including the use of bioassay and biological monitoring techniques. We look to the agricultural and silvicultural portions of the 208 plans to assess and solve any toxic pollutant problems from these activities.

EPA does not have the technical expertise, resources or delivery mechanisms to impact all those who must apply NPS controls, and does not intend to seek this capability. We will continue to utilize the expertise, resources and delivery mechanisms of the other federal agencies and levels of government to help us solve these water quality problems.

The Rural Clean Water Cost-Sharing Program, otherwise known as the Culver Amendment, and the Integrated Federal Water Quality Program in Section 304(k) are both designed to concentrate on priority water quality problems first. We cannot do everything at once. Nonpoint source control is an iterative and continuing process. As progress is made in one area, we will then move to the next problem. Section 208 agencies, state- and areawide, have prioritized their water quality problems and are to develop control plans for the most severe problems first.

Let me give an example of this prioritization. Based on the studies of the Great Lakes Basin, 60% of the agricultural sediment load is being generated from about 30% of the agricultural land area of the basin. Obviously we should concentrate our efforts on that 30% of the basin contributing 60% of the load.

The Black Creek Project Studies indicated that treating only 80 ac of the most highly erosive areas in that basin would reduce the total sediment load by 40%. The costs of control could be minimized—if these 80 ac were treated rather than 1,800 ac; the cost would be $6,000 rather than the $135,000 which would be required to treat the 1,800 ac.

Let me go into the two legislative amendments. EPA strongly supported the passage of the Rural Clean Water Amendment to assist the agriculture and forestry community in solving their nonpoint source pollution problems. On April 25, 1978, EPA and the Department of Agriculture signed an agreement setting out the mutual program responsibilities of EPA and Agriculture under this program. I would like to discuss the concepts that have and will continue to guide EPA in the development and implementation of the Rural Clean Water Program (RCWP).

The Clean Water Act authorizes the Secretary of Agriculture, with the concurrence of the Administrator of the Environmental Protection Agency, to establish and administer a program to enter into long-term contracts, of not less than five years nor more than ten years, with rural land owners and operators for the purpose of installing and maintaining BMPs to control nonpoint source pollution and to improve water quality. Only those states or areas which have an approved portion of a 208 plan dealing with rural nonpoint problems qualify for financial assistance. For example, the agricultural problem priorities and the BMPs identified in an approved agricultural portion of a 208 plan will form the basis of the critical water quality priority project areas to be approved and the BMPs to be cost-shared. The management agency designated by the governor to implement the agricultural portion of the 208 plan must certify that the BMPs cost-shared are consistent with the 208 plan.

A RCWP project area is a hydrologically related unit with critical water quality problems that result from agricultural activities. To be designated as a project area, eligible for financial assistance, the area's water quality problems *must be related* to agricultural pollutants, such as high nitrogen and phosphorus levels, toxics such as pesticides, high total dissolved solids readings, high biochemical oxygen demand and coliform levels, or sediment. RCWP deals, as does the 208 program, with more than erosion. Sediment is not the only, nor in many areas, the most severe water quality problem. About 50 million tons of fertilizers and 1 billion lb of pesticides are used annually for agricultural purposes and may cause water quality problems. The studies in the Great Lakes Basin indicate nitrogen, phosphorus, insecticides, herbicides and animal wastes are major contributors of the water quality problems. Over 45% of the phosphorus loadings of the Great Lakes is attributable to agricultural activities. Row crops on clay soils in close proximity to the lakes, rivers and streams are thought to be the primary reason for this high loading. Thus we need the flexibility to deal with either areas or sources of pollutant problems. Not all NPS pollutant problems can be solved by controlling erosion. The Integrated Pest Management Program and soil fertility testing programs are acceptable BMPs eligible for cost-sharing assistance, and any given land unit may need to apply several different types of practices to control the problems. Let me add that not all of these are structural practices.

The size of the project area will depend on the type of agricultural activities and pollutants involved. It must be of manageable size to demonstrate results within the contract period. Generally the area must be less than 200,000 ac.

Contracts are to be entered into only in areas where the designated management agency assures an adequate level of participation. If water quality improvement is to be realized from the investment, there must be sufficient participation in the program. An adequate level of participation is when 75% of the *critical acreage or sources* of the pollution is under contract, except for those areas where the approved agricultural portion of the 208 plan provides data and analyses which indicate a *greater or lesser* percentage of the acreage or sources of pollution must be treated to attain water quality standards or water quality goals. This may be a tough requirement, but this is a different type of agricultural assistance program from those in the past.

The governor, based on the priorities in the 208 plans, will recommend, *in order of priority,* project areas for cost-sharing assistance. Initially only the highest priority project areas proposed in each state will receive consideration. Project area applications will be prepared by the State Coordinating Committee. The State Coordinating Committee is to be chaired by the State Conservationist, Soil Conservation Service (SCS), and will include the State 208 Water Quality Agency; a designated representative of the State Soil and Water Conservation Agency; other state agencies as the governor deems appropriate, such as the Department of Forestry; a designated representative of the conservation districts; and representatives of the members of the National Coordinating Committee. There are many agencies involved and there will be "dynamic tension" but there are many agency programs that have to be coordinated and issues resolved if project areas are to be funded.

The national priority areas funded will be selected from among the state priorities submitted by the governor to the Secretary of Agriculture. EPA will be actively involved in the review and selection process. By letter, EPA is to concur in the priorities selected before project areas are funded. National priorities will be evaluated according to the following criteria:

1. severity of the water quality problem impacted by the agricultural pollutants;
2. economics and technical feasibility to control the problem(s) within the life of the contract; an upper limit of a per acre or per project area cost may be required if the costs of a project are not in keeping with the water quality benefits;
3. demonstration of public benefits that occur from the project, and
4. compatibility with national water quality goals and priorities.

We cannot do everything at once, so we will focus our resources on solving the most severe problems first. We are looking for the most cost-effective controls, *not* a Christmas tree or various practices that have limited water quality benefits. The costs of each project will be carefully evaluated to ensure that the limited funds are most effectively used. Those of you working with 208 agencies to develop the practices to control nonpoint source pollutants must be cognizant of the costs of what you are proposing; otherwise you might find yourselves with a program of "gold plated" BMPs that cannot be applied, funded or accepted by the owners and operators.

The BMPs that are to be cost-shared must have a *positive effect* on water quality by either (1) preventing or abating the amount of pollutant entering the stream, or (2) reducing the amount of the pollutant that is applied to the land. It is recognized that the plans upon which the contracts are to be based may include conservation practices other than those related to water quality practices and are not eligible for cost-sharing under the RCWP. There are practices for which cost-sharing assistance under RCWP will not be provided. These include drainage, channelization and bringing additional land into crop production.

The focus of the program is on agricultural practices, which can include farm woodlot operations. Requests for cost-sharing for other than agricultural practices will be evaluated on a case-by-case basis in terms of the impact on improving water quality in the project areas.

Cost-sharing assistance will be available for only privately owned lands. Federal, state, county and other public lands are included in the 208 planning process and should be managed in the most environmentally sound manner. We recognize that this does not always occur. Executive Order 11752 and Section 313 of the Act require that federal lands and facilities meet the same standards as private lands and facilities. We are working with other federal agencies to involve them more in the 208 planning process and to develop more stringent pollution controls on their lands. The same effort must occur on other public nonfederal lands.

The primary objective of the program is to cost-share practices to the extent required to ensure installation of the practices. The act provides for a 50% cost-share rate. Recommended variances to the 50% rate must be fully documented in the applications and approved by the Secretary of Agriculture based on the following criteria:

1. a documented financial burden that would prevent participation in the program;
2. substantial offsite water quality benefits that require a long-term farm investment or result in a substantially reduced farm income;
3. severity of the water quality problems to be abated;
4. compatibility with rates of other cost-sharing programs; and
5. acceptability of the practices in the area. Commonly applied practices would receive less cost-sharing assistance than innovative BMPs.

We want to avoid the game of shopping around for the best deal and pushing the rate higher than is absolutely essential. The cost-share rates at the national level will be coordinated through the National Rural Clean Water Coordinating Committee. Some rates will have to be adjusted to meet the particular circumstances of an area, but we assure you that the Office of Management and Budget will keep a watch on our cost-share rates, even if we do not.

To assist the Administration and Congress in determining the overall effectiveness of the RCWP, the Secretary of Agriculture will require administering agencies to submit reports on the accomplishments of the program, both practices planned and applied. *In selected areas,* EPA and USDA have agreed that comprehensive water quality monitoring, evaluations and analyses will be needed. The areas selected for detailed analyses will be representative of agricultural nonpoint source pollution problems, will be used to evaluate the cost and effectiveness of alternative BMPs, and to provide better information on the impact of the BMPs on water quality. This will follow-through on the seven Model Implementation Plan (MIP) project areas that have been initiated with existing Department of Agriculture funds. This analytical effort will be coordinated with the State Rural Clean Water Coordinating Committee and state water quality monitoring programs.

In summary, EPA looks for the RCWP to accelerate the implementation of nonpoint source control programs and the application of best management in those areas and for those sources of pollutants that contribute to an area's water quality problems.

The second legislative area is a reauthorization of funding for 304(k), which provides money for federal agencies to assist and implement 208 plans. In conjunction with the RCWP, EPA will use the 304(k) Integrated Federal Water Quality Management Program to utilize the technical resources and delivery mechanism of other federal agencies to assist in the implementation of water quality management control programs. Successful implementation of many 208 programs is dependent upon the expertise and support of federal agencies.

The Section 304(k) program is authorized at $100 million per year for Fiscal Years 1979-83. We are working with other federal agencies to obtain funding for the program in Fiscal Year 1980. This funding is to serve as a catalyst to bring federal agency expertise to state and local agencies to help them solve critical water quality management problems and to expedite the implementation of water quality management control programs. Funds will not be used to replace existing federal resources but will be used to increase and accelerate the ongoing programs. The 304(k) monies can fund activities that are not eligible for funding under the Section 208 Continuing Planning Process or Section 208(j), RCWP. Funds may be used by the federal agencies to contract for services with nonfederal agencies, consultants or temporary personnel where this is the most effective method for accomplishing the task. Activities selected for funding are to be based on an approved 208 plan. EPA, the federal agency and the state involved must agree on the activities to be funded. Detailed work programs and interagency agreements will be the basis of 304(k) funded activities.

Some of the activities that relate to agriculture and silviculture pollutant problems and which may be funded when appropriations are received include the Integrated Pest Management Programs and the nutrient control training programs of the Extension Service and the Irrigation Management Services (IMS) Program of the Bureau of Reclamation and the Soil Conservation Service, which would be expanded to include 300,000 ac irrigated by Bureau projects. The IMS program can already increase water efficiency by an estimated 10%, and a further increase to 65% is attainable with some additional labor and costs. We would also like to accelerate the state and private technical assistance programs of the Forest Service and have them develop and demonstrate innovative nonpoint source controls for forest and grazing areas. Finally we expect that the Fish and Wildlife Service can assist 208 agencies utilize instream flow criteria in their planning and evaluation process; help 208 agencies determine the effect of different pollutant levels on stream biota; and use water quantity/water quality modeling methods to evaluate the impact of agricultural water quantity needs on the water quality of an area.

The Rural Clean Water and the Integrated Federal Water Quality Management programs are an integral part of EPA's water quality management strategy to control pollutants from agricultural and silvicultural activities. You are the key element of the strategy for, although I have talked at some length about federal assistance programs, EPA has not lost sight that the program is most fundamentally a state and local effort. It is not my program, but your program. The answers you provide us in forums such as this will enable us all to better achieve the water quality goals that have been set forth and offer solutions to the complex problems that exist.

2

IMPROVING WATER QUALITY IN AGRICULTURE AND SILVICULTURE

D. G. Unger
    Acting Deputy Assistant Secretary of Agriculture
      for Conservation, Research and Education
    U.S. Department of Agriculture
    Washington, D.C.

I appreciate the opportunity to help open what may be the most important conference in your series.

The term "Best Management Practices" (BMPs) is a new way of describing land management systems that improve water quality. Some of them are familiar, proven practices, easily adapted to meet the needs of a specific tract of land, an individual land user, and the adjacent water resource. Other practices or issues related to them require a good deal more study.

USDA and others will be evaluating BMPs more thoroughly over the next several months. This conference provides us all with a good chance to discuss these practices in the context of our needs. It could not be more timely.

The overall job of upgrading water quality will require an effective combination of good management, technical assistance, economic incentives, research and education. It will need participation by individual land users as well as local, state and federal governments.

USDA aleady has done a great deal to help improve the quality of our Nation's water and related resources. With new legislation—and new emphasis in our continuing programs—we will do even more.

I want to discuss with you today some of America's water quality needs and the programs designed to meet them.

Sediment is America's number-one degrader of water quality, by volume. It is a multiple threat because it often carries with it pesticides, fertilizers and other chemicals.

Much of the total sediment load in streams and lakes comes from agricultural land. Just how much, and how this affects water quality, are questions still being explored.

Over the past four decades, millions of farmers, ranchers, and forestland owners and managers have undoubtedly improved water quality considerably by applying conservation practices. Yet, in areas such as the Obion-Forked Deer River Basin of western Tennessee, soil still erodes at an annual rate of 25 tons/ac from more than 1.5 million acres. Critical problems like this must be detected and treated. USDA programs can help local people and public land managers do just that.

We may not be able to cut soil erosion to zero—and may not want to—but we can reduce it considerably, through familiar conservation practices and some that are not so familiar.

Minimum tillage, for example, does a good job of reducing sediment. Contour plowing, terracing, stripcropping, windbreaks, grass seeding and sodding, and special treatment of badly eroding areas also can contribute to good water quality.

On forestlands, we can reduce sediment by improving fire management, adjusting the method and timing of timber harvest, and placing logging roads more carefully. On rangelands we can reduce sediment by adjusting the number of livestock to the forage supply and by reseeding pastures and rangelands and disturbed areas.

Pesticides and plant nutrients pose different problems that call for still other kinds of treatment. Applying these materials in proper amounts at proper times should reduce their effect on water quality.

No single technique is a panacea. All can help.

Across the country, community after community faces difficult decisions about water resources. USDA tries to help through several programs:

1. We give land owners and others a wealth of technical information, help them prepare conservation plans, and help them manage their resources and apply conservation practices.
2. We compile new information through research programs and through soil surveys, vegetation studies, snow surveys, and other inventories.
3. We have several cooperative programs for protecting watersheds, preventing floods, controlling fire, conserving and developing natural resources, and managing forestlands.
4. We serve as members of policy or advisory committees established by state and area-wide water quality agencies and provide them with vital data.
5. We manage national forests, national grasslands, and other units of the national forest system, which make up some of the most important watershed areas in many parts of the nation. We collect, analyze and use a wide variety of resource data—much of it directly related to water quality.

6. We provide information to land owners and the public about water quality problems and solutions.

Many of you have worked with USDA in carrying out these assistance programs. We need your continued support.

Looking ahead, there are four new federal laws which will strengthen our role in water quality management: The Clean Water Act of 1977; The Forest and Rangeland Renewable Resources Planning Act of 1974, as amended, better known as the RPA; The Soil and Water Resources Conservation Act of 1977, better known as the RCA; and The Surface Mining Control and Reclamation Act of 1977.

In his Environmental Message last year, President Carter pointed out the importance of state and local planning in controlling water pollution from farms and ranches, forestlands, mines, urban streets, and other sources.

The new Clean Water Act directs attention toward high-priority areas where sediment, nutrients, pesticides and other pollutants pose the greatest threat to streams, waterways and lakes. These sources of nonpoint pollution will be identified in Section 208 plans prepared by local and state agencies and approved by the EPA. Under Section 35 of the Act, Congress authorized $200 million in Fiscal Year 1979 and $400 million in Fiscal Year 1980 for USDA technical and cost-sharing aid to establish best management practices in these critical areas. We are calling the effort the Rural Clean Water Program. It will be administered by the Soil Conservation Service (SCS) in cooperation with other USDA agencies and the EPA. The USDA/EPA agreement on procedures for the program was signed April 25.

1. The Rural Clean Water Program, assuming that funds are requested and appropriated, will be carried out on a project basis to deal with critical sources of agricultural nonpoint pollution justified in approved Section 208 plans. It is not a wall-to-wall program. Whenever practicable, USDA will enter into agreements with state water quality agencies, state soil conservation agencies, or local soil conservation districts to administer the program. Where this is not practicable, and the federal government retains administration, it will be conducted by the Agricultural Stabilization and Conservation Service (ASCS).

2. In every case, priorities for long-term contracts with land owners and operators in these project areas will be set jointly by local soil and water conservation district boards and by county Agricultural Stabilization Conservation committees, based on technical information from the SCS.

The heart of the program will be contracts, for not less than five nor more than ten years, which will provide cost-sharing and technical assistance to land owners for best management practices or systems to reduce soil erosion and maintain water quality.

The cost-sharing rate is up to 50% and may be more, if the main benefits are improved offsite water quality and the land owner cannot afford his share of the cost. Thus, we will be able to adjust agreements and payments to meet local needs.

The secretary will announce in the *Federal Register* by September 30, 1978, the guidelines for operating the Rural Clean Water Program. Contracting with land owners in areas approved for the program can get underway shortly thereafter—assuming that funds are requested and appropriated.

Secretary Bergland will be assisted by a Rural Clean Water Coordinating Committee, chaired by Administrator Mel Davis of the SCS. The committee will include representatives of the ASCS; Forest Service; Farmers Home Administration; Science and Education Administration; Economics, Statistics and Cooperatives Service; and the EPA. The Association of State Soil Conservation Agencies, Association of State and Interstate Water Quality Agencies, the National Association of Conservation Districts and other groups will work closely with the committee.

The RPA, administered by the Forest Service, and RCA, administered by the SCS, direct USDA to make periodic assessments of America's basic natural resources and to take actions that will protect and improve these resources.

The RPA, bolstered by the National Forest Management Act of 1976, seeks to help all Americans understand that the nation's renewable resources are important, that long-term plans are needed for managing and using resources, and that citizens can participate in designing and evaluating the plans. Under the RPA, the Secretary of Agriculture must prepare: an *assessment* of renewable resources of U.S. forests and rangeland under all ownerships; and a long-range recommended *program* for Forest Service activities in research, in management of national forests and national grasslands, and in aid to states and private forestland owners.

USDA sent the first assessment and the first program to Congress in 1976. Both will be updated in 1980. After that, the assessment will be updated every ten years and the program every five years. This process insures that annual decisions on funding Forest Service programs will relate to the long-range program. It also insures that water quality and other important factors will be part of land mangement planning.

The RCA offers a better-than-ever chance to detect the really tough soil and water conservation problems, to develop new and more effective solutions to them, and to monitor how well the job is done. Under RCA, USDA must:

- make a continuing appraisal of America's natural resources;
- develop a five-year strategy, based on the appraisal, for dealing with soil and water conservation problems; and

- evaluate annually the progress of conservation programs in carrying out the five-year strategy.

Through RPA and RCA, local people and state agencies will blend their own plans and ideas with national multiyear programs. The RCA also will make it easier to relate soil and water conservation to the natural resource functions that the RPA assigns to the Forest Service.

The RCA could turn out to be the most significant soil and water law since the origins of our national program in the 1930s. Together, RPA and RCA provide us with an unparalleled opportunity to evaluate what we are doing and to make needed adjustments.

The Surface Mining Control and Reclamation Act of 1977, in Section 406, established what we call the Rural Abandoned Mine Program.

This is a voluntary program, designed to aid land users in reclaiming, conserving and developing coal-mined lands that are either abandoned or inadequately reclaimed. Many of these lands are contributing to pollution problems. Cost-sharing and technical help under the program are aimed at: protecting people and the environment from the adverse effects of past coal mining; and promoting development of soil and water resources in unreclaimed areas. We expect the program to become operational by late July or early August. More than a million acres could be eligible for treatment, in 377 counties in 29 states.

Several other efforts underway include:

- The Agricultural Conservation Program (ACP) of ASCS also contributes to better water quality. Under the ACP, federal cost-sharing helps farmers establish measures to control sediment and strengthen agriculture.
- USDA's new Science and Education Administration (SEA) conducts research and educational programs that support water quality efforts. SEA combines the Agricultural Research Service, the Cooperative State Research Service, the Extension Service, and the National Agricultural Library. Along with the Forest Service, these units of SEA have compiled an outstanding track record in developing and helping apply new technology.
- USDA and EPA also have begun a Model Implementation Program, or MIP, to demonstrate how a coordinated program of conservation systems can reduce nonpoint source water pollution. Our team has selected MIP project areas in seven states, including New York. The soil and water conservation work needed in these areas will begin within six months and most will be completed within three years.

USDA will continue to do everything it can to help upgrade water quality. The job will call for the best efforts of all of us.

Assistant Secretary of Agriculture Rupert Cutler has said that rural water quality management must be built on four principles:

1. a *voluntary approach* with farmers and ranchers that builds on existing working relationships to get the job done while preserving as much freedom of choice for private citizens as possible;
2. *economic incentives* to land owners to reflect the benefits that accrue to society as a whole—economic growth, improved water quality, better wildlife habitat, aesthetic values, and others;
3. *team efforts* that will accomplish more than the sum of individual programs of universities, federal and state agencies, farm groups, and conservation and environmental organizations; and
4. *recognition* of how many programs on *land* use—particularly on prime farmlands and timberlands—affect *water* quality.

I would like to add one more. We need more facts. Research and monitoring are critical. We are going to be spending a good deal of time and money on this problem and will be asking for much from farmers, ranchers and other land owners. We need to refine our information, gather new data on the effects of those practices selected to help achieve cleaner water, and constantly review and revise our practices and programs in accordance with what we know.

The biggest influence in upgrading water quality may be your influence. You have skills and ideas to contribute. You have demonstrated your willingness. I am confident that this conference will take us further toward wiser use of natural resources and a better environment for all Americans.

3

A STATE PERSPECTIVE ON NONPOINT SOURCE MANAGEMENT

P. A. A. Berle
    Commissioner
    New York State Department
       of Environmental Conservation
    Albany, New York

I would like to thank the sponsors of this conference for providing me with the opportunity to address this issue of "Best Management Practices for Agriculture and Silviculture." The states have been dealing with the nonpoint source management provisions of Section 208 for almost two years and are to submit their initial nonpoint source management plans later this year.

In New York we have contracted with the College of Agriculture and Life Sciences at Cornell University to develop the framework for a system of Agricultural Best Management Practices for the state. We have also contracted with the Applied Forestry Research Institute at State University of New York College of Environmental Science and Forestry, Syracuse, to conduct a similar study on Silvicultural Management Practices. Additionally, we have contracted for studies in three example watersheds in the state, and we have been working closely with our State Soil and Water Conservation Committee.

The input from these studies, the state committee, and our statewide system of policy, technical and citizens advisory committees has sharpened our awareness of the key problems and issues in the area of nonpoint source management. Thus it is timely to share some thoughts based on our experience of the last two years and offer some ideas on where we should be going.

We are entering a new era in water quality management which recognizes the need to control both point and nonpoint discharges. The broad perspective embodied in Section 208, encompassing point and nonpoint sources of water pollution as well as many other facets of water quality management, is very timely.

Preventing water pollution from rural nonpoint sources is an important public challenge. In New York, this challenge can best be met through a three pronged approach:

1. participation by every land owner in the existing cooperative soil and water conservation program;
2. identification of serious nonpoint water quality problems for special priority attention;
3. development of "Best Management Practices" (BMPs) tailored to meet New York's unique physical, economic and agricultural conditions.

We in New York have been concerned that, from our perspective, much of the rhetoric at the national level has tended to "oversell" the severity of nonpoint source problems. Additionally, we have a number of other types of water quality management problems—particularly toxics, combined sewer overflows, urban stormwater, industrial pretreatment, and residual waste disposal.

However, agriculture and silviculture are clearly part of the total water quality picture. They undoubtedly are contributing to degradation of water quality in many instances. As such, they cannot and should not be ignored. It is incumbent upon us to pursue an orderly program which will identify the problems and find the appropriate solutions, while at the same time maintaining a proper overall perspective.

Since the enactment of the state's Water Conservation Law in 1940, conservation of precious soil and water resources has been a keystone of our conservation ethic. This is just sensible resource management—and it also is good for water quality.

The state's agricultural community showed great statesmanship in initiating action which resulted in the 1975 amendments to the state's law requiring soil and water conservation plans for all agricultural and silvicultural land holdings over 25 acres. This initiative provides a sound basis for further actions which must be taken to meet the nonpoint source requirements under Section 208 of the Clean Water Act. The recommendations we will be developing in conjunction with the state's soil and water conservation community will undoubtedly build upon this strong foundation as we strive to meet federal water quality goals.

A substantial public information and extension effort is required by all of us in the soil and water conservation business, as only a fraction of New York's land owners are effectively participating in the existing program.

Much remains to be known about the nature of the nonpoint source pollution in New York. Therefore, in addition to participation of land owners in soil and water conservation, New York's 208 program has two major goals:

1. to assess the magnitude and geographic extent of nonpoint source pollution problems and to identify areas for priority attention; and
2. to develop systems of BMPs consistent with the ethic of good conservation which can be implemented to control nonpoint source pollution.

While we are aware of some of the nonpoint source problems in the state, we recognize that more thorough and systematic programs of problem identification are needed. We are making our initial evaluations currently within the 208 program and these will have to continue on a systematic basis over the long-term.

In developing systems of BMPs and the necessary institutional mechanisms to implement them, we are confronted with a major technical and institutional challenge which is the subject of this conference. BMPs are to be systems of land management practices which are the most economical, practical and reasonable, all factors considered, in achieving water quality goals.

We are very comfortable with and endorse the basic concept of BMPs for nonpoint sources, as an abstract concept. Instinctively, we believe that materials which belong on the land should be kept on the land and out of the water. We also believe that reasonable, practical and economical systems of land management practices can be identified which will do this. In practice, however, the implementation of this concept is extremely difficult.

The first major issue to be dealt with is the meaning of the term "water quality." Some will say that *any* contaminants entering the water unnecessarily via runoff are undesirable and should be controlled. Others will say that controls should be instituted only when there is an identified water quality problem.

At the present time, for purposes of establishing program priorities and for utilization of any available funding, we are leaning heavily toward the view that the system should respond to identified water quality problems. This will place a heavy burden on the Department of Environmental Conservation to define appropriate standards and to develop sound, systematic means of problem identification. It will force us to be more rigorous in our approach.

Although we intend to place our emphasis on clearly identified water quality problem areas at present, we should not forget that our conservation ethic, as well as the federal law, also embodies the concept of nondegradation. We will also work toward control of obviously bad practices which may contribute to the degradation of water quality.

A second major issue is the relationship of traditional land conservation practices to water quality. Some will say that all of the traditional conservation practices benefit water quality in the broad sense. Some even tend to equate all conservation practices with BMPs. On the other hand, there are those who insist that too little is known regarding the effects of conservation practices to be able to draw *any* conclusions regarding their relationship to water quality.

There are difficulties in uncritically accepting the viewpoint that all soil conservation practices relate directly to improved water quality. Research is showing us that much remains to be known about the precise relationships. Equally as important, specific types of conservation practices will have differing relationships with different water pollutants. A practice which helps to reduce problems with one pollutant may serve to increase problems with another.

A second difficulty involves the very important factors of cost and practicality. It must be remembered that when we are advocating the development and implementation of BMPs, whether through voluntary programs or regulatory means, we are potentially impinging on New York State's agricultural and silvicultural economy. The concept of BMPs includes the idea that BMPs be reasonable, practical and affordable. When we are too quick to equate the spectrum of all conservation practices with BMPs, these very important elements tend to become lost.

It is certainly true that the array of available management practices from which a system of BMPs might ultimately be developed is not new. By and large, it is the set of well known management practices which has been advocated and utilized in soil and water conservation programs for years. What is new, however, is the way in which these practices must be viewed if they are to be considered as candidates for inclusion in a system of BMPs for water quality.

I want to make it clear that we strongly support soil conservation to protect the natural resource and for purposes of other related environmental values. Our attempts to be rigorous in the way we view conservation practices for purposes of water quality should in no way diminish, but rather should enhance our appreciation of the other purposes and values associated with soil conservation.

I mentioned earlier that while some tend to be too loose in their interpretation of BMPs, others are inclined to say that too little is known to be able to make any judgments concerning the relationship of conservation practices to water quality. This point of view implies that nothing can or should be done relative to BMP development until we have the benefit of several more years of research. We cannot accept this viewpoint. We believe that the interests of all concerned will best be served by proceeding with the development of BMP systems, but with extreme caution utilizing careful analysis of the available research. Admittedly, there is much that is not known. However, a thorough and careful analysis of the available information should be able to identify those particular types of management practices which have substantial positive effects in relation to a particular type of water quality problem; those practices about which not enough is known to make a reasonable judgment; and those practices which in all probability will not have positive water quality effects.

We are seeking low-cost practices with a high probability of positive water quality effects for inclusion in an initial BMP system. Those practices selected must also meet the test of being reasonably practical within the context of agricultural or silvicultural operations. This approach seeks to minimize the risk of imposing unnecessary costs while still making the best use of what is known.

The study being conducted at Cornell to develop the framework for a system of agricultural BMPs for the state is to do the type of analysis I have described, with particular reference to agricultural conditions in the state. This is also the case with the Applied Forestry Research Institute study for Silvicultural Best Management Practices. I should note, however, that while many of the same basic principles apply, there are differences between forestry and agriculture. While the potential need for agricultural management practices can be analyzed relative to the presence or absence of an *existing* water quality problem, we see management practices for forestry as being primarily concerned with the *prevention* of *potential* water quality problems. Such problems are potentially quite severe, but are generally localized and relatively short-term in nature and consist of excessive erosion and sedimentation from badly managed harvesting operations. The way to solve these problems is with proper planning for the harvesting operation before it takes place.

The major conclusion to be drawn from what I have said so far is that technically we are dealing with a new and relatively unexplored area of concern, for which the information is simply not available to give us all the answers in the short-term. Therefore, our initial BMP systems must of necessity be rudimentary and carefully designed to stay within the limits of our current knowledge.

On the institutional side, we have a long-standing historical commitment to soil and water conservation programs which can form much of the basic framework for finding and implementing the technical solutions. This goes back to 1940, when the initial legislation was passed. As mentioned earlier, the soil and water conservation movement in New York showed great statesmanship in initiating legislation, passed in 1975, which requires that conservation plans be done for all agricultural and silvicultural operations of over 25 acres.

It is up to the various concerned interests and agencies in the state to follow this initiative. Doing this, however, is a very difficult and complex undertaking.

There are a variety of agencies at the federal, state and local levels which have a stake in the structuring of a nonpoint source management program. Generally, each of these agencies is currently performing a function which would logically fit within an overall program of nonpoint source management. Thus the basic elements upon which to build a program are largely in place. Yet at the same time, none of these existing agencies are currently performing their particular functions in exactly the manner which would ultimately be necessary under a properly administered nonpoint source management program. Each is faced with new roles, new responsibilities, altered perspectives, and unfamiliar interagency relationships.

Our job is to forge a mutually agreeable set of program functions and relationships among these agencies. In New York State we are fortunate in that the State Soil and Water Conservation Committee has accepted the role of a technical and policy advisory body to the department on nonpoint sources. This committee is, by statute, the policymaking body for soil and water conservation programs in New York and provides leadership for the state's 57 County Soil and Water Conservation Districts. Its membership includes five lay (voting) members appointed by the Governor and five advisory members:

1. the Commissioner of Environmental Conservation;
2. the Commissioner of Agriculture and Markets;
3. the Dean of the College of Agriculture and Life Sciences (Cornell);
4. the Director of the State Cooperative Extension Service; and
5. the State Conservationist (USDA, SCS).

Since membership on the committee includes representation from most of the agencies and interests directly concerned with nonpoint source management, the committee provides an excellent forum for interaction which will ultimately answer the difficult institutional questions. Perhaps the key question to be answered is that of the role of County Soil and Water Conservation Districts in nonpoint source management.

It is obvious that no state can specify a list of agricultural BMPs which automatically apply to every farm within the state's jurisdiction. Farming situations are too variable, as are physical conditions on the land and in the water. The only workable approach is a *system* of BMPs, somewhat flexible and keyed to the important physical, economic and social variables. This system must be applied through a site-specific planning process. An appropriate process already exists in the farm conservation planning done by County Soil and Water Conservation Districts with SCS technical assistance.

Historically, farm conservation planning has been done for purposes of soil conservation, productivity, and better farm management, not specifically for improved water quality. These are all important objectives which should and will not be lost. Thus, a careful definition and sorting of the roles served by County Soil and Water Conservation Districts will be necessary before an additional role is assigned.

Additionally, and perhaps more important, the County Soil and Water Conservation Districts have historically been service-type agencies, relying on the voluntary cooperation of land owners for their effectiveness. The assignment of a regulatory function to the County Districts would be to assign them a function for which they are not well equipped, which most do not want, and which might well destroy the relationship to land owners that is the basis for their effectiveness. This is one of many reasons why we in New York State, as the preferred course of action, are inclined to rely heavily on voluntary cooperation in our initial approach to nonpoint source management.

I would now like to turn briefly to some thoughts on the federal administration of the program. We recognize that many of the federal initiatives in nonpoint source control are desirable in that they tend to force necessary program development which otherwise might not occur. On the other hand, states must act wisely and carefully in response to federal requirements, so as to develop the necessary programs in ways that best fit the needs of the state and its people.

Over the last two years it has been very difficult to manage our program in this manner. Inadequate funding levels; uncertainty created by a variety of administrative decisions, court decisions, changing planning requirements, and new legislation; and a lack of adequate basic data and research have all impacted adversely on the effort to maintain an orderly and productive planning process. We will seek to work cooperatively with the appropriate federal agencies to overcome these difficulties and chart an orderly path for the future. The state/EPA agreement, recently developed with the purpose of guiding our Water Quality Management Program over the next five years, will be a key vehicle for accomplishing this.

One of the most significant items on the agenda for cooperative action among all concerned is the programming of Culver Amendment money. This provision of the 1977 Clean Water Act calls for $200 million in Fiscal Year 1979 and $400 million in Fiscal Year 1980 for agricultural cost-sharing on BMPs identified in approved Section 208 plans. Although admittedly our information base on nonpoint sources is not good, we believe that we have learned enough over the last two years to utilize this money intelligently. Our effort to program Culver funding will involve an initial, rudimentary identification of waters adversely impacted by nonpoint sources, and will utilize the results of the BMP studies being conducted by the College of Agriculture and Life Sciences and the Applied Forestry Research Institute.

In conclusion, acting in the best interests of the people of our state requires that in proceeding with nonpoint source program development we remain constantly aware of the factors mentioned earlier.

Recognizing the limitations of current knowledge, the initial development of BMPs must be cautious, and the initial systems must reflect careful analysis designed to stay within the limits of what is known.

We must encourage additional research to further our technical understanding of the problems, and develop monitoring systems to better define the magnitude and extent of problems. This must be done within an orderly process of program development over several years.

Recognizing the past record and the recent initiatives of the soil and water conservation movement in the state, we must adjust and strengthen soil and water programs so that they can effectively implement water quality solutions. At the same time we must remember our belief in the broader conservation ethic, and continue to support the other resource management objectives of these programs.

I am confident that we can do these things, and that the cooperative efforts of all concerned will ultimately result in reasonable yet effective programs for nonpoint source management.

4

RESEARCH NEEDS AND CURRENT ACTIVITIES

T. A. Murphy
   Deputy Assistant Administrator for
      Air, Land, and Water Use
   Office of Research and Development
   U.S. Environmental Protection Agency
   Washington, D.C.

The last Cornell Waste Management Conference I attended was four years ago. The agenda for that conference was broader than this year's, covering a wide range of topics on agricultural wastes. However, prominent on the agenda was the subject of nonpoint source pollution. This was a topic which had suddenly become of great concern to the agricultural community.

American agriculture was at that time apprehensive about the new agency called EPA. Looking back it is easy to see why. Agriculture had traditionally been exempt from federal regulation. Suddenly, at the beginning of this decade all this changed. First came OSHA with its forays into slippery barnyards, up rickety ladders and onto "hazardous" toilet seats. EPA was another new federal regulatory agency telling farmers they could not use certain pesticides they had been using for years. In addition, in 1972, EPA had been given a very ambitious water pollution statute (PL 92-500) with apparent authority to get involved in almost every aspect of agriculture.

EPA's first action under this new law was to develop an effluent guideline for feedlots. Although this turned out in the longer run to be fairly reasonable, EPA's early actions raised all sorts of fears. What would these wild young federal lawyers and engineers do next?

Thus, four years ago there was great apprehension that EPA would move hastily and unreasonably to control agricultural nonpoint source pollution. Specific concerns included:

- would EPA try to issue a permit for every farm?
- or impose uniform national standards for agricultural operations?
- or force major pollution control actions without justifying the need in terms of water quality improvement?

The agricultural community had historically been left to itself and liked it that way. But to its great credit, and with considerable foresight, USDA made a vigorous effort to get EPA and agricultural interests together. Since EPA's operational programs in this area were just beginning, we were approached by USDA through our research program. Out of this came: research coordinating committees, joint research projects and a general mobilization of both EPA's and USDA's research capability to address this nonpoint source problem. Frankly, it was one of the best examples of cooperation between federal agencies I have ever seen.

We in EPA, along with USDA and the land grant universities, have done considerable research on agricultural waste management in the past four years. In view of this, I would like to look briefly at what has been learned in that period, as well as what we have yet to learn. This will be done largely in terms of personal impressions.

The first change I note is that agricultural nonpoint source management has come of age. Four years ago we were a group of researchers talking to each other about what policies might emerge and how research might help out. This conference, however, is full of doers—policy officials, government regulators, technical experts who will get the job done, legislative staff, etc. There also is a greater diversity of technical disciplines, including water resources management and economics. All of this confirms our prediction in 1974 that by 1978/1979 we would see "major decisions regarding control of nonpoint sources."

What progress have we made? Considerable, in my view. First let us look at overall policy and conceptual progress.

1. We have developed the BMP concept that agricultural pollution can best be controlled by management practices adapted to specific site conditions rather than by uniform national or regional effluent requirements. This seems entirely obvious now, but was not so four years ago.
2. We have realized that BMPs are not necessarily free—somebody has to pay. This recognition is clear in the Rural Clean Water Program (RCWP) described to you in the paper by Dr. Walter Groszyk. This too was not obvious to everybody in 1974.
3. We have learned that agricultural water quality management cannot be addressed independently, but only in the context of overall crop, soil and water management. In recognition of this, we have brought to bear, through the Section 208 planning process, a significant portion of this nation's agricultural technical expertise, to assist in the development and application of BMPs. This remarkably rapid mobilization of agri-

cultural talent is a real tribute to the foresight, flexibility, and public spirit of the American agricultural community, as well as to the open-mindedness of my operational colleagues in EPA.

4. We have produced many research products during this period which should be useful in selecting and implementing BMPs.

I would like to discuss some of these in more detail, but first let me touch on those areas where we have not done as well in the past four years.

Most important, I do not think we have adequately recognized that agricultural water pollution control has to be viewed as much more than merely application of soil erosion controls. If we equate water quality control with erosion control, I see many dangers. At the operational level, this diverts us from asking the most important and absolutely essential first question in selecting BMPs—what is the pollutant we want to control? It may not always be sediment or a pollutant bound to or transported by sediment. Moreover, it tends to divert our attention from what may, in some cases, be the most cost-effective BMPs—namely "nonstructural" or cultural changes rather than erosion controls. For instance, in some cases, our most effective means for improving water quality may be through managing the timing of application of fertilizers or pesticides. In other cases, the most cost-effective method for achieving water quality goals may not be "on-farm" practices, but actions taken elsewhere in the watershed. I was discussing this whole matter recently with a state conservationist who described a drainage area in the southeast U.S. where he was working with the 208 process. In this area there was a high rate of soil loss from cropland, but also very good water quality. The apparent reason was the abundance of wetlands along the stream which presumably were filtering out sediments during high flow. Naturally one is led to ask if management of these wetlands may not be as or more cost-effective than on-farm erosion control in achieving water quality objectives.

At the national policy level, equating soil erosion control with water quality management could have very serious consequences. If it turns out that erosion controls are not as effective as we might assume against key pollutants—such as soluble nutrients or toxics—or that the major source of nutrients in some watersheds may not be cropland—but rather runoff from dairy barnlots—erosion controls may not solve the problems we are facing. Thus, we may create false expectations on the part of the public, which could lead subsequently to a serious backlash. Moreover, we could be contributing to inefficient use of public funds. Either of these possibilities could create serious political problems for both EPA and the agricultural community. Dr. Groszyk pointed this out in his paper when he said "The RCWP deals, as does the 208 program, with more than erosion. Sediment is not the only, or in many areas, the most severe water quality problem."

With this caution, I do not mean to imply that erosion controls will not play a key role or, possibly, even the dominant role as agricultural BMPs. However, it is at our peril that we oversimplify this role.

The second area that has not been given sufficient attention is our understanding of the role of sediment in degrading water quality. As far as I can tell, we know little more today than we did four years ago about how sediment either directly or indirectly affects the integrity of a body of water. This is important not only for determining the need for erosion controls beyond, of course, where we have obvious sediment problems—but also for selecting the most effective BMPs for sediment-bound pollutants such as some toxics.

Where then does this leave us in regard to what we can or cannot do today? Where do we have the technical tools available to take action, and where are these tools inadequate? Unfortunately, the answers to these questions are not entirely clear, but let me try to address them.

First, we in EPA's research program have been busy developing a number of technical tools during the past four years. This has been largely the result of two very hard-working and competent groups at our Athens, Georgia, and Ada, Oklahoma, research laboratories supported by some excellent grantees and contractors and liberal help from USDA. Many of these people are making presentations at this conference. Over the past four years we have tried to get as far as we can in our capability to relate BMPs to water quality. Our approach to this has been through the use of models. Here I use the term "models" in the broadest sense, meaning any tool for relating pollutant loading or water quality impacts to soil, water or crop management practices at a given site. These may vary from simple loading functions to highly sophisticated, computer-based simulation models. Some of the products of this effort are:

- a management guide for planners on how to evaluate, select and use water quality models;
- some very promising demonstrations of the potential for integrated pest management;
- a manual for estimating current water quality problems in a basin and projecting changes likely to result from pollution control strategies, using manual calculation techniques;
- a publication of loading functions for nonpoint sources;
- mathematical models for evaluating nonpoint source contributions under a variety of land use and land surface conditions, as well as specifically for agricultural watersheds;
- computer programs for evaluating feedlot runoff control under a wide range of climatic conditions; a design model for feedlot waste storage requirements; and a model to evaluate the ability of cropland to accept feedlot runoff;

- a "cropland runoff manual" jointly developed and published with the Agricultural Research Service (ARS), USDA, which presents detailed advice on how to develop guidelines for control of pollution from cropland;
- a series of computer models of varying complexity and scale for estimating the pollutant load reduction and water quality benefits from various BMPs; and
- other similar outputs soon to be available, *e.g.,* water quality management handbook for silviculture.

We think these are some of the best, if not the best, tools available, in terms of both their design and the data upon which they are based. How useful they really are will only be known after individuals such as those at this conference have tried them in the real world.

All of this gives us the capability to do a great deal now. We are quite confident that in many areas we can identify major pollution problems attributable to agricultural sources, and select BMPs which can significantly reduce the pollutants causing the problem.

Thus, there are a number of areas where we can move with confidence and should move vigorously in reducing significant agricultural water pollution problems—as Dr. Groszyk said this morning, "to focus available resources on solving the most severe problems first."

However, we have much less confidence in our capability to quantitatively predict the reduction of pollutant load or water quality benefits and the costs of BMPs. This will become increasingly important as we move beyond the more obvious problem areas where we will have to have a much better capability than we have now to quantify costs and benefits.

This means two things to me. First we must use our early efforts at implementing BMPs as a means to get smarter about their water quality benefits. It would be a shame if we found ourselves several years from now very expert in how to implement BMPs but not knowing whether or not they are worthwhile in terms of water quality benefits.

The second message is that we cannot ease off on our research efforts. However, we need to shift our approach somewhat. We have had our opportunity for developing a series of models and tools. Now it is time to get out and find out how good they are—both in terms of their validity and their utility—in supporting real-life decisions.

What specifically do we need to do? Most urgent, in my view, is to come up with a set of criteria for using the early RCWP projects and comparable efforts as a useful means for learning what we can apply elsewhere. This means defining what types of data collection and analysis are required to evaluate how much the BMPs that are put into practice reduce pollutant loadings and improve water quality. Some of these projects specifically should be designed to provide for this evaluation.

We also need to try some of the models and analytical tools that research has produced in the past four years. They have the potential for being of great assistance. However, their use will require patience and understanding. They are not substitutes for onsite judgment. They cannot make decisions for us. However, if we can find out how good they are, and correct their deficiencies as we go along—as we in EPA are committed to do—they can greatly improve our capability to deal with BMPs in quantitative terms in the next few years.

Another item needing our attention is possible alternative BMPs. What potential do integrated pest management, fertilizer management, or integrated watershed management have? They appear on the surface to be very attractive. We ought to evaluate them as quickly as possible.

We also need to look more closely at the economics of BMPs. Are some BMPs of benefit to the farmer in their own right? Minimum tillage and integrated pest management are possible examples. It would make life easier for all of us if the farmer saw a net benefit to his operation from applying BMPs.

Finally, it is critical that we keep working together. We have many more agricultural scientists than water quality experts to apply to this problem. Yet we face critical water quality questions. We are woefully short of people who are experienced in both aspects of the problem—agricultural and water quality. That means we have got to keep working together with a tolerance and willingness to understand each other's viewpoint. There are benefits to both of us in doing so.

SECTION II

APPROACHES FOR BMP SELECTION

5

BEST MANAGEMENT PRACTICES FOR AGRICULTURE AND SILVICULTURE: AN INTEGRATED OVERVIEW

G. W. Bailey, T. E. Waddell
    U.S. Environmental Protection Agency
    Environmental Research Laboratory
    Athens, Georgia

INTRODUCTION

This paper provides a framework for viewing the identification, selection and implementation of best management practices (BMPs) for controlling nonpoint source pollution from agriculture and silviculture. The discussion touches on the various perspectives from which BMPs must be viewed by the planner/decision-maker whose goal is improved water quality. In reaching this goal, the planner must not only consider the agriculture/forestry production system but also the diverse exogenous forces that impact this system.

Our approach takes a systems view of the agro-forestry system by integrating the many technical, economic, social and institutional factors. Such a view is absolutely necessary to achieve the goal of selecting and implementing BMPs and to assure their maintenance over time.

Agriculture and forestry are important to modern society. These economic sectors have provided food, clothing and shelter for the increasing population of this country along with a substantial portion of the world over the last three decades. This increased productivity in agriculture and forestry has resulted from the fruits of scientific research and the application of new and improved technology. Germ plasma with high productivity potential, new tillage and improved cultural practices, chemical weed and insect controls, irrigation scheduling, and more efficient and varied implements have contributed to this significant increase in productivity of food, fiber, feedstock, livestock and wood products.

The same technologies that have helped revolutionize agriculture and forestry have also aroused public concern that their application may have detrimental effects on the environment. For example, for every ton of grain we export, we may be exporting several tons of topsoil to the Gulf of Mexico. The generation of residuals like topsoil, which is removed, transported and deposited downstream, is viewed here as an output external to the agroforestry system. The intent of the Federal Water Pollution Control Act, As Amended (PL 95-217) is to internalize the cost of such residual generation by implementing BMPs.

One must first understand the nonpoint source pollution problem, however. It must be viewed in light of the legislative mandates and the related environmental laws. Next, one must define and understand the system being managed—the inputs, how the system works internally, and the outputs. The goal is to impose on this system an environmental strategy that comprises some mix of physical methods, implementation incentives, and institutional arrangements to provide pollution control. Monitoring of performance indicators dictates whether the strategy is successful or in need of adjustments. It is this sequence of logic the rest of the paper will follow.

DEFINITION OF PROBLEM

Why are we concerned about agricultural and silvicultural nonpoint sources in the first place? What is the problem? We suggest there are several plausible answers. We believe that the nonpoint source pollution problem must be viewed as a physical/technical problem, an economic problem, and a social problem.

The Physical/Technical Dimension

The National Commission on Water Quality[1] estimated that after 1983, nonpoint discharge and urban runoff will be major obstacles to attaining water quality objectives. (A nonpoint source is any unconfined source, such as a diffuse land area. A point source is a confined source, such as an industrial concern or municipal waste treatment plant.) Nonpoint source discharges primarily occur during rainfall events when storm runoff from the land surface carries sediment, pathogens, sediment-adsorbed chemicals, dissolved chemicals (such as nutrients and pesticides), heavy metals, and easily oxidizable organics into adjacent waterways. Dissolved chemicals may also percolate through the soil to interflow regions and/or groundwater, and be discharged in subsurface flows. Salinity is a major nonpoint source pollutant in the West as irrigation return flows carry dissolved salts into waterways.

Nonpoint source pollution, in addition to being characterized by its stochastic nature, is dynamic and has multimedia dimensions. It is dynamic in the sense that land uses and configurations change over time making the pollutant mix vary both spatially and temporally. The pollution is also characterized by the fact that a chemical released in one medium may have serious implications in another environmental medium. Figure 1 illustrates the movement, transformation and degradation of a pesticide among the soil, water and air media. Although the direct discharge of some chemicals may have no apparent adverse impacts in the original medium, intermedia transfers, together with degradation products, may be cause for concern.

Once in the watercourse, nonpoint source pollutants may degrade water quality and impair water use. Estimates of pollutant contributions to surface waters from selected nonpoint sources are provided in Table I. Note that approximately one-fourth of the contribution of most pollutants is attributed to the category "Natural Background." Although this table lends some insight into the relative significance of different sources, it does not sufficiently define "the problem."

The concern is not with loadings of residuals per se; rather the concern is their impact on water quality and its associated implications on public health and welfare. Obviously, the relative significance of a ton of sediment reaching a stream during a runoff-producing rainfall event may be quite different from a comparable amount of barnyard runoff continually trickling into a "dry weather" perennial stream. Not only is concentration an important component in this problem assessment but also exposure for some receptor—man, fish, aquatic organisms, etc. The concern rests on those situations in which nonpoint pollutant loadings exceed the assimilative capacity of the water ecosystem. As Bower and Spofford[3] state, "The natural environment has a capacity to assimilate, in some degree, all forms and types of residuals through the mechanisms of transport, transformation, and storage. In effect, the environment acts as a buffer between the discharger and the receptor, that is, it dissipates, absorbs, dilutes, and degrades or modifies residuals. However, the capacity of the environment to assimilate residuals varies from place to place and from time to time, depending on both local conditions and the stochastic nature of some component of the environment, such as stream flow, temperature, and sunlight." Uncertainty that surrounds what we do know about the process suggests the real dimensions of the physical aspects of the problem.

The Economic and Social Dimensions

The elevated consciousness of man concerning his environment that can be attributed to the "environmental movement" since the late 1960s has had the effect of altering social attitudes and resultant consumption patterns. Many

36 BMPs FOR AGRICULTURE AND SILVICULTURE

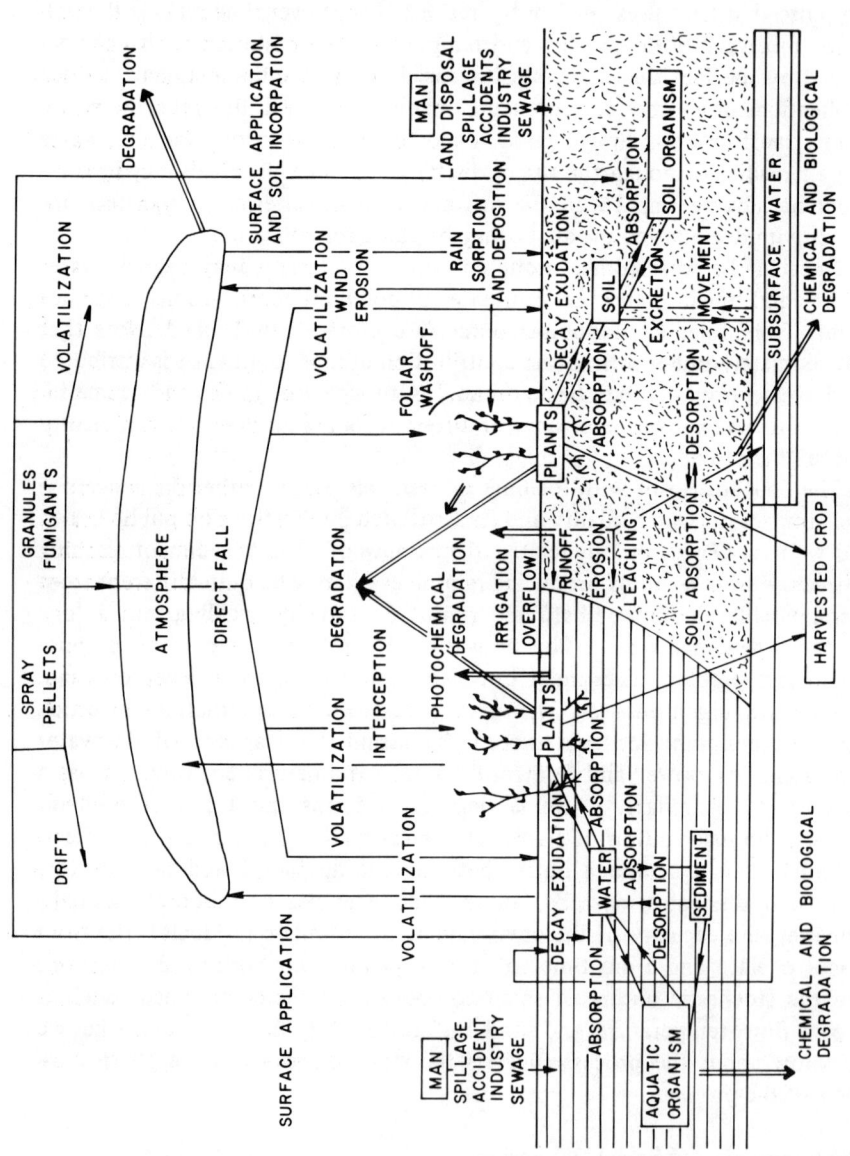

Figure 1. The pesticide cycle in the environment.[2]

Table I. Estimated pollutant contributions to surface waters from selected nonpoint sources in the contiguous 48 states[a]
(average pollution loads in million tons per year).

Nonpoint Source Category	Sediment	BOD	Nitrogen	Phosphorus	Acids[b]	Salinity[c]
Cropland	1870	9	4.3	1.56	--	57.3
Pasture and Range	1220	5	2.5	1.08	--	--
Forest	256	0.8	0.39	0.089	--	--
Construction	197	--	--	--	--	--
Mining	59	--	--	--	3.1	--
Urban Runoff	20	0.5	0.15	0.019	--	--
Rural Roadways[d]	2	0.004	0.0005	0.001	--	--
Small Feedlots	2	0.05	0.17	0.032	--	--
Landfills	--	0.3	0.026	--	--	--
Subtotal	3626	15.8	7.4	2.8	3.1	57.3
"Natural Background"	1260	5.0	2.5	1.1	--	--
Total	4886	20.8	10.0	3.8	3.1	57.3

[a] 83.8 million ha (207 million ac) in public lands (14% of contiguous U.S.), mostly in Rocky Mountain Region, were excluded due to inadequacy of information.
[b] As $CaCO_3$.
[c] From irrigation return flow.
[d] Deposition from traffic-related sources.
Sources: Midwest Research Institute, 1975. National Assessment of Water Pollution from Nonpoint Sources (Draft of Final Report), U.S. Environmental Protection Agency, Washington, D.C.

debates over social conflicts in the use of our natural and renewable resources, including water, forest land, wilderness, etc., reflect this change in social preferences. Just recently it was reported that by a three to one margin Americans are still opposed to any policy that would slow environmental clean-up programs to improve the economy or to increase energy production.[4] There is still a strong social preference for improved environmental quality. This preference can be attributed in some part to increased disposable income, increased leisure time, altered job preferences, as well as other sociodemographic factors.

This shift in attitudes and preferences has resulted in individuals making choices between increased wilderness opportunities and increased timber production (and thus reduced lumber prices). Society values environmental amenities more highly relative to other goods and services now than it did a decade ago. This behavior has placed increased pressure on the use and management of our renewable resources.

38 BMPs FOR AGRICULTURE AND SILVICULTURE

The social dimension of the problem relates to resource scarcity and the uncertain future. Although debates about the limits to growth and alternatives to growth are still carried on, the real issue that is emerging is the direction and kind of growth that should take place.[5] Issues of population growth, food and energy production, and environmental quality may cloud the distant horizon, but it seems plausible that the continued depletion of exhaustible resources will lead to greater social concern over the management and use of renewable natural resources. The uncertainty posed by the future has forced society to consider the future. This concern lends further support for a need to understand the nature of the nonpoint source problem relative to renewable resource questions.

LEGISLATIVE MANDATES AND WATER QUALITY REQUIREMENTS

These societal concerns led to the passage of a host of environmentally related legislation by the Congress of the United States and by the states. Provisions of major legislative actions that relate to nonpoint sources of pollution, including agriculture and forestry, and to watershed planning and management are discussed below.

Federal Water Pollution Control Act of 1972 (PL 92-500)

Those sections that directly pertain to the goals of the Act with respect to pollution control in agriculture and forestry are summarized.

The objective of the Act, as stated in Section 101(a), is to restore and maintain the chemical, physical and biological integrity of the nation's waters. The Act establishes as national goals that the discharge of pollutants into the navigable waters be eliminated by 1985 and that an interim goal of water quality that provides for the protection and propagation of fish, shellfish and wildlife and for recreation in and on the water be achieved by 1 July 1983.

Section 105 covers grants for research and development. Among other things it authorizes the Environmental Protection Agency (EPA) Administrator, in consultation with the Secretary of Agriculture, to make grants with respect to new and improved methods of preventing, reducing and eliminating pollution from agriculture and to disseminate such information.

Section 208 provides for areawide waste treatment management. Although this section impacts heavily on municipal and industrial pollution sources, Subsection (b)(2)(F) calls for areawide waste management plans to include a process to identify, if appropriate, agriculturally and silviculturally related nonpoint sources of pollution, including runoff from manure disposal areas, and from land used for livestock and crop production, and to set forth procedures and methods (including land use requirements) to control such sources to the extent feasible.

Section 304(e) requires the issuance of guidelines for identifying and evaluating the nature and extent of nonpoint source pollutants, and methods to control pollution resulting from, among other things, agricultural and silvicultural activities including runoff from fields, crops, and forest land.

Federal Water Pollution Control Act, As Amended (PL 95-217)

Major amendments to PL 92-500 with regard to agriculture and forestry occur in Section 208.

Subsection (b)(2)(F) redefines return flows from irrigated agriculture and their cumulative effects as nonpoint source pollution.

Subsection (j)(1) authorizes the Secretary of Agriculture with concurrence of the EPA Administrator and acting through the Soil Conservation Service and other U.S. Department of Agriculture (USDA) agencies to enter into contracts of not less than five and no more than ten years with rural landowners and operators to install and maintain BMPs to control nonpoint sources of pollution for improved water quality. Such contracts are valid only in those states or areas in which the areawide waste management program has been approved by the EPA Administrator and where the practices are first certified by the areawide or statewide management agency to be consistent with such plans and will result in water quality improvement.

Subsection (j)(2) requires the Secretary of Agriculture to furnish technical assistance to the land owner or operator for carrying out the conservation practices and measures approved by the authorized implementing agency developed under the areawide waste management plan. Cost-sharing for the installation of water quality management practices is also authorized.

Subsection (j)(3) allows the Secretary to modify earlier contracts if it is advantageous to execute the program to facilitate program administration or to accomplish equitable treatment in regard to other conservation, land use, or water quality programs.

Subsection (j)(4) requires that priority in terms of assistance be given to those areas and sources that have the most significant effect upon water quality.

Subsection (j)(5) requires the Secretary to, where practicable, enter into agreements with Soil Conservation Districts, State Soil and Water Conservation Agencies, or State Water Quality Agencies to administer all or part of the established programs and pay for that portion of the cost incurred in the administration of the program.

Subsection (j)(6) stipulates that the contracts shall be entered into only if area or state management agencies assure an adequate level of participation by land owners and operators in that area. The local Soil Conservation District and the Secretary will determine the priority of assistance rendered to assure that the most critical water quality problems are addressed.

In summary, this legislation calls for the installation of BMPs that are effective in improving water quality, economically feasible, equitable, and implementable.

Soil and Water Resources Conservation Act of 1977 (PL 95-192)

The Soil and Water Resources Conservation Act of 1977 also affects the long-term use of BMPs to improve water quality. Under this Act, a continuing appraisal is to be made of the soil, water and related resources to include, among other items: (1) analysis of the nation's soil, water and related resource problems, and (2) identification and evaluation of alternative methods for conservation, protection, environmental improvement and enhancement of soil and water resources and an analysis of their costs and benefits. A report to Congress on this appraisal is required every five years.

In summary, agriculture has received significant emphasis and attention in recent legislation. Areawide, statewide and river basin planning for water pollution control is no doubt in our future. Management systems encompassing all land uses (agriculture and silviculture, among others) to prevent or control nonpoint source pollution that is not suitable to collection and treatment must be developed. Understanding and designing these management systems will depend upon various kinds of analytical models for their identification and selection.

THE AGRO-FORESTRY SYSTEM

Agricultural and forestry resource managers are concerned with the efficient production of food, fiber, feedstuff and wood for the use and benefit of man. Multiple-objective forest management also produces amenities associated with recreation, wildlife, aesthetics, option values, watershed protection, etc. The increase in demand for these outputs stems from world population increases and shifts in consumption patterns, which, in turn, increase conflict over resource use including environmental quality. This increased pressure on environmental quality is the result of introducing more environmentally sensitive lands into production and more intensive management of current lands under production.

These production activities are complex from a physical, social and economic standpoint. A systems analysis approach is needed in order to better understand the many facets or parts, how they are interrelated, the interaction between these factors, and the dynamic nature of certain of these factors.

APPROACHES FOR BMP SELECTION 41

The systems view is a holistic one[6] which implies that an isolated study on the parts of the system will not permit the understanding of the complete system because the separate parts are linked in an interacting manner. A perturbation in one part of the system will affect not only that one part but the system as a whole.

A systems viewpoint of the agro-forestry system is shown in Figure 2. The scale of the system is arbitrary and could consist of a field, a watershed, a state, a river basin, or a national economic sector. The landscape at the field level is determined by the combination of physical characteristics of soils, topography, vegetation and drainage network. Inputs to and outputs from the system are enumerated. Inputs can be classified as controlled (deterministic) or uncontrolled (stochastic). The physical/environmental inputs are uncontrolled and contribute to the somewhat stochastic nature of the outputs: (1) supply of commodities such as food, fiber, wood and feedstuff, as well as other amenities like recreation and wildlife; and (2) pollutants/residuals generated during production operations.

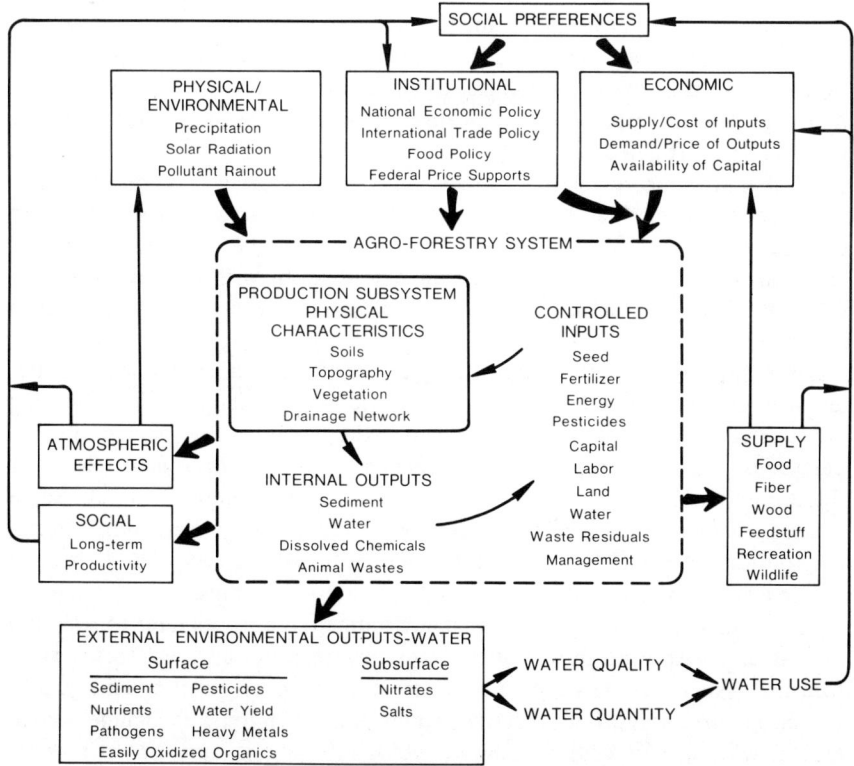

Figure 2. Systems view of agriculture and forestry.

The farmer or forest manager, although subject to external economic forces like supply/demand and availability of capital and institutional considerations such as international trade policy, federal price support programs, etc., exercises great control over a host of inputs. Such inputs include the production function components of land, labor and capital, as well as more specific inputs of chemicals (fertilizers and pesticides), irrigation water, waste residuals (such as animal wastes), levels of energy intensity, seed and management information. The latter includes information about pest dynamics, meteorology, soil/water/plant relationships, etc.

Residuals are those outputs, either material or energy, whose value is typically less than the cost of collecting, processing and transporting them for use. Because of their relatively low values, there is no incentive for the manager to prevent, collect or reuse such residuals. A pollutant, then, is a residual that adversely affects the use of one or more environmental media.

Internal outputs result from the response (or physical characteristics) of the production subsystem. Internal outputs such as sediment, animal wastes, crop residues, or irrigation return flows may be recycled or may be discharged externally into the atmosphere or into the receiving water. Internal outputs may also impair soil quality which, in turn, may affect quantity or quality of yields. In the case of pesticides, there are potential short-term effects on nontarget soil microorganisms important in the soil/plant relationship. Buildup of salts in irrigated soils may also impair productivity. The loss of topsoil is an obvious concern relative to maintenance of long-term soil productivity. This social concern has manifested itself through passage of legislation and establishment of public institutions whose theme is the conservation of our soil and water resources. To a certain extent, even without environmental controls, these internal output effects are factored into or internalized into the agro-forestry system.

External environmental outputs may affect directly and/or indirectly either air or water. The integrity of the atmosphere may be damaged from burning agricultural wastes and forest seedbeds, the volatilization/degradation of pesticides and the transport of fertilizers to the upper atmosphere. Some of these pollutants are, in turn, removed from the atmosphere in rainfall. Some residuals generation, especially those related to visual or long-term effects, may initiate a social concern that could eventually be expressed through public or legal institutions, *e.g.,* restrictions on open burning.

In the case of water, the external environmental outputs refer to residuals moved or transported into subsurface or surface water external to the agro-forestry system. These outputs may affect both water quantity and quality which, in turn, may affect different water uses. Such impacts include human health, wildlife and wildlife habitat, ecological systems, agricultural production, recreation and aesthetics. Such impacts may have direct or indirect economic dimensions or may have implications for the health and welfare of future generations. Such a social concern may then be expressed through public, legal and political institutions.

These economic and institutional considerations, exogenous to the agro-forestry system, impinge on the decision-making of the farmer or forestland manager, relative to the inputs he controls. From this systems view, outputs can be controlled by: restricting and managing inputs, changing the production system itself, and managing outputs. BMPs, therefore, may be selected to affect inputs, the projection system and/or outputs.

BMP IDENTIFICATION, SELECTION AND IMPLEMENTATION

National water quality goals dictate that programs be established to control nonpoint-source-generated pollutants. Now that the agro-forestry system has been defined in the larger social context, an environmental quality management strategy can be formulated to realize these goals. The basic interrelated components of an environmental quality management strategy are shown in Figure 3. Physical methods are those actions or practices that reduce or modify the discharge of the pollutant(s) into the environment or directly improve the assimilative capacity of the environment or both. Implementation incentives are the positive and negative inducements that stimulate the installation, maintenance and continued operation of the physical methods (*i.e.*, BMPs) or discourage other kinds of action. The institutional arrangement provides the authority to impose the implementation incentives and assigns the responsibilities for executing the activities and functions of environmental quality management.

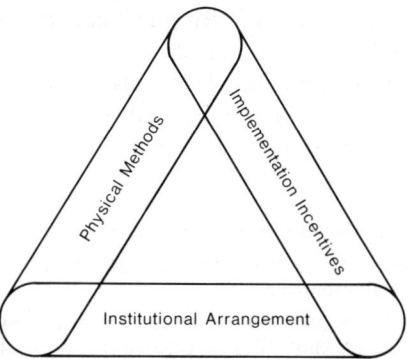

Figure 3. Components of an environmental quality management strategy.[7]

A major problem is still to rigorously define the reationale and methodologies required to select BMPs.

44 BMPs FOR AGRICULTURE AND SILVICULTURE

Section 208 requires that a process similar to that displayed in Figure 4 be followed for the effective implementation of BMPs. The water quality goals as defined by PL 95-217 have been discussed. The next step is an environmental assessment to identify actual or potential problems given a land use configuration scenario. Once defined, the potential solutions (candidate BMPs) to the problem are filtered through a "sieve" of criteria to determine the most appropriate BMPs for a particular site. Next, the implementation question must be addressed: How can the implementation and maintenance of BMPs be insured?

Environmental Assessment

The first step in this identification/selection/implementation process is to conduct an environmental assessment to determine the nature and extent of the nonpoint source pollution problem. Many different approaches or assessment methodologies are currently available. They range from quite subjective or judgmental approaches utilizing, at most, a desk-top calculation analysis, *e.g.*, Tetra Tech,[8] to more sophisticated simulation or predictive models, *e.g.*, NPS.[9] These assessment methodologies also enable the evaluation of alternative nonpoint source controls/management practices—candidate BMPs.

The Concept of BMPs

The approach embodied in PL 92-500 and PL 95-217 for the control of nonpoint source pollution from agricultural and silvicultural activities is source management rather than collection and treatment of pollutants. This approach means that management practices or combination of such practices are the vehicle to use to provide a cost-effective means to prevent or control pollution from agricultural and silvicultural activities. The term "best management practices" has been used extensively over the last two and one-half years and still is a puzzle to many. What are *best* management practices best for?

- Who—the farmer? society? or the environment? or all three?
- What purpose—erosion control? enhancement of water quality? optimization of national resource goals?
- What scale—national prescriptions or site specific prescriptions? What crop production systems? What combination of climatic, edaphic, or topographic conditions?

It is hoped that this conference will contribute to answering some of these questions. The definition for BMP as given in the Federal Code[10] is as follows:

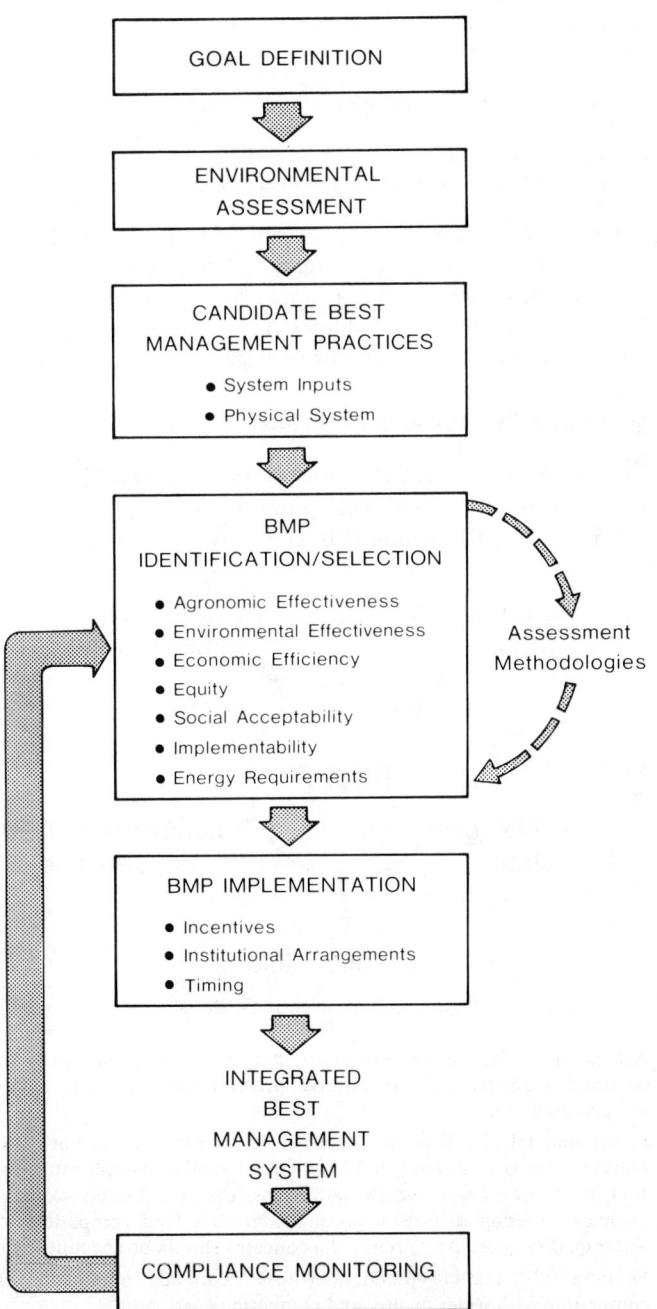

Figure 4. Schematic for agro-forestry nonpoint source control.

"Best Management Practice (BMP) means a practice or combination of practices that is determined by a State (or designated area-wide planning agency) after problem assessment, examination of alternative practices and appropriate public participation to be the most effective practicable (including technological economic and institutional considerations) means of preventing or reducing the amount of pollution generated by nonpoint sources to a level compatible with water quality goals."

Management practices are the key to source control management. Pragmatically, that is, from a farmer's standpoint, the categories of management practices for crop production (after a farmer has decided what crop he is to grow) include those displayed in Table II. These categories are crop management, soil/water management, nutrient management, and pest management.

Criteria for Defining Best Management Practices

Earlier in this paper, certain questions were raised regarding "best" management practices for whom and what purposes. To help answer these questions, the definition of BMPs implicitly comprises the following criteria (see Figure 5).

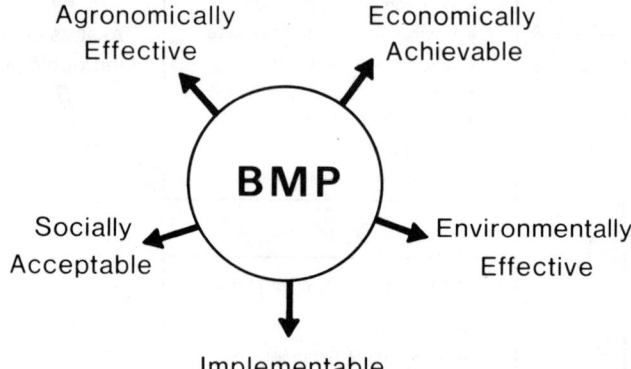

Figure 5. BMP selection criteria.

1. Agronomic effectiveness—practices that promote productivity and commodity quality and that conserve, preserve and, if possible, enhance soil productivity.
2. Environmental effectiveness—practices that prevent, reduce or control pollutant loads (dissolved or bound on particulates)—sediment, nutrients, pesticides, heavy metals, pathogens, toxics and easily oxidizable organics—entering surface or groundwater to a level compatible with water quality goals. Any intermedia concerns should be accounted for.
3. Economically feasible—practices whose economic implications are compatible with water quality and competing goals.

Table II. Crop management methods.[11]

A. *Crop Management*
1. Tillage
   - Conventional - moldboard plow, disc, narrow
   - Timing - fall, spring
   - Chisel plowing
   - Conservation - minimum, no-till
2. Crop Sequencing
   - Mono-crop
   - No-meadow crop
   - Relay cropping
   - Double cropping
3. Seed/Plant Improvement
   - Weather resistance
   - Salt tolerance
   - Production efficiency
   - Early or late maturation

B. *Soil/Water Management*
1. Runoff and Erosion Controls
   - Contouring
   - Terraces
   - Cover crops
   - Grassed waterways
   - Tile drains
   - Diversions
   - Land forming
   - Row spacing
   - Harvesting and planting times
2. Moisture Conservation Practices (*e.g.*, fallow cropping)
3. Wind Erosion Controls
   - Strip cropping
   - Barrier rows
   - Windbreaks

C. *Nutrient Management*
1. Formulation, Granular, Liquid
2. Species (*e.g.*, NH^4 vs. NO_3 form of N, animal *vs* municipal)
3. Amount Applied
4. Application Methodology
5. Timing of Application

D. *Pest Management*
1. Scouting
2. Pesticides
   - Application methodology
   - Amount applied
   - Timing of application
3. Pest-Resistant Crops
4. Integrated Controls
5. Cultural/Mechanical Methods
6. Biological Controls

4. Socially acceptable—an effective public participation process insures that practices are reasonable and will have public support.
5. Implementability and other institutional considerations—this insures that practices are legal and that institutions are in existence or can be formulated in a way to provide flexibility and, to the extent feasible, simplicity in administration in achieving implementation.

A wide array of management practices is available to control the various types of pollutants and thus reduce pollutant loads entering receiving waters. Practices for controlling erosion, reducing runoff, and managing agricultural chemicals have been discussed extensively elsewhere.[1,2] Some practices are specific for one type of pollutant; others help control two or more different types of pollutants.

Pollutants can be divided into two classes: conservative and nonconservative. Those falling into the conservative class are sediment, pathogens, and heavy metals. Those that fall into the nonconservative class are pesticides, nitrogen and phosphorus, and easily oxidizable organics. Management systems can be selected to control the two different classes of pollutants. Those practices that affect the physical system, *e.g.*, tillage, structural controls, and land use, influence erosion and runoff. They influence erosion by affecting slope, slope length, soil erodibility and cover; they influence runoff by affecting the above functions as well as soil infiltration capacity.

The nonconservative pollutants can be controlled by managing the input of these materials with regard to the timing and level of application and the nature of the chemical (environmental properties), formulation and application methodology. The timing or temporal management of these controllable inputs is crucial and should consider (1) probability of runoff event with respect to application date and the degradation or transformation behavior of the chemical (for example, greatest potential for pesticide loss in runoff occurs at application day and the amount available for runoff decreases exponentially thereafter), (2) degree of cover, (3) receptor life cycle stage at which greatest impact can occur, and (4) stream flow (greater potential impact usually would be expected at low flow than a higher flow).

Therefore, "best" management practices from an environmental standpoint:

1. provide optimum soil cover (crop residues, canopy development, and/or soil surface roughness) to dissipate raindrop impact and reduce runoff velocity;
2. provide optimum soil infiltration and flow path to minimize erosion through soil detachment and transport and reduce runoff volumes through enhanced infiltration during maximum periods of runoff producing rainfall events; and
3. minimize or reduce soil solution concentrations of soluble pesticides, heavy metals, toxics, plant nutrients, heavy metal-organo-complexes at the soil surface or within the root zone during periods of high runoff

producing rainfall events to minimize or control runoff to surface waters and movement and leaching to interflow regimes and groundwater.

Integrated with an assessment of the agronomic and environmental effectiveness of management practices is an evaluation of the economic impact on the farm or forest owner—how his net revenue may be affected by adopting BMPs. The approaches include farm budget and linear programming models. The various applications of these approaches that have been made are examined and discussed in these proceedings.

BMP Selection and Implementation

Once a practice or set of physical methods has been evaluated and deemed to be candidate BMPs, the planner is faced with other questions related to economics, fairness, legality—the whole implementation issue. The spectrum of alternative control instruments[7] ranges from education in a purely voluntary program to public purchase of the land area causing concern. Which combination of implementation incentives will most likely produce the desired result in the most cost-effective way? Technical solutions to problems are proposed all too often without attention being given to how to get the solution implemented. A solution sitting on the shelf is no solution at all. Fortunately, increasing attention is being given to how to get technology off the shelf and how to effectively transfer technical information.[13]

One way to view a taxonomy of implementation incentives is displayed in Table III. Implementation incentives, contrary to widespread belief, are broader than purely economic in nature. Incentives may be economic or noneconomic, including regulatory, administrative, judicial or educational/informational methods. Incentives may be positive, *e.g.*, tax credits, cost-sharing subsidies, education programs, etc., or they may be negative[14] to discourage certain kinds of conduct—such as soil loss tax, user charges, effluent charges, fines and permit fees.

The following questions apply in choosing among alternative implementation incentives:

1. Are there time constraints that would rule out any alternative from consideration in the short-term?
2. What are the relative equity aspects of each alternative?
3. Which alternative is most socially acceptable, *i.e.*, politically feasible?
4. Which institutions can be utilized in the implementation process? Is it realistic (legally, politically, etc.) to talk about new institutional arrangements?
5. Are there serious legal barriers that make the alternative unrealistic in the relative time frame?
6. What are the costs of implementing alternatives—including both private and social costs? The latter would include informational, contractual

Table III. Classification of implementation incentives for agriculture and forestry nonpoint sources.[7]

A. *Regulatory* (by law, ordinance, or permit)
 1. Specification of a physical method.
    - specify characteristics of raw material input, *e.g.*, no more than 5% phosphorus in fertilizer concentrate.
    - specify production process as a BMP, *e.g.*, minimum tillage with double cropping.
    - specify residuals modification or handling process, *e.g.*, sediment ponds or grassed waterways; animal manure storage requirements.
    - specify product output characteristics, *e.g.*, allowable limits on heavy metal concentrations in vegetables grown on land utilizing sludge.
 2. Specification of a result or performance.
    - specify quantify of a residual discharged per unit of time, *e.g.*, 5 tons soils loss per acre per year on the average.
    - specify performance, *e.g.*, producer certification that a pesticide is environmentally "safe."
    - specify ambient environmental quality to meet, *e.g.*, specified chlorophyll-*a* concentration in downstream impoundment.
 3. Specification of limitations on location of activity, *e.g.*, no road building on lands exceeding certain slopes; land use restrictions.
 4. Specfication of extent, timing and type of activity; *e.g.*, irrigating or fertilizing only when soil/plant tests indicate a need; no fall and winter application of animal wastes or fertilizers.
 5. Specification or procedure, *e.g.*, requirement that implementation plan be prepared and filed with Soil and Water Conservation District Office; permit be issued prior to road building.

B. *Economic*
 1. Applied directly to residuals, *e.g.*, soil loss tax.
 2. Applied to inputs or product outputs, *e.g.*, agricultural chemical tax; water use pricing.
 3. Applied to activities, *e.g.*, tax credit or allowance for accelerated depreciation for conservation tillage equipment like chisel plow; cost sharing for BMPs.
 4. Direct public investment, *e.g.*, purchase of land.

C. *Administrative* (by order within governmental agencies or other organizations)
 1. Applied directly to residuals, *e.g.*, forest management requirement for buffer strips.
 2. Applied to products used, *e.g.*, limits on chemicals used in forest management.
 3. Applied to activities, *e.g.*, limits on clearcutting, prescribed burning.
 4. Applied to services, *e.g.*, increased technical support by SCS for BMP guidance.

D. *Judicial*
 1. Court and/or administrative law review and action, or threat thereof, to compel compliance; civil and/or criminal suits.

E. *Educational/Informational*
 1. Programs can be mounted to inform groups of the implications of their activities with respect to generation of nonpoint source pollutants and adverse impacts on ambient quality, and alternative behavior patterns that would reduce such impacts, *e.g.*, use of Integrated Pest Management.

and policing or monitoring costs as well. Which alternative appears to be most economically efficient in terms of social welfare?

The Section 208 agencies have considered these questions to different levels of sophistication. It is obvious that there is still much to learn. Some will probably be learned by trial and error, an approach that can be costly to society.

Research is contributing to a better understanding of the equity aspects of alternative approaches to controlling agricultural and forestry nonpoint source pollution. Who gains? Who loses? There may be comparative advantages when one state implements more stringent nonpoint source control laws than another.[15] Several questions remain unanswered, however. Should only those who benefit pay for improvements in water quality? Should the costs of implementing the nation's nonpoint source control program be shared by all equally? Should society opt for the least-risk criterion in consideration of the temporal/intergenerational issues? Such questions are fraught with value judgments that decision-makers must bring to bear on dealing with problems having social dimensions.

Even within the agriculture/forestry sector, there are equity issues that cannot be ignored. How are farmers on more erosive soils affected relative to those on less erosive soils? How should we deal with the small woodlot owner relative to the larger land holder? To what extent should the selection of BMPs be sensitive to the size of the operation and then what incentives may be necessary to achieve compliance? Such equity issues are, unfortunately, not straightforward. To formulate policies for BMP implementation, information is needed on how different groups are affected by the policies under consideration. The valuation of such differences must largely be left to the political process and the value judgments an individual may bring into the situation.

Work with the agricultural conservation program over the past years have given us insight into the attitudes of farmers toward various soil and water conservation practices. Preliminary studies[16] have shown that farm community acceptance of such policies will be better if traditional agricultural agencies are involved in the implementation of policies than if new agencies are developed or if existing nonagricultural agencies are used. Work has shown that policies that allow the farmer flexibility in selecting the means of controlling erosion were viewed by farmers as more fair than those which impose uniform prohibition, *e.g.*, certain tillage practices.

The thrust of the current legislation relative to controlling agricultural and forestry nonpoint sources is to build to the greatest extent practicable on existing institutions. The Section 208 agencies were instituted to conduct areawide planning in an integrated sense. In addition to the role assigned to

EPA, the Clean Water Act amendments for 1977 place the responsibility for BMP implementation on the USDA, as well as the State Soil and Water Conservation Districts.

These are well established institutions with grass roots support in the agriculture sector. In the forestry sector, state and private forestry will likely play a key role in the implementation of BMPs. These institutions, especially those in agriculture, are well equipped for addressing nonpoint source problems that may be alleviated by the application of soil and water conservation practices. As our understanding of what constitutes a BMP grows, including possibilities of input management and the role of different mixes of incentives, flexibility by those institutions historically targeted toward structural solutions will be a necessity.

If, with time, new or more integrated institutional arrangements are desired to address water resource issues, then the costs of establishing such must be considered along with the perceived benefits. The establishment and effective operation of new institutions is time-consuming and usually not expedient politically.

Implementation incentives must be legal.[17] Generally, compulsory programs are more subject to legal constraints. The following questions should be important in designing implementation schemes that are legally defensible:

1. Are state water rights laws violated?
2. Does a restriction on agriculture constitute a taking?
3. Are the restrictions reasonable?
4. Is there a public purpose for the regulations?
5. Are there due process restraints upon state action?
6. What are the pros and cons of statutory vs administrative restrictions/ standards?
7. Does the policy require an exercise of policy powers?
8. Are there equal protection rights that should be considered?
9. Are there constraints upon the use of proceeds from a tax?

The economics of alternatives should consider both the private and social costs. The milieu of economic costs include effects on:

1. production costs, thus net revenues;
2. yield or outputs, including commodities and recreation opportunities;
3. prices of outputs;
4. employment;
5. informational costs, including environmental assessments;
6. public investments, including land purchase, subsidies, cost-sharing arrangements;
7. lost tax revenue related to investment credits;

8. costs associated with administration, including planning, monitoring and enforcement; and
9. external costs associated with adverse impacts on water quality and ultimately water use.

Both private and social costs combined provide a measure of social welfare. To the extent analysts can measure or decision-makers can even consider this array of economic costs, the solutions may optimize social welfare in some sense. The more thorough the analysis considers the components in this calculus, the more confidence we may have that our decisions are "reasonable" and "practicable"—the old nemeses of Best Available Technology (BAT), Best Available Technology Economically Achievable (BATEA), and Best Practicable Technology (BPT).

Economic models are available to assess, at both the national and regional levels, the economic efficiency of alternatives for controlling nonpoint pollution from agriculture. Watershed-scale models are also available to evaluate the economics of alternatives. Although no model quantitatively examines all those private and social costs about which the theorist demands information, available tools may bring sufficient insight into decision-making to understand whether the economics of alternatives may or may not have some significance at various scales. An integrated physical-economic systems analysis of irrigated agriculture is being conducted in California. A general resource planning model for the evaluation of forest management alternatives, including multiple objectives and their environmental impacts, is also under development. From this conference, information on assessment methodologies at various levels of sophistication that have been or could be used to address the issue of "what is most economic," should be forthcoming.

POLICY ISSUES AND RESEARCH NEEDS

The complexity of the problem in controlling nonpoint source pollution from agriculture and silviculture raises many issues. The following list of policy issues and research needs serves to highlight and summarize the major concerns raised in this paper:

1. The control of nonpoint source pollution associated with the agroforestry sector will require unique institutional arrangements—a considerable challenge for environmental managers.
2. The first round of BMP implementation appears to be focused on erosion control practices. Recent EPA guidance that has been issued indicates that greater attention be given to evaluating alternatives as well as looking at the relative trade-offs. The old ways of doing business may not be adequate—not adequate in the sense of being the most cost-effective way of achieving improved levels of water quality.

The cost of implementing BMPs across the nation will likely be significant. EPA Administrator Douglas Costle has said, "We must proceed in a balanced, reasonable fashion." In a broader vein, the nation must be aware of the trade-offs of increasing food production and environmental quality. Information suggests that up to a moderate level of food and fiber production, the two goals may not necessarily be incompatible.

3. The American public should not expect the farmer to voluntarily adopt a management practice that is not to his economic benefit. It is not expected in other economic sectors. The effective implementation of a nonpoint source control program or strategy will necessitate the proper mix of implementation incentives. Modified or innovative approaches for implementing incentives may be needed.
4. Nonpoint source control programs must be sensitive to the site-specific nature of the problem and to the dynamics of economics and land use. The technological solutions must be appropriate in terms of the problem identified, including energy considerations.
5. The long-term environmental implications of wide-scale applications of new and emerging technologies and the long-term impacts of environmental loadings of residuals must be recognized.
6. A more integrated view of resource management must be taken so there is not significant suboptimizing at the farm, local or regional levels. Policy-making seems to be moving in this direction, as evidenced by the major role of USDA in PL 95-217 as well as designation of pass-through monies to the U.S. Fish and Wildlife Service under the same law.
7. The ultimate fate of chemicals in the environment must be understood. This understanding must extend beyond the primary chemical impacts to the second- and third-order degradation products' impacts in a multimedia sense, especially as they relate to human exposure of chemicals and other toxic substances. As agriculture becomes more intensively managed, this issue may become even more important.
8. Agriculture/forestry nonpoint source control policies must be equitable. Environmental managers need to build trust in the agriculture community.
9. The control of nonpoint source pollution will not be without cost. Somebody will have to pay. Typically such costs will be borne by the consumer. Tough choices between "cheap food" and environmental quality will no doubt be increasingly debated. This is the same kind of debate that advocates "cheap food" now and long-term productivity or "reasonable" price of food in the future. This is not a new debate, but the rationale for the goals is different. Social preferences change, however, and no doubt will be different ten years from now.
10. The integration of BMPs into a best management system and the physical/technical linkage with water quality must be better understood.
11. Methods for achieving food production goals in concert with an environment that is compatible with sustained levels of human health

and welfare must be developed. Such an understanding will no doubt provide useful guidance to those countries relying on the agricultural/forestry sector for a significant share of their economic activity.

12. In terms of control, what is the rank and priority of the various pollutants as to their impact on water quality? A BMP may control one or more pollutants to achieve water quality goals but probably will not control all the pollutants. What criteria will be used to trade-off cost and benefits of the various combinations or practices that will control pollutants?
13. What BMPs should be eligible for cost-sharing? Only those requiring capital investment? If so, how will maintenance costs be calculated and paid? Are nonstructural practices like minimum tillage eligible and, if so, what criteria will be used to calculate such payment—net income differentials?
14. Will land use changes be allowed and cost shared?
15. Is there a role for land use regulations? Will crops be banned on Class IV to VIII lands?
16. What is the philosophy for compliance monitoring for BMP implementation and maintenance? What approaches and systems will be used to assess compliance? What statistical tests will be used in compliance?
17. What cost-sharing levels will be necessary to provide the economic incentive to install and maintain BMPs?
18. What is the degree of sensitivity, precision, accuracy and resolution required for decision-making with the use of environmental assessment methodologies? Will mathematical models be used to evaluate loadings as a function of alternate crop management systems?

This list can only be suggestive of the kinds of issues that demand attention. Many of these are addressed in these proceedings.

REFERENCES

1. National Commission on Water Quality. Staff Report, Washington, DC (April 1976).
2. Crawford, N. H., and A. S. Donigian. "Pesticide Transport and Runoff Model for Agricultural Lands," U.S. Environmental Protection Agency, Athens, GA, EPA-600/2-74-013 (1973).
3. Bower, B. T., and W. O. Spofford, Jr. "Environmental Quality Management," *Natural Resources J.* 10(4), (October 1970).
4. *Air/Water Pollution Report* 16(12), (20 March 1978).
5. Gordon, L. "Limits to Growth Debate," *Resources,* No. 52 (Summer 1976).
6. Dent, J. B., and J. R. Anderson. *Systems Analysis in Agricultural Management* (New York: John Wiley & Sons, 1971).
7. Bower, B. T., C. N. Ehler and A. V. Kneese. "Incentives for Managing the Environment," *Environ. Sci. Technol.* 11(3):250-254 (1977).
8. Zison, S. W., K. F. Haren and W. B. Mills. "Water Quality Assessment: A Screening Method for Nondesignated 208 Area," U.S. Environmental Protection Agency, Athens, GA EPA-600/9-77-023 (August 1977).

9. Donigian, A. S. "Simulation of Nutrient Loadings in Surface Runoff with the NPS Model," U.S. Environmental Protection Agency, Athens, GA, EPA-600/3-77-065 (June 1977).
10. Code of the Federal Regulations 130.2, Title 40, Protection of the Environment (1976).
11. Unger, S. G. "Environmental Implications of Trends in Agriculture and Silviculture. Volume I - Trend Identification and Evaluation," U.S. Environmental Protection Agency, Athens, GA, EPA-600/3-77-121 (October 1977).
12. U.S. Department of Agriculture and U.S. Environmental Protection Agency. "Control of Pollution from Cropland. Volume I - An Overview," (June 1976), Washington, DC, EPA-600/2-75-026a, -026b.
13. House, P. W., and D. W. Jones. *Getting it Off the Shelf: A Methodology for Implementing Federal Research* (Boulder, CO: Westview Press, Westview Special Studies in Public Systems Management, 1977).
14. Irwin, W. A., and R. A. Liroff. "Economic Disincentives for Pollution Control: Legal, Political and Administrative Dimensions," U.S. Environmental Protection Agency, Washington, DC. EPA-600/5-74-026 (July 1974).
15. Heady, E. O. "Effect of Export, Environmental and Soil Conservancy Measures on Productivity, Land Use and Income in Iowa," *Proc. Iowa Acad. Sci.* 84(3):163-167 (1977).
16. Seitz, W. D., et al. "Alternative Policies for Controlling Nonpoint Agricultural Sources of Water Pollution," U.S. Environmental Protection Agency, Athens, GA, EPA-600/5-78-005 (April 1978).
17. Environmental Law Institute. "Legal and Institutional Approaches to Water Quality Management Planning and Implementation," U.S. Environmental Protection Agency, Washington, DC (March 1977).

6

CONSERVATION DISTRICT INVOLVEMENT IN 208 NONPOINT SOURCE IMPLEMENTATION

R. E. Williams, J. E. Lake
    National Association of Conservation Districts
    Washington, D.C.

As representatives of the National Association of Conservation Districts (NACD), we are very pleased to participate in this conference. Many conservation districts throughout our country, as well as NACD, have been very much involved in water quality planning and expect to play a very significant role in implementing "Best Management Practices for Agriculture and Silviculture" which is the theme of this conference.

The activities of conservation districts have been on-going for more than 40 years in the United States. The conservation movement began in 1937 when model legislation was furnished to the states by President Roosevelt providing for the creation of conservation districts by state law. Since that time all states, Puerto Rico and the Virgin Islands have adopted such laws. Some 3,000 conservation districts have been created throughout our nation.

Most state district laws provide for establishment of districts as political subdivisions of the state. Although state laws governing conservation districts vary in some respects, their purposes are the same everywhere; that is, to focus attention on land, water, and related resource problems, to develop programs to solve those problems, and to enlist the support and cooperation from all public and private sources to accomplish district goals.

Conservation districts are managed by local citizens who know their local problems. Usually districts have from five to seven officials who are either elected or appointed depending on the laws of the particular state. There is a growing trend to provide for the election of these governing bodies at the general election. Over 17,000 men and women now serve as district officials.

Originally, conservation districts primarily served agricultural cooperators. Cities and towns were not included within most districts' boundaries. However, in recent years, conservation districts have either by amendment to the district laws or by the redefining of district boundaries included the entire soil and water resource areas encompassing urban and city dwellers as well.

Most conservation officials are farmers and ranchers, although they are being joined more and more by bankers, home owners, sportsmen, businessmen, county officials, and many other citizens concerned about natural resources. An increasing number of states require representation on district governing bodies by urban and nonfarm interests.

In every district, officials develop and continually maintain a long-range plan which contains facts about the soil, water and related resource problems of their district. The long-range plan also outlines measures that can be taken to correct the problems identified. The long-range plans must continually be updated in order to provide current resource information that is needed to assess current problems and to provide a base for setting new priorities. All districts prepare an annual plan of operations to guide the current year's activities. To accomplish the goals spelled out in the long-range plan and the annual plan of operations, district officials have developed working agreements with many local, state and federal agencies.

Through a memorandum of understanding, districts receive federal assistance from the USDA's Soil Conservation Service (SCS) to provide technical assistance to individual land owners and land users for planning and installing conservation practices needed on their lands. Districts also have memorandums of understanding and cooperative arrangements with many other federal, state and local agencies.

There are now over two million district cooperators throughout the nation. These cooperators have been working with conservation districts voluntarily to apply conservation practices (many are synonymous with BMPs) on their land for the last 40 years. However, with all these indications of success, the fact still remains that there is a tremendous job to be accomplished in soil and water conservation. New problems continue to arise, and millions of acres of our valuable cropland are still unprotected and are eroding at a rate accelerated by man's activities that will deplete the soil resource if it continues. Furthermore, the resulting sediment is recognized as the largest single polluter of our streams by volume. In addition, it is recognized that water quality can be further degraded by the excessive nutrients and pesticides carried by the sediment.

Just last year the General Accounting Office reported on a survey of the effectiveness of conservation work throughout our country. The report indicated that the SCS estimated an average of 9 tons of soil per acre per year was being lost from our nation's croplands, and that a significant amount of cropland losing soil in excess of the tolerable soil loss limits has not been protected by the application of erosion control practices. In fact, the report indicated that 42% of the 335 million acres of cropland harvested in 1975 did not have adequate erosion control techniques applied.

In recent years, attention has turned toward the effects of erosion and related pollutants on water quality. Several major events over the past few years have led to the involvement of conservation districts in 208 water quality planning. In 1970, a National Sediment Conference identified sediment as a serious polluter of our nation's waters. Conservation districts became more concerned about those water quality problems that might be created by agricultural activities. In 1972, the National Association of Conservation Districts, EPA, the Council of State Governments, SCS, and others worked to develop a Model State Act for Soil Erosion and Sediment Control to be considered for use throughout the country. The Model Act was published by the Council of State Governments in its 1973 Suggested State Legislation. Following this, NACD received a grant from EPA to assist individual states hold sediment control institutes. The purpose of these institutes was to discuss the problems related to sedimentation and water quality, to discuss potential legislation and sediment control programs that could be implemented to reduce these problems, and to educate individual district officials as to the seriousness of erosion, sediment and related water quality problems. Forty-five sediment institutes were held in cooperation with state soil and water conservation agencies, SCS, state associations, conservation districts, and others throughout the country.

SEDIMENT CONTROL LEGISLATION

As of 1977, 15 states, the Virgin Islands and the District of Columbia had adopted various forms of sediment control legislation. The legislation in these states is quite diverse and may vary a great deal from the model legislation introduced in 1972. However, the control of erosion and sediment is an important feature of all of these laws. A brief summary of the sediment control laws in five of these states follows:

Virginia

The efforts of Virginia's Soil and Water Conservation Commission and the Erosion and Sediment Control Task Force of the Governor's Council on the Environment in 1971-72 resulted in the 1972 enactment of a bill for erosion and sediment control on land-disturbing projects other than agricultural or silvicultural.

The purpose of the law was to establish and implement a statewide coordinated program to control erosion and sediment and to conserve and protect the land, water, air and other natural resources of Virginia. The State Soil and Water Conservation Commission was assigned responsibility for administering the law.

Guidelines, standards and criteria were adopted by the Commission and became effective July 1, 1974. Local control programs consistent with the state program are developed and carried out by (1) the soil and water conservation district; (2) where appropriate, by counties, cities and incorporated towns; or (3) by a joint venture between a district and a county, city or town. These local programs are approved by the Commission.

If any county, city, town or district fails to fulfill these requirements, the Commission develops and adopts a program to be carried out by the district, or if there is no district, by the county, city or town.

The local programs require an erosion and sediment control plan approved by the local government before land-disturbing activities can begin. The local authority can require an applicant to insure that emergency measures for appropriate conservation be taken at the applicant's expense. To ensure this, the authority can require a letter of credit, cash escrow, performance bond, or other legal arrangement before issuing the permit.

Iowa

Iowa's erosion and sediment control law requires abatement of erosion when a complaint is filed with the commissioners of a conservation district, provides for adoption of soil loss limit regulations by districts, and provides for state financed cost-sharing for installing needed measures. Penalties are imposed when the land owner fails to initiate necessary work within specified time limits.

Iowa was the first state in which districts experienced this new responsibility governing agricultural lands. A key stipulation in the Iowa law is that cost-sharing and technical assistance must be available before a land owner can be required to install measures to meet the requirements of the law.

Maryland

Maryland's Statewide Sediment Control Act was adopted in 1970 by the Maryland General Assembly. The Department of Natural Resources is the responsible agency. The Act requires that before land is cleared, graded, transported or otherwise disturbed for any purpose (except agriculture and single-family dwelling construction), the proposed earth change shall first be submitted to and approved by the appropriate soil conservation distrct. State projects, federal projects or projects on state-owned lands are approved by the Department of Natural Resources. Under the Act, each county and municipality is required to adopt grading and sediment control ordinances and have them approved by the Department of Natural Resources. All 23 counties and Baltimore City adopted ordinances by the end of 1972. The Maryland Attorney General has ruled that "protective stormwater measures may be imposed by the Soil Conservation District" under the 1970 Sediment Control Law.

Pennsylvania

On September 21, 1972, regulations for erosion and sediment control were adopted by the Pennsylvania Environmental Quality Board pursuant to the existing Clean Streams Law. Under the regulations all earth-moving activities on all lands, regardless of size, must have an erosion and sediment control plan. In addition to the plan, most earthmoving activities greater than 25 ac must have an erosion and sediment control permit from the Department of Environmental Resources. Inspection and enforcement activities are handled by the Department of Environmental Resources. Local conservation districts provide technical support, plan review, and monitoring assistance to the program. Included in the operating procedures is a provision that the Department may delegate portions of the enforcement program to local jurisdictions including conservation districts.

Montana

Montana's Natural Streambed and Land Preservation Law S.B. 310 requires review and approval by the local conservation districts of all proposed projects which affect perennial streams. Procedures provide for team review of projects by representatives of districts, Department of Fish and Game and the land owner. If a team cannot reach agreement, the conservation district requests the district court to appoint an arbitration board to handle the situation. The law was effective January 1, 1976.

EFFECTS OF SECTION 208

In 1972, when Congress passed the Water Pollution Control Act Amendments, it possibly enacted the most significant legislation involving conservation districts since their creation. Never before in the 40+ years of conservation district activities in this country have the challenges and opportunities been greater than they are today as a result of Section 208 of that law. Section 208, as you know, requires that each state develop state or areawide plans for controlling pollution from both point and nonpoint sources.

Nonpoint sources include such areas as agriculture, silviculture, surface mined areas, and construction sites. Districts, because of their experience, became directly involved in nonpoint planning for these activities in many states. Some of the key provisions of Section 208 that have provided the opportunity for district involvement are: the emphasis on local involvement, the requirement for identification of water quality problems by source, and the need for development of best management practices that will help solve the identified nonpoint source water quality problems. The provisions also require that the agency or agencies to manage the nonpoint program be designated by the governor. All of these provisions led very naturally to the involvement of conservation districts.

The language of Section 208 also spells out that the programs are to be carried out at the local and state levels with local public participation playing a major role in formulation and implementation of the 208 plans. Soil conservation districts are the key local agency for involving rural land owners and concerned citizens. As local land owners themselves, district officials provide the grass roots contact between government at all levels and the local people.

In addition, districts have perfected working arrangements which allow the integration of federal, state and local governmental agencies. Through this cooperation, conservation districts also have the technical expertise to provide land owners assistance in making decisions affecting nonpoint source pollution control on their land. They also have a tremendous amount of necessary resource information such as soil surveys, resource maps, Conservation Needs Inventory data, and soil loss information (Universal Soil Loss Equation) that are needed to identify the critical areas where water quality problems do exist.

In addition, districts with the technical assistance of SCS have the expertise to assist land owners with the development of plans outlining BMPs on their lands. Many existing and well-known conservation practices that have been used for years such as grassed waterways, terraces, erosion control structures, minimum tillage, pastureland management and many others are BMPs whenever they are identified as the best known means of control for agricultural nonpoint source water quality problems addressed in a 208 plan. Just because we have developed a new term which describes those measures to be applied to solve water quality problems related to agriculture doesn't mean that we scrap all the existing technical methods that we have used in the past. Instead, we will be focusing on how to use our technical experience more efficiently in addition to searching out new methods of control which will also be recognized as BMPs to improve water quality.

Districts have some real challenges to meet, and in some cases changes to make, in their own organization in order to accomplish the objectives of the nonpoint source control efforts under Section 208. To meet these challenges districts will need to reassess their priorities. The days of the "first-come-first-serve" approach for assistance are numbered. Setting priorities for conservation planning and application is a responsibility of conservation districts. Not only is this an important aspect of 208 planning but of ongoing district programs as well. SCS has agreed to provide technical assistance in accordance with the priorities set by district officials. This means that technical assistance should, and will be, available to land owners and operators on a "worst-first" basis in the future. It will mean, that instead of working with the most aggressive land owners who request assistance for relatively minor problems, SCS and other district cooperating agencies such as the Cooperative Extension Service, must concentrate on working with the less progressive operators who usually have the more difficult problems but are more hesitant to request assistance. As a result of this approach, implementation will be accomplished in the critical areas first in order to have the greatest and most immediate impact on water quality.

With the growing responsibilities conservation districts are being asked to assume, the need for additional district administrative and technical staff is critical. In many states, county and state governments provide funds to enable districts to fill at least part of this manpower need.

Federal personnel ceilings limit the number of SCS and other agency personnel available to districts. If some additional manpower needs can be met from state and local sources, better use of SCS technical assistance can be made in solving critical land protection and water quality problems.

Districts will need to continually improve their educational and informational programs in the future in order to show the need for additional support.

Districts are demonstrating their ability to make these adjustments as well as their ability to manage programs for the installation of BMPs in several programs already underway in the country. The following programs are illustrative of districts' abilities to manage programs in the future. The three examples that will be briefly discussed are the Pennsylvania Clean Streams Program, the Montana National Streambed and Land Preservation Law, and the Black Creek Demonstration Project in Indiana.

Pennsylvania Clean Streams Program

Several developments in Pennsylvania revealed the need for an expanded program for erosion and sediment control. These included the erosion and sediment problems created by industrial development and urbanization, a growing interest in, and citizen support for, total watershed management programs and the general recognition that sediment was the largest single pollutant, by volume, of water resources.

On September 21, 1972, following study by the Environmental Quality Board (EQB) and public hearings, rules and regulations for erosion and sedimentation control were adopted by the EQB pursuant to the existing Clean

Streams Law. Under the regulations, all earth-moving activities, regardless of size, must have an erosion and sedimentation control plan. In addition to an erosion and sedimentation control plan, earth-moving activities greater than 25 ac must, with certain exceptions, have an erosion and sediment control permit from the Department of Environmental Resources (DER).

The DER developed an operating procedure that would utilize conservation district expertise in the program. The staffs of the Bureau of Water Quality Management, the Bureau of Soil and Water Conservation, and the Bureau of Litigation and Enforcement jointly developed this procedure.

On projects requiring departmental permits, an application or an erosion and sedimentation control permit is submitted to the conservation district along with an erosion and sediment control plan. The conservation district has 45 days during which to act upon the application. Following technical review, the conservation district board, at an official meeting, takes action to recommend to the department that a permit should either be issued or denied. This recommendation is forwarded to the department's regional office where the permitting process takes place.

Through a department policy established by the Secretary of the DER, the Bureau of Soil and Water Conservation is to provide technical support on erosion control matters to other bureaus within the Department. Inspection and enforcement activities are handled by the Office of Deputy for Protection and Regulation and Deputy for Enforcement within the Department. Included in the operating procedures is a provision that the Department may delegate portions of the enforcement program to local jurisdictions.

The resources management portion of the program has been assigned to the Bureau of Soil and Water Conservation and the 66 conservation districts. The Bureau's Division of Soil Resources and Erosion Control implements the Department's program through informational, training, administrative and liaison activities. Districts provide information, planning assistance, plan review, and land-use monitoring assistance to the DER. Twenty-three districts have requested and have been delegated authority in the inspection portion of the program to date.

Montana National Streambed and Land Preservation Law

In 1975 the Montana Legislature passed the Natural Streambed and Land Preservation Act, referred to as S.B. 310. This law provides that conservation districts must review and approve all proposed projects which affect perennial streams such as channel changes, new diversions, riprap, jetties, new dams and reservoirs, commercial, industrial and residential developments, snagging, dikes, levees, debris basins, grade stabilization structures, bridges and culverts, recreation facilities, commercial agriculture, and certain farming, grazing and recreation activities. Conservation districts have the option of modifying this list to meet local needs.

When a district receives a proposed project, the Department of Fish and Game (DFG) is notified. If the DFG or the district requests it, a review team consisting of representatives of the district, DFG, and the private land owner examines the site of the proposal. If agreement is not reached, the District Court is asked to appoint an arbitration board. Technical assistance is provided by the SCS to all members of the team.

Under S.B. 310, the conservation districts held hearings on their proposed rules and regulations. There was substantial publicity on the new program in the newspapers, and special articles appeared in farm and livestock magazines. In 1976, the first year the law became effective, Montana districts processed some 2,000 proposals.

The Black Creek Study, Allen County, Indiana

The Black Creek Study was undertaken in 1972 by the Allen County Soil & Water Conservation District as a result of a grant from the Environmental Protection Agency, Region V, Chicago. Technical assistance was provided by the SCS, and research support was supplied by Purdue University, the Agricultural Research Service, and the University of Illinois.

The project demonstrated the ability of a soil and water conservation district to efficiently administer an extensive program for nonpoint source pollution control. The reliance on the local conservation district for the administration was shown to be a very important aspect of public acceptance and voluntary participation.

The Allen County Conservation District also demonstrated the ability of a district to efficiently handle cost-sharing funds and to carry out long-term contracts with private land owners.

Some of the major points substantiated and highlighted by the Black Creek Study were that:

1. The cost of achieving treatment on every acre of land to improve water quality would be extremely high. It probably would not be physically possible regardless of cost. Therefore, water quality improvement must be approached by treating the critical areas first. It is therefore obvious that the critical areas must be identified for any watershed before treatment efforts begin.
2. Once critical areas are identified, BMPs need to be selected for treating the critical areas. BMPs for the Black Creek Watershed were identified by the District Board of Supervisors with assistance from SCS staff. These included: field borders, grade stabilization structures, grassed waterways, livestock exclusion, pasture planting, sediment control basins, terraces, limited channel protection, and tillage methods which increase crop residue and surface roughness.
3. Farm-by-farm conservation plans were found to be essential in programs of water quality improvement. The plans should be simple in

format and selective in approach. Obligations of participating farmers must be clearly delineated.
4. A voluntary program with sufficient incentive payments and technical assistance can achieve significant land treatment aimed at improving water quality. Regulations or the threat of regulation may be required to achieve treatment on land owned by the relatively small number of noncooperators.
5. Traditional cost-sharing programs, based on a fixed percentage payment for every practice, are not adequate to sell BMPs for water quality improvement. While an overall average might be set, local districts should have the responsibility to set the rate for individual practices within the limitations.
6. Public information is critical to a successful land treatment program. Land owners and the general public should be kept up-to-date on all phases of a program from conception through planning to implementation.

A recent significant opportunity for district involvement in BMP implementation arises out of the new amendments to the Clean Water Act signed by the President on December 15, 1977. The agricultural cost-sharing section introduced by Senator Culver of Iowa authorizes $200 million in Fiscal Year 1979 and $400 million in Fiscal Year 1980 to be used for cost-share assistance for implementation of BMPs in rural areas having significant nonpoint water problems identified in the 208 water quality plan.

The amendment passed the Senate and House with very little dissent. Districts are identified in the law as the local governmental agency responsible for determining (in cooperation with the Secretary of Agriculture) priority among individual land owners and operators requesting assistance to assure that the most critical water quality problems are addressed first, and for approving cooperator's plans outlining BMPs to be installed on their land with cost-sharing pursuant to long-term contracts. This important legislation has specifically named conservation districts for direct involvement in carrying out the law.

The program which is being developed pursuant to this legislation will be called the Rural Clean Water Program (RCWP). The Secretary of Agriculture has designated the SCS as the lead agency responsible for carrying out this program.

In order for land owners to be eligible for participation in the program, their land must be identified as part of the critical areas addressed in a 208 plan certified by the governor of that state and approved by EPA. Since this program is directed at designated critical areas with significant water quality problems, it is necessary that priorities be set and funds assigned accordingly both on a national and state basis. For this reason, not every district or county will be included in the program.

The RCWP provides four options to the Secretary of Agriculture through SCS for carrying out the program at the state and local levels. These include entering into agreements for administration for all or part of the program with: (1) soil conservation districts, or (2) state soil conservation agencies, or (3) state water quality agencies, or, if none of the above, then (4) transfer of funds from SCS to Agricultural Stabilization and Conservation Service for administration of the program. Regardless of the option selected, district officials will be jointly responsible for setting the priorities for assistance as well as be solely responsible for approving plans on which contracts for cost-sharing will be based.

Districts have been working with state and areawide agencies to develop the nonpoint source phase of 208 plans for some time now. In fact, in over half the states, the state conservation agencies are preparing the agricultural nonpoint plans under contracts from the state water quality agencies. In many other states, districts are actively assisting in the development of the agricultural nonpoint plan through cooperative agreements.

As a result of this participation, and the fact that they have the expertise and working tools to accomplish implementation, conservation districts are being identified in many plans as the management agency for implementing the agricultural nonpoint plan.

In summary, the outlook for conservation districts as a result of the 208 water quality effort is excellent. The opportunity for districts to get conservation on the land has never been greater. The changes taking place in district operations are all positive changes toward meeting modern needs, more efficient use of resources, people, and tax dollars to protect both our soil and water resources.

7
THE ROLE OF CONSERVATION PRACTICES AS BEST MANAGEMENT PRACTICES

J. S. Johnson
U.S. Department of Agriculture
Soil Conservation Service
Washington, D.C.

BACKGROUND

The goal of the 1972 Federal Water Pollution Control Act Amendments (PL 92-500) is to restore and maintain the chemical, physical and biological integrity of the nation's waters. The legislation calls for achieving water quality that is suitable for fishing and swimming by 1983 wherever it is practical to do so. The question of whether it is practically, economically or humanly possible to accomplish these goals within the time frame of the law looms in the minds of many people.

The national investment for cleaning up point sources of pollution has been enormous, and the federal government and industry seem committed to continuing this investment. As progress is made in controlling municipal and industrial pollution, the problem of nonpoint source pollution becomes more apparent. A concerted effort must be made to control this pollution from diffuse sources if 1983 clean water goals are to be achieved.[1]

Nonpoint source pollution comes from many varied and diffuse sources. It is basically influenced by land use activities along with climatic, hydrologic, soil and geologic factors. Because of the relationship of activities on the land to nonpoint source pollution, land management practices will play a major role in the cleanup. Practices that are looked to for performing this role have become widely accepted as best management practices (BMPs).

The term "best management practices" originates with the Environmental Protection Agency (EPA) guidelines for the continuing planning pursuant to Section 208 of the 1972 Federal Water Pollution Control Act Amendments. The definition of BMPs in these guidelines follows:

> "The term 'Best Management Practices' (BMP) means a practice or combination of practices that is determined by a State (or designated areawide planning agency) after problem assessment, examination of alternative practices and appropriate public participation to be the most effective, practicable (including technological), economic, and institutional considerations) means of preventing or reducing the amount of pollution generated by nonpoint sources to a level compatible with water quality goals."

BMPs to control nonpoint source pollution are being incorporated into the water quality management plans being developed in accord with the requirements of Section 208. The plans address nonpoint source control needs, and identify and evaluate measures and practices that will produce a desired level of control. The practices that are selected are placed in the 208 plan as BMPs. The basic philosophy is to treat the source of problems as they occur on the land.

Because of the variability in sources, climate, topography and soils, the criteria for BMP selection must be tailored to meet the needs of the site and the nature of water quality problems. The goal of BMPs is to ensure that activities on the land do not pollute the nation's streams, lakes and rivers. This approach is far less complex and is more fair to all land owners than for pollution control experts to study a stream and try to assess portions of responsibility. A treatment plant operator can adjust his operation to meet a given effluent standard. But it is somewhat unreasonable to tell a farmer plowing his fields or a logger harvesting his trees that his operation must never exceed a given effluent or downstream water quality standard.[2]

Instead of trying to regulate agricultural and silvicultural activities with effluent standards, compliance can be based on using proper land management practices and monitoring or inspecting these activities. Research eventually should be able to provide us with the link between downstream water quality and upstream land activities, so that management practices that are consistent with downstream water quality goals can be defined.

SOIL CONSERVATION MOVEMENT

The soil conservation movement, spearheaded technically by the Soil Conservation Service (SCS) and operated locally by conservation districts, has tailormade resource management systems to fit virtually every acre of land in the U.S. For more than 40 years districts have been concerned about land and water use within their boundaries, and they have developed programs to solve land and water resource problems.

Each field office of SCS is equipped with a technical guide that is specifically adapted to the local field office area. These guides provide information to help soil conservationists and land users in considering alternatives and arrive at decisions on how land is to be used and treated to accomplish soil and water conservation objectives. The field office technical guide is the basic SCS document that provides quality standards for planning and applying soil and water conservation practices.

The SCS is an action agency. Its soil conservationists are trained and experieneced in providing onsite technical assistance to land users for planning and carrying out soil and water conservation programs. This person-to-person relationship is essential in achieving soil and water conservation objectives with our nation's farmers and ranchers. Each agricultural operation is unique. A soil conservationist, equipped with his knowledge, experience and the SCS technical guide, has a special ability to consider alternatives with farmers and ranchers and help them find acceptable solutions that will be carried out. This same approach is needed to plan and apply BMPs in rural areas.

SOIL AND WATER CONSERVATION PRACTICES COROLLARY TO BMPs

Many soil and water conservation practices and resource management systems are corollary to BMPs. They are among the practices that are being examined and assessed for their effectiveness in improving water quality in 208 planning that is underway throughout the nation. Land treatment to improve water quality is site-specific. Because a single practice will seldom achieve all the desired effects, a combination of practices is usually selected to meet the needs of each land and water problem. SCS has called this combination of practices a resource management system. The practices and systems that 208 planners select for their contribution to improved water quality are BMPs.

BMPs may consist of a single conservation practice or may incorporate several resource management systems. They may also include practices that help control nonpoint source pollution but that have not been thought of traditionally as soil and water conservation practices. Road-salt management, street sweeping, and pet litter management practices are some examples of this latter category.

Conservation cropping systems that include sod crops in sequence with row crops are excellent soil and water management systems. The protective and filtering effects of sod and its residues reduce the loss of sediment and agricultural chemicals from the land. Contour strip cropping, which provides for sod strips alternating with cultivated strips, is twice as effective in controlling erosion as contouring alone.[3]

Grassed waterways are used extensively as outlets for terraces and diversion channels and for disposal of runoff from contoured rows and natural depressions. Well-sodded waterways also filter out sediment, thus further helping to improve water quality.

Terraces have been used from the time of early civilizations to control erosion where sloping land is intensively cultivated. They are effective in reducing runoff and the transport of eroded soil to stream channels and they contribute to moisture conservation. According to SCS records, over 2 million kilometers of terraces have been installed nationally.[4]

Where crop residues are adequate to provide nearly complete surface cover, no-till planting can be a very effective year-round erosion control method.[5] This and similar conservation tillage practices have shown good results in controlling the movement of fine soil particles. There are certain problems with no-till methods that need considerable attention for resolution. They usually require more herbicides and insecticides than conventional systems. Also, nitrogen and possibly phosphorus can be leached from residues left on the surface. But the increased hazard of pollution from these soluble nutrients is usually offset by the much greater reduction in soil losses and runoff. This will become increasingly true as less persistent pesticides are used, integrated pest management practices are adopted, and fertilization timing and management practices are improved.

CONCERNS FOR EFFECTIVENESS

There are some questions unanswered about the total effectiveness of soil and water conservation practices in improving water quality. As water quality becomes more significant, the dynamics of the effects of applying a practice become more complex. More research is needed to improve our ability to design and apply practices and to predict with greater certainty their total effect on water quality. Problems vary considerably from site to site, and reliable research efforts are vast and expensive. Error analyses for a broad study of loading functions indicate that the margin of error could run as high as 50%.[6]

Still, it is logical to use soil and water conservation practices and resource management systems to improve and maintain water quality. It is widely accepted that sediment, in terms of volume, is the greatest pollutant in our surface waters. Sediment pollution muddies the water, inhibits photosynthesis, clogs fish gills, and increases oxygen demand. Also, removing sediment from public water supplies is costly. Soil and water conservation practices have a proven ability to reduce sediment pollution by controlling soil erosion. The amount of sediment reduction that takes place as erosion is reduced depends on the sediment delivery ratio.

A good example of water quality improvement through soil and water conservation practices is the reduction of sediment rates in Georgia's Jackson Lake. In 1939 the lake's sedimentation rate was 107 ac-ft/yr. By 1967 the rate had dropped to 29 ac-ft/yr, and when a planned watershed project is completed, the rate should drop to 10 ac-ft/yr.[7]

Conservation practices have also helped to reduce the turbidity of intake waters at the Atlanta, Georgia, waterworks on the Chattahoochee River. In the early 1930s total suspended solids averaged 400 ppm. As conservation practices were installed in the watershed, this declined steadily to less than 25 ppm.[8]

These two examples illustrate the beneficial effect of using a diversified soil-conserving system of grass, livestock, trees and row crops in conservation rotations supplemented by a complete water management system. They exemplify a change from land misuse to wise use and its beneficial effects on water quality.

Sediment is one of the significant transport mechanisms for nutrient and pesticide pollutants. More than half of the kinds of pesticides that are used on corn, cotton and soybean row-crops are transported by sediment. Paraquat, one of the more persistent pesticides used with no-till planting techniques, is tranported on sediment.[9] Sediment carries most of the phosphorus and organic nitrogen that gets into water. The kind of soil substantially influences pesticide and nutrient transport by sediment. More research is needed on the properties of individual soils, how well erosion control practices control the movement of fine particles, and the transport mechanism. Until this information is available, it will be difficult to predict precisely the beneficial effect of erosion control practices. The available research and information provides reasonable assurance that practices that control erosion and sediment delivery also have some effect on the reduction of pesticide and nutrient pollution. Practice design should take into account that controlling the movement of fine particles is most important because these particles have a much greater capacity than coarse particles for transporting adsorbed pollutants.

Much of the eutrophication of our lakes by nutrient enrichment is the result of increased levels of phosphorus from domestic and industrial wastes and the land. In the Great Lakes it has been estimated that diffuse sources of phosphorus may account for about 50% of the total tributary loading. Phosphorus entering the lakes from diffuse sources is primarily sediment bound since soil particles have a very strong affinity for phosphates. Practices that reduce the sediment load will, therefore, contribute to reduced phosphate loadings.[10]

The seriousness of nonpoint problems must be carefully assessed in 208 planning and the "hot spots" clearly identified. Instead of widespread general treatment, the greatest beneficial effect on water quality is likely to be accomplished by carefully designing treatment to take care of the hot spots first. A study in a 700-ha cultivated drainage area in the Black Creek Watershed in Indiana has shown that a 40% reduction in stream sediment could be achieved by concentrating treatment on only 30 ha.[11] A study in the Great Lakes Basin by the International Joint Commission's Pollution from Land Use Activities Reference Group[12] indicates that 60% of the agricultural sediment load may be generated in 30% of the agricultural area of the basin. This study stresses the need for concentrating on remedial measures in "hydrologically active zones."

DEVELOPING AREAS

SCS expertise has been drawn on to an increasing extent for control of runoff and erosion in urbanizing areas. SCS technical guides have been expanded to include soil and water conservation practices that meet the needs of these developing areas. Guides and handbooks for erosion and sediment control in developing areas are also available from many conservation district offices.

The sediment load from construction sites and developing areas comes from only a small part of our total land area. Sediment problems can be very critical, however, in those areas where development and construction activities are concentrated. The polluting effect of large volumes of sediment has been readily recognized in developing areas. It is not unusual for construction sites to produce 10 to 20 times more sediment per unit area than agriculture. Conservation districts have actively participated with state and local governments in programs to control erosion and sediment in developing areas. The National Association of Conservation Districts has provided leadership for strengthening these programs through a Model State Act for Erosion and Sediment Control.[13]

The timely application of practices to control erosion and reduce sediment delivery from developing areas is especially important. Because excessively high rates of erosion usually do not continue very long after construction is completed, action must be immediate to be effective.

Vegetation is the first line of defense against erosion and is especially useful in developing areas. Site planning to maintain as much of the natural vegetation as possible has proved effective. Proper timing of construction operations and temporary seedings are also good management practices for construction sites. Permanent vegetation should be established as soon as conditions permit. Mulches are valuable for controlling erosion, conserving moisture, and providing stability until new plantings are established.

Structural practices, such as land grading, diversions, stream bank stabilization, and grade stabilization structures, can be used where vegetative measures are not adequate for erosion control. Debris or sediment retention practices can be installed to reduce the amount of sediment delivered from construction sites. These may consist of vegetative filter strips and various structures and filters that hold water long enough for sediment to settle out.[14]

Properly designed stormwater management programs and practices also contribute to the control of nonpoint pollution from developed and developing areas.

AGRICULTURAL WASTE MANAGEMENT

Agricultural waste management systems for safely handling manure and other agricultural wastes will play an essential role as BMPs in many areas. The SCS technical guides and the waste management handbook are readily available tools for planning and installing these systems.

Waste management practices should be designed and applied to meet site-specific needs. In northern climates, for example, successful land application must rely more on the absorptive capacity of the soils and less on microbes and plant roots.[15] Time of application is important to prevent loss of nitrogen.

A study in New York State indicates that control of barnyard runoff could provide the greatest reduction in dissolved phosphorus loadings entering Cayuga Lake.[16] This same study also indicates that, next to sewage treatment, controlling barnyard runoff is the least costly means of reducing the dissolved phosphorus loading in Central New York State.

Pastures that are well located and properly managed are not usually the source of serious nonpoint pollution problems. These pastures should be large enough and grazing distributed to keep wastes from becoming concentrated. Fencing to limit the access of livestock to stream banks will reduce aggravated stream bank erosion and control the amount of defecation going directly into the stream. Fertility and pesticide management practices that provide for the proper type, timing and rate of application are also important aspects of pasture management systems for water quality.

The hazard of runoff pollution increases when animals are crowded into lots devoid of vegetation. Practices that control runoff from manure accumulations and livestock feeding areas and the incorporation of manure into the soil immediately after spreading have proven effective. Storage facilities must be designed with adequate capacity to store drainage from silos, wash water, and other wastes from the livestock operation for the time when spreading on land endangers water quality. Practices such as diversions that direct clean water around the site can help reduce the storage capacity that is needed.

IRRIGATED AGRICULTURE

Irrigation has made possible some of the most productive agricultural regions in the United States. Irrigation water management systems need careful design that is based on a complete understanding of the regional hydrology. Salinity and nutrient pollution from irrigated land are two of the major problems threatening water quality.

Proper timing and the rate of irrigation water application is important for both efficiency and water quality. Runoff recovery systems can reduce polluted runoff and raise existing 50% efficiency ratings to 85%. Specific causes and relationships need to be identified to determine with greater certainty the beneficial effects of irrigation water management systems.

COST-SHARING TO APPLY PRACTICES

Meeting water quality goals is going to be costly. SCS has estimated that it will cost $7 billion from now to the year 2000 to treat 20% of the most critical areas to improve water quality. The bulk of this expenditure falls on the owners and operators of our nation's firms and ranches, and most of the benefits will be offsite. The nature of farm commodity pricing makes it very difficult, if not impossible, for farmers and ranchers to recover such expenditures. To avoid placing an insurmountable economic burden on American agriculture, a cost-sharing program is needed to help fund the installation of practices for controlling nonpoint source pollution in rural areas.

Section 35 of the Clean Water Act of 1977 (PL 95-217) authorizes this kind of program. This new legislation calls for USDA to administer a Rural Clean Water Program (RCWP) through SCS and other departmental agencies. This program will provide technical assistance and cost-sharing for measures and practices to improve water quality. It will apply only to areas that are identified in 208 planning as high-priority.

SCS will administer this program. This agency has the technical and administrative expertise that is needed to carry out the provisions of Section 35 of PL 95-217. Technical and financial assistance will be provided to land owners and operators to help them install measures and practices to improve water quality in areas that are identified as critical in 208 plans. Cost-sharing will be provided only on the basis of long-term contracts of 5 to 10 years duration. Agreements will be signed to provide grants to conservation districts, state water quality agencies, and state soil conservation agencies to administer all or any part of the program locally.

The RCWP is being coordinated with other USDA agencies and the EPA. In accord with PL 95-217, the program is to have the concurrence of the EPA Administrator. There is a national coordinating committee of the administrators of USDA agencies and EPA. State coordinating committees are to have members from USDA, state soil conservation agencies, conservation districts, and state water quality agencies.

SUMMARY

The soil conservation movement, operated at the local level by conservation districts, has a significant role to play in efforts to control water pollution from nonpoint sources. Sediment is the greatest pollutant, by volume, of our nation's streams, lakes and rivers. Eroded soil particles, especially fine clay particles, serve to transport nutrients and pesticides that can further damage water quality. Conservation districts with SCS technical assistance have more than 40 years of experience in controlling soil erosion. The benefits of conservation practices and resource management systems are being examined in water quality management planning, and those practices and systems that are deemed effective are being selected by water quality management agencies as BMP or components of BMP.

Planning and application of site-specific practices are essential. SCS conservationists are experienced and equipped with the knowledge and technical guides to perform and train others to perform this task. The best approach is to concentrate on the critical areas or "hot spots" that are sources of nonpoint pollution. Soils, geology, hydrology and climate, along with land use activities, are among the major items to consider in locating the hot spots and in designing the control practices.

Using conservation practices with the specific emphasis on water quality improvement is a new approach. Additional research and monitoring are needed to improve practice design and application techniques for this purpose and to predict with greater certainty the total effect on water quality.

Because the expense of installing BMP will be great and most of the benefits will be offsite, a cost-sharing program is needed to make the installation economically feasible, especially in rural areas. Section 35 of the Clean Water Act of 1977 is intended to serve this need.

REFERENCES

1. Basta, D. J., and B. T. Bower. "Point and Nonpoint Sources of Degradable and Suspended Solids: Impacts on Water Quality Management," *J. Soil Water Conser.* 31(6):252-259 (1976).
2. Holtje, R., and W. C. Harper. "Is There a Point Nonpoint?" *J. Soil Water Conserv.* 30(6):262-263 (1975).
3. Beasley, R. P. *Erosion and Sediment Pollution Control* (Ames, IA: The Iowa State University Press, 1972).
4. Agricultural Research Service, U.S. Department of Agriculture. "Soil and Water Management Systems for Sloping Land," ARS-S-160, Washington, DC (1977), pp. 65-66.

5. U.S. Department of Agriculture. "Erosion and Sedimentation and Related Resource Considerations," prepared as appendix material for U.S. Water Resources Council Second National Water Assessment (1977).
6. McElroy, A. D. Regional Overview of the Impact of Land Use on Water Quality, *in* "Fluvial transport of Sediment-Associated Nutrients and Contaminants," 1976 Workshop Proc. pp. 105-113, Int. Jt. Comm. Res. Advis. Board and Pollut. Land Use Act. Ref. Group. Windsor, Ontario (1977).
7. Barnett, A. P. "Agriculture and a Quality Environment," *J. Soil Water Conserv.* 27(3):104-108 (1972).
8. Albert, F. A., and A. H. Spector. "A New Song on the Muddy Chattahoochee," *in Water 1955 Yearb. Agric.* (Washington, DC: U.S. Government Printing Office, 1955), pp. 205-210.
9. Agricultural Research Service, U.S. Department of Agriculture, and Office of Research and Development, Environmental Protection Agency. "Control of Water Pollution from Cropland. Volume I - A Manual for Guideline Development," ARS-H-5-1, Washington, DC (1975).
10. Mulkey, L. A., and J. W. Falco. "Sedimentation and Erosion Control Implications for Water Quality Management," Proc. Natl. Symp. Soil Eosion and Sedimentation by Water. Amer. Soc. Agric. Eng. St. Joseph, MI (1977).
11. Morrison, J. "Managing Farmland to Improve Water Quality," *J. Soil Water Conserv.* 32(5):205-208 (1977).
12. International Joint Commission, Pollution from Land Use Activities Reference Group. "Remedial and Preventive Strategies," *in* Task C summary report (draft), pp. 4-1-4-16, Windsor, Ontario (1978).
13. Garner, M. M. "Regulatory Programs for Nonpoint Pollution Control: The Role of Conservation Districts," *J. Soil Water Conserv.* 32(5): 199-204 (1977).
14. Highfill, R. E., and L. W. Kimberlin. "Current Soil Erosion and Sediment Control Technology for Rural and Urban Lands," Proc. Natl. Symp. Soil Erosion and Sedimentation by Water, Amer. Soc. Agric. Eng., St. Joseph, MI (1977).
15. Soil Conservation Service, U. S. Department of Agriculture. "Agricultural Waste Management Field Manual," U.S. Government Printing Office. Washington, DC (1975).
16. Bouldin, D. R., *et al.* "Lakes and Phosphorus Inputs," New York State College of Life Sciences, Cornell Univ. Inf. Bull. 127. Ithaca, NY (1977).

8

RELATIONSHIPS BETWEEN
AGRICULTURAL LAND AND WATER QUALITY

D. R. Coote, E. M. MacDonald
  Land Resource Research Institute
  Central Experimental Farm
  Agriculture Canada
  Ottawa, Ontario

R. DeHaan
  Engineering and Statistical Research Institute
  Central Experimental Farm
  Agriculture Canada
  Ottawa, Ontario

INTRODUCTION

Improved water quality depends on the reduction of loadings of pollutants to stream waters from land use activities. It is necessary to understand how present loadings are affected by different land uses before discussing methods for their reduction. It was with this principle in mind that the Pilot Watershed Studies of the International Reference Group on Great Lakes Pollution from Land Use Activities (PLUARG) of the International Joint Commission (IJC) were initiated in 1974. While these Pilot Watershed Studies recognized the need to separate the effects of agricultural land from those of forested, urban and specialized uses of the land, there was little distinction made between types of agriculture or types of land form associated with them.

A number of studies have reported on the effect of agricultural activities on water quality.[1,2,3,4] Others have presented land use/water quality relationships based on the separation of agricultural land from urban and forested land.[5,6] Few data exist from studies in which uniform methods have been applied to a range of soil and management systems in agricultural watersheds large enough to take account of overland and drainage network transport losses. Existing loadings are either not representative of those reaching streams in large watersheds, or contain unmonitored nonagricultural influences.

The study described here was developed as an adjunct to the PLUARG Pilot Watershed Studies with the objective of determining the relationships among water quality, pollutant loadings and land use data gathered in watersheds representative of different agricultural areas. The approach taken was to: (1) identify and survey a number of purely agricultural watersheds (nonagricultural land use essentially absent) representative of different agricultural systems in the Canadian Great Lakes Basin; (2) monitor the water quality at the watershed outlet; and (3) examine statistical relationships between the characteristics of these watersheds and the water quality observed at their outlets. By so doing it was possible to gain some insight into the major controlling factors responsible for the differences which are observed in water quality conditions in different agricultural environments in this diverse region. The philosophical and mathematical bases for the analytical approach used in this study are similar to those described by Haith.[6] A further objective of the study was to provide a basis for predictions of water quality in unmonitored agricultural areas of the Canadian Great Lakes Basin.

METHODS AND MATERIALS

Prior to the initiation of field studies, data were assembled by which rational selections could be made of the major agricultural areas of the Canadian Great Lakes Basin and of sites which would represent them.[7] These data included interpretations of soil properties in terms of their potential to transport pollutants to surface and groundwaters, climatic zones, and crop and livestock production activities.

An initial selection of 17 suitable monitoring sites was made from air photo and ground observation, and sampling and flow measurement was carried out weekly from April to September 1974. The sites chosen were essentially free of nonagricultural activities and were of a size considered to be hydrologically stable, with few periods of zero or excessively high (unmonitorable) flow. They were also small enough to permit the definition of agricultural characteristics in terms of reasonably homogeneous areas representing the delineated agricultural regions. Sampling, laboratory analyses and flow monitoring were carried out by the Water Resources Branch of the Ontario Ministry of the Environment (OMOE), using standard procedures used by the Ministry's Hydrology and Monitoring and Laboratories Branches.[8]

The initial data were analyzed and considered relative to the objectives of the study and the available resources. A selection was then made of 11 sites which were instrumented with continuous water level recording equipment. Six of these sites were also equipped with automatic samplers.[8] Each of the watersheds was intensively surveyed using topographic and soils maps, air photos, and farmer-by-farmer interviews to gather detailed cropping and livestock data.[9] Table I summarizes the major characteristics of the watersheds which were used in subsequent statistical analyses.

The water quality and quantity data gathered were extensive. All sites were sampled at least weekly during the 2 year period from April 1975 to April 1977, with some sites being sampled three to four times per week and up to six times per day during runoff events. Flow:stage-height rating curves for each site were prepared by the OMOE over the period of the study. Loadings of water quality parameters were calculated using a flow-concentration function[10] applied at each sample time and extrapolated to the next sample time or to the mean daily flow, assigned a time of 12.00 hr, whichever came first. Flow and water quality data were plotted and regressions of concentration on flow used to estimate missing data at times when major unsampled changes in flow occurred. The daily (or shorter time period) loads were accumulated over the 2 years of the study, divided by the total flow and a flow-weighted mean concentration of each parameter calculated. The loads were also divided by the area of each watershed and averaged over the 2 years to give a mean annual unit-area load. One watershed was dropped from this latter calculation because of inconsistent flow data. Since the characteristics of this watershed closely approximated the mean of all of the watersheds, and since flow-weighted mean concentrations for the period of available flow compared favorably with other similar watersheds, it was clear that the omission of this site would not greatly affect the results. Table II presents a summary of the water quality unit-area load data for those parameters discussed in this paper. These parameters were selected because they represent major groups of pollutants, *e.g.*, nutrients, currently used pesticides, trace elements and diffuse industrial organic contaminants.

The untransformed water quality parameters and watershed characteristics were input to a multiple linear regression program developed by the Engineering and Statistical Research Service of Agriculture Canada. Output from the program consisted of correlation coefficient matrices and listings of variables in order of regression sum of squares using one, two and three watershed characteristics at a time. From these data, combinations of characteristics which significantly increased the explained variability were selected. These were then compared with the correlation coefficient matrix to establish the independence of the variables. Selected combinations were then processed by a separate program to calculate intercepts and regression coefficients for the regression equations.

Table I. Selected watershed characteristics—11 study watersheds.

Characteristic	Description	Units	Range	Median
1. (AR)	Watershed area	ha	1860-7913	4504
2. (SC)	Soil clay content - area weighted mean[a]	%	6.6-40.0	25.0
3. (EP)	Erosion potential - from estimates based on USLE[b]	D/L[g]	0.4-2.9	1.7
4. (PS)	Pollutant transfer potential to surface water[c]	D/L	0.20-1.00	0.60
5. (PG)	Pollutant transfer potential to groundwater[c]	D/L	0.00-1.00	0.38
6. (SD)	Stream density - perennial streams from 1:50,000 topographic maps	km km^{-2}	0.50-2.16	0.65
7. (CL)	Cultivated land (row crops + cereals[d]	%	21.6-89.3	54.0
8. (RC)	Row crops (corn + beans + toabacco + vegetables[d]	%	9.5-63.5	18.7
9. (CA)	Corn area[d]	%	9.5-42.3	16.2
10. (VA)	Vegetable area[d]	%	0.0-27.8	0.1
11. (WA)	Woodland area[d]	%	3.9-37.6	9.4
12. (NA)	Nonagricultural area (excluding woodland)	%	2.0-16.9	3.7
13. (RR)	Rural residences[d]	No. km^{-2}	1.3-17.3	3.2
14. (AU)	Animal units[d]	No. ha^{-1}	0.01-0.77	0.48
15. (AG)	Average gradient - from 1:50,000 topographic maps	m km^{-1}	1.25-10.96	3.81
16. (ES)	Exposed stream bank[e]	km km^{-2}	0.030-0.216	0.065
17. (TD)	Tile drained area[d]	%	0-99	20
18. (SP)	Soil extractable phosphorus[f]	D/L	10.8-41.2	18.3
19. (FN)	Fertilizer nitrogen applied[d]	kg ha^{-1}	8.1-67.0	15.5
20. (MN)	Manure nitrogen applied[d]	kg ha^{-1}	1.1-48.1	28.9
21. (TN)	Total nitrogen applied (FN+MN)	kg ha^{-1}	29.4-89.3	57.8
22. (FP)	Fertilizer phosphorus applied[d]	kg ha^{-1}	4.9-40.5	10.1
23. (MP)	Manure phosphorus applied[d]	kg ha^{-1}	0.3-14.5	7.7
24. (TP)	Total phosphorus applied (FP+MP)	kg ha^{-1}	9.4-40.8	19.9

[a] From soil survey reports and personal communication, G. Patterson, Ontario Soil Survey, University of Guelph.
[b] From Wall et al.[11] (USLE = Universal Soil Loss Equation).
[c] From Coote et al.[7]
[d] From Ripley and Frank.[9]
[e] Personal communication, K. Knap, Ontario Ministry of Natural Resources.
[f] Estimate based on crop type and county average soil-extractable phosphorus, by crop type, from research of the Ontario Soil Testing Service (Personal communication, M. H. Miller, University of Guelph).
[g] D/L = dimensionless.

Table II. Unit-area loadings of selected water quality parameters—
range found in study watersheds.[a]

Parameter		Description	Units	Unit-area Loads	Median
1.	(SS)	Suspended solids	kg ha^{-1}	60-961	235
2.	(PT)	Total phosphorus	kg ha^{-1}	0.16-1.51	0.79
3.	(PO)	Soluble ortho-P	kg ha^{-1}	0.02-0.46	0.28
4.	(NT)	Total nitrogen	kg ha^{-1}	3.54-29.28	17.23
5.	(NN)	Nitrate (plus nitrite) N	kg ha^{-1}	2.42-25.97	13.17
6.	(NK)	Total Kjeldahl N	kg ha^{-1}	1.10-6.70	3.53
7.	(CU)	Copper	kg ha^{-1}	0.017-0.049	0.024
8.	(ZN)	Zinc	kg ha^{-1}	0.027-0.134	0.043
9.	(AT)	Atrazine	kg ha^{-1}	0.02-4.47	1.03
10.	(EN)	Endosulphan	g ha^{-1}	0.000-0.054	0.004
11.	(PG)	PGB	g ha^{-1}	0.086-0.195	0.120

[a]Water quality and quantity data supplied by Ontario Ministry of the Environment, Hydrology and Monitoring Section, Water Resources Branch under their Task C, Activity 1 involvement, International Reference Group on Great Lakes Pollution from Land Use Activities, I.J.C., except pesticides and PGB—supplied by R. Frank, Ontario Pesticide Research Testing Laboratory, Guelph.

RESULTS AND DISCUSSION

Simple Correlation Analysis

Correlations between watershed characteristics and water quality parameters for unit-area loads and flow-weighted mean concentrations were compared and found to be very similar, in spite of differences in flow from site to site over the 2 yr of data collection. Only the unit-area loads are discussed in this paper. These were preferred as they can be readily applied to source area comparisons without the requirement of supplementary flow data.

Table III shows the significant correlations between watershed characteristics and unit-area loads of the selected water quality parameters. Most noticeable in this table is the tendency for the parameters to fall into three groups: the first are those which are correlated with suspended solids (total phosphorus, total Kjeldahl nitrogen) or which appear to be similarly influenced by the same watershed characteristics as those which are correlated with suspended solids (soluble ortho-P, atrazine, zinc); the second group consists of those which are primarily in a dissolved form and appear to be related to source material availability (total nitrogen, nitrate plus nitrite-nitrogen, endosulfan); and the third are those which appear to bear little relationship with sediment and/or have no apparent source (copper, PCB).

Table III. Significant correlations between water quality parameters (unit-area loads) and watershed characteristics.

Parameter	Correlations[a]
1. (SS)	7(CL)0.74*; 8(RC)0.67*; 11(WA)-0.64*; 19(FN)0.65*
2. (PT)	2(SC)0.72*; 4(PS)0.65*; 11(WA)-0.72*
3. (PO)	2(SC)0.78**; 4(PS)0.77**; 11(WA)-0.83**
4. (NT)	7(CL)0.75*; 8(RC)0.71*; 9(CA)0.88**; 11(WA)-0.68*; 17(TD)0.81**; 19(FN)0.65*; 21(TN)0.90**; 24(TP)0.79**
5. (NN)	7(CL)0.71*; 8(RC)0.70*; 9(CA)0.87**; 17(TD)0.80**; 19(FN)0.64*; 21(TN)0.83**; 24(TP)0.81**
6. (NK)	2(SC)0.80**; 4(PS)0.65*; 11(WA)-0.65*
7. (CU)	– – – – –
8. (ZN)	2(SC)0.66*
9. (AT)	2(SC)0.86**; 4(PS)0.83**; 11(WA)-0.63*
11. (EN)	8(RC)0.68*; 10(VA)0.93**; 12(NA)0.84**; 13(RR)0.90**; 18(SP)0.75*; 19(FN)0.76*; 20(MN)-0.68; 22(FP)0.85**;
12. (PC)	2(SC)-0.81**; 4(PS)-0.79**; 11(WA)0.71*

[a] X(Y)Z* where X = characteristic no., Y = symbol, Z = correlation coefficient, * = significant at 5% level, ** = significant at 1% level.

There were two watershed characteristics which were primarily influential in the first group. These were the amount of land area in cultivated crops (or in row crops) and the soil clay content. The latter clearly had a greater effect on the quality of the sediments than it had on the total quantity of sediment (suspended solids), as can be seen from the correlations of soil clay content with phosphorus, total Kjeldahl nitrogen and atrazine. This would be expected from the high surface area:mass ratio associated with clay-sized particles. Cultivated land is generally implicated in the yield of sediment to streams because of the higher erosion rates compared to pasture or woodland. The index of pollutant transport potential to surface water (a function primarily of soil texture, structure, slope and drainage class) was closely correlated with clay content and appeared to be similarly related to quality rather than quantity of sediment. The inputs of materials as fertilizers or manures were not significantly correlated with this group of pollutants; even atrazine was not significantly correlated with the area of corn—the only crop on which it was used.

The second group of water quality parameters (total and nitrate plus nitrite-nitrogen, and endosulfan) were strongly influenced by material availability. For example, nitrate-nitrogen (which was the dominant form of nitrogen) was strongly correlated with total nitrogen input (fertilizer and manure) and the crops to which it is applied at the highest rates, *i.e.,* row crops and, most notably, corn. Though not listed in Table I, it is interesting to note that none of the crops such as tobacco or soybeans, which require low nitrogen inputs, were correlated with nitrate loads. The insecticide endosulfan was strongly correlated with the vegetable crops on which it is used and, to a lesser extent, with the other watershed characteristics which were associated with vegetable crops in these watersheds (rural residences, fertilizer use, etc.).

The third group of water quality parameters (copper, PCB) did not appear to be sensibly correlated with anything. PCB was significantly negatively correlated with clay and the pollutant transfer potential to surface water index, and positively correlated with woodland—which is certainly not a source area.

Regression Analysis

The significant correlations among watershed characteristics are shown in Table IV. For the purposes of this paper, the desirability of independence of regression variables led to the omission of regression equations containing watershed characteristics correlated at $\geqslant 95\%$ probability. The independent equations resulting from applying multiple linear regression analysis to the water quality parameter loads and watershed characteristics are presented in Table V. All characteristics contributed significantly ($p \leqslant 0.05$) to the regression sum of squares. Only those equations which contained at least one watershed characteristic which was uncorrelated with any of those contained in equations with higher r^2 values were included in Table V.

Suspended solids (SS) is seen in Table V to have had no readily understandable multiple linear regression equation capable of explaining more variability than was explained by cultivated land alone. Two significant equations containing soil clay content (SC) and soil-extractable phosphorus (SP), and woodland (WA), soil-extractable P and total P inputs (TP), respectively, were obtained. Since it is difficult to hypothesize cause-effect relationships in which phosphorus availability influences sediment loss, these equations are discussed no further.

Total phosphorus (PT) variability was well explained by including either soil-extractable phosphorus (SP), row-crops (RC), or rural residences (RR) together with soil clay content (SC), in the equation. By the further addition of watershed area (AR) to the equation containing soil clay content and soil-extractable phosphorus, an extremely high value of 99% of the variability in total phosphorus unit-area loads was accounted for. This suggests a transmission factor related to watershed area, but special caution would need to be exercised if any attempt was made to use the equation for extrapolation

Table IV. Significant correlations between watershed characteristics.

Characteristic	Correlations[a]
1. (AR)	- - - - -
2. (SC)	4(PS)0.95**
3. (EP)	8(RC)0.64*; 12(NA)0.75*; 17(TD)0.69*; 19(FN)0.71*
4. (PS)	2(SC)0.95**; 11(WA)-0.64*
5. (PG)	14(AU)-0.86**; 16(ES)0.67*; 18(SP)0.94**; 20(MN)-0.86**; 23(MP)-0.79**
6. (SD)	- - - - -
7. (CL)	8(RC)0.92**; 17(TD)0.71*; 19(FN)0.85**; 22(FP)0.75*; 24(TP)0.73*
8. (RC)	3(EP)0.64*; 7(CL)0.92**; 9(CA)0.63*; 17(TD)0.86**; 19(FN)0.98**; 22(FP)0.86**; 24(TP)0.77**
9. (CA)	8(RC)0.63*; 17(TD)0.79**; 21(TN)0.91**
10. (VA)	12(NA)0.88**; 13(RR)0.94**; 17(TD)0.63*; 19(FN)0.71*; 22(FP)0.86**; 24(TP)0.73*
11. (WA)	4(PS)-0.64*
12. (NA)	3(EP)0.75*; 12(VA)0.88**; 13(RR)0.88**; 18(SP)0.65*; 22(FP)0.66*
13. (RR)	10(VA)0.94**; 12(NA)0.88**; 22(FP)0.75*
14. (AU)	5(PG)-0.86**; 18(SP)-0.87**; 19(FN)-0.63*; 20(MN)0.97**; 22(FP)-0.67*; 23(MP)0.86**
15. (AG)	- - - - -
16. (ES)	5(PG)0.67*
17. (TD)	3(EP)0.69*; 7(CL)0.71*; 8(RC)0.86**; 9(CA)0.79**; 10(VA)0.63*; 19(FN)0.89**; 21(TN)0.77**; 22(FP)0.70*; 24 (TP)0.70*
18. (SP)	5(PG)0.94**; 12(NA)0.65*; 14(AU)-0.87**; 19(FN)0.72*; 20(MN)-0.85**; 22(FP)0.79**; 23(MP)-0.78**
19. (FN)	3(EP)0.71*; 7(CL)0.85**; 8(RC)0.98**: 10(VA)0.71*; 14(AU)-0.63*; 17(TD)0.89**; 18(SP)0.72*; 22(FP)0.88**; 24(TP)0.73*
20. (MN)	5(PG)-0.86**; 14(AU)0.97**; 18 (SP)-0.85**; 23(MP)0.88**
21. (TN)	9(CA)0.91**; 17(TD)0.77**;
22. (FP)	7(CL)0.75*; 8(RC)0.86**; 10(VA)0.86**; 12(NA)0.66*; 13(RR)0.75**; 14(AU)-0.67*; 17(TD)0.70*; 18(SP)0.79**; 19(FN)0.88**; 24(TP)0.88**
23. (MP)	5(PG)-0.79**; 14(AU)0.86**; 18(SP)-0.78**; 20(MN)0.88**
24. (TP)	7(CL)0.73*; 8(RC)0.77**; 19(VA)0.73*; 17(TD)0.70*; 19(FN)0.73*; 22(FP)0.88**

[a] X(Y)Z* where X = characteristic no., Y = symbol, Z = correlation coefficient, * = significant at 5% level, ** = significant at 1% level.

Table V. Results of regression of selected water quality parameter unit-area loads on watershed characteristics.

Parameter	Units	No. of Variables	Equation	Explained Variance (r^2)	F^a
SS	kg ha^{-1}	1	-148 + 8.90 CL	0.55	9.62*
		1	548 - 13.68 WA	0.41	5.56*
		2	-762 + 22.46 SC + 24.59 SP	0.75	10.46**
		3	577 - 24.25 WA + 23.20 SP -18.43 TP	0.85	11.67**
PT	kg ha^{-1}	1	0.13 + 0.029 SC	0.51	8.40*
		2	-1.08 + 0.047 SC + 0.035 SP	0.91	33.81**
		2	-0.26 + 0.029 SC + 0.012 RC	0.80	13.72**
		2	-0.21 + 0.034 SC + 0.053 RR		12.79**
		3	-0.65 + 0.043 SC + 0.034 SP -0.000072 AR	0.99	249.72**
		3	-0.95 + 0.037 SC - 1.051 AU -0.00013 AR	0.90	17.75**
PO	kg ha^{-1}	1	0.005 + 0.011 SC	0.61	12.60**
		2	-0.198 + 0.012 SC + 0.0084 TP	0.90	32.50**
		2	0.310 - 0.010 WA + 0.14 SD	0.88	26.89**
		2	-0.108 + 0.012 SC + 0.017 RR	0.87	23.78**
		3	-0.197 + 0.012 SC + 0.0057 TP +0.010 RR	0.96	44.48**
		3	-0.030 + 0.012 SC + 0.0069 FP -0.000021 AR	0.95	41.43**
		3	-0.174 + 0.33 PS + 0.0087 TP +0.078 SD	0.94	29.18**
NT	kg ha^{-1}	1	-6.65 + 0.42 TN	0.81	33.17**
		1	1.77 + 0.29 CL	0.56	10.05*
		2	-7.73 + 0.47 RC + 0.35 MN	0.92	42.97**
		2	-9.27 + 0.33 TN + 0.15 CL	0.92	37.95**
		2	-4.02 + 0.45 TN - 5.19 SD	0.90	29.90**
NN	kg ha^{-1}	1	-0.20 + 0.66 CA	0.75	24.89**
		1	-0.55 + 0.26 CL	0.51	8.27*
		2	-4.15 + 0.40 TN - 6.53 SD	0.86	21.69**
		2	-8.41 + 0.42 RC + 0.29 MN	0.82	16.22**
NK	kg ha^{-1}	1	-1.11 + 0.13 CL	0.62	13.17**
		1	5.64 - 0.097 WA	0.42	5.86*
		2	8.49 - 0.32 AG - 0.00068 AR	0.64	6.24*
		3	0.13 + 0.16 SC + 0.080 SP -0.00033 AR	0.88	14.14**
CU	kg ha^{-1}	2	0.00037 + 0.00039 CL + 0.11 ES	0.64	6.23*
		3	-0.032 + 0.00088 SC +0.0013 SP + 0.15 ES	0.90	18.57**
		3	-0.011 + 0.00046 CL +0.0087 SD + 0.12 ES	0.77	6.88*

Table V (continued).

Parameter	Units	No. of Variables	Equation	Explained Variance (r^2)	F^a
ZN	kg ha^{-1}	1	0.013 + 0.0022 SC	0.43	6.11*
		3	-0.135 + 0.0036 SC +0.14 PG + 0.57 ES	0.96	48.61**
AT	kg ha^{-1}	1	-0.0006 + 0.00011 SC	0.73	21.86**
		1	0.0030 - 0.000072 WA	0.40	5.39*
		2	-0.0012 + 0.0033 PS +0.000024 CL	0.83	17.61**
		3	-0.0023 + 0.0034 PS +0.000061 CA + 0.00000025 AR	0.90	18.84**
		3	-0.0002 + 0.0028 PS +0.000054 CA - 0.00011 AG	0.87	13.30**
EN	g ha^{-1}	1	0.0041 + 0.0018 VA^b	0.86	48.11**
		1	-0.0182 + 0.0012 SP	0.56	10.12*
		1	-0.0068 + 0.00054 RC	0.46	6.84*
		2	0.0033 - 0.010 SD +0.0035 RR^c	0.90	33.08**
PC	g ha^{-1}	1	0.189 - 0.0025 SC	0.65	14.91**
		1	0.102 + 0.0020 WA	0.51	8.33*
		2	0.210 - 0.0028 SC - 0.0031 RR	0.82	15.78**
		2	0.149 - 0.0021 SC +0.0000073 AR	0.81	15.10**
		2	0.204 - 0.0021 SC - 0.26 ES	0.81	14.50**
		3	0.237 - 0.0027 SC - 0.30 ES -0.0035 NA	0.98	81.72**
		3	0.120 - 0.0023 SC +0.0012 CA + 0.0000093 AR	0.93	27.02**
		3	0.200 - 0.0028 SC - 0.0048 RR +0.00061 FN	0.91	20.22**

[a] F values followed by ** are significant at $p \leq 0.01$; F values followed by * are significant at $p \leq 0.05$.
[b] Orchards were also strongly associated with endosulfan loads but were omitted from this study because only one watershed had significant orchard area.
[c] There were also some significant multiple regression equations for endosulfan which included animal units and/or manure inputs, which were ignored.

purposes due to the relatively small and uniform size of the watersheds studied.

Soluble orthophosphorus (OP) variability was, like total phosphorus, dominated by soil clay content (SC). Improvements in r^2 values occurred when either total phosphorus inputs (TP), or rural residences (RR) were added, with the greatest variability accounted for (96%) when both were added together. Watershed area (WA), stream density (SD) and pollutant transport potential to surface waters (PS) contributed significantly to variability in soluble orthophosphorus in one or more of the regression equations, suggesting that this parameter is especially subject to transmission factors.

Total nitrogen and nitrate (plus nitrite) nitrogen (NT and NN) can be considered together since the latter dominates the former, and the regression results were almost the same for both (see Table V). Most variability (92%) was accounted for by either row crops (RC) combined with manure nitrogen inputs (MN), or cultivated land (CL) with total nitrogen inputs (TN). Total nitrogen inputs combined with stream density (SD) accounted for 90% of the variability in these parameters. No equations containing three watershed characteristics were significant improvements over those listed above.

Total Kjeldahl nitrogen (NK) was closely related to total phosphorus (TP) and consequently soil clay content (SC), soil-extractable phosphorus (SP) and watershed area (AR) together accounted for the greatest amount of variability (88%), as they did for total phosphorus. Manure nitrogen (MN) could be substituted for area in this equation with little reduction in r^2 (85%).

Copper (CU) variability could be accounted for to some degree (64%) by cultivated land (CL) and exposed stream banks (ES). Substituting soil clay content (SC) and soil-extractable phosphorus (SP) together for cultivated land increased the variability explained to 90%. Stream density (SD) could also be used with cultivated land and exposed stream banks to improve the r^2 values, which further supports the concept of an apparent native soil influence on copper due to the absence of anthropogenic sources in agricultural watersheds.

Zinc (ZN) is more insoluble than copper, and this may account for the greater influence of the soil clay content (SC) on the regression equations for this parameter. The role of exposed streambanks (ES) appears to be similar to that seen with copper loadings. Zinc, like most trace elements, has few sources in agricultural environments.

Atrazine (AT) is a widely used herbicide in corn. However, corn did not enter into the regression equations significantly until three characteristics were added simultaneously. Clay content and the index of pollutant transport potential to surface waters were dominant factors, indicating that soil properties are exerting a strong influence on atrazine movement into streams.

Endosulfan (EN) is an insecticide used primarily in vegetables, tobacco and orchard crops. Loadings were strongly correlated with vegetable area. The presence of tobacco and orchards in the watersheds was limited to too few sites to permit inclusion of these crops in the analysis. No characteristic could be added to vegetables so as to add significantly to the regression sum of squares.

PCB (PC) loadings could not be explained by way of regression equations. Those which were significant, and there were many, contained variables that were uncorrelated with soil or water movement. Since there are no known sources of PCB in agricultural watersheds, these results suggest that the suspected atmospheric input route is probably relatively unaffected by watershed characteristics.

The regression analysis approach to the explanation of variability in water quality loadings to streams draining agricultural watersheds is a useful step in understanding the dominant relationships which are active in these areas. These relationships can lead to the selection of prediction equations for the estimation of loadings from unmonitored areas. They can also be usefully applied in the selection of remedial management practices which will have the highest probability of success in reducing contaminant loads on a watershed scale.

The equations presented in Table V, and discussed above, are not a complete listing of all possible regressions which can be obtained with the available data. Equations which contain correlated characteristics, if reasonable representations of known cause-effect relationships, can also be helpful in explaining the impact of agriculture on water quality.

CONCLUSIONS

There has been a persistent gap between the two most generally available forms of information on the effects of agriculture on water quality. These two forms of information are the results of small-scale "plot" studies and large-scale river basin or lake loading analyses. The former are usually limited in terms of variability of soil, management and climatic conditions. The latter generally fail to distinguish between even distinctly different types of agricultural environments. The results of this study help to bridge the gap between these different approaches.

The role of soil particle size as a major influence on phosphorus, organic nitrogen, zinc and atrazine loadings from agricultural land has been clearly demonstrated. This holds important implications in terms of the efficacy of remedial measures selected without regard to soil texture. In cultivated areas, reduction of current loadings may be difficult to achieve with standard remedial programs. Furthermore, losses of some of these materials from fine-textured soils may be unavoidable regardless of the land use practice employed.

The influence of source material availability on stream loadings of the more water-soluble contaminants in streams is evidenced by the loadings of soluble orthophosphorus and nitrate-nitrogen which can be accounted for to a considerable degree by the inputs (fertilizer and manure) of phosphorus and nitrogen, respectively. Endosulfan, an example used here of the currently used pesticides, was present in relation to the crops on which it was used. Reductions for all of these materials can probably be expected if inputs are reduced, and/or availability to the water system is controlled, *e.g.*, by avoiding contact with surface runoff and by better timing of nitrogen applications to match crop uptake.

Finally, the results indicate that some materials, such as PCB and copper, are essentially unrelated to any aspect of agriculture. Control or reductions should not be expected through any remedial programs applied to agricultural activities.

ACKNOWLEDGMENTS

This study was carried out as part of the activities of Task Group C of the Pollution from Land Use Activities Reference Group, an organization of the International Joint Commission, under the Canada-U.S. Great Lakes Water Quality Agreement of 1972. Funding was provided by Agriculture Canada's Research Branch, the Ontario Ministry of Agriculture and Food and the Ontario Ministry of the Environment. Findings and conclusions are those of the authors and do not necessarily reflect views of the Reference Group or its recommendations to the Commission.

REFERENCES

1. Neilsen, G. H., and A. F. Mackenzie. "Relationships Between Soluble and Sediment Nutrient Losses, Land Use and Types of Soil in Agricultural Watersheds," Proc. 12th Can. Symp., Water Pollution Research Canada (1977), pp. 121-134.
2. Armstrong, D. E., K. W. Lee, P. D. Uttormark, D. R. Keeney and R. F. Harris. "Pollution of the Great Lakes by Nutrients for Agricultural Land," U.S. Task A Report, Vol. 1, International Reference Group on Great Lakes Pollution from Land Use Activities, I.J.C., Windsor, Ontario (1974).
3. Porter, K. S., Ed. *Nitrogen and Phosphorus: Food Production, Waste, and the Environment* (Ann Arbor, MI: Ann Arbor Science Publishers, Inc., 1975).
4. Hore, F. R. and A. J. MacLean. "Agriculture Canada Task Force Report on Implementation of the Great Lakes Water Quality Agreement," Research Branch, Agriculture Canada, Ottawa (unpublished, 1973).
5. National Eutrophication Survey. "Relationships Between Drainage Area Characteristics and Nonpoint Source Nutrients in Streams," Working Paper No. 25, Pacific Northwest Environmental Research Laboratory, National Environmental Research Center, Corvallis, Oregon (1974).
6. Haith, D. A. "Land Use and Water Quality in New York Rivers," *J. Env. Engr. Div., Proc. Am. Soc. Civ. Engrs.* 102 (EE 1):1-15 (February 1976).

7. Coote, D. R., E. M. MacDonald and G. J. Wall. "Agricultural Land Uses, Livestock and Soils of the Canadian Geat Lakes Basin," A Report of the Engineering Research Service and the Soil Research Institute, as part of Agriculture Canada's Contribution to the Implementation of the Geat Lakes Water Quality Agreement, 1973-74; Task C (Canadian Section), PLUARG, I. J. C., Windsor, Ontario (June 1974).
8. Ontario Ministry of the Environment. Work Plan, Task Group C (Canadian Section), Activities 1, 3 and 4 Studies, International Reference Group on Great Lakes Pollution from Land Use Activities, International Joint Commission (January 1976).
9. Ripley, B. D. and R. Frank. "Land Use Activities in Eleven Agricultural Watersheds in Southern Ontario, Canada, 1975-76," Ontario Ministry of Agriculture and Food (March 1977).
10. DeMayo, A., and E. Hunt. "NAQUADAT Users Manual," Inland Waters Directorate, Water Quality Branch, Environment Canada, Ottawa (1975).
11. Wall, G. J., L. P. J. Van Vleit and W. T. Dickinson. "Soil Erosion Investigations, Section II, Annual Report 1974-75, Agricultural Watershed Studies, Task C (Canadian Section), International Reference Group on Great Lakes Pollution from Land Use Activities, I.J.C., Windsor, Ontario (1975).

9

INTEGRATING WATER QUALITY AND BEST MANAGEMENT PRACTICES

A. F. Taverni
> Research Staff, The New York State Assembly
> Albany, New York

R. F. Dworsky
> Lake Champlain Basin Study
> Burlington, Vermont

INTRODUCTION

Various sections of PL 92-500 set forth methodologies to solve a water quality objective. Certainly all of us have been deluged with a veritable barrage of new terms—201, BAT, 303, BMP, 208 and 404, among others. The specific reference for this conference is Section 208 and a review of best management practices (BMPs) for the control of agricultural and silvicultural nonpoint source water pollution.

Although the federal government, through legislation, has directed that nonpoint sources of pollution be identified and controlled, there is currently little information on water pollution caused by land use activities. Little appears to be known about the linkage between land use and water quality. Although the water quality of a stream can be determined, there is currently no way of measuring accurately nonpoint impacts—either by cultural activities or natural processes—due to a number of variables including land use, topography, climate, and the chemistry and physical transformation of material in land and water transportation. The Council of State Governments concluded that "little or no land data exist by which management and regulatory programs can be justified. Efforts at managing diffuse source pollutants will continue to be impeded until greater emphasis is placed on providing adequate baseline data."[1]

In addition to baseline water resource data, when discussing integration of water quality and best management practices, one is struck by the fact that these topics are very limited in terms of the breadth of water resources planning. If a broad view is taken, we might find that to secure full value for investments in water quality or BMPs, a more comprehensive approach is needed. This paper has several focus areas which might explain the reasons for broader planning and management of water resources.

First, we will briefly discuss the problems of the Act as they impact proper assessment of BMPs and water quality. Second, we discuss the problems of the public's rising expectations with regard to water quality management and the potential problems inherent in a large-scale single-function act. Third, we will outline the problems of federal, state and local consistency with regard to various sections of the Act and potential management/planning options to gain full national benefit from clean water. Fourth, three potential integrated methodologies will be discussed: (a) the Genesee River Study, which provides a macroapproach for BMPs and water resources management; (b) the Hudson Level B Study, which provides a policy approach for BMP and water resources; and (c) the Lake Champlain Level B Study, which provides a microapproach to land and water planning.

BACKGROUND

At the risk of not being totally in step with the objectives of this conference, this section of the paper will provide an advocacy role in discussing BMPs. It is our belief that the implementation of PL 92-500 and the coordination required at the various federal, state and local levels is inconsistent. Consequently, we fear a disasterous public response due to rising expectations for clean water, coupled with a hasty distribution of limited public funds for water quality versus the development of a national water resources program.

We do not mean to denigrate the efforts for solving nonpoint problems. Historically they have formed a basis for environmental management at multiple levels of government. Further, as we are all aware, there are many applicable aims associated with BMPs in agriculture and silviculture.

The potential benefits which could accrue from BMPs are endless. BMPs could provide public health benefits, recreation benefits, and improved aesthetic quality. Reduction in sediment and erosion could concomittantly aid drinking water supplies by reducing the amount of toxics and organics delivered to public waters. BMPs could reduce sediment and erosion which, in itself, would achieve a considerable environmental goal. BMPs might reduce flooding and increase groundwater infiltration and water table management.[2]

However, we really do not know what the relationship of BMPs is to water quality improvements. In addition, we do not know the costs—either social or economic—that must be borne to attain this environmental "benefit." The fact is that *no* studies have been undertaken to determine what constitutes good or bad water quality.

The Missouri River was not named "Big Muddy" for nothing and Green Bay, Wisconsin, certainly was named due to algal blooms, caused by sedimentary conditions. In a study at the University of Massachusetts, it was found that if all the waters in New England could be classed A, B or C (Class A being the best), only 14% of all New England's waters would have been classed as A prior to the Pilgrims' landing. The explanation of this is fairly obvious. Most of the headwaters started in wetlands and flowed through large bogs. This increased the biochemical oxygen demand (BOD) loading due to detritus and plant material. Much of the runoff was from granite rock or shallow soil that caused pH problems. The cyclic climatic pattern produced large variable deposits which flowed into streams periodically increasing organic loadings. The percentage of Class A waters has subsequently been reduced from about 14% to 5%. Perhaps this is what BMPs should concentrate on.

As stated above, innumerable benefits can accrue through a BMP strategy. However, the theory that BMPs will necessarily improve water quality deserves scrutiny. BMPs must take water quantity into consideration as it regulates our capacity to understand our goals. BMPs should also reflect exogenous conditions which might dictate high background levels of pollutants.

The inherent assumption in PL 92-500 is one of natural productivity and diversity in water. To some extent, what is one man's pollution could be another's integral part of the water ecosystem. Improving water quality must look at the basic resources as well as its quantity and then apply BMPs. PL 92-500 did not look at the relationships of water quality and biological productivity. It failed to provide a water quality strategy in terms of natural assimilative capacity, and the assimilative capacity in terms of water quantity.

As we know, different flows yield different qualities. The classifications of receiving waters vary as to time and space. Many areas had significant nonpoint pollution problems prior to modern technology, and this provided a specific biological response. Unless an appropriate vehicle is used to manage conflicts or future water allocations, then public investments and confidence in water quality control will be undermined.

A possible framework for providing effective BMP might focus on the following triage:

1. bodies of water where capital investment in pollution control or land treatment will produce little or no environmental response because of low assimilative capacity of the receiving waters (in these cases it might be suggested that little extra public money be invested);

2. bodies of water where capital investment will have a positive response in terms of water quality; and
3. bodies of water where capital investment will make no difference because of the high assimilative capacity of the receiving waters.

This type of system would simplify the allocation and distribution of resources to both point and nonpoint problems and would stratify receiving waters so that a determination of "responsiveness" could be made.

We feel that a single-function view of water quality as advanced by the Environmental Protection Agency is not effective in the real world. A mechanism has not been set up for resolution of use conflicts at the local, state or federal regional level. The mechanism which does exist solves only water quality problems and not the more far-reaching problems of flow management or future water allocation.

A definition of clean water/water quality needs to be tempered with an appropriate view of the resource and the ecosystem response. One way is to examine the receiving waters and determine if best management practices will be effective in improving water quality or if quality must be integrated with quantity planning and management. This will require research and analysis of water quality standards as well as BMPs. Subsequently, an analysis must be made of BMPs in terms of costs and benefits, not only with regard to water resources, but also with regard to agriculture or forestry.

While the public is footing the bill for this very limited objective of water quality improvement, it does so within the perspective of promised quality improvement. The second section will look more closely at the case of New York State and the problems of rising expectations.

NEW YORK STATE

PL 92-500 established a ringing set of societal goals:

- it is the national goal that the discharge of pollutants into the navigable waters be eliminated by 1985;
- it is the national goal that wherever attainable, an interim goal of water quality which provides for the protection and propagation of fish, shellfish, and wildlife and provides for recreation in and on the water be achieved by July 1, 1983.[3]

It has been claimed that these are the most ambitious set of national goals ever established. Section 208 establishes the foundation for their attainment with its point and nonpoint source control provisions.

Consider what these goals mean. We have created within society a set of rising expectations in anticipation of the elimination of discharge and improvement of fishable/swimmable waters. We have created these expectations against a backdrop of administration confusion, nebulous guidelines, and fragmented communications.

The potential consequence of spurned expectation could be catastrophic. Consider the following New York State example. Although a leading industrial state, New York has taken vast strides to preserve its resources and protect its environment. Prior to 1974, among a host of legislative enactments were a Wild, Scenic and Recreational River System, a Tidal Wetland Protection Act, a billion-dollar Pure Waters Bond Issue (approved by the voters), creation of the Adirondack Park Agency and other pieces of salutory legislation. In 1974 and 1975 the State of New York enacted three major pieces of enviromental legislation which are related to the control of nonpoint sources:

1. the Freshwater Wetlands Act which required the inventorying of all freshwater wetlands in the state in excess of 12.5 ac and instituted a series of regulatory measures to protect this valuable resource;
2. the Mined Land Reclamation Act which provided for the control of mining practices especially as they contribute to erosion and sedimentation; and
3. the State Environmental Quality Review Act (SEQR) which requires all public agencies, local governments and developers to prepare impact statements for projects of certain magnitudes.

Of these three programs, the only one which has been substantially implemented is the Freshwater Wetlands Act. Amendments to SEQR in 1976 and 1977 effectively delayed its implementation and modified its scope. The Mining Reclamation Act has never been implemented and presently languishes in a bureaucratic half-life.

The reasons for this lack of implementation sound all too familiar: confusion over legislative intent, lack of knowledge and appropriate tools, lack of implementing powers, lack of coordination and communication with local governments, differing executive priorities, and a lack of financial resources.

The Department of Environmental Conservation, which was given administrative responsibilities for these programs, has been saddled with additional responsibilities without benefit of a real budget increase in four years. Presently it is under attack and efforts are being made to split off its fish and wildlife responsibilities and give them to an independent commission. Reasons for this attempt include dissatisfaction with the department's environmental-ecologic orientation and a general uneasiness with the department's overall effect over the past several years.

The correlations which could be drawn between this example and a possible federal scenario over the next several years remain to be seen. However, if the public's rising expectations are not met after the expenditures of $12 billion for facilities and BMPs, if there is no appreciable increase in water quality because it is difficult to weigh any improvements in the absence of a benchmark, or if PL 92-500 has only served as a massive public works/capital construction boondoggle, then the analogies will become apparent.

Groups like the New York State Conservation Council, whose expectations have been frustrated, may attempt to dismantle programs and reconstitute policies to their benefit. Existing coalitions may fragment and competing constituencies may emerge because various planning groups have set out with half-a-basket-full of tools in search of a theory.

NONPOINT SOURCE CONTROL

This section will briefly examine the problems of setting the priorities needed to achieve the elusive goal of nonpoint source control. To begin with, agriculture and silviculture are only two of the several nonpoint control components identified in Section 208. The water quality impact of agriculture and silviculture may vary in relative importance with construction, acid mine drainage, and/or urban runoff, depending upon a host of economic, geographic and social factors. In fact, any of the above frequently contribute more to negative water quality than either agricultural or silvicultural problems. Clearly, BMPs in Vermont are not the same as those in Iowa, either in terms of scale or magnitude.

Given this type of differing demands and a lack of national consistency for review and approval, what selection criteria exist to identify critical problems and determine priorities? The quagmire of governments and regulations makes the evaluation of identified nonpoint problems at the local level elusive; at the state level difficult; at the federal regional level intractable; and at the national level implausible. Without such criteria, it is virtually impossible to determine which problems have larger impacts than others, which can be solved in terms of environmental response, and which may never provide an ecosystem response, just a dollar sink.

The problem is exacerbated by the many levels of government, each doing its own thing (Table I).[4] As we know, one level submits grant applications which are approved by the others above it in the hierarchy. Basically all levels can plan, but only one has the constitutional standing to zone. Conflicting demands precipitated by differing constituencies clamoring for a share of finite resources breed confusion in the absence of a national strategy.

Further, EPA has never said that there will be a common set of elements, evaluated by a common standard, which will relate to water quality improvement in terms of costs and benefits. No evaluation of the impact on social cost, energy or recreation is made. This seems to preclude the evaluation of Section 208 BMPs in terms of opportunity costs or revenues foregone, water quality, or other more manageable social goals.

APPROACHES FOR BMP SELECTION 99

Table I. Water quality management as provided for by the Federal Water Pollution Control Act (PL 92-500).

Government Level	Section 208 Basin Plans	Section 303 State Plans	Section 208 Regional Management	Section 201 Project Supplementation
Federal	Federal grants for plans development	Federal grants to states	Federal grants to states and designated local agencies	Federal grants for treatment works
	Plans to be integrated analysis of water quality and water resources issues on a basin or sub-basin level	Approval of state planning process	Approve certified 208 plans	Approve local grant applications: - project included in applicable 208 plan - project in conformity with state 303 plan - project entitled to priority re: Section 303 plan - federal requirements complied with re: project
	Plan to focus on water quality and water resources matters in designated section 208 area			
	All plans to be completed by January 1980			
State	Substantial state participation visa representation on river basin commissions	Develop and implement statewide planning process to result in plans for all navigable waters of state	Designate 208 local/regional areas and organizations	Certify project priority
			Certify each areawide plan as being consistent with applicable 209 and 303 plans	Approve local grant applications
		Plan to include: - effluent limitations, water quality limitations, and compliance	Areawide waste treatment management planning for balance-of-state non-	Contribute to local share of funds

Table I (continued)

Government Level	Section 208 Basin Plans	Section 303 State Plans	Section 208 Regional Management	Section 201 Project Supplementation
		schedules; total maximum daily load for pollutants; - inventory and ranking in order of priority of needs for construction of treatment works - incorporation of all elements of applicable 208 areawide waste treatment management plans	designated area Implementation of 208 plans for nondesignated areas Statewide regulatory program for nonpoint sewer controls	
Local/ Regional	Little or no local/ regional involvement in plan development	Limited local involvement in plan and process development	Areawide waste treatment management Planning for designated 208 areas Implementation of 208 plan for designated areas	Develop grant application Project level decisions Local share of funds

In a paper entitled "After Pollution, What?,"[5] Carlos Stern makes the following point concerning our commitment to clean up the water. If the public is to reap full benefits of this multibillion-dollar investment, it will be essential for related lands to be planned and managed. To expand this single-point function, elements such as water quality management or agriculture or silviculture BMPs are not enough. There must be a larger commitment to plan and integrate other resource problems such as recreation, urban and rural development, and agricultural land preservation. To do less means that the full value of this labor will not be available to the public.

There is another mechanism available for consolidating BMPs and resource management. It is provided in PL 89-80 which states[6]:

> It is the policy of Congress to encourage the conservation, development, and utilization of water and related land resources of the United States on a comprehensive and coordinated basis by the Federal Government, states, localities, and private enterprise with the cooperation of all effectual Federal agencies, states, local governments, individuals, corporations, business enterprises, and others.

This establishes the framework for the Level B studies later provided for in Section 209 of PL 92-500 as a process for integrated resource management. As will be demonstrated in the next section, it may be that Level B offers the remaining hope for integration of water resources and water quality elements.[7]

Section 209 is a vehicle for the integration of water resources and water quality issues at the river basin level. Section 303 provides for the development of statewide planning processes for the development and implementation of effluent limitations and water quality standard compliance schedules, and the development of priorities for the construction of waste treatment works. Section 208 provides for the development of continuous areawide waste treatment management processes at the local and regional levels. Section 201 provides for the development and implementation of specific treatment works projects. Therefore, 201 provides for the smallest level of detail, the actual project, while 209 plans represent the macrolevel of integrated water resources and water quality management. The conceptual relationship between these authorities is illustrated by the pyramid in Figure 1.[4]

"Level B," or river basin plans, are intended to solve complex land and water resource problems. The planning process requires that a broad range of alternative measures be considered, and that short- and mid-term action plans be suggested. These alternative measures are valued against the Water Resources Council's Principles and Standards for planning and evaluation, or against similar criteria to simulate a range of alternatives through consideration of multiple objectives. This methodology would be useful in optimizing the use of federal dollars and resources to improve (1) economic development, (2) environmental quality, and (3) overall water quality.[7]

102 BMPs FOR AGRICULTURE AND SILVICULTURE

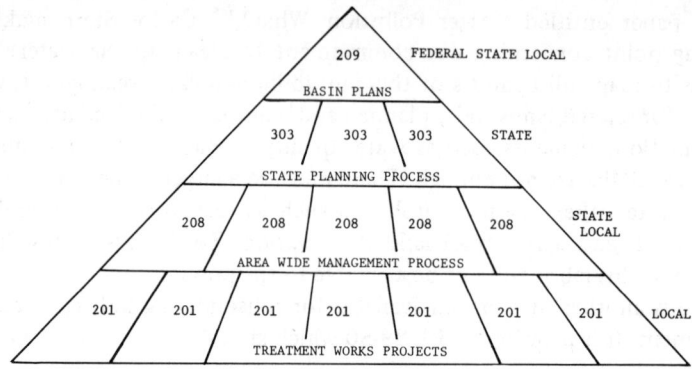

Figure 1. Conceptual relationship between planning authority sections in PL 92-500.

At a broader level, it is possible to integrate water quality and water quantity management through the use of the Comprehensive Coordinated Joint Plans (CCJP) mechanism. CCJPs are integrated with the data and planning objectives developed by the Water Resources Council in its national assessment, and identified as priority needs and problems by basins, presumably with input from state and 208 agencies as to integrating nonpoint and other water resources management. The potential utility of CCJPs in resolving water quality/water quantity problems and inserting community and state water plans into a basin framework of a federal water policy is still being resolved.

Several case studies of the potential integration of water and land resources have been made in the concluding part of this paper.

THE GENESEE RIVER BASIN STUDY

The Genesee River Basin Study* (Figure 2) was published by the Coordinating Committee of the Corps of Engineers in 1969 and completed prior to PL 92-500. While not a Level B plan as identified by the Water Resources Council, it provided the traditional elements of a water resources planning effort. "The objective of the study was to devise a sound program for the development of water and related resources to meet the immediate and long-range needs of the basin in an orderly, efficient, and timely manner."

The principal needs were determined to be flood protection, water quality control, recreation, fish and wildlife enhancement, irrigation, and agricultural

---

*A subsequent report by the Genesee River Basin Board and DEC entitled "A Comprehensive Water Resources Plan In the Genesee River Basin" will be reviewed at a public hearing this spring.

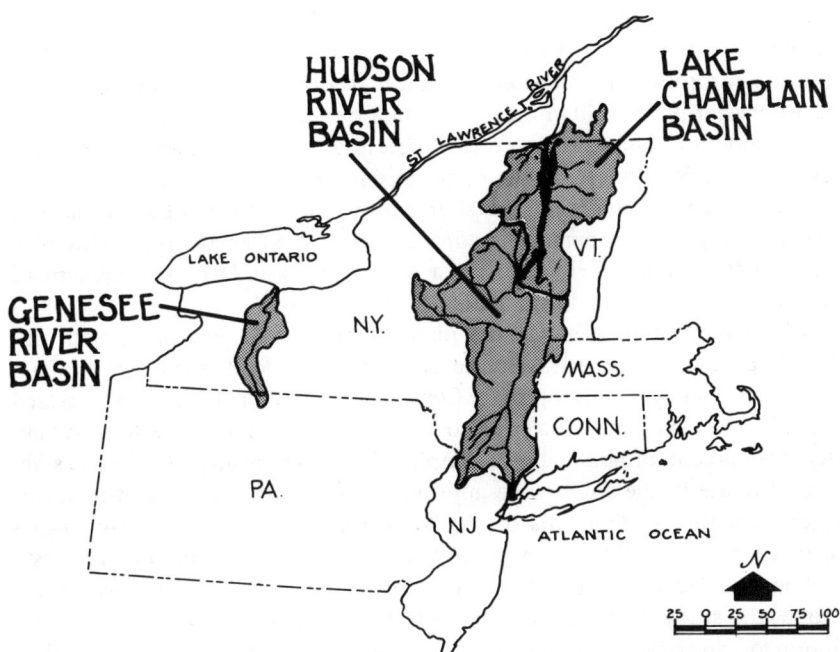

Figure 2. Study areas.

land and water management. The most practicable means to provide for these and other needs of the basin was identified through a comprehensive plan of structural and nonstructural measures. The basic analysis for decision was that tangible benefits exceeded project costs.

Without describing the details of the basin, a summary would show the Genesee River Basin covers 2,479 mi^2, mostly in western New York, with a small portion, 96 mi^2, in northwestern Pennsylvania. It is roughly elliptical in shape, with a north-south major axis of approximately 100 mi and a maximum width of about 40 mi. The river rises in the Allegheny highlands in Potter County, Pennsylvania, at an elevation of about 2,500 ft, flows approximately 157 river mi in a generally northward direction to its mouth at Rochester Harbor on Lake Ontario, at an elevation of about 247 ft. The topography of the southern portion, the Upper Basin, upstream of Mount Morris Dam, is steep and rugged, while the northern portion, the Lower Basin, is gently rolling. The two major divisions of the basin also closely parallel the two land resource areas which comprise the basin: the Allegheny Plateau and the Ontario Lake Plans Service Area, a region of about 750 mi^2 north and west of the Basin lying between Rochester and Lockport, New

York. This area was included because of its rich agricultural potential and the possibility of supplying it with irrigation water from the Genesee River Basin via the New York State Barge Canal, which traverses the region from west to east and crosses the Genesee on grade at Rochester. The dominance of the Rochester metropolitan area in population, employment, income, industrial production, and even in certain sectors of the agricultural economy, is the most significant factor in the basin's economic base.[8]

It was clearly stated that the short-range phase of the study would not meet all the projected demands and/or needs of the Genesee Basin. The plan was expected to serve, however, as a guide for the orderly development of water and land resources in the Basin.

It is important to examine the planning framework associated with this plan. The water resource was considered as the vehicle for the larger program of natural resources development. Consequently, water quality was assessed not only as a functional area but also as part of a concomitant water and land use/development program. For example, the water quality standards established for use in the study were approved by New York State and in accordance with the requirements of the Water Quality Act of 1965. By using a dissolved oxygen (DO) concentration as the standard and assuming that waste loading and treatment levels would improve BOD point source waste discharge, the study determined that the DO requirement could be met by "exclusion or diversion of wastes, by advanced treatment of wastes or by flow augmentation."[8]

Waste loads and subsequent management did consider land source problems other than point. These included combined sewer systems and agricultural and other land sources that add to surface water pollution problems.

Major souces of water pollution from farm activities are fertilizer, pesticides and sediment. Agriculture land runoff and waste discharge contributed about 322,000 lb (1965) of soluble phosphates which produced excessive algae production in lakes and streams. No determination of pesticide problems was discussed.

Sediment and erosion studies indicated that heavy sediment loads are delivered during high runoff periods, and dredging of the lower Genesee for navigation requires the removal of 82,000 tons annually.

It was generally estimated that the solution to land problems could be estimated in terms of land treatment needs (Table II). Land treatment measures included those to reduce erosion, eliminate excessive water conditions, improve unfavorable soil conditions, and protect forest lands. In addition to the existing program of funding, a specific reference was made to current and recommended program changes for forest land treatment measures (Table III) which would specifically improve water quality through reduction of sediment and erosion.[8]

Table II. Needs and cost summary for conservation treatment, total program and program to 1980, Genesee River Basin, 1967.

Land Use	Total Program		Program to 1980	
	Ac Needing Treatment	Estimated Cost	Ac to be Treated	Estimated Cost
Cropland	312,000	$20,053,000	111,000	$7,126,000
Pasture	154,000	10,117,000	42,000	2,755,000
Forestland	271,000	11,000,000	74,500	3,030,000
Total	737,000	$41,170,000	227,500	$12,911,000

Because flow needs were determined as a measure for DO, numerous reservoir sites were determined to meet the needs of water-based recreation, water supply, fish and wildlife needs, and most importantly the projected sites had the potential to yield a high enough supplemental water to meet water quality standards. The economic feasibility of these sites was determined by estimating the annual alternative cost of advance treatment and comparing this with the cost of constructing, maintaining and operating the potential site. Where cost of augmentation was less than alternative cost of treatment, a site was determined to be feasible. Subsequently, EPA rejected flow augmentation as a quality control method.

A concluding comment would show that the comprehensive development of the basin met not only the water quality needs—as determined in 1969—but also reduced sediment for navigation purposes as well as water quality, provided for fish and wildlife development, increased production of food and fiber, and used a basic benefit/cost analysis.

Table III. Current and recommended program for forest land treatment measures, Genesee River Basin, 1966.

Practice	Annual Accomplishments Under Current Programs	Recommended Total Program to 1980	Acceleration Above Current Programs
Management Plans	60 plans	1,650 plans	870 plans
Tree Planting	1,800 ac	35,500 ac	12,100 ac
Hydrologic Stand Improvement	1,100 ac	23,000 ac	8,700 ac
Woodland Grazing Control (fencing)	500 ac 2 mi	15,000 ac 60 mi	8,500 ac 34 mi
Erosion Control (skid trail and logging road)	40 ac 1 mi	1,000 ac 25 mi	480 ac 12 mi

THE HUDSON RIVER BASIN STUDY

Instituted under a different authority than the Genesee Study, the Hudson River Basin Study focus was not on "comprehensive" problems but on significant land and water problems facing the basin within the next 15-25 years. The objective of the study is to develop implementable solutions and/or recommend additional studies required to meet the anticipated problems. Specific focus areas are: water supply, water quality, recreation, flood damage reduction, disposition of dredged spoil, and access. The Hudson Study relied on the Water Resources Council's Principles and Standards for economic analysis, environmental management, and policy and program development.

The Hudson River Basin extends over an area of 13,365 mi^2, 95% of which is in New York State, with small portions in New Jersey and Connecticut.

The river begins on the southern slopes of the high peaks country of the Adirondack Mountains and flows more than 300 mi through some of the most beautiful lands in the state of New York. The Mohawk River, which joins the Hudson at Cohoes, drains about 3,500 mi^2 of the east central portion of New York State. Much of the Mohawk mainstream is incorporated in the Barge Canal System extending about 100 mi from Cohoes to Rome. The lower Hudson extends from Cohoes to the Battery in New York City and flows through some of the most urbanized land in the country.

The Hudson and its tributaries carry an important part of the commerce and the wastes of the more than 2 million people who live in the valley. It is today a source of water supply for many municipalities and is a potential water supply for as many as 12 million people in the New York City Metropolitan region. There are tremendous conflicts over multiple objectives and these are exacerbated by large and varied economic, political and social groups.

Critical water quality problems have been associated with the Hudson River for some time. The purpose of the Hudson Study is to integrate water and land resources management with federal and state programs which will provide for the most responsive enhancement of resources. Some water quality study focuses are generally described below.[9]

Nonpoint Sources

In the Metropolitan New York area, the control of nonpoint sources of pollution is a major problem. Wet weather discharges through combined sewer overflows are a major water quality problem. For most areas of the basin, stormwater runoff problems exist.

In other locations, sources of nonpoint pollution are creating water quality problems accentuated by the impairment of normal flow by dams and locks.

Waste Heat

The Hudson River Basin is a highly industrialized area, with many industries using the surface water of the region as a heatsink. The heat dissipation capacity of the Hudson estuary is limited.

With respect to power plants and energy usage, there will be a consumptive use problem due to conflicts between these and other uses, depending upon future federal policy on cooling water requirements. Lesser use of cooling towers will increase thermal loading, while greater use of cooling towers will increase consumptive water usage for cooling.

Pumped storage power projects will conflict with withdrawal and instream uses for other purposes.

Salinity Levels

Preservation of existing salinity levels is required to maintain the present water intakes on the river downstream from Poughkeepsie to fulfill needs for potable water supply, fish and wildlife management, and maintenance of existing estuarine and wetland ecosystems.

In the event there is extensive use of cooling towers, it is possible that low flow augmentation may be needed in order to maintain the present location of the salinity front.

Industrial

Chemical and biological pollutants will be of significance to municipal water supplies, commercial and sport fishing, and recreational pursuits. Discharges of solid wastes and toxic industrial discharges such as PCBs further degrade water quality. Oil spills and improper handling of hazardous materials are adversely affecting water quality, visual appearance, and recreational uses. Excessive acid precipitation has caused fish kills and affected 11,500 ac of Adirondack lakes and ponds.

Dredging and Navigation

With increase in vessel size, channel enlargement will be necessary along with improvement to shoreline and transportation facilities, and maintenance dredging will continue. Thus dredge spoil disposition and effects on fish and wildlife habitat present potential problems. Control of harbor drift is especially costly.

Transportation

Considerable maintenance dredging is required because of channel filling. Disposition of dredged spoil conflicts with navigation as well as fishing and shoreline interests. Transportation corridors usages conflict with other land uses. Major sediment sources should be identified and a determination made whether it is economical to control the source and dredge the channel less frequently.

In this study specific efforts are being made to tie sediment and erosion problems [using Conservaton Needs Inventory (CNI)] to the cost of removal and value of soil losses. These two approaches are being used to calculate a basinwide set of costs which could translate into new or expanded programs. Section 208 plans would suggest BMPs to meet water quality objectives; however, the level B approach is to link value and costs of removal to BMPs— certainly a more politically supportable and analytical method.

Table IV identifies Agricultural Stabilization and Conservation Service expenditure in the Hudson Basin for practices alleviating nonpoint pollution.[10]

Table IV. Hudson Basin expenditure for practices alleviating nonpoint pollution.

Land Use Practices	Cost/Ac Average of All Costs	1970 Quantity	1970 Cost[a]	1972 Quantity	1972 Cost	1975 Quantity	1975 Cost
Best Identifiable Practices	$526	40,893	$3,318,517	62,343	$5,217,681	94,028	$9,574,900
Cost per Ac			$81.15		$83.69		$101.83

[a]Cost = Quantity x cost/ac of each practice cost.

Table V identifies the cost of land treatment needed to solve nonpoint problems.

Table V. Cost of land treatment on the Hudson Basin.

Reach	Acres Needing Treatment	Cost at 106/ac[a]
Lower Hudson	112,330	$11,906,980
Upper Hudson	19,118	2,026,508
Mohawk	123,476	13,088,456
Schoharie Ck.	76,887	8,150,022
TOTAL	331,811	35,171,966

[a]Based on expenditure made in basin in 1973, 1974 and 1975.

Table VI is perhaps the most interesting based upon acres needing treatment on which sediment loads were calculated. The value of topsoil lost from cropland pasture and orchards was estimated at $5 per ton. This value represents the cost of fertilizer needed to replace nutrients. Ten percent of total gross erosion from all sources was estimated to be delivered to streams. The cost of removal to keep channel capacity in areas of deposition was calculated at $2.31 per ton for removal of sediment that would affect stream flow that results in increased depth and frequency of flooding.

Table VI. Magnitude and damage from sedimentation in Hudson Basin.

Reach	Sediment (Tons)	Cost of Removal	Value of Fertility Lost
Upper Hudson	202,983	$346,476	$1,754,960
Lower Hudson	343,184	774,275	6,957,935
Mohawk	295,887	679,127	2,940,285
Schoharie	200,730	333,226	4,219,665
TOTAL	1,042,784	$2,133,104	15,872,845

This points out that there are other costs associated with sediment and erosion, and the benefits accrue not only through improved water quality but also through revenue not spent for other values. The primary objective might not be, then, water quality, but navigation benefits, flood protection or recreation—none of which can be estimated if a single-function goal of water quality is pursued.

Through the use of The Principles and Standards, a cost-effective multiobjective analysis can be made to integrate water quality with other national and environmental objectives.

THE LAKE CHAMPLAIN STUDY

The Lake Champlain Basin Level B Study Plan found that while land and water problems existed and these needed solutions, few were considered "critical." Because of this fact, as well as the belief that the public deserves good products from planning efforts, a closer integration of 208 planning and Level B planning and several joint publications were undertaken. The Champlain Plan will seek to improve and demonstrate the relationships between planning under Sections 208 and 209. It was hoped that the examples here could integrate water quality and BMPs as well as water quantity concerns.[11]

Two major studies were undertaken by the Level B program and subsequently used by the Vermont 208 program. The first was a forestry study entitled "Nonpoint Pollution from Forest Lands." This study concludes that no research information on the effects of timber harvesting on water quality in Vermont is available. However, it points out that only 2% of the total forest land in the Basin is harvested each year, and only about 10% of the harvested acres are served by roads or ski trails. To impair water quality, it was estimated that over 50% of the forestland in a watershed would have to be harvested in one year to impact a stream larger than a headwater stream. At present this was not expected.[12]

However, it was noted that increased forest harvesting would be taking place due to an increase in wood being used for home use. In addition, Burlington Electric has plans to build a 50-MW electric wood generating plant. The potential for increased wood demands was emphasized because of other studies conducted. In the energy and power report, it was determined that wood might provide for some of the anticipated increase in energy demand.

It was estimated that forest practices do not significantly contribute to the nonpoint source loading in water quality management. It was estimated, however, by the energy and power task force that increased use would be made of wood for energy generation. A method to assess the water quality impacts of wood demands was needed. The Lake Champlain Basin Study sponsored a study called "Modeling Vermont's Forest Resources in the Lake Champlain Basin." Using the methodology of system dynamics, a computer model is developed to forecast alternative futures for Vermont forest regions. The model projects volume and net growth over the next 25 years under the impact of removals for wood-utilizing industries, and for a wood-fired power plant in the basin. This provided an example of integrated resource and water quality planning.[13]

A second plan, a Farm Practices Study, was conducted using a micro-approach toward solving water quality problems.

The LaPlatte River is the largest tributary to Shelburne Bay, a bay experiencing accelerated eutrophication. The river, approximately 15 mi long, drains an area of approximately 29,056 ac.

Beginning in July 1975 and ending in June 1976, the State of Vermont, Department of Water Resources, began an intensive study of Shelburne Bay to determine the quantity of nutrients coming from point and nonpoint sources. A summary of the results of the study indicates that:

1. Almost 40% of the annual total phosphorus loading in Shelburne Bay comes from nonpoint sources.
2. Phosphorus is the primary nutrient that will have to be managed to control the rate of eutrophication in Shelburne Bay.

3. Sections of the Shelburne Bay drainage basin which are dominated by agricultural land use and the lack of a protective buffer zone are associated with high nonpoint loading rates.

The Lake Champlain Level B Study funded the U.S. Soil Conservation Service office in Burlington, Vermont, to carry out a 1-year study of the LaPlatte watershed. The purpose of the study was to compute erosion from various land uses, streambanks, roadbanks and construction sites, and to estimate sediment delivery to Shelburne Bay from the LaPlatte River watershed. Also, erosion problems are to be examined in relation to land management practices.

Study findings indicate two major problems associated with farming. The foremost problem appears to be excessive soil loss associated with poor farm conservation practices. Contributory practices include continuous corn production and inadequate rotations on lengthy slopes, lack of buffer zones on stream borders, and the lack of contour plowing and stripcropping. Although stream bank erosion is highly visible, the study indicates that in quantity, it is not the most significant when compared with sheet and rill erosion. According to this study, 42% of the cropland is eroding in excess of 12 ton/ac/yr.

The other problem identified in the watershed is manure management. Study data[14] show 3363 animal units in the watershed sample plots.* Extrapolating this data to the entire watershed, 269,040 tons of manure will be produced (this is based on 16 tons production per animal unit per year). Measured cropland on the LaPlatte watershed is 3398 ac. Hayland and pasture account for 12,314 ac. If the manure is applied only to cropland, the application rate is 79 ton/ac. If the total amount is applied to cropland and one-half the hayland and pasture, the rate of application is 28.1 ton/ac. The study goes on to indicate that nearly one-half the soils in the watershed are in the low productivity group for which 20 ton/ac is a more appropriate application rate.

The study found that problem practices are found on both cooperator and noncooperator fields.†

In summary, it was found that while cropland represents only 10% of the land acreage, it generates 86.2% of computed soil loss.

Rather than go into the technique for BMPs, it would be appropriate to discuss what this study means in terms of other study elements and planning. The outputs include the development of a phosphorus model for Lake

---

*An animal unit is defined as one dairy cow (1000 lb) or equivalent size animal.
†A cooperator is a farmer who belongs to the local soil conservation district. Before a farmer or a landowner can receive conservation planning assistance, he/she must be a cooperator.

Champlain to assess in numerical form the water quality improvements. This model utilizes all phosphorus loadings into the Basin. As a result, we can see if improvements in water quality will exist with BMPs and where our investments will be most effective.

In another study on the economic viablity of agricultural lands, it was found that the agriculture development had low or medium economic viability in the foreseeable future. Because of the expanding Burlington area, this area was under extremely high development pressure. In a rational planning process it would be expected that additional growth and transition from agriculture to housing would modify the role and expectation of BMPs. In other words, why invest in something that will be gone shortly?[15]

In summary, this is only an element of an integrated approach for water quality management. Clearly no decision with regard to nutrients could be made on Lake Champlain without considering all political boundaries. Even within a basin a procedure was needed to identify those areas where BMPs were practiced, and lastly a procedure was needed to determine changing land uses where the use of BMPs would be the most effective in the long run as well as the short. In forestry not only were sediment problems estimated, but a prediction model was designed for wood generation industries that could be expected. The Lake Champlain Study is a vehicle for coordinating bi-state efforts for planning consistency.

CONCLUSIONS

To tie together a report such as we have presented, several points must be made, and these are listed below.

1. It is plain to see that EPA has taken on the attempt to resolve water quality issues in a vertical profile. As with all vertical organizations, these have relatively thin intersecting elements with broad horizontal environmental issues (Figure 3). Water quality planning and management are extremely subject to flow regulations, and natural environmental system, and social uses of land outside the planning framework of Section 208.

2. We think it is safe to say that not having reached desired planning efficiencies in water resources development, we cannot expect too much in terms of definitive succinct management programs to solve water quality management problems. Continuing new innovative projects in planning can and must be used to meet the public objectives of clean water.

3. Public objectives, public participation and public awareness have been heightened to such a degree in terms of water quality management that any faltering or lack of success will probably reflect negatively on the entire national pollutant abatement strategy—in other words, we owe it to our clientele, the public, not to fail. Because of this and because we want to secure full

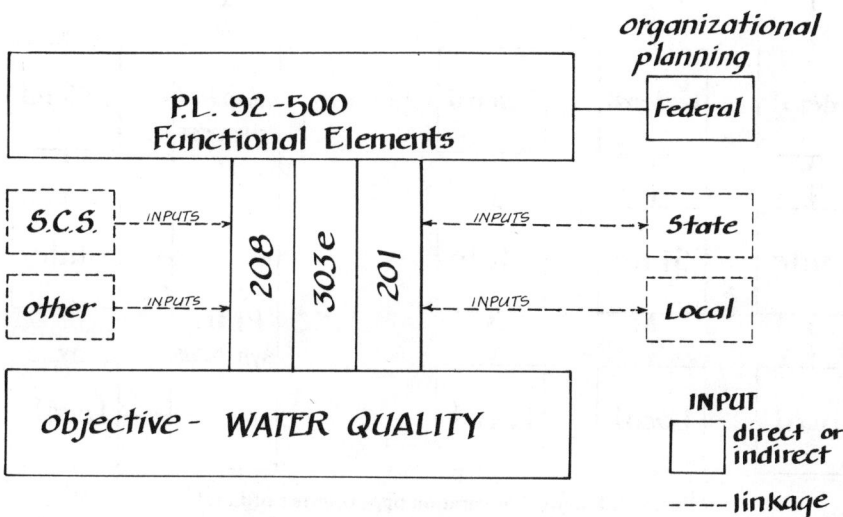

Figure 3. Vertical functional planning.

public value as a desirable goal in a limited objective strategy, it is necessary to broaden planning and management and to provide *reasonable* expectations for advanced water and related land values.

4. One of the federal mandates calls for consistency in planning and management. Federal consistency is defined as: Federal water activities are carried out in agreement with all principles and objectives of approved regional water resources management plans. This apparent coordination is more far-reaching than the limited objectives stated in water quality management by EPA. In addition, consistency for evaluation must be required at all federal levels, especially when reviewing limited political objectives called for in 208 plans.

5. The opportunity to integrate water quality management and water resource (quantity) management is stated in PL 92-500. Close departmental/agency coordination for policy and planning is more likely to be achieved than basic reorganizational functions. Level B studies and CCJP activities are possible methods to integrate various governmental levels and functional programs (Figure 4).

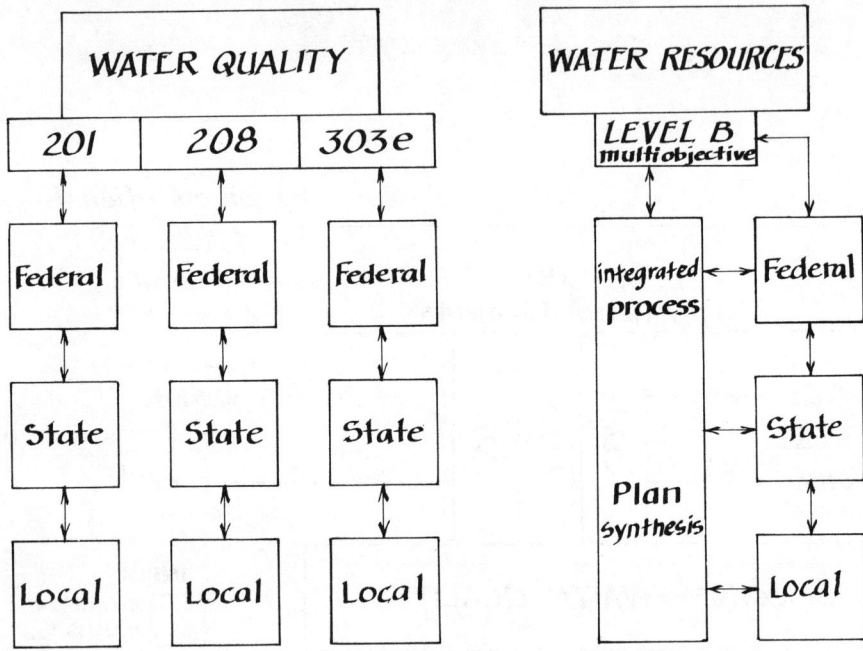

Figure 4. Potential integration opportunities of Level B.

6. The last element of this paper shows that BMPs are workable and desirable when placed in a context of multiobjective planning. Until a closer integration of water quality and water quantity is undertaken by the federal agencies, it is not certain that local responses can be made. If this is the case, then public investment will not be made in the national interest or even in a local interest. If this happens, certainly another issue of government credibility for solving problems will be raised.

REFERENCES

1. Agency of Environmental Conservation. "Vermont 208 Program," Montpelier, VT (1978).
2. U.S. Environmental Protection Agency. "Clean Water and the Land: Local Government's Role," Washington, DC (1977).
3. U.S. Environmental Protection Agency. "208 Plans and Water Quality/Quantity Conflicts," (1977).
4. "Integrating Water Quality and Water and Land Resources Planning," Cornell University, Ithaca, NY (1976).
5. Stern, C. "After Pollution, What?," Cornell University (1966).
6. Public Law 89-80. The Water Resources Planning Act, U.S. Government (1965).

7. U.S. Environmental Protection Agency. "Relationship of Level B Planning and Water Quality Management Planning," Washington, DC (1976).
8. U.S. Army Corps of Engineers. "Genesee River Basin Study," (1969).
9. New York State Department of Environmental Conservation. "Hudson River Basin Plan of Study" (1976).
10. Hudson Level B Study. Department of Environmental Conservation, Albany, NY unpublished data (1978).
11. "Lake Champlain Plan of Study," Burlington, VT (1977).
12. *Nonpoint Pollution from Forest Lands,* Lake Champlain Basin Study, Burlington, VT (1978).
13. Stone, B. "Modeling Vermont's Forest Resources in the Lake Champlain Basin," Agency of Environmental Conservation, Montpelier, VT (1978).
14. Soil Conservation Service. "LaPlatte Watershed Report," Burlington, VT (1978).
15. "The Economic Viability of Agricultural Land in the Shoreland Towns of Lake Champlain," Lake Champlain Basin Study, Burlington, VT (1978).

10

ENVIRONMENTAL IMPLICATIONS OF TRENDS IN AGRICULTURE AND SILVICULTURE

S. G. Unger
Development Planning and Research Associates, Inc.
Manhattan, Kansas

The environmental implications of a continued growth in the levels of output from both the agriculture and the silviculture sectors of the U.S. economy are dependent not simply upon growth, *per se*; rather they are also dependent upon those current and emerging trends which will characterize the agricultural and silvicultural production systems in the future. This paper summarizes those major trends *within* the respective sectors that will most importantly affect each sector's environmental implications in the long term.

GROWTH AND THE ENVIRONMENT

Generally accepted OBERS projections indicate that the future demand for both agricultural and forestry products will increase rapidly.* The moderate growth indices of Figure 1 indicate the outputs for 1985 and 2010 required to meet OBERS projected population, GNP, disposable income, trade, and other economic conditions. For example, by 2010, agricultural and silvicultural output levels are projected to increase from an index of 100 in 1972-74 to 171 and 192 for these sectors, respectively.

A widespread belief is that such growth will automatically result in environmental degradation. With static technology and unchanging management practices, this belief would surely be true. However, through the use of

---

*OBERS Projections for the agriculture and silviculture sectors are presented in the report "1972 OBERS Projections Regional Economic Activity in the U.S." (Supplement), U.S. Water Resources Council, Washington, D.C., May 1975.

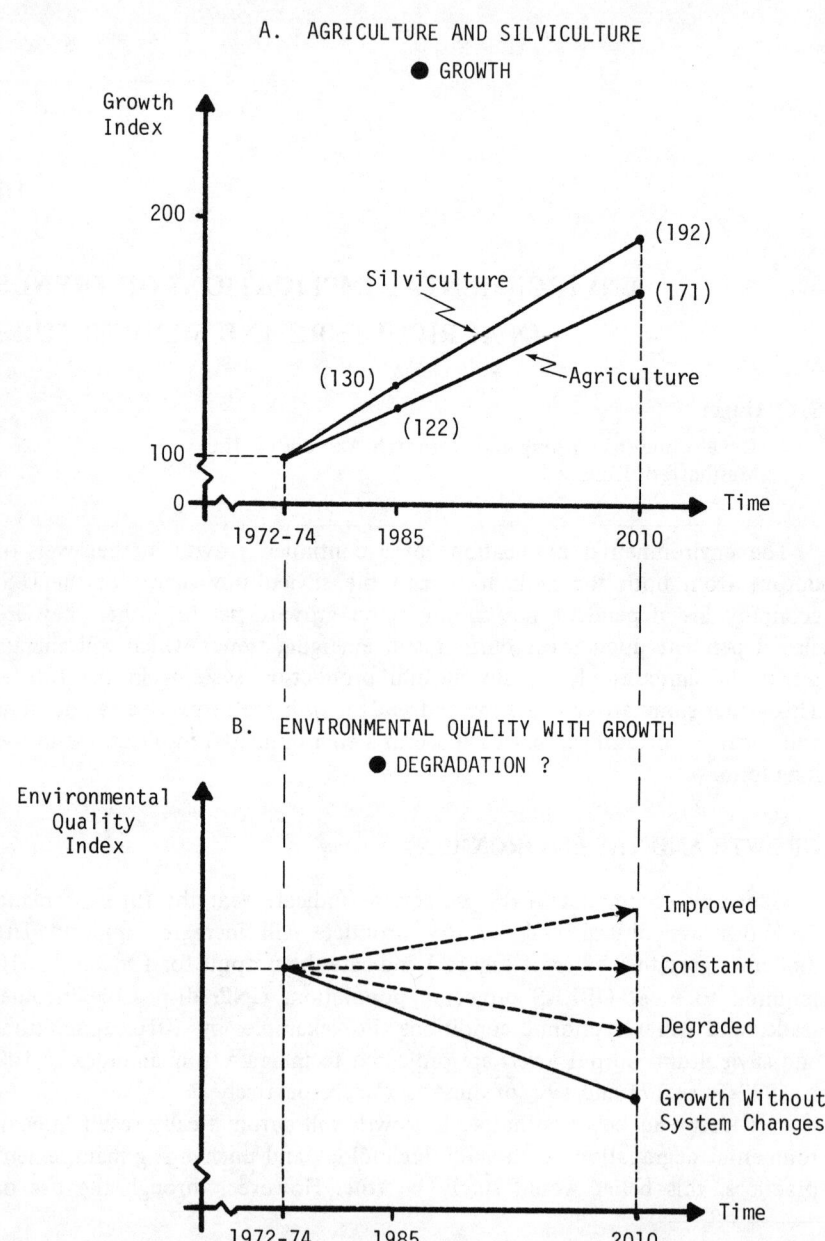

Figure 1. Projected growth indexes for agriculture and silviculture and their consequent potential environmental quality changes.

qualitatively improved inputs, advanced management practices, and appropriate residual treatments over time, environmental quality may well be maintained and even enhanced. This possibility is also depicted graphically in Figure 1 which shows that, because of the types of trend changes proposed, the future environment may not be degraded under growth conditions.

AGRICULTURE'S AND SILVICULTURE'S EXPECTATIONS

Under the research procedures from which this study stems, panels of national agricultural and silvicultural professionals participated in a workshop to assess, first in detail and then in the aggregate, the environmental effects of projected trends in agriculture and silviculture. Five subsector panels were involved:

Sector	Subsector
Agriculture	Nonirrigated Crop Production
	Irrigated Crop Production
	Feedlot Production
	Range and Pasture Management
Silviculture	Silviculture and Harvest Management

Each subsector panel had an original list of specific trends which they were allowed to modify. After completing their detailed assessments and their aggregate analyses, the fundamental conclusions of each sector's panels were that:

> *Agriculture* can have, with achievable developments, the capability to supply moderate growth output levels of food, feed and fiber products for the U.S. to 2010 while also maintaining or enhancing the environment affected by agricultural production.
>
> *Silviculture* can have, with achievable developments, the capability to supply moderate growth output levels of timber and pulp products for the U.S. to 2010 while also maintaining or enhancing the environment affected by silvicultural production.

To the layman, these general conclusions may appear overly optimistic. However, whether overly optimistic or not, the assessments and the rationale leading to them deserve a careful and thorough review. In the following sections of this paper, some of the major supportive assessment results are shown. Further details have been presented in the EPA report as cited.

TREND DEFINITIONS

Much of the agriculture and the silviculture panels' expressed optimism for meeting both growth demands and favorable environmental quality standards stems from their knowledge of the respective production systems and their beliefs that trends in "controllable variables" will be sufficient to meet both the growth and the environmental objectives. Oftentimes, those not directly involved with these sectors do not preceive that the production systems are themselves dynamic and capable of modification over time.

In short, both agricultural and silvicultural production relationships may be characterized by the general production function:

$$O(P,R) = F(I_1, I_2, \ldots, MP_1, MP_2, \ldots, RT_1, RT_2, \ldots | \text{EXOGENOUS FACTORS})$$

where
- O = Output
- P = Products
- R = Residuals
- I_i = Inputs, i=1,2,...
- MP_j = Management Practices, j=1,2,...
- RT_k = Residual Treatments, k=1,2,...

Of these variables, the inputs, management practices and residual treatments were regarded as controllable production factors that affect the sectors' product and/or residual outputs. Other exogenous factors beyond producer control were, for projection purposes, presumed to be held constant or to change in a specified manner.

This study focused in particular on the indicated controllable variables and their "expected trends" in extensiveness of use (E) in the short term (1985) and the long term (2010). Further, the intensiveness of environmental effects (I) for each trend-variable was assessed, and a multiplicative environmental effects rating (R) was defined as R = ExI. Using such a basis, the silvicultural and agricultural workshop panels determined the rankings of the major trends within each subsector and sector. [Note: The results presented here are the implications of long-term (2010) trends only.]

MAJOR SILVICULTURE TRENDS*

Within the silviculture sector, a total of 39 specific trends were assessed by the evaluation workshop panel. These specific component trends were subsequently grouped into ten trend clusters of related activities, as shown, in rank order, in Table I. Also shown are the panel's index ratings of the relative environmental importance of each of the trends. The top ranked trend has an index of 100 and all others range between 100 and 25.

As shown in Table I, two trend clusters (access to timber and site preparation practices) ranked 1 and 2 were given index ratings of 100. As such, both

---

*The trends and trend clusters described refer to the national assessment findings only. The silviculture workshop panel also assessed key regional differences (Western–mixed confier, Eastern–hardwood and softwood). Significantly different environmental implications exist among the regions assessed.

Table I. Summary of major environmentally related trends in the silviculture sector.

Trend (Practices)	Panel Ratings[a]	
	Rank	Index
Access to Timber	1	100
Site Preparation	2	100
Log Extraction	3	75
Utilization	4	50
Fire Control	5	50
Growth Enhancement	6	50
Stand Conversion	7	50
Pest Control	8	50
Cutting System	9	25
Stand Establishment	10	25

[a]Ratings established by a panel of silviculture professionals in an EPA-sponsored evaluation workshop. The rank indicates the trend cluster's rank order of environmental importance; the index is a subjective measure of each trend's relative importance compared to the top-ranked trend which has an index score of 100.

types of practices will have major environmental effects within the silviculture sector in the long term. Further, based on more detailed information, the net environmental implications of these trend clusters are expected to be beneficial relative to current practices (as are all the major trends presented herein). Access to timber practices refers more specifically to the construction and maintenance of both permanent and project road networks leading into the areas of harvestable timber resources. Site preparation practices refer to specific procedures and techniques to prepare harvested sites for reforestation. Such practices include mechanical preparation, burning prescription, chemical treatments, and soil moisture control.

The next three most important environmentally related trends in silviculture and harvest management are log extraction, utilization, and fire control practices. Improvements in extraction practices (rank 3) are expected to include better harvest-unit layout and the development of environmentally preferred logging equipment. Better utilization practices (rank 4) refer to the continuing improvements in the optimum utilization of the biomass, *e.g.*, minimum size and quality extensions, species-use enlargements, and the determination of optimal retention or removal levels of forest residues to enhance the environmental effects following harvest. Fire control practices (rank 5) refer in general to the need for improved fire prevention, detection and control practices. Perhaps unexpectedly, the need exists for a greater use of prescription burns as a fire control/prevention method.

Five additional trend clusters in the silviculture sector, ranked 6 to 10 in Table I involve other types of management practices which are expected to improve over time. In general, future growth enhancement practices (rank 6) will involve improved stocking rates, the elimination of undesirable species, prescription uses of fire for optimal growth, fertilization improvements and irrigation developments. Improved stand conversion practices (rank 7) will involve the elimination of undesirable or the placement of desired tree species. Pest control practices (rank 8) will include the selected use of chemical pesticides, increased uses of biological agents (*e.g.*, attractants, repellants, pheromones), and the use of mechanical methods to eliminate infested materials. Cutting system practices (rank 9) are to be site-specific designed and include such cutting systems as selection, group and clear-cutting. Finally, stand establishment practices (rank 10) are expected to include improvements in species selection, planting material developments, planting mechanization techniques, and stock/spacing controls.

The silviculture workshop panel evaluated not only the environmental implications of its major trends but also their aggregate environmental effects ratings from 1976 to 2010. First, the panel estimated that because of the improved efficiency of management practices, this sector's adverse environmental effects on the primary media (water, soil and air) would diminish during the 1976-2010 period. However, and secondly, the panel felt unable to judge the trends overall effects on certain ecological systems and their aesthetic factors (secondary and tertiary media). The workshop's judgments, hence, were made exclusive of these latter important considerations (important insofar as the secondary and tertiary effects of silviculture and harvest management are often the most visible and sensitive effects to be assessed and managed). Finally, the silviculture panel explicitly excluded Alaska's forest system in its assessments, a system whose forest inventory and growth capacity exceeds all that of the northern region of the coterminous U.S. The ultimate management of Alaska's timber resources, mostly federally owned, is still undecided.

MAJOR AGRICULTURE TRENDS

A total of approximately 200 specific trends were assessed within the four separate subsectors of agriculture—nonirrigated crops, irrigated crops, feedlots, and range and pasture management. The trends were subsequently grouped into 56 trend clusters of related activities, and each subsector panel was responsible for 12 to 16 major trends. For purposes here and as summarized in Table II, only the top five trend clusters from each subsector are discussed. As shown, both the within-subsector ranks and their relative

Table II. Summary of major environmentally related trends in the agriculture sector by subsector.

Subsector and Trend (P or I)[a]	Panel Ratings[b]	
	Rank	Index
Nonirrigated Crop Production		
Runoff Control (P)	1	100
Improved Seeds (I)	2	90
Conservation Tillage (P)	3	80
Integrated Pest Control (P)	4	70
New Pesticides (I)	5	65
Irrigated Crop Production		
Water Application (P)	1	100
Runoff Control (P)	2	80
Nutrient Application (P)	3	70
Integrated Control (P)	4	60
Soil-Plant Analysis (P)	5	40
Feedlot Production		
Size (I)	1	100
Waste Management Design (I)	2	80
Residual Disposal (P)	3	70
Odor Control (P)	4	60
Ration-Efficiency (P)	5	50
Range and Pasture Management		
Grazing Management (P)	1	100
Stocking Rates (I)	2	90
Renovation (P)	3	70
Nutrients/Pest Control (P)	4	60
Structural Developments (P)	5	60

[a] P = practices (primarily); I - inputs (primarily changes in quality).
[b] Ratings established by subsector panels of agriculture professionals in an EPA-sponsored evaluation workshop. The rank indicates the trend cluster's rank order of environmental importance; the index is a subjective measure of each trend's relative importance compared to the top-ranked trend which has an index score of 100.

index ratings (rank 1 = 100) are presented. These index ratings range from 100 to 40 across the four subsectors.

Each of these subsectors is itself a major component of the agriculture sector and each has its unique environmental concerns. This uniqueness is evident by the trend descriptions shown, and, at best, their environmental pollution concerns will involve various types of policies and programs to achieve the desired environmental management. Despite this uniqueness, the workshop panels assessed the general ranking of these major trends *across* the agriculture sector, *i.e.*, the top five trends from each of the four agriculture subsectors were ranked from 1 to 20. The results of these rankings are summarized as follows:

Subsector	Score*
Nonirrigated Crop Production	"21"
Irrigated Crop Production	"48"
Feedlot Production	"63"
Range and Pasture Management	"78"

*Sum of ranks for five subsector trends. Minimum possible score equals 15.

These computed scores reflect the agriculture subsector panels' composite views of the relative importance of environmental concerns within agriculture. Clearly, the crop production subsectors, especially the nonirrigated subsector, were regarded as having trends with relatively high environmental implications in the future. This was due both to the *extensiveness* of use and to the *intensiveness* of effects of the component activities comprising the trends. In general, crop production activities were seen as rather intense practices which occur on a regular, generally annual, basis. In contrast, range and pasture management, while widespread and considerable, is less intense and does not normally involve annual cultivation practices.

Nonirrigated Crop Production

As indicated in Table II, the runoff control practices (rank 1) trend group was rated highest in expected environmental importance in the future. Improvements in runoff control and their consequent erosion reduction are expected to result from increased terracing, grass waterway construction, contour farming, narrow-row planting and the optimizing of the timing of operations. Improved seeds and plants (rank 2) will also have major beneficial environmental implications due not only to higher yielding potentials but also to such factors as improved weather resistance, disease resistance, insect resistance, and salt tolerance.

The third-ranked trend, conservation tillage practices, includes various forms of reduced tillage that utilize crop residues to better control the environmental effects of crop production. Integrated pest control practices (rank 4) involve both improved scouting and the use of appropriate pest control techniques, whether chemical, biological or mechanical, to best control observed or predictable pest problems. Finally, the expected use of new pesticides (rank 5) anticipates the continued development of improved inputs such as systemics, microencapsulated pesticides, surfactants, biodegradable pesticides, and alternative formulations.

Irrigated Crop Production

While irrigated crop production requires management practices similar to those for the nonirrigated crops subsector, its primary distinguishing feature is the inclusion of water management practices. As indicated in Table II, improvements in water application practices (rank 1) are expected to have the greatest environmental implications in the future. Improvements are expected in irrigation scheduling, in the differentiation of timing and amounts of irrigation on alternate crops and soils, in the recycling and control of tailwater, and in the use of alternative application systems, *e.g.*, sprinklers. The next most important trend, runoff control (rank 2), is expected primarily to involve more extensive use of land grading. Other forms of runoff and erosion control such as contour farming, terraces, and grass waterways were also considered.

The nutrient application practices trend (rank 3) is important in irrigated crop production because of the generally intensive nutrient application rates with irrigation. Methods to improve the efficiency of plant uptake and to reduce nutrient losses are important. In particular, multiple applications, improved nutrient placement, and irrigation applications are practices that are expected to increase productivity and enhance environmental effects. Integrated controls (rank 4), primarily pest controls but also integrated nutrient applications, will involve the integration of biological, chemical and mechanical methods to achieve desired controls while minimizing their environmental effects. The use of soil-plant analysis (rank 5) is expected to be an increasingly important practice as a means to balance the soil supplies of inputs (nutrients, water) with the plant needs throughout the plant growth stages. A continual monitoring system is needed, and trend improvements in this analysis procedure are expected.

Feedlot Production

This subsector of agriculture is not directly a major land-using subsector. In fact, the feedlots themselves (beef, dairy, swine, sheep and poultry lots) are classified as point sources of pollution rather than nonpoint sources. Indirectly, however, the solid waste disposal requirements cause this subsector to have major nonpoint source environmental implications.

As is also indicated in Table II, feedlot size (rank 1) was regarded by the agricultural feedlot panel as the most important controllable trend variable in feedlot environmental management. Increased feedlot sizes are anticipated in the future, but these size changes will be accompanied by improved waste management designs (rank 2) and improved residual (solid and liquid) disposal practices (rank 3). These three top-ranked trends are interrelated for purposes of effective environmental management, and the feedlot panel foresees improved environmental effects via integration of these input and management practice trends.

Odor control practices (rank 4) and ration-efficiency practices (rank 5) complete the top five major trends in feedlot production. Odor pollution is seen as a major concern in humid and semiarid regions, but steady improvements in odor control are expected and will include improved waste management facility designs. The ration-efficiency practices trend is also expected to generally enhance waste management control—increased feeding efficiencies will effectively lower the wastes generated per unit of production. A negative effect of ration-composition changes may also occur: if higher roughage levels are used, a longer feeding cycle will be required and greater waste loads will consequently result.

Overall, the feedlot subsector expects to improve its aggregate environmental effects. Not directly assessed, however, was the problem of disposing of solid wastes on cropland or other land sites. This land disposal option is generally acceptable, and the panel believed that feedlot locations and sizes could be and would be designed with a predetermined waste disposal plan to account for this environmental concern.

Range and Pasture Management

The range and pasture management subsector accounts for the largest proportion of total land use in the U.S., *i.e.*, about 40% or 571 million acres (nonfederal). Environmentally this subsector has the advantage that ranges (and pastures to a lesser degree) are not intensely cultivated. However, where moisture, soil quality and favorable topography are inadequate, this is also a limitation. In such areas range renovations and forage improvements are presently neither practical nor environmentally feasible.

Ranges and pastures are primarily used directly for livestock (and other animals) production. The two top-ranked trends—grazing management practices (rank 1) and stocking rate inputs (rank 2)—most directly affect the use of range and pasture resources, and these two trends are often integrated. The grazing practices trend includes both continuous and specialized (with rotations) systems and, more recently, the use of complementary forage seedings to extend or improve grazing potentials. Stocking rates can be controlled, and improved management will involve determining the proper numbers of animals for the grazing system used and the vegetative conditions which may vary annually and seasonally.

The range and pasture renovation practices trend (rank 3) will increase in use to improve the establishment of grass cover and thus to better stabilize the soil. These practices will include mechanical and chemical methods and prescribed burning. Improved nutrient and pest control practices (rank 4) will increase the quality, quantity and productivity of ranges and pastures and will beneficially affect the environment, especially over time. The structural developments trend (rank 5) was seen as an important means of improving ranges and pastures. Such developments are expected to include structures such as ponds, catchment basins, drainage structures, and even fencing to better manage range and pasture use. These practices will aid in grazing management and will improve forage quality.

A variety of additional favorable trends in input use and management practices was foreseen by the range and pasture management workshop panel. As is true within other subsectors, many potential developments which have favorable environmental implications are possible and generally expected. However, much more development is needed to assure that these expected trends do materialize.

PRINCIPAL ENVIRONMENTAL CONCERNS

Although both the silviculture and agriculture sector panels were generally optimistic about the overall environmental implications of their major trends in the long term, a wide range of environmental concerns still exists. Primarily, a large gap potentially exists between what *can* be done and what *will* be done regarding environmentally preferred input developments, management practices, and residual treatments. Four of the principal environmental concerns expressed by the workshop panels are summarized next.

Concern 1. Government Coordination

Foremost, confusion exists currently about the respective roles of federal, state and local governments in environmental quality management. Basically, state and local governments may adopt legislation or policies of their choice so long as they meet minimum federal requirements. However, such still creates problems at the local implementation level because (1) the minimum federal standards are yet often unknown (*e.g.,* Section 208, PL 92-500), and (2) state or local governments may adopt even more stringent controls and, consequently, limit the adoption of even known federal guidelines.

A related and probably even more complex coordination problem is that involving multimedia interaction. As is more frequently being realized and discussed, the environmental problems and concerns of one primary medium (water, air, soil) cannot be resolved independently of the others. Furthermore, secondary media effects such as the impacts of pollution on ecosystems (aquatic life, terrestrial life and, ultimately, human health) may be a resultant complication of pollution control measures, *e.g.,* the dispersement of municipal sludges on agricultural lands may result in heavy metal concentration that will affect terrestrial and aquatic life. Even further, certain tertiary environmental effects such as aesthetics and wildlife preservation goals need to be integrated and coordinated by governments. These latter concerns are particularly evident in forestry and range management areas where multiobjective production and environmental goals are regularly sought. The requisite coordination needs are so complex that improved guidelines are needed soon to foster improved management practices.

Concern 2. Technology Development

Much of the optimism for the potential improvements in the long-term environmental effects of agriculture and silviculture, particularly for the more intense agriculture production sector, stems from the expected development over time of improved technology. Examples of anticipated improved technology to become available by 2010 are (1) improved germ plasm (higher yielding, disease-resistant, weather-resistant, pest-resistant, salt-tolerant), (2) improved chemicals such as pesticides and nutrients (less persistent, less toxic, and improved nutrient availability), (3) breakthrough in biological nitrogen fixation sources, especially for nonlegumes, (4) improvements in the photosynthesis capability of plants, and (5) the designing of improved environmentally beneficial equipment. Many other examples of improved technology could also be cited, but the main point is that such improvements, while anticipated, will not necessarily be forthcoming. Thus, increased attention and support should be given to fostering such developments in order to enhance the environment through an improved quality of inputs.

More broadly, the agriculture and silviculture workshop professionals expressed their conviction that a renewed emphasis be given to the general support of basic science research. A fear existed among the professionals that too much reliance was being given to the development and implementation of past basic science discoveries without adequate support for further fundamental plant and animal science research. In essence, the feeling existed that the past "capital stock" of scientific knowledge is being expended without an adequate funding of basic science research to increase the supply of "capital stock" for future generations.

Concern 3. Land Use Policy

Within the time frame of this study, the present to 2010, both silviculture and agriculture perceive a threat to their production capacities because of inadequate land use policies. In general, both sectors rely fundamentally upon the availability of the land resource and its associated soil productivity. The 1975 land use estimates by the Soil Conservation Service, USDA, indicated that about 40% of the nonfederal U.S. land was in ranges and pastures,

28% in cropland, 26% in forests, and 6% in urban and other use categories. Despite the relatively low current percentage of land in urban and other uses, increasing concern exists that prime farmland and prime timber resources are steadily shifting away from "commercial" use in agriculture and silviculture. By 2010, a significant proportion of the current cropland base may be impacted by urban sprawl, transportation networks, federal installations or developments, and other uses (including private holdings for noncommercial purposes). Such uses, whether publically preferred or not, will indirectly increase the environmental implications of growth on the remaining land base, for its full utilization will require a greater intensification of production.

Besides farmland losses in the agriculture sector, the silviculture sector may also be affected. Increased private holdings and federal acquisition or control of forests may frequently, indeed almost inevitably, result in a loss or underutilization of timber resources. Hence, both agriculture and silviculture professionals are increasingly concerned about land use and associated land use policies which now inadequately preserve these sectors' capacities for future growth.

Furthermore, particular attention should be given soon to those land use decisions that are effectively irreversible. For example, urban developments on prime farmland are unlikely to revert to their original agricultural use for economic reasons even though the same land will have a much higher relative agricultural value in the future. Finally, land use policy should have a long-term assessment perspective. By 2010, agriculture and silviculture land needs will more closely approach (or exceed in some regions) the available supplies, but by then it may be too late to increase effectively the useable supply of the most arable lands.

Concern 4. Public Investments/Cost-Sharing

Agriculture and silviculture representatives generally anticipate that some environmental management problems will require more than an individual producer's actions. Such concerns as watershed or river basin problems that affect multiple producers' lands will require public commitments to and investments in actions protective of environmental quality. Additionally, the fact that an individual producer's improved management practices can have both private and public benefits may make it desirable to induce better management practices through cost-sharing methods reflective of such public *vs* private benefit distributions. Such implementation methods would again represent a form of public investment in environmental management. This form of public investment would also likely involve more thorough benefit/cost assessments to ascertain both the costs and benefits (and their distribution) of alternative management practice alternatives.

RESEARCH AND PUBLIC POLICY ISSUES

In conclusion, the above trends assessment summary and the principal concerns statement lead to further research needs and prospective policy issues which are expected to become more evident as environmental quality management in agriculture and silviculture progresses. The main factors affecting research and policy issues, as expressed by the agriculture and silviculture professionals in this study's workshop assessments, can be summarized as follows.

Regional Differences

Both the agriculture and silviculture sectors (and the agriculture subsectors) reflect important regional differences in their environmental concerns, differences that affect major pollutants, major media effects, and potential best management practices. Such differences require regionally applicable research programs and policies to achieve improved regional environmental management.

The U.S. comprises vast resources over large geographic areas. The associated climates, topographic features, and soil characteristics differ importantly and thereby affect the environmental results of agricultural and silvicultural production. The primary media effects of production differ relatively on a regional basis, and hence the associated research and policy options should be planned and implemented accordingly.

High vs Moderate Growth

The above trend assessment results were based upon moderate growth assumptions to 2010. One of the alternative growth scenarios proposed by the Economic Research Service, USDA, depicts a high growth scenario for agriculture positing a 15-20% increase in agriculture's output over the moderate growth case—primarily for food and feed grains exports. A further study of this high growth case for agriculture has shown that the types of trends in input use and in management practices would be much the same as under moderate growth, except that their environmental implications would be worse than those under moderate growth. For example, as more marginal farmland is utilized and input-use intensified on existing farmland, the environmental effects would be toward degradation within the same time peiod, *i.e.,* 1976-2010.

Interaction Effects

The alternative high growth scenario also has differential environmental implications by region because the projected high growth scenario requires relatively more food and feed grains, outputs which are concentrated in the Cornbelt and Great Plains states. Further, spillover production effects will occur, involving all regions to some degree, especially the southeastern region. Overall, the high growth scenario raises serious environmental quality management questions.

As suggested, the high growth scenario is largely dependent upon higher export demands. U.S. trade policies are involved which further relates to such other economic policies as balance of payments and energy. In brief, with the obvious circularity of these implications, our Nation's agricultural (and silvicultural) trade policies (exogenous factors) may have more significant environmental implications for the U.S. agriculture and silviculture sectors than the "controllable variables" within these sectors, *per se*.

Despite the growth scenario chosen, much must yet be done—in part through research and policy controls—to achieve the generally favorable potentials for environmental management in agriculture and silviculture. If high exports are desired and needed, then even greater attention and support will be required to maintain or improve the relatively favorable environmental effects of these sectors. Finally, the regional environmental implications differ, and their importance would be recognized by research programs and policy options regardless of the growth alternative projected.

ACKNOWLEDGMENTS

This paper is based on research for the U.S. Environmental Protection Agency by DPRA, Manhattan, Kansas 66502, and the Tuolumne Corporation, Corte Madera, California 94925. In particular, more detailed results are presented in the EPA report: "Environmental Implications of Trends in Agriculture and Silviculture, Volume I: Trend Identification and Evaluation," EPA-600/3-77-121, Athens, Georgia, October 1977.

11

BEST MANAGEMENT PRACTICES FOR FERTILIZER USE

W. C. White
    The Fertilizer Institute
    Washington, D.C.

H. Plate
    Agway Inc.
    Syracuse, New York

To manage is to direct or to control. In regard to best management practices (BMPs) for fertilizer use, the objective is to direct or control practices which maximize nutrient uptake by plants with the end result of increased crop yields, while simultaneously minimizing loss of nutrients to the environment. At best, the route to this objective, when viewing fertilizers in equipment hoppers or tanks aboveground, is circuitous and subject to natural and cultural practices.

Failure to fully achieve this objective of complete nutrient recovery results in a portion of the applied nutrients not being absorbed by plants and, hence, subject to loss to the environment. From the short-term view, nutrients absorbed by plants are deducted from the pool considered subject to loss to the environment, either to ground water, surface water, or to the atmosphere. The fact that nutrient recovery by the crop to which applied is never complete establishes the basis of concern about losses to the environment of plant nutrients.

Several fundamentals of plant growth deserve emphasis in reference to nutrient recovery from fertilizers. First, plant roots are inherently ill-fitted for high nutrient recovery. In the case of corn, for example, roots occupy only about 1% of the soil volume. Additionally, much of the root surface is inoperative for nutrient absorption. Secondly, nutrients are absorbed only when present in the soil solution bathing root surfaces. Concentrations of nutrients in soil solutions as shown by Barber[1] are on the order of magnitude of 0.05 ppm P, 4 ppm K, 30 ppm Ca and 25 ppm Mg. Replenishment rates of these levels in the soil solution are relatively slow, partly because of the short distance nutrients diffuse (about 1 cm for N, 0.02 cm for P and 0.2 cm for K).[2]

Compound these limitations with the point that the average distance between corn roots in the plow layer is 0.7 cm,[3] and one can begin to appreciate why 100% recovery of applied nutrients to a given crop is not "in the books." However, for purposes of maximizing economic returns on fertilizer investments, as well as environmental protection, increased nutrient recovery is an objective of agronomic management practices.

IMPLICATIONS OF FERTILE SOILS

Without reversing laws of nature, such as the law of gravity, fertile soils will have high nutrient levels. And with the latter case, water percolating through soils to tile drains or aquifers, or surface water running downhill will contain nutrients at one concentration or another. In either case, water is the transport agent, and "control points" for improving nutrient recovery by plants center largely on water movement through or across soils with its sediment load.

Annual loss in the U.S. from surface runoff is estimated to be about four billion tons of sediment,[4] with about three-quarters of this from agricultural lands. Of this total, three billion settle in reservoirs, rivers and lakes, and one billion goes to the oceans. Most of the nutrients lost in this runoff are associated with sediment particles, and representative losses for Maryland cited by Miller[5] are:

	Cropland (lb/ac/yr)	Forests (lb/ac/yr)
Inorganic N	4 - 8	0.4 - 4
Inorganic P	0.1 - 1.5	0.03-0.8

Recognizing that water will continue running downhill and typically carry 10 to 12 ton/ac of sediment annually, it is obvious that with fertile soils there will be some nutrient losses to the environment, even with the best management practices.

The job for BMPs is to minimize these losses. Nutrients absorbed by plants are not subject to environmental losses, at least for the time being. Leaks or losses to the environment will occur, however, as the nutrients continue step-by-step in their cycles. BMPs only attenuate the inexorable leaks to the environment.

FOUR BMP TECHNIQUES

Rate of Application

In 1977 U.S. consumption of fertilizers amounted to 10.6 million tons of N, 5.6 million tons of P_2O_5 and 5.8 million tons of K_2O. Per acre rates of application varied widely from crop to crop, but it is safe to assume that the "fit" between crop needs and applied rates could have been improved in many cases. At best, the process of determining accurate application rates is one of approximation. Absolutely perfect fits are hardly, if ever, achieved in view of unpredictable weather, soil variation, etc.

Management practices should minimize the error between rates applied, whether deficient or excessive, with the optimum for a given soil-crop combination. Many tools are used for this purpose, from accumulated experience to computerized soil analyses, or soil tests. There is no substitute for basic knowledge of soils in properly matching rates with needs, but there are supplemental aids that can cut the margin of error between optimum and actual use.

Soil testing is one of the BMPs to minimize the error between optimum and actual rates. Without going into a full explanation of this practice, it is worthwhile to examine the extent of use of this practice.

Records of USDA of soil samples tested since 1968 are shown in Table I.

Table I. Soil samples tested.

	Government (Million)	Commercial (Million)	Total
1968	1.3	2.2	3.5
1972	1.3	1.2	2.5
1974	1.2	0.8	2.0
1975	1.3	0.9	2.2
1976	1.7	1.4	3.1
1977	1.7	1.5	3.2

Several years ago, The Fertilizer Institute began focusing attention on ratios of acres per soil sample. This comparison is a better indicator of the extent of soil sampling than number of samples alone. Examples for 1975 and 1977 are given in Table II.

The ratios in Table II show intensive sampling in the urban northeast, probably distorted with nonfarm samples, and in several states with large acreages of specialty crops such as California and Florida. There are wide differences between the midwest states with states operating state labs generally having wider ratios than where commercial labs are the main source

Table II. Harvested crop acres per soil sample.

	1975	1977
Connecticut	5	6
California	41	33
Florida	26	20
Massachusetts	28	52
Illinois	110	68
Iowa	343	122
Kansas	852	519
Texas	630	586
U.S.	144	104

of testing. In the extensive grain areas of the plains states, ratios shoot up to 500 to 800 harvested acres per soil sample. Recalling general instructions to include no more than 10-15 ac per sample, and to sample once at least every three years, one gets a range of 30 to 45 ac annually per sample as a reference point for the ideal. Information above shows that in 1977 the national ratio of 104 was 2.3 to 3.5 times these "ideal" ratios (45 and 30 ac per sample, respectively). Thus, one could project, using these same multipliers, that the number of soil samples in the U.S. (assuming harvested acreage the same as in 1977) should be 7.4 to 11.2 million annually.

Obviously, as a BMP, there is much room for increased soil sampling as a means of fitting application rates as closely as possible to individual crop/soil needs. An associated diagnostic tool, or BMP, is plant tissue testing. According to USDA, the maximum number of plant analyses reported was 508,500 in 1973, falling to 277,100 in 1977. Combined, soil and plant testing provide the best available diagnostic means for assessing the fit between nutrient requirements of a crop and nutrient supplies.

Time of Application

Each plant has a unique nutrient absorption pattern, beginning with a few ions per day for a seedling to 3 to 5 lb/ac per day of nitrogen and potassium during peak growing periods. If nutrient supplies are deficient because of poor timing of application during these specific periods, yields are cut and nutrient losses to the environment increased. Thus, *timing* of application is extremely important to maximizing nutrient uptake by plants, and, conversely, minimizing losses to the environment.

Many items affect optimum timing of nutrient application—crop, soil, date of planting, climate, etc. This bundle of items has to be assembled field by field, and generalizations are dangerous.

One impressive example of how these factors vary, not only from farm to farm but field to field, is shown by work of Kamprath et al.[6] The experiment showed different patterns of nitrogen retained and lost by several soils when applied in December ahead of a corn crop. An example of differences reported for several soils is shown in Table III.

Table III. Nitrogen remaining at corn planting time from nitrogen applied[a] in December.

Depth (in.)	Soil		
	Goldsboro Nitrogen, lb/ac	Portsmouth Nitrogen, lb/ac	Georgeville Nitrogen, lb/ac
0-6	5	49	50
6-12	9	24	43
12-18	15	13	10
18-24	17	--	
24-30	11	--	
30-36	14	--	
36-42	13	--	
Total	84	86	103

[a] 150 lb/ac N as ammonium nitrate.

In Table III one has a wide range of soil textures, exchange capacity, clay content, horizon thickness, etc. The Goldsboro soil is most subject to leaching losses, as shown by the depth of nitrogen movement. In contrast, the Portsmouth and Georgeville soils are such that percolation is minimized. Incidentally, conditions for denitrification would be favored in the poorly drained Portsmouth soil located in the Tidewater region.

The North Carolina work illustrates the complications of determining optimum timing of fertilizer application. Soils differ widely, even with a given field. Management by the individual farmers is critical in matching nutrient applications to meet peak demands of individual crops unique to their growth cycles on each soil.

For phosphorus and potassium, elements much less mobile than nitrogen, timing of application is less critical. Retention of these by soil, and benefits of built-up fertility levels, present entirely different aspects to the issue of timing. Preplant applications, even preyear applications can be beneficial. One example of the latter is provided by results of a "Time Deposit Fertilization" program of Agway Inc.[7] It involves applying phosphorus and potassium at seeding time for a 3-yr stand of alfalfa. During a 3-yr test period, the program gave higher alfalfa yields on seven of eight farms. This program, however, was not as good as conventional annual application programs on sandy soils with low capacities for holding potassium, nor is this practice recommended for these soils.

Balanced against the agronomic aspects of application timing are the pressures of labor, equipment and fertilizer supply availability. With larger tonnage involved and stresses on delivery systems increasing, there has been a general trend toward more preplant fertilizer application. Evidence of this is provided by TVA data,[8] showing the following:

Percent of Fertilizers Applied in the Fall

	FY '55	'60	'65	'70	'73	'74	'75
Percent	27	28	29	31	29	37	39

Increasing the lapse of time between application and uptake in some cases increases the chance of loss to the environment, especially in the case of nitrogen. Some practices such as subsurface application, nitrification inhibitors, selection of nutrient source, etc., can reduce these losses, if not eliminate them in most cases.

Method of Application

Whether nutrients are applied to the soil surface, localized in the row, broadcast and mixed with the soil, or in irrigation systems can significantly affect nutrient recovery by the crop and, likewise, the extent of possible loss to the environment. The implications of minimum tillage, for example, impact methods of application. In the case of minimum tillage where nutrients are largely applied over the surface, there obviously will be an increased surface loss of such nutrients compared with conventional methods of application where the nutrients are mixed to the plow depth.

Placement, or method of application, for phosphorus is particularly important. Localized placement is favored on soils of high phosphate fixation capacity. Generally, methods of application for nitrogen and potassium have little, if any, significant effect on their uptake by plants, except in the case of special forms of nitrogen where improper method can result in volatilization losses.

With increased use of irrigation in the U.S., there also has been a marked increase in the application of nutrients in irrigation water. Again, differences in nutrient utilization can be observed, and in crop yields as well as possible losses to the environment. One example is shown by an irrigation experiment in Washington[9] where the application of nitrogen in sprinkler irrigation gave yields nearly twice those from the same quantity of nitrogen applied in furrow irrigation. The soil used was sandy, and apparently there was a major leaching loss with the furrow irrigation system, which required approximately four times as much water as the sprinkler system.

As with other aspects of nutrient use, optimum methods of application vary widely between locality, crops, etc. Combining agronomic information available for a given location can point to those methods of application that maximize nutrient recovery by plants and, hence, constitute BMPs from the environmental point of view.

Forms of Nutrients

Little difference is shown between forms of fertilizer phosphate and potash in regard to recovery by plants or loss to the environment. The focus on form of nutrients will be on nitrate *vs* ammonium and organic sources of nitrogen.

At the core of the question of nitrogen form is the degree of solubility and, hence, the extent to which nitrogen may be lost by percolation or by surface runoff. Additionally, the rate at which a given form of ammonium or organic nitrogen converts to nitrate enters the picture regarding susceptibility of the material to environmental loss. Nitrate is the mobile form and, with all other factors equal, there will be greater losses to the environment of nitrate forms than from others.

With the sharp increases in ammonium sources of nitrogen in U.S. agriculture, such as anhydrous ammonia, nitrogen solutions, urea and ammonium phosphates, there is a smaller percentage of nitrogen being applied today in the nitrate form. However, most ammonium forms convert to nitrate-nitrogen in a few weeks unless prevented by low temperature or by nitrification inhibitors.

Agronomic recommendations regarding delayed application of ammonium nitrogen until some soil temperatures go below 50°F. have been stressed for years. The point is to delay application of ammonium nitrogen until the nitrification rate, a function of temperature, becomes so low that there is little transformation of the ammonium nitrogen until warmer conditions—about planting time. Stipulations for fall application will vary from state to state depending on soil conditions and constitute BMPs at the local level.

In the past few years the use of bactericides, such as N-Serve, has increased sharply. The purpose for such a product, generally applied with anhydrous ammonia or nitrogen solutions is to inhibit the bacterial oxidation of ammonium to nitrate-nitrogen. Again, as long as nitrogen can be retained in the ammonium form, it is in most situations immobile and not subject to removal by percolation or surface runoff, except with associated soil particles. Such products only delay the eventual oxidation of ammonium nitrogen, but in the case of corn can delay the oxidation in some situations to the extent that leaching losses or denitrification losses are reduced, and yields increased.

SUMMARY

The few examples presented illustrate the wide range of factors to consider in controlling nutrient levels in soils so as to maximize plant recovery and to minimize loss to the environment. Even with the best available matrix mathematics, all individual factors cannot be quantitatively evaluated for every single eventual possible field condition. The judgment of the farmer must come into play in exercising BMPs for nutrient use in crop production.

At the outset, the fact has to be recognized that nature precludes eliminating all losses of applied nutrients to soils to the environment. Fertile soils will have relatively high nutrient concentrations and with these, in turn, there will be increased losses to the environment. The objective of agronomic management in this regard is to minimize these. However, we cannot realistically expect to have nutrient losses from fertile productive soils reduced to what many environmentalists consider background levels of natural conditions where nutrient levels are low. One has to recognize the positive aspects of high fertility in producing increased plant cover that, in turn, cuts erosion.

There are means, however, of reducing losses of nutrients from fertile, high-nutrient soils to the environment. Matching accurately as possible the rate of application of nutrients to a given crop is probably the first step. If the mark is missed with the rate of application, percent nutrient recovery by the crop will be decreased and, conversely, the risk for losses to the environment increased. There are diagnostic tools to help the farmer in his management practices for determining proper rates of application such as soil testing and plant analysis.

Probably next in the order of importance in ranking BMPs for nutrient management is that of time of application. Through timing, supply of nutrients can be fitted fairly closely to peak demands by the crop. Again, this is fully a responsibility of the farmer taking into account local soil, climatic and crop conditions.

We have also shown that even methods of application can affect nutrient recovery and loss. Increasing use of irrigation systems and minimum tillage will impact this aspect of nutrient management directly. Lastly, the form of nutrient applied in many cases is for selection by the farmer. There are instances where the form of nutrients, particularly for nitrogen, will affect susceptibility to loss and constitutes another BMP.

In final analysis, reducing adverse nutrient losses to the environment from fertile soils centers on the transport system—either water moving through a soil or water moving across a soil. Water is the transport agent for nutrients, either directly as in the form of a solvent or as a means for physically moving soil particles with adsorbed nutrients. The point to stress in nutrient conservation from the environmental point of view, in many cases, rests squarely with the dual objective of soil and water conservation.

REFERENCES

1. Barber, A. "Water—Essential to Nutrient Uptake," *Plant Food Review*, Vol. 10 (1964).
2. Barber, A. "Efficient Fertilizer Use," Agronomic Research for Food, ASA (1976).
3. Mengel, D. B., and S. A. Barber. "Rate of Nutrient Uptake Per Unit of Corn Root Under Field Conditions," *Agron. J.* 66:399-402 (1974).
4. Wadleigh, E. H. "Waste in Relation to Agriculture and Forestry," USDA Mis. Pub. No. 1065 (1968).
5. Miller, F. P. "Agriculture and Nonpoint Source Pollution: Nutrients," Univ. of Md., Agron. Dept. Memo.
6. Kamprath, E. J., et al. "Nitrogen Management, Plant Population and Row Width Studies with Corn," N.C. State Agr. Exp. Sta. Tech. Bul. 217 (1973).
7. *Agway Cooperator*, Vol. 14 (February 1978).
8. Fertilizer Summary Data, TVA (1976).
9. Middleton, J. E., et al. "Irrigation and Management for Efficient Crop Production on a Sandy Soil," Wash. State Univ., Agr. Res. Bul. 811 (1975).

SECTION III

NUTRIENT MANAGEMENT

12

ANIMAL MANURE MOVEMENT IN WINTER RUNOFF FOR DIFFERENT SURFACE CONDITIONS

D. B. Thompson
  Department of Agricultural Engineering
  University of Minnesota
  St. Paul, Minnesota

T. L. Loudon, J. B. Gerrish
  Department of Agricultural Engineering
  Michigan State University
  East Lansing, Michigan

INTRODUCTION

Land application appears to be the most economical means for utilization and handling of animal waste. The goals for land application of manure are to maximize the use of nutrients available and at the same time minimize any pollution potential. Any time animal waste has been applied to the soil surface without incorporation, there is potential for nutrients to be transported by snowmelt or rainfall to surface waters. By using proper management and conservation practices, the amount of runoff and erosion can be reduced, thus reducing nutrient loss.

Protecting surface water quality and preventing the loss of valuable plant nutrients are two very practical reasons for developing management practices to minimize the nutrient content of surface runoff. Soil and water conservation practices have been developed to minimize soil erosion. Similarly, practices to control nutrient loss are being studied and tested. The results of one such study are presented in this report.

Animal manure is spread on frozen and snow-covered fields during the winter in the northern states when manure storage is not available or is insufficient to store the entire season's manure production. Winter spreading does not allow the opportunity for incorporation of the manure with surface soil. Therefore, when snowmelt or rainfall occurs, there is a greater potential for nutrient loss than under normal spreading conditions.

BUFFER ZONES

The purpose of this study is to examine the quality of winter runoff under various surface conditions and determine whether a hazard exists due to winter spreading of animal manure. Runoff from manured areas was examined to determine if the nutrient load was significantly different from that of control areas. The influence that a buffer zone has on runoff quality downslope from a winter spreading area was observed for three common agricultural field conditions.

A buffer zone utilized as a management practice may be defined as an area situated between two areas which are in possible conflict. The objective of the buffer zone is to lessen the possibility of adverse impact from land application areas on surface waters. Runoff coming from a field spread with animal manure not incorporated with the soil may contain a high concentration of soluble nutrients, soil and organic particulates. Soluble nitrogen and phosphorus may be leached from the manure and held in solution while organic matter and soil particles are carried in suspension. The function of the buffer zone is to provide an area where nutrients and particulates can be removed from runoff prior to entering the surface water system.

The buffer area may have a permanent vegetative cover or a cultivated surface. The objective of the buffer zone is achieved through adsorption of nutrients, decreasing surface runoff velocity and volume, and increasing surface detention capacity. Other mechanisms such as infiltration, dilution and filtration of particles are taking place in varying amounts on different surface conditions. Soluble nutrients are adsorbed on the surface of soil particles (mostly silt and clay) and thereby removed from the runoff solution. The velocity of surface runoff can be decreased by various types of surface vegetation and crop residue. Increasing surface roughness with fall plowing, or discing, will decrease runoff velocity and increase surface detention time by creating numerous pockets and depressions where runoff can be detained. Large numbers of macrosized surface depressions are created by the tillage tool and these surface configurations are especially important to water management.[1] Values for random roughness created by different tillage practices are given by Burwell et al.[2]

The quantity of nutrients carried by surface runoff is dependent in part upon the transport capacity of the runoff. Transport capacity is an expression of the energy associated with moving water in runoff. As the transport capacity increases, the amount of sediment and particulate matter which can be carried increases. A reduction in the transport capacity of the runoff causes sediment to drop out of suspension and be deposited. By reducing the runoff velocity, the transport capacity is reduced and the amount of material carried in suspension in the runoff is reduced.

LITERATURE REVIEW

Doyle et al.[3] found that 4-m buffer strips were effective in reducing levels of nitrogen, phosphorus, potassium and fecal bacteria in surface runoff from manure-treated plots. Their data indicated that the greatest reduction in nutrient levels took place rapidly over a relatively short distance of buffer area. Phosphorus loading rates for manured plots remained higher than control plots after 4 m of buffer but showed a 62% reduction in concentration from runoff collected at the edge of the manured area. Concentrations of indicator bacteria (fecal coliforms, fecal streptococci) were significantly reduced after 4 m of forest and grass buffer strips.

The number of studies evaluating the effectiveness of buffer zones downslope from manure application areas is limited. Results of similar research on overland flow for treatment of animal waste and feedlot runoff can be used for comparison. In an overland flow system there are frequent manure loadings, compared with a single application (per season) in manure applications; however, the principles of nutrient removal as feedlot runoff or diluted manure flows over the soil surface are the same. Several examples of nutrient reduction during overland flow are presented to illustrate the capacity of the soil to remove nutrients from runoff.

Overcash et al.[4] reported an overland flow pretreatment of poultry waste. The waste which flowed over a 2.4-ha terrace system carried approximately 20 kg N/day or about 3000 kg N/ha/yr. With a 15-m flow distance, Overcash reported a 60-70% mass reduction in nitrogen. Eighty to ninety percent removal of nitrogen was obtained by increasing the flow distance to 30 m (4.9 ha) while maintaining a constant waste load.

Willrich and Boda[5] reported that COD, PO_4 and inorganic-N showed mass reductions of 67, 62 and 62%, respectively, with an average application rate of 0.48 kg COD, 0.07 kg PO_4, and 0.24 kg NH_3 per 5-hr period on 30.5-m overland flow plots.

The quality and quantity of runoff is somewhat dependent upon the season of application and subsequent weather conditions. The physical condition of the soil (frost content, soil texture and structure) will influence the amount of infiltration and runoff.[6] Two terms are used to describe the structure of frozen soil for the north central states region. Concrete freezing is observed most frequently in cultivated fields or areas with sparse vegetative cover. Honeycomb freezing is characterized by a loose, porous structure easily broken into pieces. It is found most frequently in grassland, meadows and pastures. As little as 1 in. of concrete frost prevents infiltration of rain or snowmelt while infiltration may be good in the case of honeycomb freezing.

Soil texture is determined by the proportion of sand, silt and clay particles. Soil texture influences infiltraton, and the number of adsorbtion sites, and consequently the quantity of nutrients which can be adsorbed from the runoff. Soil organic matter and manure aid in the formation of soil aggregates. Poor soil structure (limited aggregation) will decrease the amount of infiltration and increase runoff. Zwerman et al.[7] found that a single application of 13.5 ton/ha of solid dairy manure increased soil infiltration by 27% in a continuous corn culture.

The season and method of manure application have been shown to have a large-scale effect on amount of manure remaining on the soil surface. Midgley and Dunklee[8] found that the amount of nitrogen lost in runoff from surface-applied manure during the winter was inversely related to the amount previously lost to the air through volatilization. Immediate incorporation with the soil has been shown to be the most effective means of reducing nutrient loss through volatilization and runoff. Hensler et al.[9] investigated the influence of the season of application on the nutrient loss from dairy manure. Winter application on frozen, snow-covered ground resulted in a three-fold increase in the annual average nitrogen and phosphorus losses as compared with control areas. Much of this loss resulted from one storm event which occurred only a few hours after the manure was applied to frozen soil.

Manure application on melting snow or just prior to a rainfall event represents the worst possible case for nutrient loss. Klausner et al.[10] investigated surface runoff losses of inorganic nitrogen and total soluble phosphorus from field-spread dairy manure and found losses were increased when manure was spread during active thaw periods. Losses were minimized when manure was applied and then covered with snow, which melted at a later date. Klausner found that with a snow-covered, 35-ton/ha application rate, nutrient losses differed little from control areas. Zwerman et al.[11] reported a similar conclusion. Nutrient loss from control plots or watersheds originate primarily from natural soil fertility, leaching of organic material on the surface, and precipitation.[10,12]

Witzel et al.[13] found that nutrient losses from winter and spring runoff from four small watersheds were the same even though some of the watersheds had winter-spread manure while others did not. On one watershed where fertilizer applications on a per-acre basis were double that of the others, the loss of N and K was lower. Thus, it is evident that runoff characteristics may vary from one location to another independent of manure or fertilizer application and may more often be due to variation in the topography and physical soil properties.

Zwerman et al.[11] presented data indicating that nitrogen and phosphorus in runoff was reduced by leaving plant residue on the soil surface. Young and Mutchler[14] indicated that there was only a slight increase in nutrient loss from manured over unmanured corn plots. Higher nutrient losses occurred on manured alfalfa plots than unmanured alfalfa. The characteristics of certain surface conditions make them more likely to retain nutrients from manure applications than others. Upon investigation, Young and Mutchler found very little difference in thawing rates of manured and unmanured alfalfa plots. Data from this experiment and others indicate that there are variations in runoff quality and/or quantity from different surface conditions with an equal amount of animal manure applied.

Manure and plant residue have both been indicated as having effects on runoff. Converse et al.[15] observed the average runoff from plots for three years and reported that runoff from the check plots was significantly greater than from manured plots. Similar observations have been made by other investigators.[13,14] Doyle[16] concluded that the concentrations of nutrients (NPK and Na) were dependent on the number of rains previously leaching the manure but was independent of total rainfall and the amount of runoff collected.

EXPERIMENTAL PROCEDURES

Twelve plots (3 x 60 m) were established on a moderately well-drained sandy loam (Hillsdale Sandy Loam) with a slope of 4% (Figure 1). Three sample sites were located on each plot. The first site was located in the manured area 12 m from the upper plot end. The second sample site was located 36 m downslope from the upper end or 24 m downslope from the first sample site. The third sample site was located at the lower end of the plot at 60 m.

Fresh dairy manure was applied to the upper 24 m of plot surface. Two lengths of overlapping buffer zone were tested downslope from the manured area. The second sampling location was located 12 m downslope from the manured area, and the third was 33 m downslope. Runoff collected at the second location had flowed over 12 m of buffer zone and was compared to the quality of runoff which flowed over 36 m of buffer area.

Fresh stanchion barn dairy manure (80% moisture), with a moderate amount of wood shavings and straw bedding, was applied at the rate of 63 ton/ha. The manure was applied uniformly with a pitchfork on a 10-cm snow cover in early January while temperatures were below freezing. During the two years of the study, the manure application was covered by snow which fell within two days after spreading. The first significant amount of snowmelt and runoff did not occur for approximately 30 days after application.

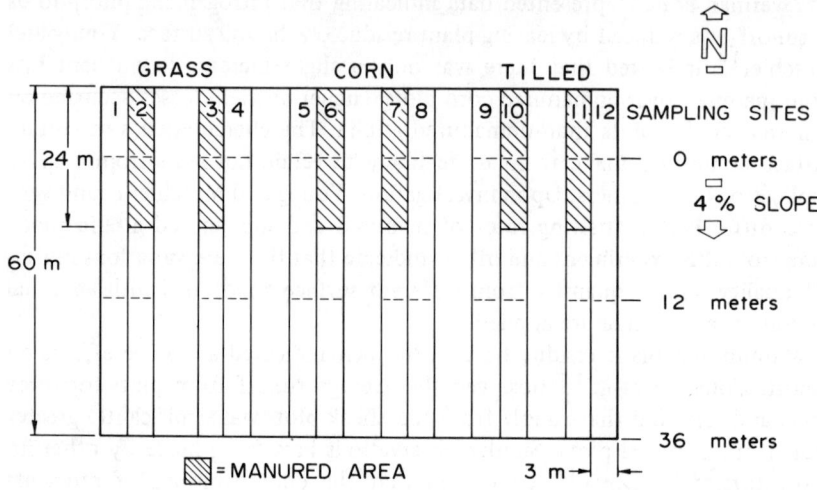

Figure 1. Plot layout.

A proportionate amount of runoff was collected from the plot using troughs across the plots the first year and dust pan collectors the second year and diverted to 200-liter reservoirs which were buried in the ground outside the plot at each sample location. The contents of the reservoir were mixed for one minute with a centrifugal pump before a sample was taken. All samples were refrigerated at 4°C until laboratory analysis was completed.

DESCRIPTION OF SURFACE CONDITIONS

Three surface conditions were studied. Each surface condition with manure-treated and control plots had two replicates, totaling four plots for each surface cover. Grass surface (orchard grass) was selected to simulate effects of winter application on a continuous vegetative cover. Field corn was planted across the slope in 93-cm rows (47,000 plants/ha) in preparation for the other two surface covers. Manure was applied in the spring prior to tillage and planting operations at 34 ton/ha. The corn plot area was sprayed with herbicide prior to corn emergence for weed control. In the fall of the year, the corn was harvested for silage leaving approximately 20 cm of stubble. Four plots were left in this condition and hereafter will be referred to as corn stubble. The third surface condition was created by discing the remaining corn stubble. The area was worked over twice, first parallel to the slope and the final time perpendicular to the slope. Each plot was bordered by galvanized sheet metal. The border material was 20 cm wide and placed in the soil vertically 10 cm deep in the soil with 10 cm above the soil surface.

RESULTS AND DISCUSSION

Winter runoff was monitored for changes in water quality which result from application of animal manure on different surface conditions and the reduction of nutrients in surface runoff as it moves through the buffer zones downslope from the manured area. Data were collected during both snowmelt and rainfall events in the winter and early spring of 1976 and 1977. The results presented here are based on 112 and 95 surface runoff samples taken the first and second year, respectively. Samples were taken at three locations on the plots described earlier. The average values given in Table I for control plots are the seasonal averages of six sample locations from the two replicates of unmanured plots for each surface condition. Values for manure-treated areas are the average of all samples from two replicates of each respective buffer zone. Concentrations in runoff for all the parameters tested were higher in magnitude during the second year of the study. The differences between years can be attributed to variation in meteorological factors and timing of the runoff events. Twelve runoff events were sampled in 1976 and four events were sampled in the 1977 season. More of the 1976 samples were taken later in the season after some initial leaching and decomposition of the manure had occurred. As a result average concentrations for 1976 are biased downward. The increase in concentrations the second year is not likely a result of manure nutrient accumulation in the test plot soil since runoff concentrations from control plots show a similar, though not proportional, increase. The first year of the study the grass plots consistently show a lower concentration of nutrients than the other surface conditions, with the exception of phosphorus which was similar for all. The sampling technique was changed in 1977 to avoid snow and ice accumulation near the sample point which caused some dilution of initial samples. Lower runoff volumes were noted on grass plots, less than 30% of other surface conditions, due to greater infiltration. The second-year results show concentrations in runoff are similar at respective sample sites for all three surface conditions. For each year, concentrations of nutrients in control-plot runoff are similar for all the surface conditions, suggesting little difference in background levels.

The results presented in Table I indicate that average nutrient concentrations decrease as runoff moves downslope from the manured area. The sample site designated as "0 meters" was located within the manured area; concentrations observed at this location represent runoff at the downslope edge of a manured area. Nutrient levels in runoff are highest at this point and represent potential pollution if allowed direct access to surface waters without management. Concentrations at the second sampling location are reduced considerably after flowing over only 12 m of buffer zone. Average nutrient concentrations in runoff which passed over 36 m of buffer zone show that nearly all of the

Table I. Average nutrient concentrations in surface runoff[a] (mg/l).

		COD		NH_3		NO_3		Phosphorus		TKN	
		1976	1977	1976	1977	1976	1977	1976	1977	1976	1977
Grass											
Treated Buffer	0 m	155	911	5.0	50.2	—	6.8	1.7	10.7	23.5	115.0
	12 m	81	633	1.9	28.8	—	3.7	0.6	6.0	14.5	62.9
	36 m	65	333	0.7	17.2	—	2.6	0.7	3.2	12.8	35.5
Control		56	139	1.2	3.3	—	1.7	.84	2.6	10.2	14.3
Corn Stubble											
Treated Buffer	0 m	500	1103	18.0	44.6	—	6.1	1.6	8.6	41.9	97.7
	12 m	182	350	4.7	12.7	—	2.0	1.2	5.3	24.6	35.5
	36 m	71	193	0.4	6.6	—	2.1	0.5	2.8	10.2	31.8
Control		63	179	1.8	8.4	—	3.2	.69	2.6	16.2	25.8
Tilled Surface											
Treated Buffer	0 m	586	1006	17.2	60.8	—	6.4	1.9	15.4	42.3	130.0
	12 m	186	596	3.5	13.0	—	2.5	0.8	4.1	18.4	45.0
	36 m	67	171	0.4	5.4	—	1.5	0.9	3.1	11.0	14.7
Control		71	165	.76	3.6	—	2.4	0.5	2.8	15.7	14.0

[a]From data collected in January, February and March, 1976 and 1977.

manure-contributed nutrients present in the runoff at the 0-m location have been removed.

The degree of variability between runoff events causes large standard deviations to result when a large number of samples from different types of runoff events are analyzed. The variations in concentration are so large that it becomes difficult to draw conclusions without more sophisticated statistical analysis. The variation which tends to mask the actual nutrient removal can be avoided by tabulating data for each runoff event separately. Differences caused by climatic factors and variable plot conditions can be minimized by calculating the percentage reduction of each nutrient which takes place as the runoff moves downslope. The nutrient reduction can be most accurately calculated when the numbers utilized are from the same plot and runoff event. These individual percentage reductions were averaged to arrive at the average percent reduction shown in Table II.

Table II. Percent reduction of nutrients in winter runoff[a].

	Length of Buffer Zone	
	12 m	36 m
Overall Average Reduction (three surface conditions, 2-yr average)	62% (56, 68)	73% (69, 77)
Average for Three Surfaces—1975-76	68% (60, 76)	65% (56, 74)
Average for Three Surfaces—1976-77	60% (53, 67)	77% (72, 82)
Two-year Average by Surface Condition		
Grass Cover	63% (54, 72)	72% (67, 77)
Corn Stubble	55% (39, 69)	68% (59, 77)
Tilled Surface	66% (58, 74)	78% (68, 88)

[a] Average reduction for ammonia, total Kjeldahl nitrogen, total phosphorus and chemical oxygen demand; numbers in brackets give the 95% confidence interval.

The effectiveness of the buffer zone can be judged by its ability to reduce the nutrient concentration in surface runoff. Nutrient concentrations are greatly reduced after runoff has flowed through a 12-m buffer area. The overall average of total nutrient reduction for the three surface conditions, two-year average, was 62% for the 12-m buffer zone and a 73% nutrient reduction for the 36-m buffer zone. The percent reductions for each of the surface conditions are given in Table II. The tilled surface had the highest average reduction with a 66% reduction at 12 m and 78% after the 36-m buffer zone. The grass surface had the next most efficient buffer zone with 63 and 72% reduction on the 12- and 36-m buffers, respectively. The corn stubble plots removed 55% of the nutrient load in the 12-m buffer zone and 68% in the 36-m buffer zone. The confidence intervals for each group of numbers are given to illustrate the amount variation.

All of the surface conditions compared in this study did an equally satisfactory job of reducing nutrient concentrations in runoff. The numbers used in the calculations to this point included background levels of nutrients. If background concentrations are subtracted, the percentage reductions appear much higher, sometimes going over 100%, suggesting that runoff from manured plots may frequently have lower nutrient concentrations than unmanured-plot runoff.

The average concentrations for control plots (*i.e.*, background concentrations) of each surface condition were subtracted from the concentration of each sample at the corresponding sampling location; these numbers were subsequently averaged to arrive at the values given in Tables III and IV.

Background concentrations are those nutrients contributed to runoff from the same surfaces but without the addition of manure. Background nutrients may possibly be reduced through soil and water conservation practices which were not a part of this study. The objective of the buffer zones was to reduce nutrient concentrations in runoff as a result of surface-applied manure.

Table III. Average percent reduction of nutrients with background concentrations subtracted (2-yr average).

Buffer Zone Length	COD	NH_3	NO_3	P	TKN	Average
12 m	78%	84%	92%	68%	88%	= 82%
36 m	96%	109%	106%	83%	93%	= 97%

Table IV. Average nutrient reduction by winter runoff year (percent)[a].

January February March	COD		NH_3		NO_3		TKN		P	
	12 m	36 m	12 m	36 m	12 m	36 m	12 m	36 m	12 m	36 m
1976	87	94	93	125	–	–	106	104	55	69
1977	69	96	75	93	92	106	69	91	80	97

[a] Background concentrations subtracted for calculations.

Nitrate levels in winter runoff were measured only in the second year of this study. Nitrate concentrations averaged less than 7 mg/l and were reduced to background levels rapidly by an unexplained mechanism. Nitrate levels were so low that they were often lower than background levels at the 0-m sample location. Total phosphorus concentrations occasionally increased as runoff moved downslope as a result of increased soil erosion during rainfall events which carried sediment and attached phosphorus. Total phosphorus was generally reduced to background levels within 36 m. Removal of nitrogen forms was generally higher than phosphorus removal (Table III). From Table I it can be seen that NH_3 and TKN are more readily reduced on tilled and corn-stubble surfaces than on the grass surface condition. COD values were generally the same for all surface conditions except for less COD reduction on the grass surface the second year.

The overall nutrient concentrations are reduced by approximately 60% within the first 12 m of buffer zone with background concentrations not subtracted. With the background subtracted, the percentage reduction should represent the actual reduction of manure-contributed nutrients in the runoff and increase to near 80%. The data collected under the conditions of this study indicate that for any of the three surfaces studied, a 36-m buffer zone will remove between 80 and 100% of the nutrients added by the winter application of animal manure.

CONCLUSIONS

Based upon the conditions of this study our data would lead us to conclude the following:

1. Nutrient concentrations decrease as runoff water moves downslope from a manured area.

2. Nutrient concentrations contained in runoff leaving a manured area are greatly reduced as the water moves across a 12-m buffer strip. In our study, this was equivalent to a buffer strip equal to one-half of the length of the manured area. The extent to which the buffer area would have to be lengthened for longer manured areas is uncertain.

3. On a sandy loam soil with a 4% slope, a buffer zone 36 m long reduced nutrient concentrations in runoff from manured plots to levels equal to unmanured plot runoff.

4. The overall average nutrient concentration reduction for a buffer zone 12 m long was 62% compared with a 73% overall reduction after a 36-m buffer zone if background concentrations are not subtracted.

5. With background concentrations subtracted, buffer zones removed an average of 82 and 97% of the manure-contributed nutrients from winter runoff with a 12- and 36-m buffer zone.

6. Runoff volumes from manured areas have been shown to be less than from unmanured areas by past researchers. Based on the concentrations reported in this study and runoff volumes from manured areas less than, or equal to, those from nonmanured areas, nutrient loading rates in winter runoff from manured areas followed by adequate buffer zones should not exceed those from unmanured areas.

7. The quality of winter runoff from unmanured areas was essentially the same for the three surface conditions observed in this study.

REFERENCES

1. Burwell, R. E., R. R. Allmaras and M. Amemiya. "A Field Measurement of Total Porosity and Surface Microrelief of Soils," *Soil Sci. Soc. Am. Pro.* 27(6):697-700 (1963).
2. Burwell, R. E., R. R. Allmaras and L. L. Sloneker. "Structural Alteration of Soil Surfaces by Tillage and Rainfall," *J. Soil Water Conser.* 21(2):61-63 (1966).
3. Doyle, R. C., G. C. Stanton and D. C. Wolf. "Effectiveness of Forest and Grass Buffer Strips in Improving the Water Quality of Manure Polluted Runoff. ASAE Paper No. 77-2501. St. Joseph, MI 49085.
4. Overcash, M. R., F. J. Humenik, P. W. Wesserman and J. W. Gilliam. "Overland Flow Pretreatment of Poultry Manure," ASAE Paper No. 76-4517, St. Joseph, MI 49085 (1976).
5. Willrich, T. L., and J. O. Boda. "Overland Flow Treatment of Swine Lagoon Effluent," ASAE Paper No. 76-4515, St. Joseph, MI 49085 (1976).
6. Storey, H. C. "Frozen Soil and Spring and Winter Floods," *The Yearbook of Agriculture.* (Washington, DC: U.S. Government Printing Office, 1955).
7. Zwerman, P. J., A. B. Drielsma, D. J. Jones, S. D. Klausner and D. Ellis. "Rates of Water Infiltration Resulting from Applications of Dairy Manure," Proceedings of 1970 Cornell University Conference on Agricultural Waste Magement (1970), pp. 263-270.
8. Midgley, A. R. and D. E. Dunklee. "Fertility Runoff Losses from Manure Spread During the Winter," Agricultural Experiment Station Bulletin No. 523, University of Vermont (1945).
9. Hensler, R. F., R. J. Olsen, S. A. Witzel, O. J. Attoe, W. H. Paulson and R. F. Johannes. "Effect of Method of Manure Handling on Crop Yields, Nutrient Recovery and Runoff Losses," *Trans. ASAE* (1970), pp. 726-731.
10. Klausner, S. D., P. J. Zwerman and D. F. Ellis. "Surface Runoff Losses of Soluble Nitrogen and Phosphorus Under Two Systems of Soil Management," *J. Environ. Qual.* 3:42-46 (1974).
11. Zwerman, P. J., S. D. Klausner and D. Ellis. "Land Disposal Parameters for Dairy Manure," Proceedings of 1974 Cornell Waste Conference (1974), pp. 211-221.
12. Timmons, D. R., R. F. Holt and J. J. Latterell. "Leaching of Crop Residues as a Source of Nutrients in Surface Runoff Water," *Water Resources Res.* 6(5):1367-1375 (1970).

13. Witzel, S. A., N. Minshall, M. S. Nichols and J. Wilke. "Surface Runoff and Nutrient Losses of Fennimore Watersheds," *Trans. ASAE* (1969), pp. 338-341.
14. Young, R. A., and C. K. Mutchler. "Pollution Potential of Manure Spread on Frozen Ground," *J. Environ. Qual.* 5(2):174-179 (1976).
15. Converse, J. C., G. D. Bubenzer and W. H. Paulson. "Nutrient Losses in Surface Runoff from Winter Spread Manure," ASAE Paper No. 75-2035, St. Joseph, MI 49085 (1975).
16. Doyle, R. C., D. C. Wolf and D. F. Bezdicek. "Effectiveness of Forest Buffer Strips in Improving the Water Quality of Manure Polluted Runoff," Managing Livestock Wastes 3rd Intern. Symp. on Livestock Wastes, ASAE, St. Joseph, MI 49085 (1975), pp. 299-302.

13

ESTIMATION AND MANAGEMENT OF THE CONTRIBUTION BY MANURE FROM LIVESTOCK IN THE ONTARIO GREAT LAKES BASIN TO THE PHOSPHORUS LOADING OF THE GREAT LAKES

D. W. Draper, J. B. Robinson
    Department of Environmental Biology
    University of Guelph
    Guelph, Ontario, Canada

D. R. Coote
    Land Resource Research Institute
    Ottawa, Ontario, Canada

INTRODUCTION

The Great Lakes Quality Agreement between the United States and Canada, signed at Ottawa on April 15, 1972, included a recommendation to study pollution of the Great Lakes arising from agriculture, forestry and other land use activities. Accordingly, the International Reference Group on Great Lakes Pollution from Land Use Activities (PLUARG) was established in late 1972 and produced a detailed plan outlining an extensive study scheduled for completion in July 1978.

As a major part of the work assigned to the Canadian section of PLUARG, a number of small watersheds representative of larger agricultural regions in Ontario were studied. The purpose of this work was to relate water quality at river mouths to specific land uses and practices in the river basin. Research into the effects of livestock related land uses in these Ontario watersheds was undertaken by several groups including the authors.

The objective of this study was to derive a numerical estimate of the lake loadings attributable directly to livestock. The authors have collaborated with a group headed by Dr. F. Madison of the Water Resources Center, Madison, Wisconsin, which has been engaged in similar work for the U.S. side of the Great Lakes Basin. Their work is reported separately at this conference by Moore et al.[1] Differences in farming practices and physiography of the two countries became evident during this exchange. Therefore, the results presented here should be considered representative of the livestock problem in the Canadian Great Lakes Basin only.

Of the pollutants of livestock manure origin, the most significant in terms of lake water quality is thought to be phosphorus. The oxygen-demanding carbonaceous component of livestock manure may cause local pollution problems but is attenuated very markedly during tributary transport to the lakes. The same is probably true of pathogenic microorganisms originating in livestock manure. Nitrogen of manure origin which contaminates local water supplies moves by both surface and groundwater routes, is subject to transformations and may enter permanent sinks during transport.[2] This mobility makes the movement of N to the lakes difficult to quantify.

Phosphorus occurs in natural waters in a multitude of forms,[3] but in terms of trophic effects in the lakes, particular significance is given to orthophosphate. This form, also called soluble inorganic phosphorus, dissolved reactive phosphorus or soluble orthophosphate, is referred to as the "readily available" or "biologically available" form of phosphorus. The orthophosphate which enters runoff from livestock manure sources is ephemeral in nature in that it is readily changed to nonavailable forms during transport. This change is dependent on other water quality parameters such as sediment load. It was decided, therefore, that for this study total phosphorus must be the parameter considered.

The disadvantage of this decision is that lake loadings of total-P are not necessarily related to lake trophic conditions. However, strong positive correlation between spring overturn total-P and summer mean chlorophyll-*a* concentration in a large group of southern Ontario water bodies has been demonstrated by Dillon and Rigler.[4] Rast and Lee[5] have also shown a strong positive correlation between total-P and dissolved reactive phosphorus in a number of water bodies. Estimation of the lake loadings of total-P from livestock sources may therefore be significant as an index of effect on lake quality, but at present this cannot be stated with certainty.

For the purposes of this study it is assumed that total phosphorus has a stream delivery ratio of 1.0, *i.e.*, the calculated inputs to streams and the resultant lake loadings are equivalent in the long term. This approach has been used even though there is some evidence that tributary delivery ratios are less than 1.0. However, the delivery ratio for any stream is difficult to evaluate and can be expected to vary from one basin to another. Other workers may wish to apply a factor less than 1.0 to the loadings estimated in this study.

To quantify livestock inputs of phosphorus, a simple modeling approach was used based on a synthesis of information on specific livestock management practices and the factors relating these practices to stream loading. In what follows, attention has been focused upon the main areas of livestock management which contribute to lake phosphorus loading. Values assigned to these components are not intended to apply to all areas at all times, or even to be a complete definition of the situation in Ontario. The problem of cattle entering streams, for example, has not been dealt with here, due mainly to a lack of background studies. What has been attempted is simply an estimation of the contribution of phosphorus to lake loadings on a per animal basis for the manure source areas considered.

ATTENUATION OF PHOSPHORUS IN OVERLAND FLOW

Only a small proportion of the phosphorus excreted in livestock manure ever reaches runoff receiving channels. Uptake by plants on manured croplands, accumulation in soils and retention at livestock facilities are the main alternatives. Much of the manure-P which erodes from deposition sites in storm or snowmelt runoff settles out during transport in overland flow (as distinct from channel flow which, as indicated above, we assume to be 100% effective in transport of P).

Recognition of the importance of the process of nutrient loss in overland flow, or attenuation, led to an attempt to derive an estimate of a distance, referred to subsequently as "critical distance," over which attenuation to background levels would be likely to occur on unmanured surfaces. Several studies, mainly concerned with the use of "vegetative filters" or grassed terraces for the treatment of effluent and runoff from livestock facilities, provide data useful for this estimation.

Sutton et al.[6] constructed a serpentine-type infiltration channel 800 ft long by 3 ft wide (at the bottom) and observed that at no time during the 2-yr study did runoff from a 6336-ft^2 feedlot travel further than 400 ft into the channel. Loudon[7] examined runoff water quality below winter-manured and unmanured plots on a 3.5% tilled slope and concluded that no significant difference in P concentrations existed at 120 ft downslope. An analysis of several other studies[8-12] indicated that a distance of from 30 to 1000 ft may be required for near complete attenuation of P in runoff.

The main factors resulting in this wide range of distances appear to be (1) the ratio of the volume of runoff to the treatment surface area, and (2) detention time. The confinement of runoff in an embanked area or in a defined channel increases the runoff volume to surface area ratio and reduces the amount of deposition and infiltration which can occur, thus increasing the distance required for attenuation of a major portion of the nutrients in runoff.

In general, in studies where runoff was not well channelized or confined within fixed boundaries and where flow rate was not controlled (through settling basins or detention facilities), distance of up to 400 ft would appear to be required for attenuation of P in runoff to near background levels. Under ideal conditions, as little as 30 to 40 ft over a well-vegetated surface has been shown to be effective in controlling nutrients in polluted runoff.[9,13] Considering this range of distances, it was decided that for the purposes of the model, a slightly conservative choice of "critical distance" in the range of 100 to 400 ft would be appropriate.

DATA BASE

The importance of the attenuation process in ultimately determining the delivery of phosphorus to streams led to an examination of the distances between livestock production sites and receiving channels and an attempt to determine the relative impact of livestock operations according to this distance. To obtain this information, data were extracted from a study[14] referred to subsequently as the inventory, involving the interpretation of aerial photographs (1:15,840) to give information on the size, type and location of livestock operations in a large portion of southern Ontario. The photographs used were mainly from 1971 and 1972, with a smaller part of the surveyed area photographed in 1966. All large* livestock and poultry operations were recorded, while smaller operations were recorded only if they were considered "close" to a drainage channel, water body, road or urban development. The recording procedure involved listing, for each operation included, the estimated number and type of animals housed, the distance to the closest channel or stream, water body and roadway, the type of manure management system, and the confinement and feeding management.

In all about 50% of the actual 1976 Ontario population of confined fattening cattle in the survey area[15] were recorded in the inventory and about 10% of all dairy cattle and heifers, beef cows and bulls were accounted for. This was considered a reasonable sample upon which to base assumptions as to the spatial distribution of the general livestock population relative to receiving channels in the southern Ontario Basin. In addition, this sample provided a basis for estimating the proportions of livestock of different types managed under different systems, some representing a large, others a negligible risk for runoff loading of manure-P.

Throughout the following discussion, reference to livestock populations will be made in terms of "animal units" (au). Since there is considerable variation in the production rate of manure nutrients between different species

---

*Dairy: $>$ 75 head; beef cattle: $>$ 150; swine: $>$ 300; horses: $>$ 75; poultry: (subjective).

and ages of animals, some uniform basis for comparison is necessary. The animal unit, used in Ontario in establishing waste management guidelines[16] is based on the nitrogen excretion rate of the animal. One animal unit will annually produce the amount of N normally required to bring a 1-ac corn crop to maturity (approx. 70 kg). Table I lists the animal unit equivalents for various livestock classes as applied in this study.*

Table I. Animal units of manure production in terms of nitrogen. After: Agricultural Code of Practice, Ontario.

Type of Livestock or Poultry	Annual Basis (365 days)
1 Dairy Cow (plus calf)	1 animal unit
2 Heifers (for dairy purposes)	1 animal unit
1 Beef cow (plus calf)	1 animal unit
1 Bull	1 animal unit
1 Horse	1 animal unit
4 Sows (plus litter to weaning)	1 animal unit
125 Laying Hens	1 animal unit
4 Sheep (plus lambs)	1 animal unit
	Market Basis (as marketed)
2 Beef Feeders (gain 400-1100 lb)	1 animal unit
4 Beef Feeders (gain 400-750 lb)	1 animal unit
4 Beef Feeders (gain 750-1100 lb)	1 animal unit
15 Hogs (gain 40-200 lb)	1 animal unit
1000 Broiler Chickens or Roasters (4-5 lb)	1 animal unit
300 Turkey Broilers (11-12 lb)	1 animal unit
150 Heavy Turkey Hens (19-20 lb)	1 animal unit
100 Heavy Turkey Toms (30-32 lb)	1 animal unit
40 Veal Calves (gain 90-300 lb)	1 animal unit
1000 Pullets	1 animal unit

RESULTS FROM INVENTORY

Figure 1 illustrates the inventory data on proximity to aerially definable runoff receiving channels of "large" livestock enterprises in southern Ontario. It is evident that about 60% of these animals are found within 400 ft of clearly defined channels and that animal numbers decline logarithmically with distance. No preference for the location of operations relative to channels may necessarily be inferred since it is possible that this distribution just

---

*The Ontario animal unit is not to be confused with the animal unit used in the U.S., which is based on the number of animals of a given type equivalent to 1000 lb liveweight.

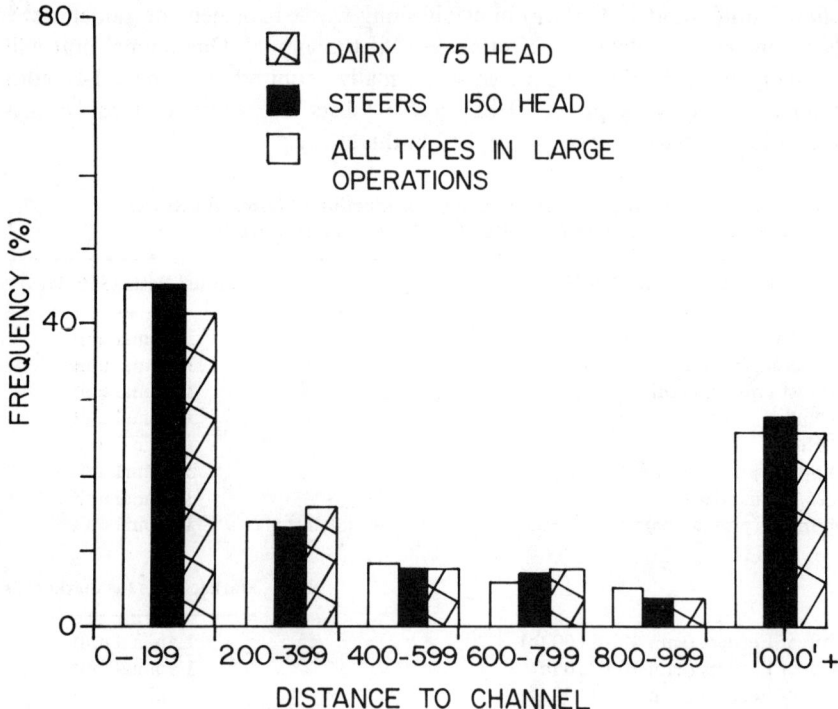

Figure 1. Proximity to runoff channels of animal units contained in large facilities in Southern Ontario based on air photo interpretation.

reflects the drainage pattern in the region. The development of artificial drainage channels on and near the feedlots and barnlots is also a factor.

The relationship of inventory animal numbers to distance from a channel was analyzed and it was found that between 0 and 400 ft the curve:

$$f_i = 0.165 - 0.025 \log d_i$$

approximates the proportion of confined animals, f_i, in the distance interval, d_i, from a channel, using 25 ft intervals (d_i = 25/2, 75/2, ... 400 - 25/2 ft). This relationship was applied to the general livestock population, to obtain an estimate of the number of confined animals at successive 25-ft distances from a channel between 0 and 400 ft.

The inventory sample also provided information essential for determining the pollution potential of various types of livestock and poultry operations in the study area. Tables II and III summarize these data. Table II indicates, for example, that only about 13% of swine were found at facilities having

Table II. Confinement of inventory livestock
(% of each livestock category).

	Dairy	Beef (mixed ages)	Feedlot Cattle	Swine	Poultry	Other
Covered Only						
Large Operations	0.2	4.9	4.0	82.4	60.0	0
Small-Medium Operations	1.3	5.3	0.1	4.7	28.7	4.8
Outside Lots						
Large Operations	41.2	39.6	93.5	11.4	7.5	50.9
Small-Medium Operations	57.3	50.2	2.4	1.5	3.8	44.3

outside barnlots or feedlots. Therefore 87% of the population of swine could reasonably be excluded as potential contributors to stream loading of phosphorus originating in barnlots. However, Table III indicates that 58% of swine are housed in operations with outside storage of solid or semisolid manure, prone to runoff, and hence this proportion of the actual swine population must be considered as contributing through this management factor.

CALCULATION OF LIVESTOCK P LOAD

Overland Flow Delivery Ratio

For any particular area, the overland flow delivery ratio is defined here as:

$$F_1 = \frac{\text{pollutants reaching receiving channels}}{\text{pollutants loaded to runoff at sources}}$$

If it is assumed that attenuation of pollutants occurs linearly with respect to distance of travel in overland flow, until at some critical distance, C_d, the pollutant level is 0 (or "background"), then over any distance, d, less than C_d,

Table III. Assumed manure management of livestock operations based on inventory samples (% of each livestock category).

	Cattle	Swine	Poultry
Separate Solid or Semisolid Manure Storage Areas	0	58.2	85.3
Liquid Manure Storage	0	41.8	14.7
Manure Storage on Open Lot Area	100	0	0

the proportion of pollutant remaining is $1-d/C_d$. The distribution of confined animals (and consequently of manure sources) in southern Ontario, restated from above, is:

$$f_i = 0.165 - 0.025 \log d_i, (d_i = \frac{25}{2}, \frac{75}{2} \ldots (400 - \frac{25}{2}) \text{ ft}) \tag{1}$$

Therefore, under these assumptions, the overland flow delivery ratio for the area is:

$$F_1 = \sum_{i=0}^{C_d} (1 - \frac{d_i}{C_d}) f_i \tag{2}$$

As discussed previously, it was decided to assume that the critical distance for phosphorus attenuation lies between 100 and 400 ft. For 100 ft, $F_1 = 0.16$, for 400 ft, $F_1 = 0.40$. Thus, if a 100 ft critical distance is chosen, 16% of the phosphorus in runoff from all feedlots and barnlots in the area is said to reach a channel.

Livestock Load Calculation

Figure 2 illustrates the main pathways through which manure phosphorus enters runoff and subsequently receiving channels. Note that excretion on pasture has not been indicated as a manure source of runoff loading. It was thought that runoff would be minimal under a forage crop. Manure spread during the growing season on croplands is also indicated as not contributing, as this disposition of manure is considered to be a cropland management problem, not a livestock problem. In any case, moderate applications of manure to some crops are believed to reduce erosion,[17] improve soil structure,[17,18] water holding capacity[19] and infiltration rate.[20]

Three component manure sources were considered as contributing to the total-P load: direct runoff from open feedlots and barnlots, direct runoff from open manure holding areas separate from lots, and direct runoff from land on which manure is spread in winter when soil is frozen. Coefficients, or proportions have been estimated for the various branch points of the flow chart of Figure 2, making it possible to estimate the contribution from each of these components. These were derived in part from the inventory data and in part from literature values. A critical distance of 400 ft is shown in Figure 2. However, calculations were based on the assumption of a possible range of critical distance of 100 to 400 ft.

Specific excretion rates were first applied to censused livestock populations to obtain an estimate of annual P production for each area studied. Frank and Ripley[21] provided livestock population estimates in the 11 PLUARG study watersheds, and provincial statistics[15] were used for population estimates in

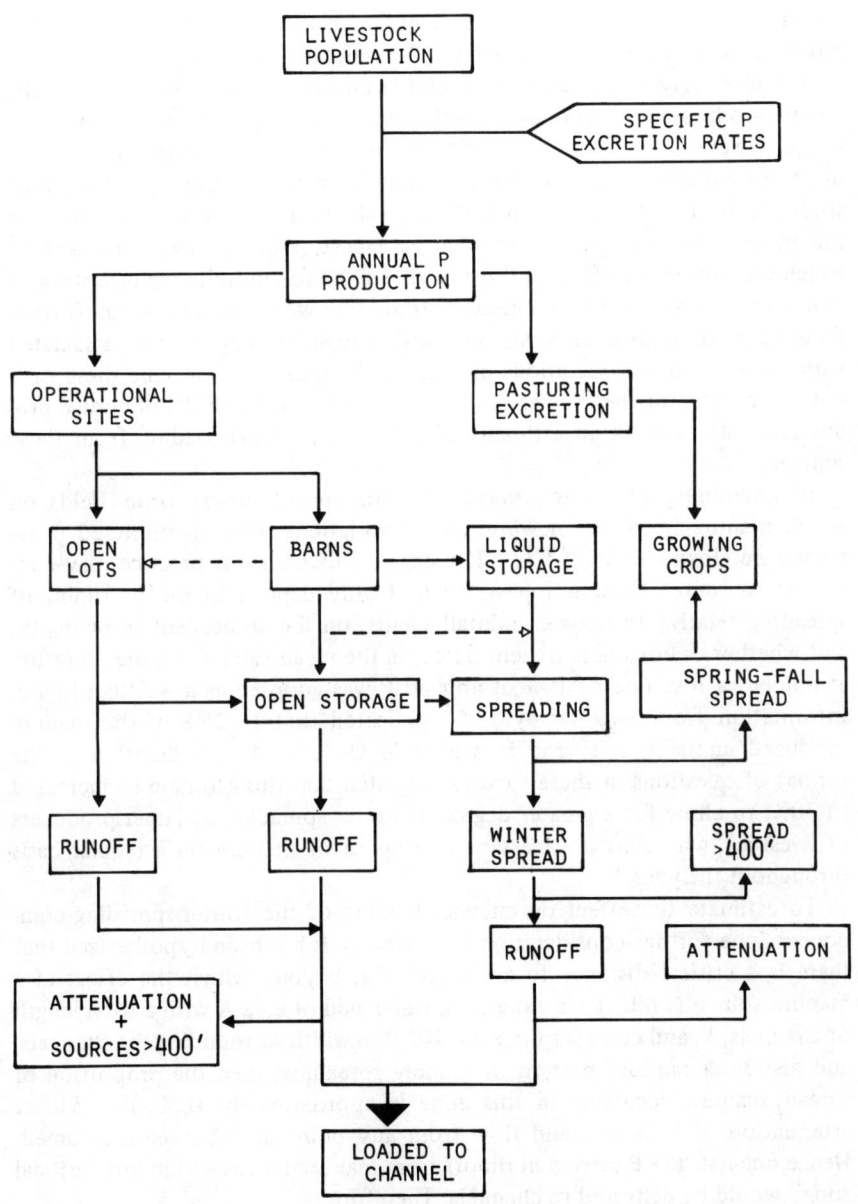

Figure 2. Manure-P pathways to runoff receiving channels.

the southern Ontario Great Lakes Basin. Excretion rates of P for different animals were generally in good agreement with those used by Moore et al.,[1] but differences exist which probably reflect dietary differences.

The inventory information of Tables II and III was used to apportion the annual production of manure-P into the "open lots" and "open storage" components. Several literature sources[22-26] were analyzed to obtain an estimate of the proportion of annual P production lost in runoff from open lots. Most studies indicated that less than 10% of production was eroded annually, and the mean value of 5% was adopted as a best estimate. Two studies[22,27] which monitored runoff from three open, solid and semisolid manure storages indicated that about 3% of annual P production would be lost in runoff from these sources. Solid manure storages with a high bedding content associated with chicken broiler operations produce little runoff, and because these were not represented in the studies referred to, a lower value of 2% of annual production was used as an estimate of the overall runoff loading from these sources.

In examining literature concerned with spring runoff from fields on which manure is spread in winter, wide variations were encountered in reported net losses of P.[2,28-33] The degree to which this practice may contribute to runoff loading appears to be highly dependent on the timing of spreading relative to thaw or rainfall events, on the antecedent snow depth, and whether or not soil is frozen. Based on the mean value from the literature examined, a loss rate of 10% of applied P was adopted as a working figure. Information from two surveys[34,35] indicated that 19-25% of the manure produced annually is spread in winter in Ontario. A consideration of the format of questions in these surveys suggested that this estimate be increased to 30%, to allow for a greater degree of winter spreading by larger producers of livestock who find it necessary to dispose of manure on a regular basis throughout the year.*

To estimate the effect on channel loading of the winter-spreading component, one further consideration is necessary. It has been hypothesized that there is a critical distance to a channel, C_d, beyond which the effect of a manure source is nil. If we examine a watershed of area A with a total length of channels, ℓ, and consider the zone $2(C_d)$ in width surrounding the channels, and assume a random pattern of manure spreading, then the proportion of spread manure occurring in this zone is approximately $\ell(2C_d)/A$. Linear attenuation of P in overland flow from any point in this zone is assumed. Hence one-half the P carried in runoff from manured areas within this "critical zone" would be delivered to channels. Therefore

---

*A complete description of the calculations used to arrive at values for runoff from lots, storages and winter-spread manure is reported in Robinson and Draper.[36]

NUTRIENT MANAGEMENT 169

$$F_2 = \frac{\ell \times 2C_d}{2A} \quad (= \text{channel density} \times C_d) \qquad (3)$$

where F_2 is an estimate of the delivery ratio to channels of P in runoff from winter-manured fields in the watershed in the same way F_1 is an estimate of the overland flow delivery ratio of P in runoff from operational sites.

Under the above set of assumptions and numerical estimates, per animal unit loads of P were calculated for the whole of the southern Ontario Basin and for the PLUARG study watersheds. The estimates range from 80-220 g P/au/yr for Ontario and from 90-240 g P/au/yr in the PLUARG watersheds. The lower estimates are derived by assuming a critical distance of 100 ft and the higher values by assuming 400 ft. Notice that increasing the distance assumption by a factor of four increases the unit loading value by less than a factor of three. This results from the decreasing frequency with which livestock occur at greater distance intervals from channels (Figure 1). Per animal unit loads are readily converted to unit area loads by multiplying by animal density.

The estimates of unit loads are more sensitive to the assumption regarding critical distance than to estimates of channel density, which affects only the contribution from the winter-spreading component. In calculating the above unit loads a channel density estimate of 1.09 km/km² based on an analysis of 1:50,000 scale topographic maps of the PLUARG study areas was used, and this is probably lower than the true density.* If in fact the channel density were twice this figure, the animal unit load for a 400 ft critical distance assumption would be increased by only about 15%. A greater proportion of the livestock unit load would, however, be accounted for under these assumptions, by the winter-spreading component.

In order to verify loading estimates made as described, field data are required from small watershed studies in which water samples are taken immediately upstream and downstream of livestock operations. One such study[37] has already been completed in part of the Little Ausable River watershed, one of the PLUARG agricultural watersheds. The 13 livestock sites studies had a total of 1263 animal units on 1449 ha and yielded an annual flux of total-P of 732 g/ha. Allowing for an estimated annual areal loading of 503 g P/ha estimated by Miller and Spires[38] for the cropland in this same area, a per animal unit load of 263 g total-P per year is estimated from these data. The estimates based on the method described in this paper ranged from 90 g P/au/yr (100 ft critical distance assumption) to 240 g P/au/yr (400 ft critical distance assumption) for the PLUARG watersheds. The unit load

---

*Detailed analyses by Moore et al.[1] found channel densities in Wisconsin of 2.3 to 4.6 km/km².

based on the field data is higher than the maximum unit load estimate in part owing to the lack of stream protection and the occurrence of manure spreading on floodplains in this study area.

DISCUSSION AND RECOMMENDATIONS

It is estimated that there are about 2,120,000 animal units in southern Ontario producing nearly 22,800 MT of manure-P annually. The present model predicts that 160-570 MT, or 0.7-2.5% of this annual production is loaded directly to streams from feed and barnlots, manure storage areas and winter-spread manure through runoff. This analysis does not include any P loss from cropland receiving manure other than in winter. Inputs of total-P to streams from fertile croplands may be quite large, but it appears that soil type and cover rather than manure application rate are the important determinants.[38]

Table IV shows the main sources of the livestock total-P loading as derived in the foregoing section. It would appear that remedial measures should first be applied to livestock operational sites, particularly feedlots and barnlots associated with beef and dairy cattle within 400 ft of channels. Berms, broad grassed waterways or runoff retention facilities with clean water diversions above lots, and eavestroughing for roofed areas will be required, depending on site conditions.

Table IV. Components of livestock total-P loading in southern Ontario Great Lakes Basin.

Category	Critical Distance Assumed	
	100 ft	400 ft
	(% of Total-P Load)	
Beef (> 150 Head) Direct Lot Input	48.7%	43.9%
Winter Manure Spreading	13.4	19.4
Dairy (< 75 Head) Direct Lot Input	15.4	14.0
Dairy (> 75 Head) Direct Lot Input	10.9	10.0
Poultry/Hog Solid and Semisolid Manure Storages	5.0	4.5
Hogs Direct Lot Input	2.0	1.8
Poultry Direct Lot Input	0.5	0.5

It should be noted that erosion may cause the formation of temporary channels below feedlots and barnlots, especially where the surface slopes noticeably and vegetation is sparse. Attenuation is greatly reduced when runoff enters even small channels, and therefore feedlots and barnlots located on or above sloping round should be protected by a strip of dense perennial ground cover immediately downslope. The width of this strip would be determined by the design and present management of the facility.

Elimination of inputs from feedlots and barnlots should decrease livestock inputs of total-P by over 50%. Successful programs to reduce winter-spreading of manure near drainage channels would have the most significant effect, next to eliminating the feedlot and barnlot runoff problem. The study conducted by our U.S. collaborators[1] found that inputs from winter-spreading appear to be of greater relative importance. Their results further draw attention to the importance of drainage density in different physiographic regions in determining the potential impact of this practice.

Where winter-spreading cannot be curtailed, alternatives for ameliorating its effects should be considered. Spreading where buffer strips of perennial vegetation intercede between fields and streams, where drainage enters field sinks, and where crop residues are left on the land reduces the impact of this practice.

Buffer strips of natural vegetation around streams should be vigorously promoted, not only in connection with livestock operations, but in any area of intensive agriculture. As little as 30 ft of natural vegetation can be effective in attenuating over 90% of the nutrients carried in runoff,[13] and has the added benefits of promoting in-stream nutrient removal[39] and recreational use potential.

It is also recommended that further research on the effects of cattle entering streams be conducted. Some preliminary estimates have been made suggesting that there is potential for as much as 160 MT of total-P to enter streams frequented by cattle in the warm summer months in southern Ontario.[36] During this season, low stream flows prevail and the potential for impairment of stream quality, and the stimulation of algal blooms in near-shore areas of receiving water bodies is great. Stream bank erosion is also known to be accelerated by the presence of cattle near streams, and the establishment of a natural stream bank and stream bed biota is prevented. Present knowledge does not permit a proper assessment of these potentials, but they appear sufficiently important in some areas to warrant a consideration of fencing of perennial streams with the provision of alternate livestock watering facilities.

DISCLAIMER

The study discussed in this article was carried out as part of the efforts of the International Reference Group on Great Lakes Pollution From Land Use Activities (PLUARG), an organization of the International Joint Commission, established under the Canada-U.S. Great Lakes Water Quality Agreement of

1972. Funding was provided through Agriculture Canada. Findings and conclusions are those of the authors and do not necessarily reflect the views of the Reference Group or its recommendations to the Commission.

ACKNOWLEDGMENTS

The authors wish to thank F. W. Madison, I. C. Moore and R. Schneider of the Water Resources Centre, University of Wisconsin for their cooperation on this project and to R. Hore of Agriculture Canada for his support and guidance.

REFERENCES

1. Moore, I. C., F. W. Madison, R. R. Schneider. "Estimating Phosphorus Loading from Livestock Wastes: Some Wisconsin Results," in: *Best Management Practices for Agriculture and Silviculture,* Proceedings of the 10th Annual Cornell University Conference, this volume (1978).
2. Hensler, R. F., R. Olsen, S. Witzel, O. Attoe, W. Paulson and R. Jonannes. "Effect of Method of Manure Handling on Crop Yields, Nutrient Recovery and Runoff Losses," *Trans. ASAE* 13:726-731 (1970).
3. Vollenweider, R. A. "Scientific Fundamentals of the Eutrophication of Lakes and Flowing Waters, with Particular Reference to Nitrogen and Phosphorus as Factors in Eutrophication," Tech. Report DAS/CSI/68.27, OECD, Paris (1968).
4. Dillon, P. J., and F. H. Rigler. "The Phosphorus-Chlorophyll Relationship in Lakes," *Limnol. Oceanog.* 19:767-773 (1974).
5. Rast, W., and G. F. Lee. "Summary Analysis of the U.S. Portion of The North American OECD Eutrophication Study Results Emphasizing Nutrient Loading-Lake Response Relationships and Trophic State Indices," U.S. Environmental Protection Agency Report (1977).
6. Sutton, A. L., D. D. Jones and M. C. Brumm. "A Low-Cost Settling Basin and Infiltration Channel for Controlling Runoff from an Open Swine Feedlot," ASAE Technical Paper No. 76-4516 (1976).
7. Loudon, T. L., and J. B. Gerrish. "Animal Manure Movement in Winter Runoff for Different Surface Conditions," in: *Best Management Practices for Agriculture and Silviculture,* Proceedings of the 10th Annual Cornell University Conference, this volume (1978).
8. Edwards, W. M., F. Chichester and L. Harold. "Management of Barnlot Runoff to Improve Downstream Water Quality," Proc. 1st Int. Symp. on Livestock Wastes. ASAE (1971), pp. 48-50.
9. Haupt, H. F., and W. Kidd, Jr. "Good Logging Practices Reduce Sedimentation in Central Idaho," *J. Forestry* 63:664-670 (1965).
10. Overcash, M. R., F. J. Humenik, P. N. Westerman, D. M. Covill and J. W. Gilliam. "Overland Flow Pretreatment of Poultry Manure," ASAE Technical Paper No. 76-4517 (1976).
11. Sievers, D. M., G. Garner and E. Pickett. "A Lagoon-Grass Terrace System to Treat Swine Waste," Proceedings 3rd Int. Symp. on Livestock Wastes, ASAE (1975), pp. 541-543; p. 548.

12. Willrich, T. L., and J. O. Boda. "Overland Flow Treatment of Swine Lagoon Effluent," ASAE Technical Paper No. 76-4515 (1976).
13. Doyle, R. C., D. Wolf and D. Bezdicek. "Effectiveness of Forest Buffer Strips in Improving the Water Quality of Manure Polluted Runoff," Proc. 3rd Int. Symp. on Livestock Wastes (1975), pp. 299-302.
14. Coote, D. R., E. MacDonald and M. Rigby. "A Selective Inventory of Large Livestock Operations in Southern Ontario (By Aerial Photograph Interpretation)," Combined report of the activities of Eng. Res. Service and Soil Res. Inst. as part of Agriculture Canada's contribution to the Implementation of the Great Lakes Water Quality Agreement, 1973-1974, Section IV (1974).
15. *Agricultural Statistics for Ontario 1976*, 95th ed., Statistics Section, Economics Branch, Ontario Ministry of Agriculture and Food (1976), pp. 32-37.
16. Agricultural Code of Practice Government of Ontario, Toronto, Ontario (1976).
17. "Management of Nutrients on Agricultural Land for Improved Water Quality," U.S. Environmental Protection Agency, Water Poll. Cont. Res. Series (1971).
18. Guttay, J. R., R. I. Cook and A. E. Erickson. "The Effect of Green Manure and Stable Manure on the Yield of Crops and on the Physical Condition of Atappan—Parkhill Loam Soil," *Soil Sci. Soc. Am. Proc.* 20:526-528 (1956).
19. Olsen, R. J., R. F. Hensler and O. J. Attoe. "Effect of Manure Application, Aeration, and Soil pH on Soil Nitrogen Transformations and on Certain Soil Test Values," *Soil Soc. Am. Proc.* 24:222-225 (1970).
20. Murphy, L. S., G. W. Wallingford, W. L. Powers and H. L. Manges. "Effects of Solid Beef Feedlot Wastes on Soil Conditions and Plant Growth," in: *Waste Management Research,* Proceedings of the 1972 Cornell Agricultural Waste Management Conference. (Washington, DC: Graphics Management Corp., 1972).
21. Frank, R. and B. D. Ripley. "Land Use Activities in Eleven Agricultural Watersheds in Southern Ontario, Canada, 1975-76," International Reference Group on Great Lakes Pollution from Land Use Activities, Ontario Ministry of Agriculture and Food, Project No. 80645 (1977).
22. Coote, D. R., and F. R. Hore. "Runoff Characteristics and Pollution Potential from Cattle Feedlots and Manure Storages in the Canadian Lower Great Lakes Basin," Report to I.J.C. International Reference Group on Great Lakes Pollution from Land Use Activities (PLUARG), Engineering Research Service, Ottawa, Canada (1976).
23. Gilbertson, C. B., J. Ellis, J. Nienaber, T. McCalla, and T. Klopfenstein. "Physical and Chemical Properties of Outdoor Beef Cattle Feedlot Runoff," University of Nebraska Ag. Exp. Sta. Res. Bull. No. 271 (1975).
24. Madden, J. M., and J. N. Dornbush. "Measurement of Runoff and Runoff Carried Waste from Commerical Feedlot," Proc. 1st Int. Symp. on Livestock Wastes, ASAE (1971), pp. 44-47.
25. Sharpley, A. N. and J. Syers. "Phosphorus Transport in Surface Runoff as Influenced by Fertilizer and Grazing Cattle," *N.Z. J. Sci.* 19:277-282 (1976).
26. Swanson, N. P., L. N. Mielke, J. C. Lorimer, T. McCalla and J. R. Ellis.

"Transport of Pollutants from Sloping Cattle Feedlots as Affected by Rainfall Intensity, Duration and Recurrence," Proc. 1st Int. Symp. on Livestock Wastes. ASAE (1971), pp. 51-55.
27. Magdoff, F. R., J. F. Amadon, S. P. Goldberg and G. D. Wells. "Runoff from a Low-Cost Manure Storage Facility," *Trans. ASAE* 20(4):658-665 (1977).
28. Converse, J. C., G. Bubenzer and W. Paulson. "Nutrient Losses in Surface Runoff from Winter Spread Manure," *Trans. ASAE.* 19:517-519 (1976).
29. Klausner, S. D., P. Zwerman and D. Ellis. "Nitrogen and Phosphorus Losses from Winter Disposal of Dairy Manure," *J. Env. Qual.* 5:47-49 (1976).
30. McCaskey, T. A., G. Rollins and J. Little. "Water Quality of Runoff from Grassland Applied with Liquid, Semi-liquid and 'Dry' Dairy Waste," 1st Int. Symp. on Livestock Wastes, ASAE (1971), pp. 239-242.
31. Midgeley, A. R., and D. Dunklee. "Fertility Runoff Losses from Manure Spread during the Winter," University of Vermont Ag. Exp. Sta. Bull. No. 523 (1945).
32. Minshall, N. E., S. Witzel and M. Nichols. "Stream Enrichment from Farm Operations," *J. San. Eng. Div. Am. Soc. Civ. Eng.* 96(SA2):513-524 (1970).
33. Philips, P. A., A. J. Maclean, F. R. Hore, F. J. Sowden, A. D. Tennant and N. K. Patni. "Soil, Water and Crop Effects of Selected Rates and Times of Dairy Cattle Liquid Manure Applications under Continuous Corn," Agriculture Canada, Engineering Research Service Report No. 7522-540 (1975).
34. Bangay, G. Ontario Agricultural Practices Survey, unpublished information (1977).
35. Ketcheson, J. W. "Farm Manure Disposal Practices during Winter in Ontario—A Survey," Report for Water Pollution Control Directorate, Environmental Protection Service Canada Dept. of the Environment, Contract No. 565526 (1975).
36. Robinson, J. B., and D. W. Draper. "A Model for Estimating Inputs To the Great Lakes From Livestock Enterprises in the Great Lakes Basin," Final Research Report to Agriculture Canada, Ottawa, Canada (1978).
37. Beak Consultants Ltd. "Evaluation of Remedial Measures for the Agricultural Management of Water Quality," Final Report to Agriculture Canada for the Int. Ref. Group on Gt. Lakes Poll. from Land Use Activities (1977).
38. Miller, M. H., and A. C. Spires. "Contribution of Phosphorus to the Lower Great Lakes from Agricultural Activities in Ontario," Phosphorus Integration Report, Task Group C (Canadian Section) International Reference Group on Great Lakes Pollution from Land Use Activities (1978).
39. Robinson, J. B., H. R. Whitely, W. Stammers, N. K. Kaushik and P. Sain. "The Fate of Nitrogen in Small Streams and its Management Implication," in *Best Management Practices for Agriculture and Silviculture*, Proceedings of the 10th Annual Cornell University Conference, this volume (1978).

14

ESTIMATING PHOSPHORUS LOADING FROM LIVESTOCK WASTES: SOME WISCONSIN RESULTS

I. C. Moore
Water Resources Center
University of Wisconsin-Madison
Madison, Wisconsin

F. W. Madison
Water Resources Center and
 Soil Science Department
University of Wisconsin-Madison
Madison, Wisconsin

R. R. Schneider
Department of Economics
Williams College
Williamstown, Massachusetts

INTRODUCTION

The signing of the Great Lakes Water Quality Agreement in April 1972 between the United States and Canada signaled the start of a major effort on the part of both countries to quantify pollutant loadings to the Great Lakes from a variety of differing land uses. To implement this Agreement, the International Joint Commission (IJC), through the Great Lakes Water Quality Board, established the International Reference Group on Great Lakes Pollution from Land Use Activities (PLUARG).

The present study was undertaken at the Water Resources Center at the University of Wisconsin-Madison in conjunction with the EPA-sponsored Washington County Project and the IJC sponsored Menomonee River Pilot Watershed Study. Estimates of loadings from livestock wastes are integral parts of both projects.

A report done for PLUARG[1] estimates that nearly 58 million tons of manure are produced annually in the U.S. Great Lakes Basin. About 80% of this total is from cattle of all types, 10% from horses and the balance from swine, sheep and poultry. These wastes contain over 700,000 tons of primary nutrients (nitrogen, phosphorus and potassium). Quite clearly, wastes generated by livestock on the U.S. side of the Great Lakes represent a significant pollution potential.

Researchers at the University of Guelph, under the direction of Dr. J. B. Robinson, developed a method for quantifying phosphorus losses from animal wastes which served as a basis for this study. Our experience in estimating animal loadings by this method for the Wisconsin portion of the Great Lakes Drainage Basin represents the first step in determining the loadings basinwide. This areawide study was prompted by the general lack of reasonable estimates of the magnitude of the animal waste problem over large geographic areas.

Estimates of this kind are difficult to develop and their calculation, as reported in this study, is based on a number of fairly generalized assumptions. The figures generated are important as they serve to put the problem in perspective on several levels. For this reason, they are useful to planners and decision-makers both for ongoing nonpoint source planning and for setting investment priorities for control of diffuse pollution.

METHODOLOGY

The method developed by researchers at the University of Guelph for estimating phosphorus loading to Lake Ontario from livestock manure sources in the Ontario Great Lakes Basin[2] is followed here except where differences in available data or geographic characteristics warranted modification. Attention is focused in this study on the contamination of surface runoff with phosphorus (P) derived from two areas of livestock waste accumulation: barnyards and manure-spread fields. Although additional loadings of manure P result from pastured lands, estimates of loadings from this source are not included in the present study.

Barnyard/feedlot areas with the highest potential for runoff contamination are those with the following characteristics: (1) the yard area is completely open, (2) manure is allowed to accumulate or is scraped into uncovered piles, (3) water from uphill sources runs through the lot area, and (4) animals are confined to the area for considerable periods of time. Therefore, these animal concentration areas are considered to be manure P sources in this study. In Ontario, operations with solid and semisolid manure storage facilities are also analyzed as contributors of P to surface waters. However, the data available indicate that the use of this type of storage is limited in Wisconsin and that those waste storage systems used in the state are not susceptible to appreciable losses.

For the purposes of this study, manure-spread fields are considered to be manure P sources only during the 4 months of the year when the ground is frozen. When the ground is open, applied manure is considered to be a part of the soil. Manure P losses, therefore, are incorporated into total P loadings associated with soil erosion. Studies have shown that during the warmer months spreading manure can even help to reduce P in runoff by increasing the infiltration capacity of the soil and by reducing the velocity of surface water movement.[3]

The critical assumptions of the Guelph methodology concern the availability of manure P for transport by storm runoff and the delivery of that P to the Great Lakes. These are fully described in the next section. It should be noted here that all assumptions concerning phosphorus are based on *total phosphorus, i.e.,* both sediment attached and dissolved forms. While dissolved inorganic P is in the form most readily available for plant uptake, sediment P has also been shown to increase the fertility of a waterbody over the long run.[4,5]

The following description of the analysis of P loading to the Great Lakes from animal waste sources in Wisconsin is organized into four sections representing sequential phases of the analysis. Most of the discussion centers around the calculation of delivery ratios for manure P as this proved to be the most difficult factor to estimate.

PHOSPHORUS LOADING ANALYSIS

Definition of Study Areas

The State of Wisconsin is bordered on the east by Lake Michigan and on the north by Lake Superior. The Wisconsin Great Lakes Drainage Basin extends over the entire north-south range of the State (Figure 1). Considerable climatic and physiographic variability is exhibited within the Basin. For analytical purposes, the basin was divided into five study areas of more uniform physiography. These areas, shown in Figure 1, are based on an aerial division of the state found to be significant in a study of the frequency and magnitude of flooding.[6] Statistical significance was based primarily on topography, soil type, and bedrock geology. Physiographic boundaries were adjusted to county lines because most data inputs could not be disaggregated below the county level. The importance of this delineation will become obvious when techniques used to estimate the delivery of manure P to receiving waters are described below.

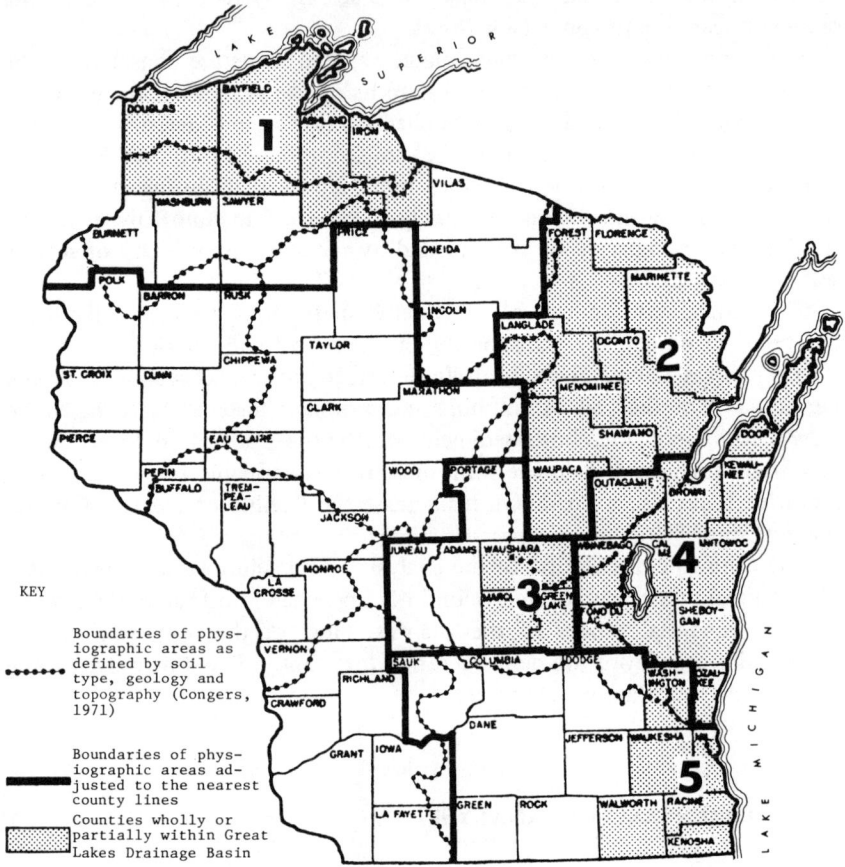

Figure 1. Map of Wisconsin showing physiographic areas and Great Lakes Drainage Basin counties.

Quantity of Total P Present at Source

Total numbers of livestock and poultry on Wisconsin farms in 1974 were taken from the Census of Agriculture.[7] Waste production and characteristics are based on values established by the American Society of Agricultural Engineers.[8] The average weight and annual total P production assumed for each of the ten census animal categories are presented in Table I. Manure P production figures used are comparable to those in the Ontario study with the exception of the figure for beef cattle. The American Society of Agricultural Engineers[8] reports an average of 18 kg of P/yr for mature beef cattle (1000 lb weight) while Canadian estimates are only 10 kg of P for the same

Table I. Phosphorus production and animal unit conversion factors by animal type.

Animal Type[a]	Average Weight (lb)	Phosphorus Production lb/yr (kg/yr)[b]	No. of U.S. Animal Units per Animal[c]	No. of Canadian Animal Units per Animal[d]
Beef cows and heifers that had calved	1,000	40 (18)	1.0	0.8
Beef heifers and heifer calves	600	24 (11)	0.6	0.5
Steers, bulls, steer and bull calves	750	30 (13)	0.75	0.6
Dairy cows and heifers that had calved	1,200	32 (15)	1.2	1.2
Dairy or milk heifers and calves	600	16 (7)	0.6	0.6
Hogs and pigs	175	9 (4)	0.175	0.2
Sheep and lambs	100	2.4 (1)	0.1	0.1
Hens and pullets of laying age	4	0.4 (0.2)	0.004	0.007
Broilers	4	0.4 (0.2)	0.004	0.01
Turkeys	20	0.04 (0.02)	0.02	0.03

[a]USDA, Census of Agriculture.[7]
[b]American Society of Agricultural Engineers.[8]
[c]One animal unit = 1000 lb liveweight.
[d]One animal unit produces 70 kg of manure nitrogen/yr.

animal per year. The reason for this difference is not entirely clear but may be related to differences in animal management and feeding.

Conversion factors used to compute U.S. and Canadian animal units from the census data are also listed in Table I. By definition one U.S. animal unit weighs 1,000 lb. One Canadian animal unit produces approximately 70 kg of nitrogen in its manure annually. Differences between the two animal units are minimal when the above definitions are strictly interpreted as shown in Table I. However, as Canadian animal units were designed to give farmers a rough idea of the fertilizer value of manure produced on their farms, they are usually defined in more general terms. For example, one Canadian animal unit equals a dairy cow and a calf or four sows plus litters. This makes comparisons between U.S. and Canadian animal units somewhat misleading. Consequently, Wisconsin P loading estimates are expressed in terms of the U.S. animal unit definition only.

Management information for Wisconsin livestock and poultry operations was obtained from two sources. Data concerning the use of barnyard runoff controls and manure storage facilities (Table II) were taken from the Wisconsin Domestic Animal Waste Inventory.[9] Information on animal housing and confinement (Table III) was provided by University of Wisconsin livestock and poultry experts. The Animal Waste Inventory was a cooperative effort of state and local resource agencies to collect data necessary for the evaluation of the environmental impacts from animal wastes. Over 1,700 sample barnyards were inventoried on randomly selected sample plots representing about 2% of the land area of each county.

Table II. Wisconsin livestock and poultry operations with waste controls.[a]

Waste Control	Type of Operation				
	Beef (%)	Dairy (%)	Swine (%)	Sheep (%)	Poultry (%)
6 Months Manure Storage Capacity	30	10	40	60	25
Barnyard Runoff Controls	10	5	45	10	5

[a]Wisconsin Domestic Animal Waste Inventory.[9]

Table III. Confinement and housing on Wisconsin livestock and poultry operations.

Management System	Type of Operation				
	Beef (%)	Dairy (%)	Swine (%)	Sheep (%)	Poultry (%)
Confined to Open Lots	100	100	30	50	0
Totally Enclosed	0	0	70	0	100
Unconfined	0	0	0	50	0

Several important assumptions were based on this information. No P loading from barnyard runoff was attributed to those situations where animals were thought to spend most of the time on pasture and/or under roofed enclosures or in barnyards with adequate runoff controls installed. The amount of manure available for spreading on frozen ground was taken to be 30% of the total manure excreted on an annual basis and no winter-spreading losses were assigned to farms with 6-mo waste storage capacity.

Phosphorus at Source Available for Runoff

Research has shown that the amount of manure-P which will enter surface runoff is dependent upon a number of site-specific factors. The most important variables appear to be the frequency, intensity and duration of rainfall and the spring thaw conditions.[10,11] The assumption is made here that the amount of manure-derived P carried in runoff from a wide range of sources can be represented as a set percentage of the total-P excreted on an annual basis. After examining a number of research reports on the movement of P during surface runoff, Draper et al.[2] made the following estimates:

1. 5% of the P excreted annually enters surface runoff from barnyard sources.
2. 10% of the P in manure spread on frozen ground is carried away in runoff.
3. 2% of the P excreted annually enters runoff from solid and semisolid storage facilities.

The 5 and 10% figures were used to estimate the quantity of P in surface runoff from barnyards and winter spread fields, respectively, in this study.

Phosphorus in Surface Runoff Delivered to the Great Lakes

Changes which occur in the volume and composition of runoff during transport to receiving waters involve complex chemical, physical and biological reactions. Factors such as the rate of soil infiltration and percolation, the type and extent of vegetative cover, the nature and distribution of native soil P, and slope are important variables to be considered in detailed manure P loading analyses.[12]

In this more general study, it is assumed that over a large area the primary determinant of P delivery ratios is the degree of contact between the transporting water and the ground surface. During the period of high water/ground contact associated with overland flow, the P in runoff is assumed to be linearly attentuated. If the flow remains unchannelized long enough, all of the manure derived P will be effectively removed. The distance at which close to 100% removal occurs is designated the "critical distance." Based on an extensive literature review, Draper et al.[2] concluded that the critical distance generally falls between 100 and 400 ft. During channelized flow, when both the time and the amount of water/ground contact are minimized, the attentuation rate is assumed to be zero. In effect then, all manure P delivered to channels is presumed to be carried ultimately to the Great Lakes.

It follows from these assumptions concerning nutrient delivery that data on the distance from manure sources to channels are critical inputs to the analysis. The "channels" that are important in this P loading analysis include perennial streams, intermittent or ephemeral streams, and small gullies, ditches and man-made conveyances which carry runoff to the larger watercourses. In Ontario, the required overland flow distances for barnyard sources were taken from a study of large livestock operations in southern Ontario in which aerial photographs (1:15,840) were used to identify barnyards and natural and man-made channels. For spread manure, the appropriate flow distances were estimated from available data on channel density for southern Ontario.[2] Comparable information was not available in Wisconsin; therefore techniques for approximating overland flow distances from the data that were available were devised.

Location of Barnyards with Respect to Channels

The only data source which relates Wisconsin barnyards to surface water bodies over an extensive area is the previously mentioned Domestic Animal Waste Inventory.[9] Data collectors in that inventory were instructed to measure distances from sample sites to receiving waters as the "distance (as water flows) to perennial and intermittent receiving waters as shown on USGS topographic maps."

As illustrated in Figure 2, the perennial and intermittent streams marked on U.S. Geological Survey (USGS) topographic maps typically make up only a small portion of the total length of channels in a given drainage basin. In Figure 2A, the upper portion of a tributary to the Pine River in Waushara County, Wisconsin, is reproduced directly from USGS topographic maps (scale 1:24,000). In Figure 2B, the same area is depicted but drainage channels indicated by contour crenulations have been added. The term contour crenulation refers to the inflections in contour lines pointing upslope which denote channels too small to be shown by stream symbols.[13] From this example, it is apparent that the inventory-measured distances are not sufficiently detailed for this analysis. It should be noted that very recent channels and artificial watercourses which are a part of the true drainage network cannot be identified from topographic maps.

The inventory barnyard to receiving water distances were adjusted for input into the analysis in the following manner. First, a small drainage basin (9 to 46 km^2) was selected for analysis in each of the five study areas. Two "channel density" values were then computed for each basin from USGS topographic maps (1:24,000):

1. *Channel Density A*. The total length of *marked* perennial and intermittent streams (Figure 2A) divided by the area of the drainage basin.
2. *Channel Density B*. The total length of marked perennial and intermittent streams *plus* the drainage channels determined from contour crenulations (Figure 2B) divided by the area of the drainage basin.

NUTRIENT MANAGEMENT 183

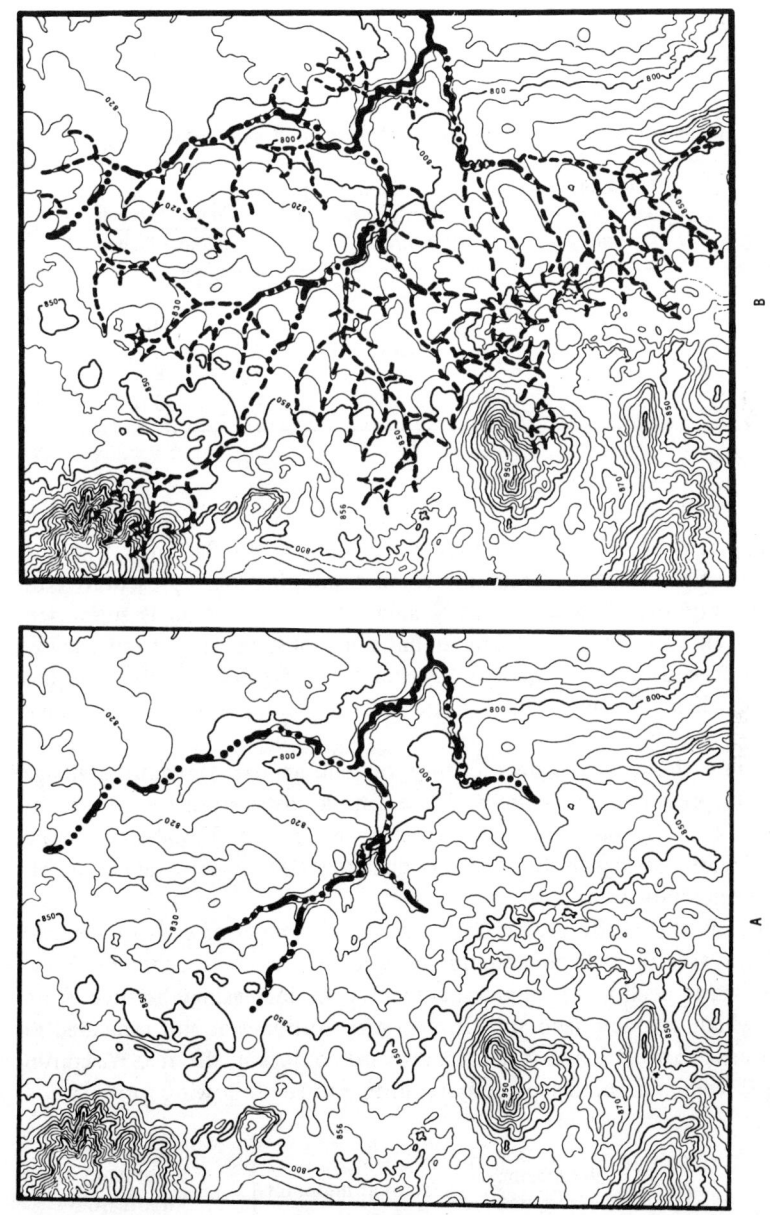

Figure 2. A. Parts of the Poy Sippi and Saxeville (Wisconsin) quadrangles showing perennial (———) and intermittent (•••———) streams (USGS, 7.5-min series); B. Same area showing additional drainage channels (—-—-—) determined from contour crenulations.

Finally, approximations of actual overland flow distances were made by dividing the inventory-measured distances by the ratio of channel density B to channel density A (Table IV).

Table IV. Channel densities determined from selected sub-basins within study areas.

Study Area	Channel Density A (km/km^2)	Channel Density B (km/km^2)	Channel Density B / Channel Density A
1	1.2	4.6	3.8
2	0.9	3.1	3.4
3	0.7	3.1	4.4
4	0.8	3.3	4.1
5	0.8	2.4	3.0

An important assumption underlying this data manipulation is that the channel densities calculated from the small sub-basins are representative of channel density over the larger study areas. This assumption is supported by the fact that channel development has been observed to be remarkably uniform over wide areas of similar topography and soil type.[14,15] As has been noted, these physiographic features were considered in the determination of study area boundaries.

Histograms showing the relationship of sample barnyards to drainage channels are presented in Figure 3. These sample distributions were used as estimators of the population distribution in each area. Variability among barnyard locational distributions is thought to reflect physiographic differences between study areas.

Findings of this study are similar to those of Draper et al.[2] for Ontario in that barnyards in both areas tend to be located close to channels. However, the sharp decline in animal numbers with increasing distance from channels which characterized the Ontario distribution was not indicated in the Wisconsin data. P loadings from barnyard sources in the five Wisconsin study areas were computed separately according to the following formula:

$$\text{barnyard manure P delivered to channels} = \left[\sum_{i=1}^{4} \text{\% of farms in distance interval (i)} \times \text{1 - attenuation factor at midpoint of distance interval (i)} \right] \times \text{total P in runoff from barnyard source}$$

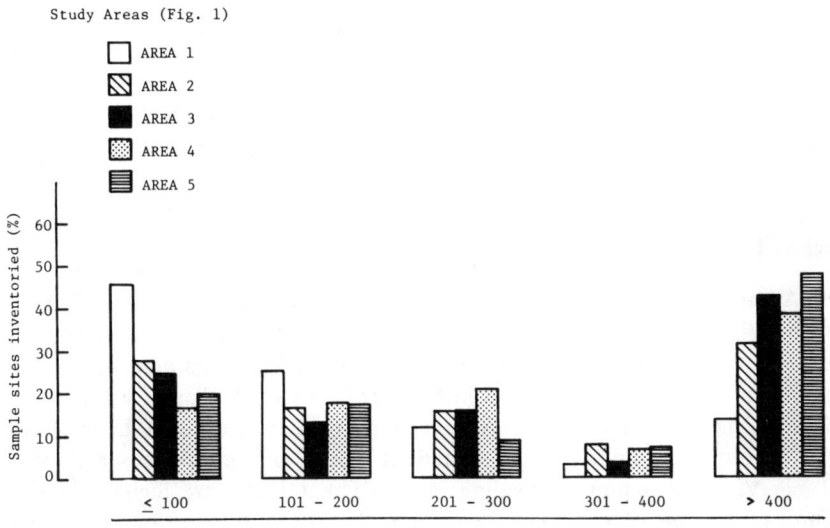

Figure 3. Location of sample sites with respect to channels.

Location of Winter-Spread Manure with Respect to Channels

Calculation of distances of overland flow from manure spread on frozen fields is less involved than that for barnyards because manure is typically spread evenly on croplands which are randomly distributed between channels. Given uniform application, an average attenuation rate of 50% can be applied to all winter-spread manure P within the critical distance of a channel. Furthermore, the percent of manure applied within the critical distance of a channel is approximately equal to the percent of land within that distance. P losses from winter-spread manure were calculated for each study area with the following series of equations:

1. $$\text{average distance of overland flow} = \frac{1}{2 \times \text{channel density}}$$

2. $$\text{\% of land or manure within critical distance of a channel} = \frac{\text{critical distance}}{\text{average distance of overland flow}} = 2 \times \text{critical distance} \times \text{channel density}$$

3. winter-spread manure-P delivered to channels = 0.50 × total P in runoff from manure spread within critical distance

Channel densities used in the above formulae were those calculated from the selected sub-basins in each study area and include channels indicated by contour crenulations.

RESULTS

Annual loadings of total phosphorus to Lake Superior and Lake Michigan from Wisconsin animal waste sources were calculated using critical distances of 100 ft and 400 ft. The results, then, should be interpreted as the endpoints of a range of loading estimates predicted by the application of the Guelph methodology to the Wisconsin portion of the Great Lakes Basin.

Wisconsin findings are compared to those of Ontario researchers in Table V. Total manure P loadings to the Great Lakes are higher from the Ontario portion of the Basin than from the Wisconsin portion. The opposite is true when results are considered on a per animal unit basis. Wisconsin manure P loading estimates in grams per animal unit per year are 38% higher under the 100-ft critical distance condition and 59% higher under the 400-ft condition. This finding primarily reflects differences in surface water hydrology and overland flow estimation techniques between the two Great Lakes Basin regions. A higher proportion of manure P sources in Wisconsin was found to be located within the critical distance of channels.

Another factor known to be contributing to the higher Wisconsin per animal unit loadings relates to the total P content of manure excreted assumed in the two studies. Wisconsin values, as reported by the American Society of Agricultural Engineers,[8] are slightly higher than those used in Ontario for most animal categories, and significantly higher for beef cows. Finally, some distortion may result from the use of a different definition of animal units in the two countries.

The percent of all manure P produced by livestock and poultry in the Wisconsin portion of the Great Lakes Drainage Basin predicted to be delivered to Lake Superior and Lake Michigan annually was found to be between 0.8 and 2.5%. This compared favorably with the Ontario range of 0.7 to 2.5%.[2]

Differences in total manure P loadings, as broken down by source, correlate well with differences in the types of operations predominating in the two Great Lakes Drainage Basin regions studied. The Wisconsin Basin is characterized by numerous small dairy operations with a smaller number of beef operations (about 5:1). Herds of greater than 50 head are uncommon

Table V. Comparison of total phosphorus loadings.

	Wisconsin Portion Great Lakes Drainage Basin		Ontario Portion Great Lakes Drainage Basin[a]		
	critical distance			critical distance	
	100 ft	400 ft		100 ft	400 ft
	Total Loadings (MT/yr)				
	130	416		150	570
	(g/animal unit/yr)				
	110	350		80	220
	Total Loading by Source (percent)				
Dairy Cattle	50	44	Beef > 150 Head	49	44
Winter-Spread Manure	30	38	Dairy Cattle	26	24
Beef Cattle	18	16	Winter-Spread Manure	13	19
Hogs	1	1	Poultry/Hog Manure Storage	5	5
			Hogs	2	2

[a]From Draper et al.[2]

on either type of farm.[16] In contrast, the Ontario Basin is rapidly becoming the center of beef feeding operations in eastern Canada. Between 1951 and 1971, the number of dairy cattle in the Ontario Basin decreased by 44% while the number of beef cattle increased by 93%.[17]

Half of the total loading in Wisconsin was found to be associated with dairy barnyards, while these operations account for only a quarter of the Ontario total P loading. In Ontario, beef feedlots of over 150 head contribute nearly half of the total loading. Beef feedlots of all sizes make up only 17% of the Wisconsin total. The fact that the winter-spreading component of the total manure P loading is higher in Wisconsin than in Ontario is primarily attributable to differences in channel density in the two regions. Direct lot input from hog operations is a small portion of the total in both Wisconsin and Ontario (1 and 2%, respectively).

In Tables VI and VII, total P loadings from manure sources in each of the five Wisconsin study areas are presented. Total loadings are typically increased by a factor of 3 when the critical distance is changed from 100 ft to 400 ft. Thus, the percent of total loading attributable to each area and to various sources within each area remains fairly constant under each condition.

Table VI. Total phosphorus loadings to the Great Lakes from animal wastes in the Wisconsin portion of the drainage basin (critical distance—100 ft).

Source	Lake Superior Basin		Lake Michigan Basin							
	Study Area 1		Study Area 2		Study Area 3		Study Area 4		Study Area 5	
	P delivered (kg/yr)	%	P delivered (kg/yr)	%	P delivered (kg/yr)	%	P delivered (kg/yr)	%	P delivered (kg/yr)	%
Barnyards and Feedlots	4,396	76	26,794	75	7,691	73	41,203	65	10,334	71
Dairy Cattle	2,389		20,152		4,436		31,079		6,619	
Beef Cattle	1,994		6,257		2,961		9,226		3,467	
Hogs	9		378		284		885		238	
Sheep	4		7		10		13		10	
Winter-Spread Manure	1,412	24	8,997	25	2,777	27	21,970	35	4,302	29
Totals	5,808	4	35,791	28	10,468	8	63,173	49	14,636	11

Table VII. Total phosphorus loadings to the Great Lakes from animal wastes in the Wisconsin portion of the drainage basin (critical distance—400 ft).

	Lake Superior Basin		Lake Michigan Basin						
	Study Area 1		Study Area 2		Study Area 3		Study Area 4		Study Area 5	
Source	P delivered (kg/yr)	%	P delivered (kg/yr)	%	P delivered (kg/yr)	%	P delivered (kg/yr)	%	P delivered (kg/yr)	%
Barnyards and Feedlots	11,657	66	66,201	64	18,460	61	127,353	58	33,073	64
Dairy Cattle	6,334		49,788		10,647		96,061		21,183	
Beef Cattle	5,289		15,462		7,108		28,516		11,096	
Hogs	24		934		682		2,737		761	
Sheep	10		17		23		39		33	
Winter-Spread Manure	5,751	34	36,888	36	11,418	39	87,882	42	17,818	36
Totals	17,408	4	103,089	25	29,878	7	215,235	52	50,891	12

Approximately 4% of the total manure P loading from the Wisconsin portion of the Great Lakes Drainage Basin is delivered to Lake Superior with the remaining 96% ending up in Lake Michigan. Livestock operations in East Central Wisconsin are responsible for over half of the manure P loading to Lake Michigan. The animal unit density in this area (Study Area 4) is twice as high as the mean animal unit density for all areas studied (0.51 animal unit/ha to 0.23 animal unit/ha). Barnyard and feedlot loadings are, on average, two-thirds of the total loading when a critical distance of 400 ft is used. Under the 100-ft condition, the proportion of the total loading attributable to barnyard/feedlots increases slightly. This reflects the increasing frequency of the barnyards at smaller distances from channels (Figure 3).

RECOMMENDATIONS

Several recommendations for reducing the total-P loading to the Great Lakes attributable to animal waste sources are suggested by the results of this study. Implementation strategies for animal waste nonpoint source pollution control should be directed at barnyard and feedlot areas first. Clean water diversion techniques (often referred to as the Environmental Eye) appear to offer promise for reducing nutrient and sediment outflow from these animal concentration areas.

With this management technique, clean water is diverted away from the feedlot area by upslope berms and diversion ditches. Roof water is moved away by means of properly sized and maintained gutters and downspouts. Water falling directly on the barnyard/feedlot area is treated by means of settling ponds, serpentine ditches or low-slope, low-gradient ditches. Management techniques like this were designed and installed in five Wisconsin dairy feedlots in 1977 at costs ranging from $1500 to $3500 per farm.

Problems arising from the application of animal wastes on frozen ground can be solved either through the provision of adequate waste storage facilities or through the implementation of management guidelines outlining judicious application of wastes on frozen ground. The former strategy has the potential of imposing severe economic constraints on many farmers, whereas the latter offers the possibility of achieving a significant nutrient reduction at little or no cost to the farmer.

Management strategies for winter application of wastes should include recommended setback distances from any channel. Additionally, spreading should be confined to low-slope lands, preferably less than 6%. Lands below steeper slopes should be protected by diversion moving upslope water away. If possible, wastes should be applied on corn fields where the stover has been left. Fields suitable for winter application of manure should be identified prior to the winter season. Those fields which do not meet the management criteria for winter-spreading should receive manure applications only while the ground is open.

In conclusion, it should be pointed out that the development of these estimates of P loadings from animal wastes has been extremely difficult because of a lack of sufficient data on waste management. Agricultural statistics at both state and federal levels are production oriented. Outside of the limited Domestic Animal Waste Inventory, no facts and figures on how Wisconsin farmers manage livestock wastes are currently available. It is recommended that detailed information on those aspects of agriculture that may have significant environmental impacts be collected in future surveys, most importantly, in the Census of Agriculture. Without these data, problem assessment is severely handicapped and investment strategies for remedial programs cannot be properly directed.

ACKNOWLEDGMENT

Funding for this study was provided in part by the U.S. Environmental Protection Agency (Grant # G005139-01) and by the University of Wisconsin-Madison.

REFERENCES

1. Doneth, J. "Materials Usage in U.S. Great Lakes Basin," Report for Pollution from Land Use Activities Reference Group, Ann Arbor, MI (1975).
2. Draper, D. W., J. B. Robinson, and D. R. Coote. "Estimation and Management of the Contribution by Manure by Livestock in the Ontario Great Lakes Basin to the Phosphorus Loading of the Great Lakes," *Best Management Practices for Agriculture and Silviculture*, Proc. 1978 Cornell Conference, Rochester, NY, this volume (1978).
3. Young, R. A., and R. F. Holt. "Winter-Applied Manure: Effects on Annual Runoff, Erosion, and Nutrient Movement," *J. Soil Water Conserv.* 32(5):219-222 (1977).
4. Sagher, A. "Availability of Soil Runoff Phosphorus to Algae," Ph.D. Thesis, University of Wisconsin-Madison (1976).
5. Wetzel, R. G. *Limnology.* (Philadelphia, PA: W. B. Saunders Co., 1975).
6. Conger, D. H. "Estimating Magnitude and Frequency of Floods in Wisconsin," U.S. Geological Survey open file report, Madison, WI (1971).
7. U.S. Department of Commerce, Bureau of the Census. 1974 Census of Agriculture, Vol. 1, Part 49, (Washington, DC: U.S. Government Printing Office, 1974).
8. American Society of Agricultural Engineers. Committee S & E-412 report AW-D-1, St. Joseph, MI (1973).
9. Wisconsin Domestic Animal Waste Inventory. Report on file Wisconsin Department of Natural Resources, Madison, WI (1976).

10. Converse, J. C., G. B. Bubenzer and W. H. Paulson. "Nutrient Losses in Surface Runoff from Winter-Spread Manure," Paper No. 75-2035, Am. Soc. Ag. Eng., St. Joseph, MI (1975).
11. Klausner, S. D., P. J. Zwerman and D. F. Ellis. "Nitrogen and Phosphorus Losses from Winter Disposal of Dairy Manure," *J. Environ. Qual.* 5(1):47-49 (1976).
12. Ryden, J. C., J. K. Syers and R. F. Harris. "Phosphorus in Runoff and Streams," *Agronomy* 25:1-45 (1973).
13. Strahler, A. N. "Quantitative Analysis of Watershed Geomorphology," *Am. Geophys. Union Trans.* 38:913-920 (1957).
14. Horton, R. E. "Erosional Development of Streams and Their Drainage Basins; Hydrophysical Approach to Quantitative Morphology," *Geol. Soc. Am. Bull.* 56:275-370 (1945).
15. Leopold, L. B., M. G. Wolman and J. P. Miller. *Fluvial Processes in Geomorphology* (San Francisco: W. H. Freeman and Company, 1964).
16. Wisconsin Statistical Reporting Service. Wisconsin Agricultural Statistics. Madison, WI (1977).
17. Bangay, S. E. "Livestock and Poultry Wastes in the Great Lakes Basin: Environmental Concerns and Management Issues," Social Science Series No. 15, Environment Canada, Burlington, Ontario (1976).

15

PHOSPHORUS—A POTENTIAL NONPOINT SOURCE POLLUTION PROBLEM IN THE LAND AREAS RECEIVING LONG-TERM APPLICATION OF WASTES

K. R. Reddy, R. Khaleel, M. R. Overcash, P. W. Westerman
Department of Biological and Agricultural Engineering
North Carolina State University
Raleigh, North Carolina

INTRODUCTION

Wastes of animal, municipal or agricultural origin are applied on the land with an objective of either supplying plant nutrients to the growing crop at recommended rates, or at higher rates for disposal purposes. Under both the conditions application rates are usually based on the available nitrogen (N) content of the wastes. The basic assumption is that N is the limiting plant nutrient in most regions of the U.S., with salt content being the land-limiting constituent in the arid regions. Consequently the waste loading rates are determined based on limiting constituent. Little attention has been directed to the phosphorus (P) in research and monitoring of the wastes applied to the land where long-term effects are considered. Besides crop removal, N applied through wastes is also removed from the soil system through various biochemical processes and transport mechanisms, whereas P applied to the soil (along with the N-containing waste) was assumed to accumulate in the soil through adsorption and formation of insoluble precipitates. Phosphorus in the soil water readily reacts with the available calcium, iron and aluminum in the soil to form insoluble compounds; and most soils have a capacity for immobilizing applied P. The P concentration in naturally occurring groundwater is low, typically 0.05 mg/l or less.[1] However, soils over periods of successive high application of wastes have a finite capacity for fixing the soluble P and may saturate most of the fixation sites in the soil thus increasing the P concentration of the soil solution. Since application rates of animal wastes on farm land are usually regulated according to the N requirements of crops to preclude nitrate leaching and groundwater contamination, there is no control over the P application rates, resulting in P application in excess of crop requirements. For example, the data presented by Westerman et al.[2] indicate that an annual swine lagoon effluent loading rate of 670 kg N/ha/yr results in a P loading rate of 161 kg/ha/yr. The Coastal Bermuda grass grown on this Norwalk loamy sand soil removed about 23% of applied P and 55% of applied N. This resulted in an excess P application of 117 kg/ha/yr. The average P removal capacity by some of the crops is shown in Table I. The ratio between the annual P removed from the croplands to P applied from all sources ranges from 0.21 to 1.85 for several regions of the U.S.[3] This excess application of P results in the increase of available P in the soil.[2,4-8]

Table I. Phosphorus removal by some of the selected crops.

Crop	Average Yield/ha	P Removed (kg/ha)
Corn:		
Grain	6343 liters	35
Stover	8960 kg	15
Cotton:		
Lint	1680 kg	
Seed	2520 kg	19
Wheat:		
Grain	2819 liters	20
Straw	8960 kg	4
Rice:		
Grain	7840 kg	20
	7840	6
Tobacco		
Leaf	4450 kg	6
Stalks and Tops	4032 kg	7
Johnson Grass	11 tons (metric)	94
Coastal Bermuda grass	16.9 tons (metric)	50
Tall Fescue Grass	3.1 tons (metric)	32
Soybeans:		
Grain	2114 liters	22

Continuous application of wastes beyond the capacity of the soil to sorb P will result in elevated levels of P concentrations in the surface runoff and leachate. Research reported by several workers[9-12] indicates high concentration of P in the surface runoff water from the land areas receiving animal wastes. To quantify the P transport in the surface runoff water and leaching from the land areas receiving animal wastes, it is important to develop the relationship of P in the plant-soil system used for land application of waste.

BEHAVIOR OF P IN THE SOIL WASTE SYSTEM

When wastes containing soluble P are applied to the soil, the most important phenomena that occur are the sorption, desorption or dissolution reactions. The removal of sorption of P from solution by soil is significantly related to the presence of amorphous oxides and hydrous oxides of iron and aluminum and is known to be rapid compared to the other adsorption sites. This kind of sorption is more predominant in acid soils. Under alkaline soil conditions, sorption of inorganic P by $CaCO_3$ has been established. This results in the formation of calcium phosphates, which occur in the soils above pH 8, and are related to apatites. The fate of applied P as a function of pH has been shown in Figure 1. Another form of P removal from the soil solution is the sorption

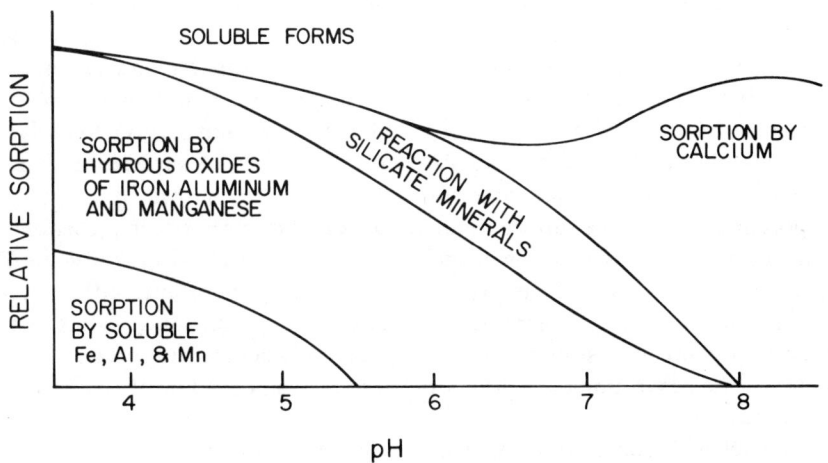

Figure 1. Fate of applied phosphorus in the soil system.

by the silicate minerals. Phosphorus in the adsorbed phase exists in equilibrium with the P solution. The capacity of the soil to sorb applied P is often described by an adsorption isotherm, which describes the relationship between the P concentration in the soil solution and the amount of P sorbed. When increasing amounts of P are added to a series of soil suspensions, the P sorbed per gram of soil can be plotted against the equilibrium P concentration. The resulting curve flattens and appears to approach a maximum at high concentration, a feature typical of chemisorption reactions. Maximum sorption capacity of the soil is generally reached upon saturation of all sorption sites by P application. This capacity varies from soil to soil and is dependent on several physicochemical properties of the soil. In recent years several adsorption models have been developed and discussion of the models are beyond the scope of this paper.[1,13,14] The most widely used adsorption isotherms are the Freundlich adsorption isotherm and the Langmuir isotherm. The Freundlich isotherm is empirical in nature and assumes that the amount of P sorbed per gram of adsorbing material would be proportional to the equilibrium P concentration, raised to some fractional exponent. The Langmuir isotherm is based on the assumption that the monolayer coverage of the sorbing surface is with phosphate ions. This equation is most often used by researchers to estimate the adsorption maximum. Adsorption of P can also be described in terms of an equilibrium P concentration (EPC), at which P is adsorbed and desorbed at equal rates and thus is in dynamic equilibrium.[15,16] An EPC value indicates the capacity of the soil to desorb or adsorb, when that soil is in contact with liquid of certain P concentration. The importance of EPC values are further discussed in the latter part of the paper. An understanding of the P adsorption-desorption process is very important in describing the quality of the surface runoff water. The total P which is generally defined in the surface waters may be divided into soluble P and suspended insoluble P. The soluble P is composed of insoluble inorganic orthophosphate (obtained by the analysis of the filtrate after 0.45-μm separation), hydrolyzable polyphosphates and organic P. Phosphorus levels of surface waters reported in the literature are generally soluble ortho P or total P. Although importance of P in the surface runoff water and streams has been recognized, little attempt has been made to differentiate between and quantify the P forms in the runoff and stream which are of potential importance with respect to their impact on the microbial productivity of the standing waters. The most important process controlling the quality of leachate and surface runoff waters is the adsorption-desorption of P from the soil or sediment transported to the stream.

The research reported in this paper is related to the behavior of P in the land application of wastes and the relationship to nonpoint source pollution from such areas. Several soil samples were obtained from the land areas receiving different types of animal waste and were evaluated for the effect of waste application

1. on the availability of P in the soil solution
2. on the adsorption-desorption characteristics of the soil waste system and the capacity of the soil to release P in the event of rainfall-runoff transport to receiving streams, and
3. on the characteristics of the P movement in the soil profile after several years of waste application.

The animal wastes examined were beef, poultry, swine, and long-term application of swine lagoon effluent (a common pretreatment for swine waste in southeastern U.S.).

CHARACTERISTICS OF ANIMAL WASTES

Some of the selected characteristics of several animal wastes used in the land application studies are presented in Table II. The P content of the wastes ranges from 0.52% in beef feedlot wastes to as high as 2.30% in swine wastes. The ratio of total P to total N content was lower for beef and dairy wastes and higher for poultry and swine wastes, indicating that at the same loading rate of N, application of swine and poultry wastes results in high P applications compared to beef and dairy wastes. Phosphorus in animal wastes is found in both organic and inorganic forms. The proportion of each of the forms is of some importance to P availability for plant uptake or for transport in the leaching and surface runoff. Paperzak et al.[17] have extensively examined P fractions in 49 samples including fresh manures from different classes of livestock, manure with and without litter, and manures of different ages. Their results indicate that approximately 27% of waste P was present in organic form and 73% in inorganic form in the fresh manure samples obtained from eight classes of livestock free of litter. The data obtained from several other sources (see Table II) indicate the ratio of ortho P to total P ranges from 0.29 to 0.70 for different animal wastes.

EXPERIMENTAL APPROACH

The soil types used include Norfolk loamy sand (typical of the Coastal Plains region), Cecil sandy loam (typical of the Piedmont regions), and Davidson clay loam. The types of animal wastes and their loading rate are shown in Table II. The details of the experiments are described by Reddy et al.[8] and Westerman et al.[2] For the field site (Norfolk loamy sand site), soil samples were also obtained at different depths up to 1.05 m, to measure the P movement and sorption characteristics at various depths. For the Cecil sandy loam site only surface 0- to 15-cm soil samples were obtained. All the soil samples were characterized for the sorption properties and equilibrium P concentration, dilute acid-extractable P ($0.05\ N$ HCl + $0.025\%\ N\ H_2SO_4$), and water-soluble P.

Five grams of soil were mixed with 100-ml volume of P solutions (made in $0.01\ M\ CaCl_2$) of known initial concentration ranging from 0 to 100 µg/ml, plus three drops of toluene, and then were equilibrated under continuous shaking for a period of 18 hr. After the equilibration the supernatant liquid was filtered through 0.45-µm filter and analyzed for P remaining in the solution on a Technicon autoanalyzer. Adsorption isotherms were constructed

Table II. Phosphorus applied in various experiments based on the N loading rates.

Type of Waste	Loading Rate Determined by N Content (kg N/ha)	P Applied Based on N Loading (kg P/ha)	P Loading Rate/N Loading Rate Ratio	Ortho P/ Total P Ratio	Remarks	Reference
Experiment I						
Beef Waste	1260	365	0.29	0.32	Laboratory study soil type - Norfolk loamy sand, 120-day decomposition period.	8
Poultry Waste	1260	374	0.30	0.29		
Swine Waste	1260	724	0.58	0.66		
Experiment II						
Poultry Waste	300	65	0.22	0.39	Laboratory study Norfolk loamy sand and Davidson clay loam, 30-day decomposition period.	29
	600	192	0.32	0.39		
	1200	301	0.25	0.39		
Experiment III						
Swine Lagoon Effluent[a]	356	81	0.23	0.70	Field study soil type Norfolk loamy sand - Coastal Bermuda grass, 5-year application.	2
	712	161	0.23	0.70		
	1425	322	0.23	0.70		
Experiment IV						
Swine Lagoon Effluent[a]	712	161	0.23	0.70	Field study soil type Cecil sandy loam - fescue grass system, 3-year application.	2
	1425	322	0.23	0.70		

[a] Loading rates shown are applied every year from April through September on Norfolk loamy sand site and from May through November on Cecil sandy loam site, each year.

and EPC values were determined as described by White and Beckett,[19] Taylor and Kunishi,[15] and Logan and McLean.[16] The EPC is defined as that concentration at which no net adsorption or desorption takes place, *i.e.*, the concentration that is supported by the solid samples when in contact with an ambient solution such that no P is either gained or lost by the solid. Acid-extractable P and water-soluble P were measured as described by Olsen and Dean.[20]

EXPERIMENTAL RESULTS

The results obtained from the soil samples of experiment I are shown in Table III. Application of animal wastes at high loading rates (disposal rates)

Table III. Effect of different types of wastes on P behavior in Norfolk loamy sand soil.

Treatment	pH	Water-Soluble P	Acid-Extractable P	P Sorbed at 1000 μg/g Added	EPC
		(μg/g of soil)			(μg/ml)
Control	6.00	2.0	43.5	140.0	0.06
Beef Waste	6.35	22.0	154.0	16.0	6.40
Poultry Waste	6.65	23.9	235.0	16.0	10.00
Swine Waste	6.45	48.9	347.0	-8.0	80.00[a]

[a]Obtained from the extropolation of the curve.

increased the water-soluble, dilute acid-extractable P, and EPC values. The increase in EPC values indicates a decrease in the adsorption capacity of the soil. The maximum N loading rate used in this study resulted in a P loading rate of 365, 374 and 724 kg/ha for beef, poultry and swine wastes, respectively. The maximum sorption capacity of the soil was approximately 315 kg P/ha. The EPC values obtained for the soil treated with swine wastes were extremely high compared to the EPC values obtained for the soils treated with beef and poultry wastes, indicating enormous capacity of this soil to release P, when it comes in contact with the liquid containing the concentration of P less than EPC values. If this situation occurs under field conditions, the P released from the surface soil will be moved into lower soil layers by the leaching water. The data on the effect of poultry waste loading rate on P sorption characteristics of the Norfolk loamy sand and Davidson clay loam are presented in Table IV. Water-soluble P, acid-extractable P, and EPC values increased with increase in loading rate of poultry waste in both soil types. The EPC values increased from 0.06 to 5.20 μg/ml, and from 0.02 to 0.35

Table IV. Effect of poultry waste loading rate on P behavior in
Norfolk loamy sand and Davidson clay loam.

Treatment	pH	Water-Soluble P	Acid-Extractable P	P Sorbed at 1000 μg/g Added	EPC (μg/ml)
		(μg/g of soil)			
Norfolk Loamy Sand					
Control	4.75	2.0	43.5	140.0	0.06
65 kg P/ha	5.00	14.5	100.0	135.0	0.40
192	6.35	28.5	180.0	145.0	1.25
301	6.90	108.0	345.0	150.0	5.20
Davidson Clay Loam					
Control	4.10	0.2	10.5	728.0	0.02
65 kg P/ha	4.60	0.4	57.0	636.0	0.03
192	5.50	4.2	148.5	616.0	0.09
267	6.00	14.3	256.5	602.0	0.35

μg/ml, in the soils treated with highest loading rates for Norfolk loamy sand and Davidson clay loam, respectively. Maximum P sorbed (measured at 1000 μg P/g of soil) was 140 to 150 and 602 to 728 μg P/g of soil, for Norfolk loamy sand and Davidson clay loam, respectively. High sorption capacity of Davidson clay loam resulted in lower EPC values. The typical adsorption isotherm as influenced by poultry waste loading rate is shown in Figure 2.

The data presented in Table V show the effect of long-term disposal of swine lagoon effluent on P sorption characteristics. The results obtained in this study were consistent with those observed in laboratory incubations. Application of swine lagoon effluent increased the water-soluble, dilute acid-extractable P, and EPC values in Norfolk loamy sand and Cecil sandy loam sites. After a 5-yr application of swine lagoon effluent on a Norfolk loamy sand site, the water-soluble P of the surface soil increased from 0.6 to 19.4 μg/g of soil, at highest loading rate, whereas dilute acid-extractable P increased from 16.6 to 152.2 μg/g of soil. The EPC values increased from 0.012 to 22.0 μg/ml at highest loading rate of P. The sorption capacity decreased from 200 μg P/g of soil to 18 μg P/g of soil, at highest loading rate. Similar increasing trends were also observed in Cecil sandy loam treated with swine lagoon effluent for a 3-yr period; however, the magnitude of increase was much lower in this soil type.

MOVEMENT OF P IN THE SOIL PROFILE OF NORFOLK LOAMY SAND

Phosphorus distribution in the Norfolk loamy sand soil profile was measured after a 5-yr application of swine lagoon effluent at loading rates shown

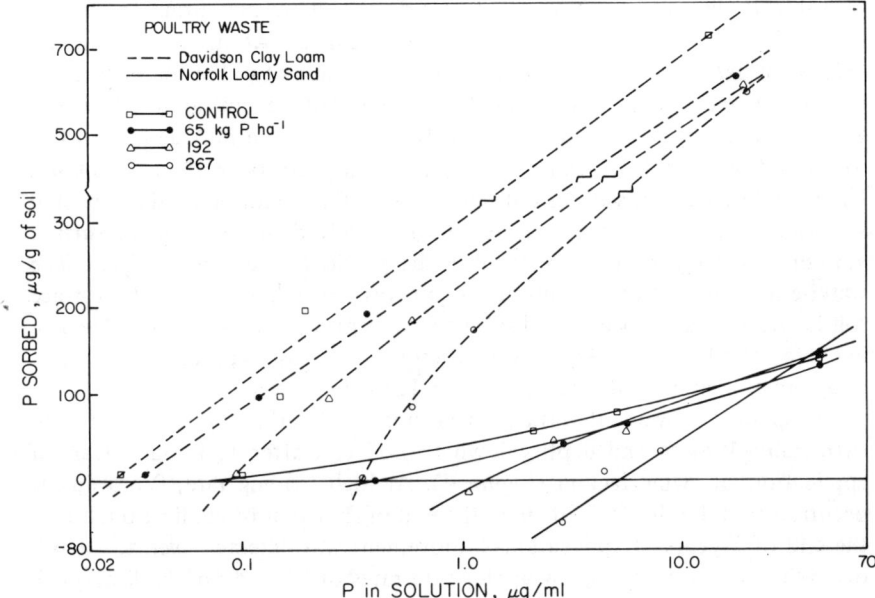

Figure 2. Adsorption isotherm of the soils treated with poultry wastes at several loading rates.

Table V. Effect of long-term application of swine lagoon effluent on P behavior in Norfolk loamy sand and Cecil sandy loam.

Treatment	pH	Water-Extractable P	Acid-Extractable P	P Sorbed at 1000 $\mu g/g$ Added	EPC ($\mu g/ml$)
		($\mu g/g$ of soil)			
Norfolk Loamy Sand[a]					
Control	4.85	0.6	16.6	200.0	0.01
81 kg P/ha/yr	4.70	12.4	91.7	57.5	1.65
161	4.60	14.4	131.4	54.5	5.30
322	4.90	19.4	152.2	18.0	22.00
Cecil Sandy Loam[b]					
Control	5.40	2.3	28.8	168.0	0.10
161 kg/ha/yr	5.35	12.3	88.0	136.0	0.78
322	5.40	19.8	130.0	104.0	2.65

[a] Swine lagoon effluent was applied on soil surface for a period of 5 yr.
[b] Swine lagoon effluent was applied on soil surface for a period of 3 yr. Surface soil samples had high organic material.

in Table II. Phosphorus movement was described by measuring the water-soluble P at different depths in the profile (Figure 3). At highest loading rate (322 kg P/ha/yr), the water-soluble P was 22.5 µg/g of soil in the surface 7.5 cm, and P movement was detected at a depth of 75 cm, at which water-soluble P content was 3.5 µg/g of soil. If we assume the bulk density of this soil as 1.5 g/cm^{3},[21] we can estimate the total porosity of the soil as 0.42. Upon a rainfall event, assuming all soil pores are filled with water, the expected soil solution concentrations will be approximately 80.4 µg/ml in the surface 7.5 cm of soil layer and 12.5 µg/ml in the profile at a depth of 75 cm. The water-extractable P of the untreated soil was 0.6 µg/g of soil in the surface soil layer, with an expected soil solution concentration of 2.1 µg/ml. At loading rates of 81 and 161 kg P/ha/yr, P movement was recorded after 5 years of application at a depth of 20-30 cm, and 30-45 cm, respectively.

Phosphorus movement was also measured by determining the dilute acid-extractable P at several depths down to 105 cm, after 1, 4 and 5 years of application of swine lagoon (Figure 4). At high loading rate, P movement occurred to a depth of 15-20 cm at the end of first year of application, but at the end of 5 years of application, P movement was detected even at a depth of 75-90 cm. Similarly at low application rates of 81 and 161 kg P/ha/yr, P movement was detected at depth of 30 and 45 cm, respectively, after 5 years of application.

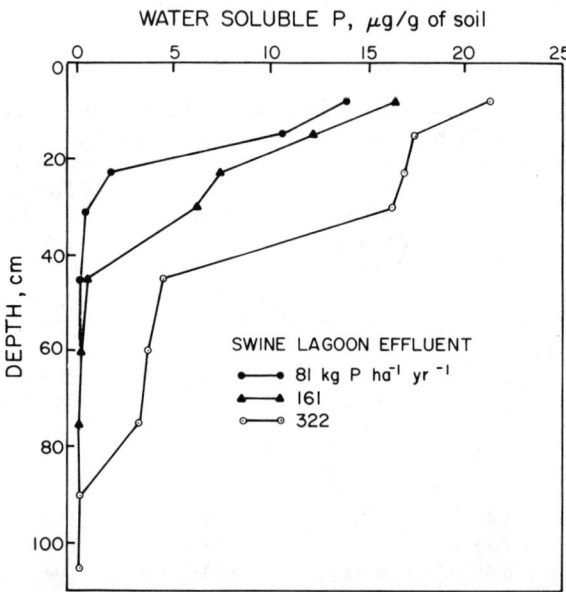

Figure 3. Distribution of water-soluble P, in a Norfolk loamy sand soil profile, after a 5-yr application of swine lagoon effluent.

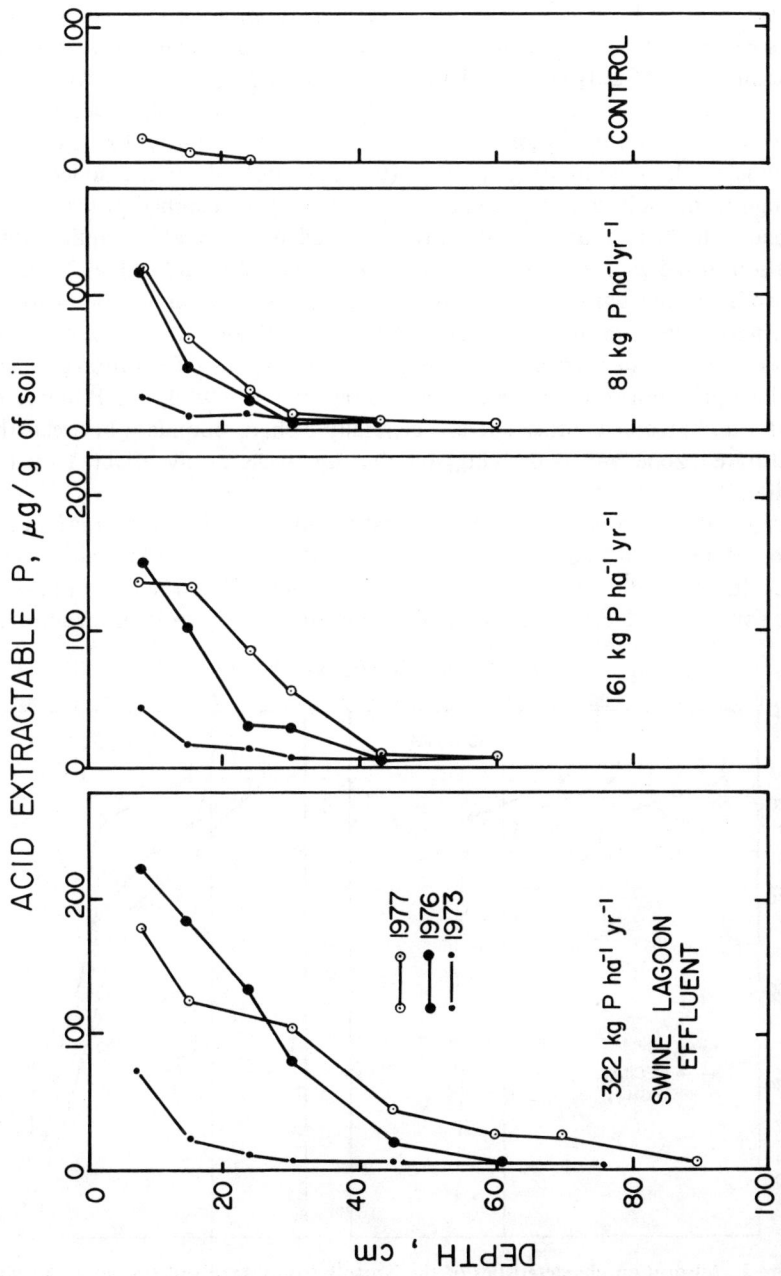

Figure 4. Distribution of dilute acid-extractable P in a Norfolk loamy sand soil profile, after 1-, 4- and 5-yr application of swine lagoon effluent.

To understand the mechanisms involved in the P movement deeper into the soil profile, P sorption capacity of the soil was measured at each depth in the soils treated with swine lagoon effluent, by equilibrating for a period of 18 hr in 0.01 M $CaCl_2$ containing 0.5, 5, 10 and 60 µg/ml, which corresponds to 10, 100, 200 and 1000 µg P/g of soil. The P sorbed or desorbed at each depth is presented in Figures 5 and 6. For the surface soils at lower concentrations in the equilibrating solution (0.5 µg P/ml), all of the added P remained in the solution, plus some of the P from the adsorbed phase was desorbed into the solution. Adsorption of added P from the equilibrating solution of 0.5 µg P/ml was observed at depths of >20, >30 and >42 cm, for the soils treated with low, medium and high rates of swine lagoon effluent. At a maximum rate of application (1000 µg P/g of soil), only 2-5% of added P was sorbed in the surface layers and the percent of added P sorption capacity of the upper portions of the soil profile become saturated, and P moves to greater and greater depths. There is generally a sharp boundary between the P-saturated zone and underlying soil that has been barely affected by the applied P.

From the adsorption isotherms constructed, at each depth, EPC values were estimated at each depth in all the treatments. The EPC values were much higher in the surface soil layers and decreased with depth, indicating that sorption of P can be expected only at extremely higher levels of solution

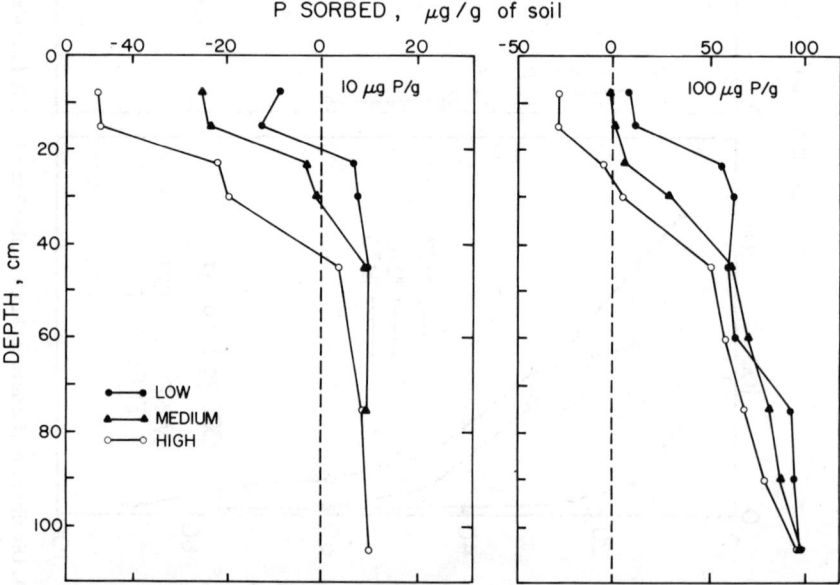

Figure 5. Adsorption characteristics of the Norfolk loamy sand soil treated with swine lagoon effluent for 5 years.

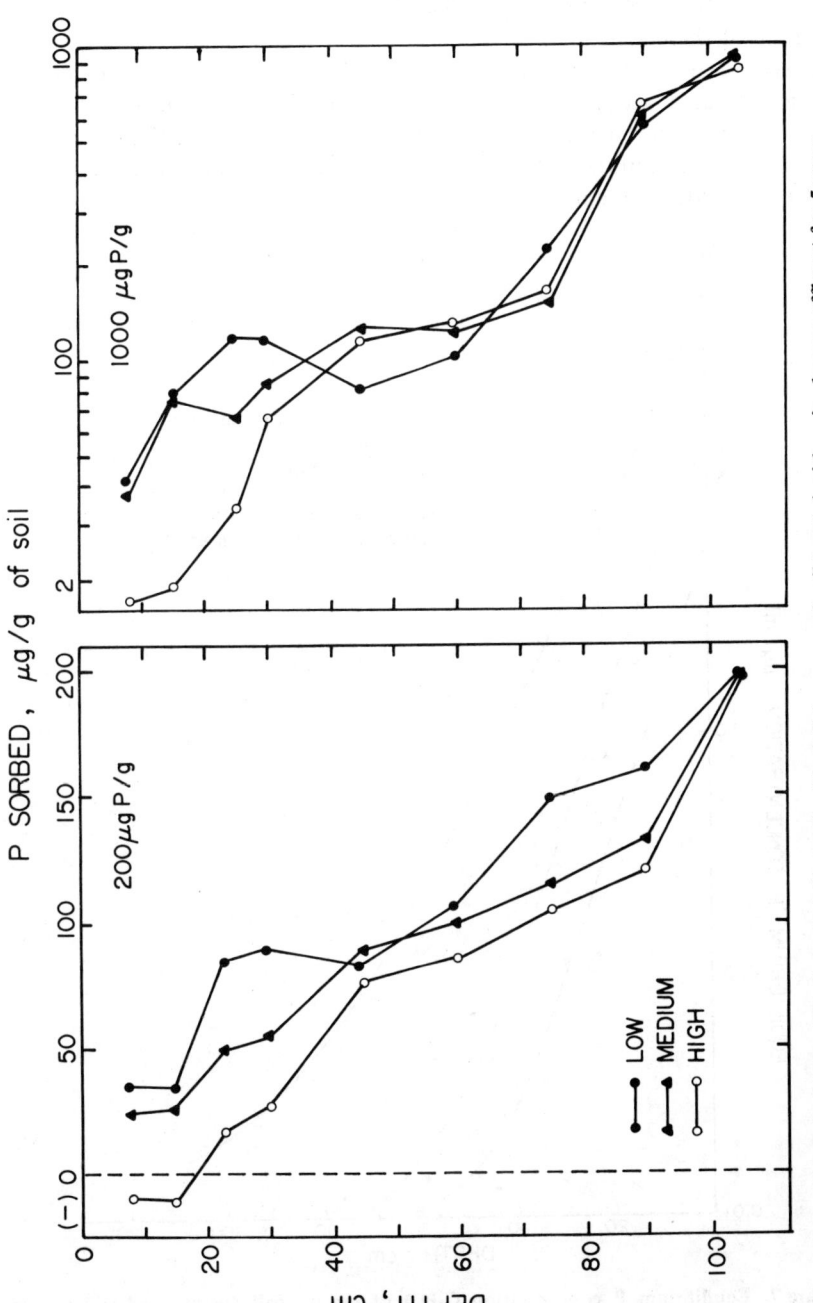

Figure 6. Adsorption characteristics of the Norfolk loamy sand soil treated with swine lagoon effluent for 5 years.

P in the surface soil layers (Figure 7). The high EPC values also indicate the capacity of these soils for releasing P in the event of rainfall. The EPC values decreased consistently from 22.0 to 0.02 µg/ml, up to a depth of 105 cm, in the soils treated with high rate of swine lagoon effluent application, and did not reach a stabilized value. At lower rates of swine lagoon effluent application EPC values approached control values at a depth of 60 and 75 cm for 81 and 161 kg P/ha/yr loading rate, respectively.

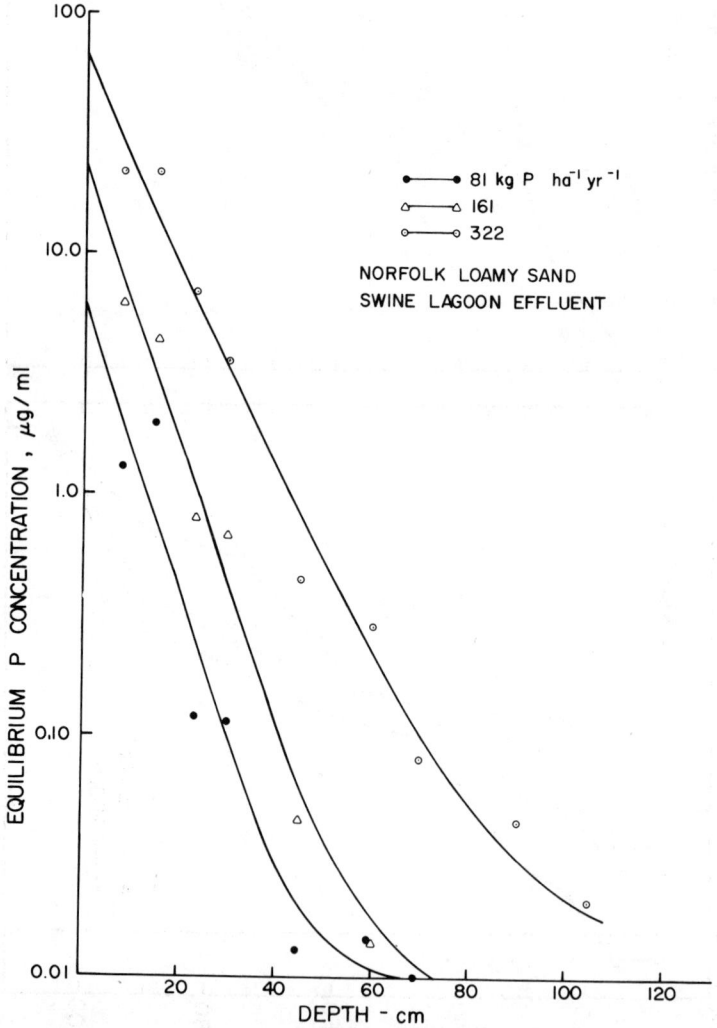

Figure 7. Equilibrium P concentration (EPC) of the Norfolk loamy sand soil profile treated with swine lagoon effluent for a period of 5 years.

A significant relationship was observed between EPC values and P extracted with dilute acid and water. The data used in calculating these relationships were obtained from all of the experiments reported in the early part of the paper. This relationship was best described by a power function as shown below:

$$\text{EPC} = 0.0014 \, (\text{AEP})^{1.58} \qquad R^2 = 0.75** \qquad (1)$$
$$\text{EPC} = 0.11 \, (\text{WSP})^{1.21} \qquad R^2 = 0.82** \qquad (2)$$

where EPC = equilibrium P concentration, μg/ml
AEP = acid-extractable (0.05 N H Cl + 0.025 N H$_2$SO$_4$), phosphorus, μg/g of soil (soil to solution ratio 1:10)
WSP = water soluble P, μg/g of soil (soil to solution ratio 1:10)
** = significant at 0.01 level of probability.

In all of the experiments reported, waste application decreased the P sorption capacity of the soil. This was due to the application of high rates of soluble inorganic P and due to mineralization of organic P during decomposition of waste, which probably saturated most of the adsorption sites. Another probable reason could be that, during decomposition of wastes, several organic acids are produced which form stable complexes with Fe and Al and consequently block P retention by them.[22-24] Thus when phosphates and an organic anion are present together, the decrease in P adsorption by an adsorbent must arise from the specific adsorption of the organic anion resulting from competition between phosphate and the organic anions for adsorption sites. The capability of organic anions decreasing P sorption would be determined by the relative stabilities of the Fe (or Al) organic anion complex and Fe (or Al) phosphate complex. Decrease in sorption capacity of the soil and increase in EPC values have great significance in relation to the transport of P in leaching or surface runoff water. All these soils showed a greater tendency to desorb high amounts of P into the water leaching through the profile or to the surface water. This explains the movement of soluble P in the profile at Norfolk loamy sand site, shown in Figure 4, after 5 years of application of swine lagoon effluent. These soils with high EPC values have greater potential to pollute surface waters because, if these soils are eroded as sediments into the receiving stream, they continuously desorb or release P until equilibrium concentration is attained.

The data presented in Table VI show the amount of P accumulated at the end of each year of application of swine lagoon effluent in the Norfolk loamy sand soil. After accounting for the crop removal and the losses in the runoff, the amount of P accumulated in the system at the end of each year was 61, 73 and 81% of applied P at an application rate of 81, 161 and 322 kg/ha/yr, respectively. From the adsorption isotherm measured on the untreated Norfolk loamy sand soil at each depth, the P sorption maximum was estimated

Table VI. Mass balance of the P applied at Norfolk loamy sand site, for a period of 5 yr.

Treatment	Total P Applied	Crop Removal	Estimated[a] Loss in Runoff	Net Accumulation in the Soil	Total P Accumulated in 5 yr (kg/ha)
	(kg P/ha/yr)				
Site I					
Swine Lagoon Effluent	81	29	2.9	49.2	246
	161	41	3.0	117.0	585
(Conc. 55 mg/l)	322	50	11.0	261.0	1305

[a]Estimated from the data presented by Overcash et al.[11]

using the Langmuir equation.[25] The maximum sorption capacity of the soil at 0-15 cm and 0-105 cm soil profile, was 488 and 6145 kg P/ha. The number of years required to reach the maximum sorption capacity of the soil at any given loading rate of waste can be estimated as

$$TS = S\,max/Pg \qquad (3)$$

where S max = maximum adsorption capacity of the soil layer at any given depth, kg/ha
Pg = phosphorus gained by the soil after each year of application of waste, kg/ha/yr, *i.e.*, applied P minus crop uptake, and surface runoff losses
TS = time required to saturate the soil layer with P, years

The estimated time required to saturate the soil profile from 0-15 cm and 0-105 cm is shown in Table VII. At high application rates of P (161 and 322 kg/ha/yr) each year resulted in excess of P sorption capacity of the soil, thus saturating the system, whereas at low rates of application (81 kg/ha/yr) approximately 5 more years of application of swine lagoon effluent was necessary to saturate the system. However, to saturate the soil profile to a depth of 105 cm it takes about 24, 53 and 125 years of application of waste at the loading rate shown above. Estimates of maximum adsorption based on isotherm almost invariably underestimate potential P immobilization observed under field conditions, particularly if the waste application results in high concentration of P in the soil solution.[26] Some researchers have found that the sorption capacity of a P-saturated soil can be rejuvenated with time, presumably because some of the surface-sorbed P forms a crystalline precipitate over a period of time, exposing a fresh sorbing surface.[27,28] Phosphorus sorption potential of a soil depends on the content of potential P-sorbing surfaces.

Table VII. Showing the number of years required to saturate the soil profile with P.

Depth (cm)	Net Accumulation of P in the System at the End of Each Year (kg/ha/yr)	Maximum[a] Adsorption Capacity of the Soil (kg/ha)	Number of Years Required to Saturate the Soil Profile (yr)	P Gained by the Soil After 5-yr Application (kg/ha)
0-15	49.2	488	9.9	246
	117.0	488	4.2	585
	261.0	488	1.9	1305
0-105	49.2	6145	124.9	246
	117.0	6145	52.5	585
	261.0	6145	23.5	1305

[a]Estimated from adsorption isotherm, using the Langmuir equation.

SUMMARY AND CONCLUSIONS

Wastes of animal, municipal or agricultural origin are applied on the land with an objective of either supplying plant nutrients to the growing crop at recommended rates or at higher rates for disposal purposes. The application rates are generally based on the available N content, thus resulting in higher loading rates of P than plant utilization. Phosphorus applied to the land was assumed to accumulate in the soil through adsorption and formation of insoluble precipitates. Recent concern about P enrichments in surface runoff water from the land areas receiving wastes raised the question of what effect wastes might have on the behavior of P in the soil-waste system. To evaluate this problem, surface soil samples were obtained from the soil areas receiving wastes to determine the P behavior in the system. Application of animal wastes increased the water-soluble and dilute acid-extractable P and decreased the sorption capacity of the soils. This resulted in increased levels of EPC in the soil solution. Significant relationships were obtained between EPC of the system and water-soluble and acid-extractable P, expressed by some power function. Continuous application of waste (swine lagoon effluent) to a loamy sand beyond the sorption capacity of the soil, resulted in high concentrations of P in the deeper soil layers and surface runoff waters. Upon using the land, serious consideration should be given to the P assimilatory capacity of the soil system without causing high levels of P contribution to the surface waters.

ACKNOWLEDGMENTS

This is a contribution of the Department of Biological and Agricultural Engineering, and North Carolina Agricultural Experiment Station, Raleigh,

N.C. The research was supported in part by the United States Environmental Protection Agency on Grant No. R-805011-01-0. The authors wish to thank the Environmental Protection Agency for funding the research, and wish to express their appreciation to Ms. June Preston and Ms. Dottie deBruyne, for their assistance during chemical analysis.

REFERENCES

1. Enfield, C. G., C. C. Harlin, Jr., and B. E. Bledsoe. "Comparison of Five Kinetic Models for Orthophosphate Reactions in Mineral Soils," *Soil Sci. Soc. Am. J.* 40:243-249 (1976).
2. Westerman, P. W., M. R. Overcash, J. C. Burns, L. D. King and F. J. Humenik. "Long-Term Fescue and Coastal Bermudagrass Crop Response to Swine Lagoon Effluent," paper presented at annual meeting, Amer. Soc. Agr. Eng. June 26-29, 1977, North Carolina State University (1977).
3. Walsh, L. M., M. E. Sumner and R. B. Corey. "Consideration of Soils for Accepting Plant Nutrients and Potentially Toxic Nonessential Elements," in *Land Application of Wastes,* Soil Conservation Society of America (1976).
4. Warren, R. G., and A. E. Johnston. "Hoosfield Continuous Barley," Rep. Rothamsted Exp. Stn. for 1966. Harpenden, Herts., England (1967), pp. 320-338.
5. Olsen, R. J., R. F. Hensler and O. J. Attoe. "Effect of Manure Application, Aeration, and Soil pH on Soil Nitrogen Transformation and on Certain Soil Test Values," *Soil Sci. Soc. Am. Proc.* 34:222-225 (1970).
6. Randall, G. W., R. H. Anderson and P. R. Goodrich. "Soil Properties and Future Crop Production as Affected by Maximum Rates of Dairy Manure," in *Managing Livestock Wastes,* The Proc 3rd International Symposium on Livestock Wastes, Amer. Soc. Agr. Eng., St. Joseph, MI (1975).
7. Manges, H. L., L. S. Murphy, W. L. Powers and L. A. Schmid. "Ultimate Disposal of Beef Feedlot Wastes onto Land," Report to EPA Grant No. R803210-01.
8. Reddy, K. R., R. Khaleel, M. R. Overcash and P. W. Westerman. "Evaluation of Nitrogen and Phosphorus Transformations in the Soil-Manure System in Relation to Nonpoint Source Pollution," *Agron. Abstracts.* (1977), p. 35.
9. Minshall, N. E., S. A. Witzel and M. S. Nichols. "Stream Enrichment from Farm Operations," *J. San. Eng. Div., Proc. Am. Soc. Civil Eng.* 96:513-24 (1970).
10. Doss, B. D., Z. F. Lund, F. L. Long and L. Mugwira. "Dairy Cattle Waste Management: Its Effect on Forage Production and Runoff Water Quality," Bull. 485. Agri. Expt. Sta. Auburn Univ., Auburn, AL (1976) p. 39.
11. Overcash, M. R. *et al.* Unpublished results (1976).
12. Young, R. A., and C. K. Mutchler. "Pollution Potential of Manure Spread on Frozen Ground," *J. Environ. Qual.* 5(2):174-179 (1976).

13. Laidler, K. J. *Chemical Kinetics,* 2nd ed. (New York: McGraw-Hill Book Co., 1965).
14. Oddson, J. K., L. Letey and L. V. Weeks. "Predicted Distribution of Organic Chemicals in Solution and Adsorbed as a Function of Position and Time for Various Chemicals and Soil Properties," *Soil Sci. Soc. Am. Proc.* 34:412-417 (1970).
15. Taylor, A. W., and H. M. Kunishi. "Phosphate Equilibria on Stream Sediment and Soil in a Watershed Draining an Agricultural Region," *J. Agri. Food Chem.* 19:827-831 (1971).
16. Logan, T. J., and E. O. McLean. "Nature of Phosphorus Retention and Adsorption with Depth in Soil Columns," *Soil Sci. Soc. Am. Proc.* 37: 351-355 (1973).
17. Paperzak, P., A. G. Caldwell, R. R. Hunziker and C. A. Black. "Phosphorus Fractions in Manures," *Soil Sci.* 87:293-302 (1959).
18. Reddy, K. R., R. Khaleel, M. R. Overcash and P. W. Westerman. "Phosphorus Adsorption-Desorption Characteristics of the Soil Utilized for Disposal of Animal Wastes," paper to be presented at 11th ISSS Congress, Edmonton, Canada (1978).
19. White, R. W., and P. H. T. Beckett. "Studies on the Phosphate Potentials of Soils, 1. The Measurement of Phosphate Potential," *Plant Soil.* 20:1-16 (1964).
20. Olsen, S. R., and L. A. Dean. "Phosphorus," in *Methods of Soil Analysis. Part 2,* C. A. Black, Ed. (Madison, WI: Amer. Soc. Agron., 1965), pp. 1035-1048.
21. Lutz, J. F. "Movement and Storage of Water in North Carolina Soils," Report WRRI, North Carolina State Univ., Raleigh, NC (1970).
22. Gaur, A. C. "Studies on the Availability of Phosphate in Soil as Influenced by Humic Acid," *Agrochimica.* 14:62-65 (1969).
23. Nagarajah, S., A. M. Posner and J. P. Quirk. "Competitive Adsorption of Phosphate with Polygalacturonic and Other Organic Anions on Kaolinite and Oxides Surfaces," *Nature* 228:83-85 (1970).
24. Singh, B. B., and J. P. Jones. "Phosphorous Sorption and Desorption Characteristics of Soil as Affected by Organic Residues," *Soil Sci. Soc. Am. J.* 40:389-394 (1976).
25. Olsen, S. R., and F. S. Watanabe. "A Method to Determine a Phosphorus Adsorption Maximum of Soils as Measured by the Langmuir Isotherm," *Soil Sci. Soc. Am. Proc.* 21:144-149 (1957).
26. Adriano, D. C., L. T. Novak, A. E. Erickson, A. R. Wolcott and B. G. Ellis. "Effect of Long-Term Disposal by Spray Irrigation of Food Processing Wastes on Some Chemical Properties of the Soil and Subsurface Water," *J. Environ. Qual.* 4:242-248 (1975).
27. Beek, J., and F. A. M. deHaan. "Phosphate Removal by Soil in Relation to Waste Disposal," Proc. International Conf. on Land for Waste Management, Ottawa Canada (1974), pp. 77-86.
28. Sawhney, B. L., and D. E. Hill. "Phosphate Sorption Characteristics of Soils Treated with Domestic Water," *J. Environ. Qual.* 4:342-346 (1975).
29. Crane, S. R. "A Laboratory Investigation of the Short-Term Chemical and Microbial Transformations in the Soil Following Surface Land Application of Poultry Manure," M.S. Thesis, North Carolina State University (1978).

16

NUTRIENT AND PESTICIDE MOVEMENT FROM FIELD TO STREAM: A FIELD STUDY

J. L. Baker, H. P. Johnson
   Department of Agricultural Engineering
   Iowa State University
   Ames, Iowa

M. A. Borcherding
   Department of Agronomy
   Iowa State University
   Ames, Iowa

W. R. Payne
   Environmental Research Laboratory
   U.S. Environmental Protection Agency
   Athens, Georgia

INTRODUCTION

The Federal Water Pollution Control Act Amendments of 1972 (PL 92-500) specified that the Administrator of the U.S. Environmental Protection Agency shall, in cooperation with state and other federal agencies, provide guidelines for identifying and evaluating the nature and extent of nonpoint sources of pollution. Section 208 of that law specifies that agriculturally related sources be identified and that procedures and methods be developed to control, to the extent feasible, such sources. The development of a better understanding of the chemistry and transport processes related to sediment, nutrient and pesticide movement is needed to make possible sound recommendations for control of agricultural pollutant sources. The present understanding of field-to-stream relationships related to sediment transport, division of sources of nutrient contribution, and pesticide transport is particularly deficient.

Objectives

The general objective of this field study is to provide a better understanding of the relationships between management of farm systems and stream water quality and to enable analysis of such systems using mathematical models. Specific objectives are

1. to collect data to refine EPA's Agricultural Runoff Model;[1]
2. to relate runoff and soil-erosion nutrient (nitrogen and phosphorus) losses to hydrologic factors as altered by land management;
3. to relate losses of pesticides in runoff to pesticide properties and to hydrologic and management factors; and
4. to relate sedimentation to hydrologic and management factors.

In this paper, the study is described and initial results are presented.

Background

The 3-yr study involves cooperation among the Agricultural Engineering and Agronomy departments of Iowa State University; Hydrocomp Incorporated, Palo Alto, California; and the EPA Environmental Research Laboratory, Athens, Georgia.

The field site for the study is a heavily row-cropped region of Tama County, which is in east-central Iowa in the Four Mile Creek Basin. The topographic features and drainage patterns are related to Pleistocene glaciation and recent modification. Glacial deposits mantle the bedrock to a maximum thickness of 110 m. Nearly all the till is covered with loess, as much as 11 m deep on the divides but thinning on the valley sides. Six silt loam soils account for 93% of the soil map units.

About 75% of the watershed is planted to corn and soybeans; 25% is in small grain, meadow and pasture. About 166 kg/ha of nitrogen and 25 kg/ha of phosphorus were applied to corn in 1977. The respective amounts for soybeans were 7 and 24 kg/ha. Most of the land is conventionally tilled for corn, *i.e.*, plow, disk and plant. Soybean land on which corn will be grown is often disked instead of plowed. Herbicides commonly used are Aatrex, Lasso, Sutan, Sencor and Treflan. Insecticides commonly used are Counter, Dyfonate, Furadan and Thimet.*

Tile drains are installed along waterways. In the more level upper reaches of watersheds and in floodplains, field tile systems may be installed. About

---

*Product names in this paper are included for the benefit of the reader and does not imply endorsement or preferential treatment of the product by Iowa State Univesity or the U.S. Environmental Protection Agency.

1% of the land is terraced. A few farmers use conservation tillage, contouring and strip cropping.

Average annual rainfall in the region is about 820 mm; average annual water yield is about 150 mm. Annual sediment yields are about 150 MT/km^2 for 50-km^2 watersheds. Natural drainage is well developed except in the upper reaches of watersheds.

Previous Studies

Hydrologic models provide a means for quantitatively defining the primary transport medium related to nonpoint pollution, *i.e.*, water. Other models are integrated with the transport model to provide prediction tools. Models related to sediment, pesticide and nutrient transport are in various stages of development and testing. One of the first sediment transport models integrated with a hydrologic model was presented by Negev.[2] His model incorporated soil splash erosion, overland flow entrainment, rilling, and channel transport. David and Beer[3] presented a similar model applied to Four Mile Creek. Onstad and Foster[4] described a model that incorporated detachment, rill-interrill contributions and transport. Onstad *et al.*[5] presented a comparison of three models developed for prediction of erosion and sediment transport. Several other sedimentation models have been developed.

Modeling chemical transport is a relatively new technique, the development of which awaited models for both hydrology and sedimentation. Crawford and Donigian[6] described a model for pesticide transport in runoff. The model simulates the loss of pesticides from agricultural lands and has components related to hydrology, sediment loss, pesticide-soil interaction, and pesticide attenuation factors. Bailey *et al.*[7] presented a review and analysis of chemical transport model components. Recent efforts by Donigian and Crawford[8] and Donigian *et al.*[1] include a nutrient component. Frere *et al.*[9] described ACTMO, an agricultural chemical transport model. Subsystems in ACTMO include water and sediment transport, degradation, volatilization, absorption, leaching, mineralization, and uptake processes. Considerable effort is presently being devoted to the development of such models. The need for good field data of chemical transport from field to stream is evident.

One reason the specific study site was selected was that some previous information was available. The U.S. Geological Survey (USGS) has gaged Four Mile Creek from 1962 to 1974. Kunkle[10] reported a groundwater hydrology study. Ruhe* and Beer† completed other hydrology and water quality studies.

---

*Unpublished report on surface water and groundwater hydrology of Four Mile Creek by R. Ruhe, Department of Agronomy, Iowa State University, Ames.
†Personal communication, C. E. Beer, Iowa State University, Ames.

PROCEDURE

Site Instrumentation and Observations

Rainfall and flow-measuring equipment are located at selected sites in the largest watershed. Recording rain gages are used to measure at five sites within the 50-km^2 area above site 4 (see Figure 1). A sixth rain gage is located nearby at the weather station on the farm cooperator's farm. Weather data taken during the period of roughly April through October, in addition to rainfall, include daily pan evaporation, daily total wind, max-min temperatures, relative humidity, and solar radiation. During the winter months, in addition to snowfall, max-min temperatures, soil temperatures (at 2.5, 7.5, 15, 30, 60 and 90 cm), relative humidity, and solar radiation data are taken. Flow is measured at eight sites as shown in Table I. Flow occurs during most of the year at sites 4, 5 and 6. Flow at the other stations is ephemeral. Flow at sites 1, 2 and 3 is measured by the use of 4-ft H flumes. Flow measured at sites 4, 5 and 6 rely on metered calibration.

Table I. Description of flow-measuring stations.

Site No.	Area (ha)	Device	Cover
1	5.6	H Flume	Corn or beans
2	7.6	H Flume	Corn or beans
3	6.0	H Flume	Grass (pastured)
4	5055	USGS Station	Mixed Cover
5	3575	USGS Station	Mixed Cover
6	345	USGS Station	Mixed Cover
7	284	Box Culvert	Mixed Cover
8	149	Headwall and Culvert	Mixed Cover

Sites 7 and 8 are useful for measuring storm runoff and rely on culvert hydraulics for stage-discharge relationships. Tile outlets for tile draining waterways in farmed land are identified in Figure 1. Discharges from the tile outlets are not measured, but water quality samples are taken from the flow.

During the winter months, an attempt is made at two locations to measure snowfall by use of large open-cylinder gages protected by windshields. One gage is located between sites 1 and 2, the second, at the weather station. Soil moisture is measured to a depth of 150 cm at eight locations (five at site 3) within the watershed above sites 1, 2 and 3 at three times during the year when cores are taken for other analyses.

Watersheds 1, 2 and 3 represent single vegetative cover conditions and a given crop management practice. Watersheds 1 and 2 are planted to corn and

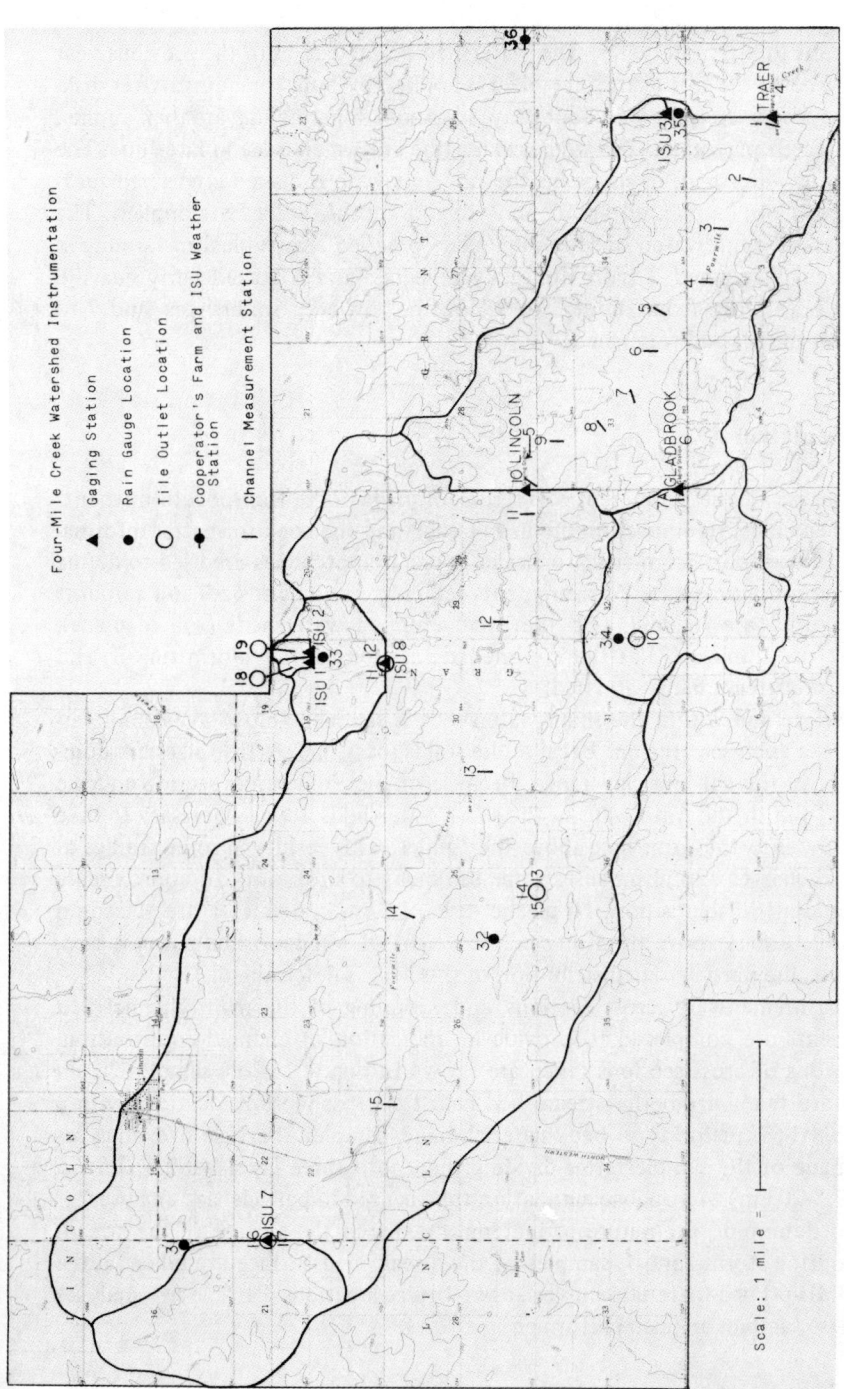

Figure 1. Locations of watersheds, sampling sites, and rain gages in study area.

soybeans in alternate years. The watershed planted to corn the previous year is plowed in the spring after fertilizer is applied by bulk spreading. After disking, soybeans are planted about May 10, and the herbicides are then applied by research personnel. The soybeans may be cultivated once in late June. The alternate watershed, in soybeans the previous year, is disked after fertilizer is applied. Corn is planted about May 1, and the herbicides are applied. The corn may be cultivated once in June. The practices are typical of farming in the area. Watershed 3 is in Kentucky bluegrass and is grazed fairly heavily. The slopes range from about 2 to 9% on the row-crop watersheds, and 7 to 18% on the grassed watershed.

Sedimentation

Because of the prominent role sediment plays in the transport of nutrients and pesticides, its impact on life in a stream, and the need for better information on transport from field to stream, several approaches are used to define the transport processes. Measurements related to sediment sizes and amounts transported are made (1) in the small cropped watersheds (soil in place), (2) at the flumes, (3) at the intermediate stations during storm runoff, and (4) in the stream banks and bed.

Surface soil within the small watersheds is sampled (plow layer), and particle size analyses are run. Because the soil is loess, the particle sizes are dominantly in the silt and clay range. Because of the broad cross section and the lack of fall in the waterways of the small watersheds, it was necessary to dike the flow into the drop box above the flumes. This results in some storage at high discharges and also causes some sediment to drop out. To approximate the amount of deposition, 10 plastic disks, 15 cm in diameter, are placed on the soil surface above the flumes. The weight of soil deposited per unit area, and also the particle size distribution of this soil, is determined.

Measurements of cross sections and sampling of the main channel bed sediments are completed to provide an indication of channel contribution. Locations of cross sections taken are shown in Figure 1. Core samples 20 cm deep are taken from the stream bed near ten cross section locations with a US BMH-53 piston-type bed material hand sampler. Particle-size analyses are made of the samples after dividing them into three layers (0-2.5, 2.5-7.5 and 7.5-20 cm). A visual-accumulation tube is used in particle size analyses.

To determine the nature of material moving at the surface of the stream bed during storm runoff, samples of the stream bed surface are taken with a US BMH-60 bed-material sampler at two bridge locations. Particle size analyses of these samples are also performed.

Surface Runoff, Creek Flow, Subsurface Flow, and Precipitation Sampling

When runoff occurs at small watershed sites 1, 2 and 3, samples are taken by three different methods. At each site an automatic pumping sampler (PS-69 interagency sampler) has been installed capable of taking 46 samples (4-liter samples at sites 1 and 2 for sediment, pesticide and nutrient analyses; 1-liter samples at site 3 for nutrient and sediment analyses) at 2-minute intervals when depth of flow exceeds 15 cm in the flume (0.04 m^3/sec). The intake for the pumping sampler is a stainless steel well-point located near the bottom of the approach box for the flume. There is a 25-cm head-wall on the approach box to facilitate mixing. The samples are pumped to refrigerated containers (glass for pesticide samples) to maintain sample integrity. In addition, ports on the flume walls for five single-stage samples at depth of 7.5, 15, 30, 45 and 60 cm are installed to provide samples when the automatic pumping samplers fail. Finally, an effort is made to have project personnel present during runoff to take grab samples at low flows and make sure the automatic samplers function.

When runoff occurs at intrabasin sites 7 and 8, samples are taken at 10- to 30-min intervals by hand, at low flows by passing an open narrow-necked bottle back and forth through the flow and at higher flows by using a hand-held, depth-integrating sampler (US D-48). At creek sites 5 and 6, where flow is nearly continuous, bottle grab samples are taken weekly. At high flows, during surface runoff, additional samples are taken (5 to 10, with the hand-held, depth-integrating sampler). At the final creek site 4, grab samples are taken on a daily basis (every other day during extremely low flows). At high flows an additional five to ten samples are also taken with a winch-operated, depth-integrating sampler (US D-74).

Subsurface drainage water, intercepted by tile drains, is sampled approximately once a week when flow occurs. Of the ten tile drains sampled, eight have open outlets where flow is grab-sampled. Flow from the remaining two, which drain the waterways of sites 1 and 2, is sampled by use of a hand vacuum pump connected by a tube to the tile drain.

Precipitation for the months April through November is measured to provide both intensity and amount data. During this period, precipitation is sampled for nutrients at site 2 with a collector that opens automatically during rainstorms. Samples of snowfall are taken at the weather station for nutrient analyses.

All water samples taken are analyzed for NH_4-N, NO_3-N, PO_4-P and Cl using the Technicon Auto-Analyzer II system. NH_4-N is analyzed by the alkaline phenol method, NO_3-N by the cadmium reduction method, PO_4-P by the ascorbic acid reduction method, and Cl by the ferricyanide method. Sediment in selected samples (with at least 10,000 ppm sediment) is analyzed for K_2SO_4-extractable NH_4-N and acid/fluoride-extractable available P (Bray-1).

Sediment plus dissolved solids concentrations are determined by drying a 30-ml portion of sample. Dissolved solids concentrations are determined by drying a 30-ml portion of sample that had previously been centrifuged for 10 min at 42,000 G. Particle size analysis of the suspended sediment in runoff samples is made with a Micromeritic Sedigraph 5000 particle size analyzer. This instrument uses X-ray attenuation to measure sediment concentration in a narrow horizontal band of the sedimentation cell and is used to determine the size distribution from about 75 to 0.5 μ. The samples are run in their native runoff water (concentrated to a 6% soil suspension, if naturally lower) with no dispersive agent added.

Portions of water and sediment samples from sites 1, 2 and 4 are taken from about April through September and air-shipped, cold in insulated containers, to the EPA Environmental Research Laboratory at Athens, Georgia, for herbicide analyses.

These water and sediment samples are analyzed for alachlor, cyanazine, metribuzin, paraquat and propachlor by the method of Payne et al.[11] This method was modified for these compounds by (1) adjusting soil-core moisture to 15 to 20% to improve extraction efficiency of atrazine and (2) using Hall model 700 electrolytic conductivity detectors in the nitrogen and halogen modes instead of the Coulson systems. Recoveries of fortified samples at concentrations of 0.5 to 50.0 ppb (water) and 0.05 to 5.0 ppm (sediment or soil) were in the range of 90.0 to 98.7% (alachlor), 93.0 to 100.0% (cyanazine), 87.6 to 102.6% (metribuzin), 91.0 to 99.0% (paraquat, in sediment or soil), and 91.0 to 100.0% (propachlor). Because of the high extraction efficiencies, no corrections for extraction efficiencies are made. Data on atrazine measurements, made for creek samples only, are not included in this paper.

Chemical Application

Granular N, P and K fertilizers at the rates of 135 kg/ha of N, 39 kg/ha P, and 37 kg/ha K are custom-applied and disked down on the corn ground; 34 kg/ha P and 56 kg/ha K are custom-applied and disked down or plowed under on the soybean ground (sites 1 and 2). Starting in 1977, granular fertilizers at the rates of 78 kg/ha N, 20 kg/ha P, and 19 kg/ha K are custom-applied to the pastureland without incorporation.

Immediately after planting, herbicides are broadcast sprayed without incorporation to the cropped watersheds by project personnel. Alachlor, metribuzin and paraquat are applied to the soybean ground at rates of 2.24, 0.56 and 1.12 kg/ha, respectively. Propachlor, cyanazine and paraquat are applied to the corn ground at rates of 2.24, 2.24 and 1.12 kg/ha, respectively. Table II lists the water solubilities and vapor pressures of the herbicides studied. To quantify the amount sprayed, the spray nozzles are calibrated, ground speed is recorded, and samples of the spray mixture are taken for herbicide analyses. In addition, sets of doubled, 12.5-cm filter paper disks are placed on the soil surface, passed over by the sprayer, and then quickly sealed in glass flasks for later analyses.

Table II. Names, water solubilities, and vapor pressures of herbicides studied.[a]

Herbicide Name		Water Solubility (mg/l)	Vapor Pressure (mm Hg at 25°C)
Common	Trade		
Alachlor	Lasso	242 at 25°C	2.2×10^{-5}
Cyanazine	Bladex	171 at 20°C	4.0×10^{-9}
Metribuzin	Sencor/Lexone	1220 at 20°C	1.0×10^{-5}
Paraquat	Ortho Paraquat	Soluble[b]	Nonvolatile
Propachlor	Ramrod	580 at 20°C	2.3×10^{-4}

[a] Data from *Herbicide Handbook,* Weed Science Society of America (1974).
[b] Even though soluble, paraquat is not found in water because of the strong bonding of the positive ions to soil.

Soil Sampling for Pesticides and Nutrients

To determine the amount of herbicide present in the soil, immediately after herbicide application at sites 1 and 2, 12 surface soil samples (423 cm^2 in area to a depth of about 5 cm) are taken for herbicide analyses. Until the first significant rain, this type of sampling is performed daily for the first week, every other day for the second week, and every third day thereafter. After the first rain, soil samples are taken with a 19-mm-diameter soil probe in four layers to 30 cm (0-1, 1-7.5, 7.5-15 and 15-30 cm) to determine both the amount and location of herbicides in the soil profile. Three sets of 16 composited soil cores are taken from each watershed at each sampling. This sampling is performed after each runoff event or periodically (roughly every 2 weeks) during the time when no runoff occurs. These samples, like runoff water and sediment samples, are refrigerated (actually frozen) prior to air shipment to the EPA Environmental Research Laboratory at Athens, Georgia, for herbicide analyses.

Nutrient soil sampling takes two forms. First, several of the pesticide core soil samples (about four through the growing season) taken to 30 cm are also analyzed for nutrients. Second, three times a year a 5- or 7.5-cm diameter core sample is taken to 150 cm in nine layers (0-1, 1-7.5, 7.5-15, 15-30, 30-45, 45-60, 60-90, 90-120 and 120-150 cm), once in the spring, once mid-season, and once in the fall. Soil samples are analyzed for water-extractable NH_4-N, NO_3-N, PO_4-P, and Cl; for K_2SO_4-extractable NH_4-N; and for acid/fluoride-extractable available P (Bray-1).

RESULTS AND DISCUSSION

Precipitation

Precipitation amounts in the 2-yr study period were above average in the initial 6 mo (through June 1976), below average in a very dry 12-mo period (384 mm versus an annual average 823 mm), and above average in the final 6 mo. Although the drought resulted in a shortage of surface runoff events and flow from the watershed, particularly during the herbicide season, crop yields were good. Of the 544 mm of precipitation received at sites 1 and 2 in 1976, 106 mm was snowfall; of the 828 mm received in 1977, 64 mm was snowfall.

One aspect of precipitation data, which is very important in trying to predict runoff and chemical losses, is the variability in intensities and amounts over short distances. On an annual basis total precipitation measured at the six gaging sites ranged from 511 to 587 mm in 1976, and from 807 to 930 mm in 1977. Variation on a storm basis is probably more important and was more severe. This is evidenced by two storms of particular interest. On 4/19/77, 20 mm of rain fell at sites 1 and 2 causing 4.5 mm of runoff to occur from recently fertilized site 1. At a rain gage site 2.4 km south, however, only 10 mm fell, and no precipitation was measured 6.6 km west. On 9/17/77, 36 mm of rain fell at sites 1 and 2 causing 0.2 mm of runoff from site 1 and no runoff from site 2. At site 3, 6.9 km away, however, 70 mm fell causing 8.4 mm of runoff from the pasture, which usually has far less runoff than cropped areas.

Flows

Flows in Four Mile Creek in both 1976 (106 mm) and 1977 (40 mm) were much less than the 13-yr average (209 mm). Table II gives data for total surface runoff for the three small watersheds, and Table IV gives runoff data for major events. The well-documented effect of pasture on reducing runoff relative to row-cropped areas is evident in the data (excluding the event of 9/17/77 when site 3 received nearly twice as much rain as sites 1 and 2). The effect of the timing and depth of tillage on runoff is also very evident in the data. In both years, storm events after primary tillage resulted in much more runoff from the disked watershed than from the plowed watershed. In addition, the relative amounts of runoff from sites 1 and 2 for 5/29/76 and 6/13/76 were reversed by cultivation of site 2.

Sedimentation

Sediment loads have been reported by the USGS for the water years 1970 through 1973 for sites 4, 5 and 6. Annual suspended sediment loads reported

Table III. Average concentrations and annual amounts of soluble nutrients and dissolved and suspended solids in precipitation runoff, tile flow, and creek flow.

	Depth (mm)	NH$_4$-N		NO$_3$-N		PO$_4$-P		Cl		TDS		Suspended Solids	
		(ppm)	(kg/ha)	(ppm)	(kg/ha)	(ppm)	(kg/ha)	(ppm)	(kg/ha)	(ppm)	(kg/ha)	(ppm)	(kg/ha)
1976													
Precipitation	554	0.78	4.2	0.76	4.1	0.054	0.29	1.5	8.2	33	179	26	144
Runoff													
Corn (#2)	51.0	0.44	0.22	5.2	2.7	0.048	0.02	2.1	1.1	172	88	10440	5380
Beans (#1)	58.4	0.09	0.05	3.5	2.0	0.028	0.02	7.3	4.3	134	78	4650	2730
Pasture (#3)	13.0	0.20	0.03	0.9	0.1	1.160	0.15	4.0	0.5	129	17	310	40
Tile Flow													
#10-19	---	0.11	---	10.2	---	0.069	---	14.4	---	330	---	---	---
Creek Flow													
#4	106.2	0.23	0.24	7.8	8.3	0.070	0.07	11.1	11.8	273	290	1350	1420
1977													
Precipitation	828	0.87	7.2	1.00	8.3	0.028	0.23	1.5	12.5	51	420	36	302
Runoff													
Corn (#1)	11.7[a]	2.30	0.27	0.7	0.1	0.720	0.08	7.4	0.9	156	18	38380	4670
Beans (#2)	0.9	0.10	0.00	0.3	0.0	0.040	0.00	1.0	0.0	100	1	20000	174
Pasture (#3)	8.6[b]	0.66	0.06	0.4	0.0	0.940	0.08	1.9	0.2	60	5	310	27
Tile Flow													
#10-19	---	0.13	---	14.0	---	0.168	---	20.2	---	352	---	---	---
Creek Flow													
#4	39.5	0.23	0.09	10.7	4.2	0.116	0.05	17.1	6.8	329	130	180	70

[a] 40% of this runoff occurred within 24 hr of fertilizer application and incorporation.
[b] A very localized rain caused 98% of this runoff.

Table IV. Runoff and sediment concentrations
for major events on the small watersheds.

1976	Soybeans (Site 1)		Corn (Site 2)		Pasture (Site 3)	
	Flow (mm)	Sediment (ppm)	Flow (mm)	Sediment (ppm)	Flow (mm)	Sediment (ppm)
Feb-March Snowmelt	45.7	360	30.7	2320	10.4	195
	(plowed 4/5)		(disked 4/2)			
4/23 (28 mm rain)	0.6	--	6.0	8080	0.2	300
	(planted 5/19)		(planted 4/28)			
5/29 (29 mm rain)	3.8	45600	11.4	34800	1.0	1800
			(cultivated 6/7)			
6/13 (37 mm rain)	8.1	9200	1.2	4300	0.2	--

1977	Corn (Site 1)		Soybeans (Site 2)		Pasture (Site 3)	
	(mm)	(ppm)	(mm)	(ppm)	(mm)	(ppm)
4/19	(disked 4/19)		(plowed 3/31)			
(20 mm, 9 mm site 3)	4.5	66700	0.0	--	0.0	--
8/15-16 (38 mm)	4.8	15800	0.8	21900	0.0	--
9/17 (36 mm, 70 mm site 3)	0.2	--	0.0	--	8.4	310
9/18 (13 mm)	0.3	1360	0.1	1510	0.0	--

ranged up to 360 MT/km² (average at site 4 was 132) with maximum daily sediment concentrations of 7800 ppm. One to three percent of the suspended sediment was in the sand size range; while the bed material sizes reported varied widely, most bed material is smaller than 500 μm.

The initial observations in this study showed, as expected, that the highest sediment loads and concentrations are associated with the small row-crop watersheds. Table III presents summary sediment data for 1976 and 1977. Although rainfall was greater than normal in 1977, most of it occurred in the fall. Thus, erosion losses were low. During the late spring and summer months, very low rainfalls were observed each year. Erosion losses from the small watersheds were, therefore, lower than expected. In spite of the low rainfall, sediment concentrations in runoff samples from the row crop areas were relatively high, the highest recorded being about 100,000 ppm at site 1 (4/19/77). No runoff occurred on that date from site 2 because the rain infiltrated or was contained within the voids in the recently plowed cornstalk stubble. Flow sampled in the flume contained sediment in the silt and clay range.

Sediment deposited on the plastic disks above the flumes at sites 1 and 2 varied in thickness, partly because of deposition from local scour. In the one event (8/15/77) for which data were obtained (disks were not installed in 1976), about 15,000 kg/ha of sediment deposited on a small area above the flume. Although the area was only a fraction of a hectare in this case, the deposition could be considerable in severe storms. The particle sizes on the plates were dominantly in the silt and clay range.

Figure 2 shows the particle size distribution as a function of time and sediment concentration as flow decreased after peak discharge for a single runoff event from site 1 (4/19/77). At 1713 hr on that date less than 20% of the transported material was smaller than 2 μm, the concentration was over 100,000 ppm, and the discharge about 0.50 m^3/sec. As flow receded at 1754 hr, the 2-μm clay size had increased to 70% of the load, concentration had decreased to 12,600 ppm, and discharge to 0.001 m^2/sec. Because particle size is related to particle transport and chemistry, these data should be useful for modelers trying to define sediment delivery.

Summary data for site 4 (50 km^2) are presented in Table III. Sediment loads were low (142 and 7 MT/km^2 for 1976 and 1977, respectively).

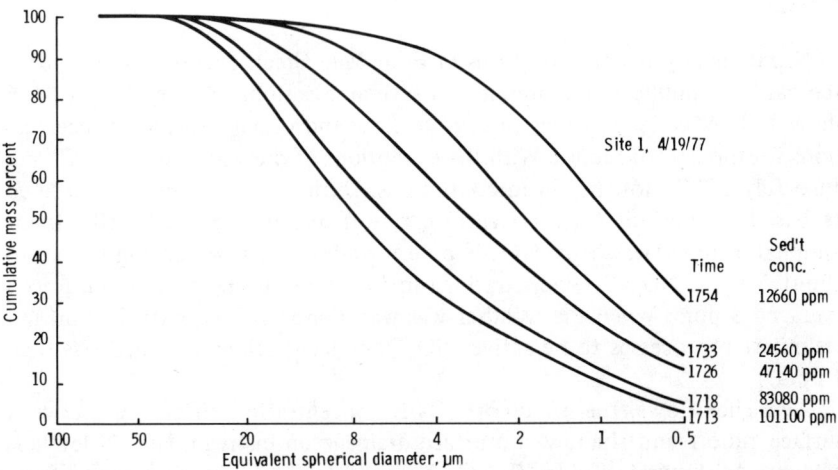

Figure 2. Particle size distribution for sediment in five runoff samples collected from site 1, 4/19/77.

Nevertheless, the weighted mean concentration for 1976 was greater than that for any of the 4 yr of USGS records. Individual samples contained concentrations up to 17,000 ppm (6/15/76). The peak concentrations precede the peak discharges at all sites where samples are taken.

Data related to the main channel are taken to assist in defining the channel contribution to sediment load and the nature of the material moving at or near the bed. Cross sections of the channel were taken at ten selected locations between gaging sites 4 and 5 in 1967 and again in 1977. The areas of the cross sections at low bank-overflow level increased from an average of 10.5 m^2 to 13.3 m^2, an increase of more than 25% in 10 yr over a length of about 3040 m. Cross sections further upstream were not taken in 1967, and thus channel cross section enlargement upstream could not be assessed. However, the average annual sediment production from the measured portion of the channel (based on a specific weight of 1.34 MT/m^3) was about 18% (24 of 132 MT/km^2/yr) of the average measured sediment discharge from above site 4.

The creek bed is often stratified. There is no observable systematic layering of the sandy and finer soil materials. Samples taken after larger flows may provide an observable pattern. Sand is not always found at the stream bed surface at present. Particle size analysis of cores removed from the bed show much scatter between cores at the same depth and location, and with depth. About half of the samples contained more than 50%, 62μm or larger, sand sizes.

Nitrate

Nitrate-nitrogen concentrations in Four Mile Creek exceeded the 10-ppm standard for public water supplies on several occasions. Figures 3 through 6 show NO$_3$-N concentrations relative to flow and other measured concentrations for the 2 yr of record. With the exceptions of the winter of 1976-77 and June-July 1977, flow has been continuous. During low-flow periods, defined as base-flow periods (*i.e.*, no surface runoff and no flow from tile drains sampled in the watershed), NO$_3$-N concentrations are low, ranging from 0 to about 5 ppm. NO$_3$-N concentrations in low flow during winter conditions averaged 3 ppm. When the weather was warm enough for denitrification and assimilation processes to be active, NO$_3$-N concentrations averaged less than 1 ppm.

At higher flows, the effect of NO$_3$-N concentration differences between surface runoff and shallow subsurface drainage on instream NO$_3$-N levels is evident. As shown in Table III, NO$_3$-N concentrations in surface runoff from row-cropped and pastured areas are generally much less (especially for pasture) than in shallow subsurface drainage from tile drains. This effect is particularly evident for two runoff events in May and June 1976, when relatively high

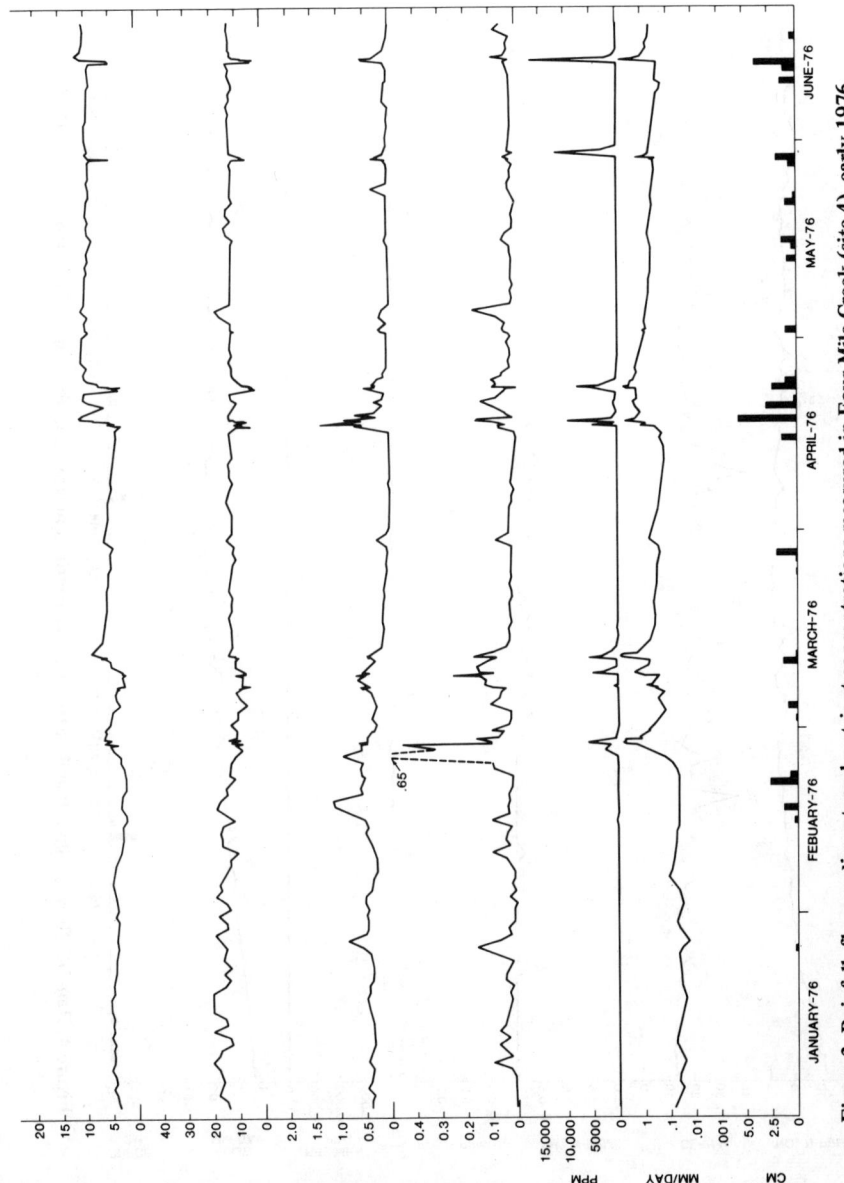

Figure 3. Rainfall, flow, sediment, and nutrient concentrations measured in Four Mile Creek (site 4), early 1976.

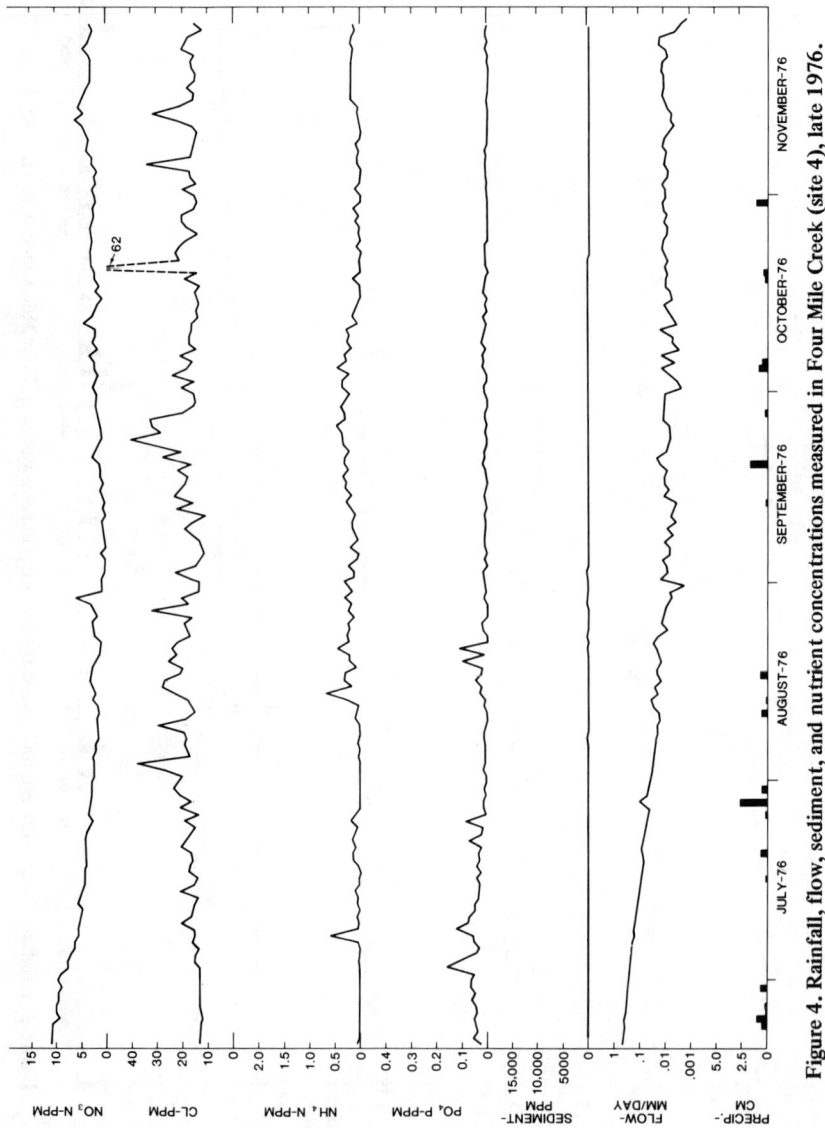

Figure 4. Rainfall, flow, sediment, and nutrient concentrations measured in Four Mile Creek (site 4), late 1976.

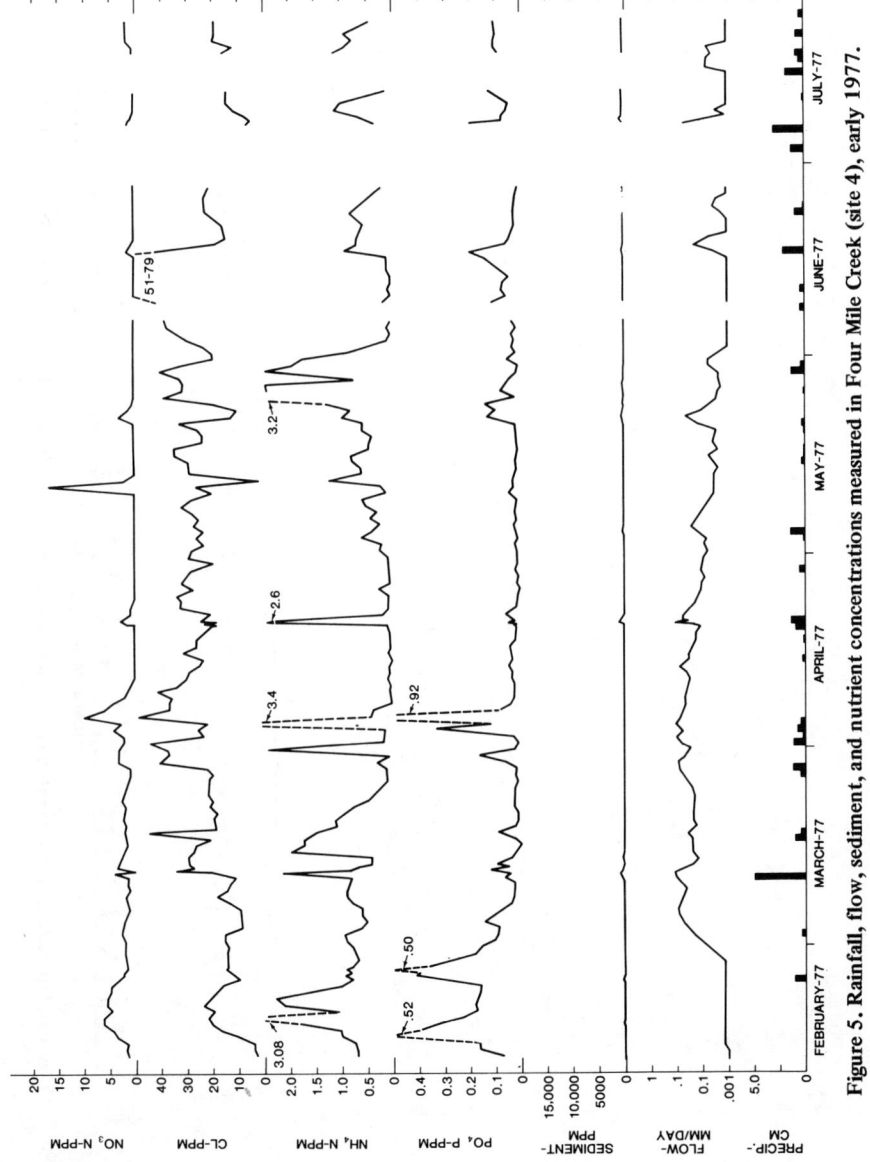

Figure 5. Rainfall, flow, sediment, and nutrient concentrations measured in Four Mile Creek (site 4), early 1977.

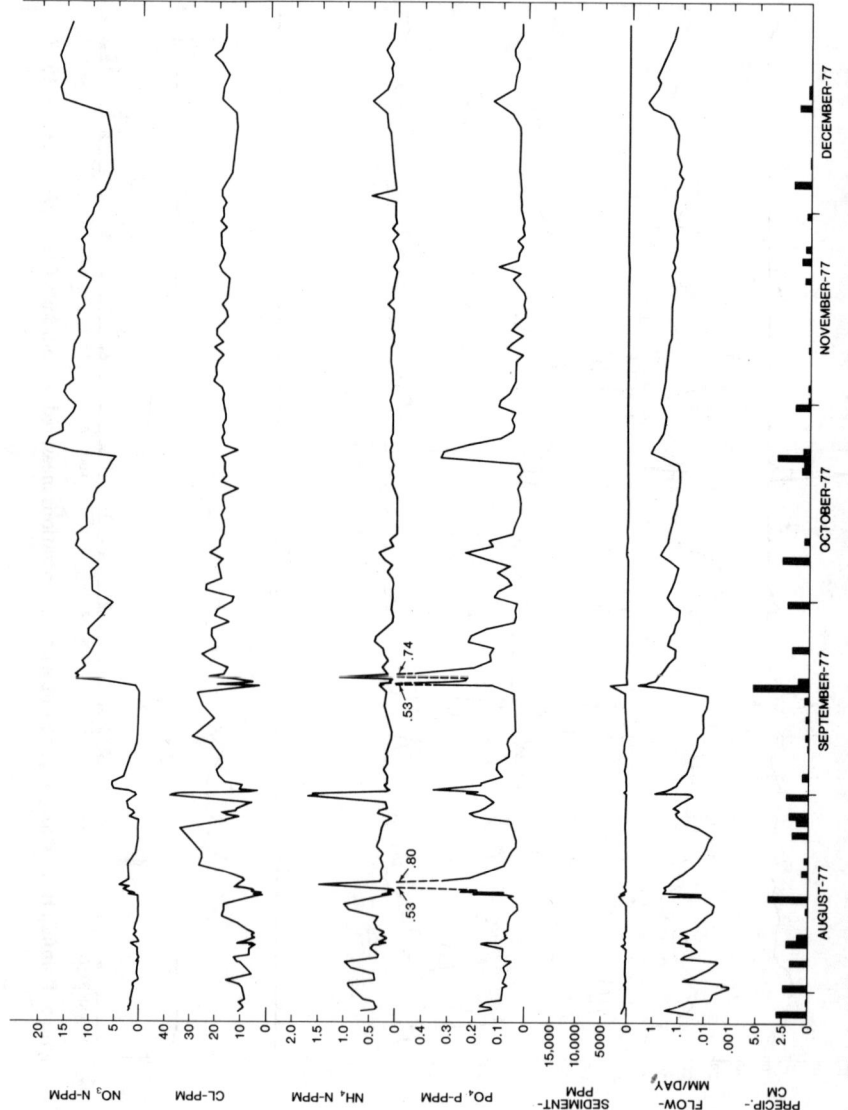

Figure 6. Rainfall, flow, sediment, and nutrient concentrations measured in Four Mile Creek (site 4), late 1977.

NO_3-N levels were temporarily decreased at peak flows because of the lower NO_3-N levels in surface runoff. As surface runoff passed, NO_3-N levels quickly returned to their previous levels, or higher, as shallow surface drainage again dominated the creek flow.

The impact of NO_3-N concentration differences between base flow and tile flow on instream NO_3-N levels is also evident, particularly for three rainfall periods in September, October and December 1977. The summer of 1977 was very dry, but rains in August and September caused the tile drains to begin to flow again. In each period instream NO_3-N concentrations increased with flow by about 9-14 ppm in a matter of 2 days (in which time any surface runoff that may have occurred had passed the gaging station). After each increase there was a gradual decline as the shallow soil profile drained. Irregularities in these declines were correlated with small rains and small increases in flow.

In addition to differences in NO_3-N concentrations between base flow and tile flow, the quality of water draining from tile lines varies. Part of the large increase in instream NO_3-N levels in October reflected the fact that NO_3-N levels in tile flow increased an average of 6.5 ppm through the week of October 19-25 during which 52 mm of precipitation fell. Because most of the tile lines in the watershed drain waterways in the draws of the rolling topograph, it is probable that these lines drain water from a range of depths (and possibly of varying water quality) depending on soil moisture conditions.

Water samples withdrawn from tile lines draining the waterways on rowcrop sites 1 and 2 (samples have been taken only in the fall of 1977) have averaged 7.7 and 18.1 ppm, respectively (site 1 received 135 kg/ha of N in 1977, none in 1976; vice versa for site 2). Results are shown in Figure 7 for average NO_3-N concentrations in soil water sampled down to 150 cm in November 1977. Data for the pasture sites are also included for sake of information. A comparison of values for tile effluent with the profiles shown indicates that the tile lines are intercepting water equivalent to that at about the 90-cm depth. The low NO_3-N levels at 120-150 cm shown in Figure 7 were also found for five other samplings over the 2-yr period. This indicates that vertical leaching of NO_3-N to depths greater than 150 cm is presently limited in extent at the three sampled sites.

NO_3-N loss from the watershed for the 2-yr study period (12.5 kg/ha) has roughly equalled that deposited with precipitation (12.4 kg/ha). However, losses, which have been primarily associated with shallow subsurface and deep subsurface (base) flow, could be expected to be much higher during a more normal rainfall period.

Ammonium

Several times in the spring of 1977 instream NH_4-N levels exceeded the 2-ppm level believed toxic to fish and, during the 2-yr study, commonly exceeded

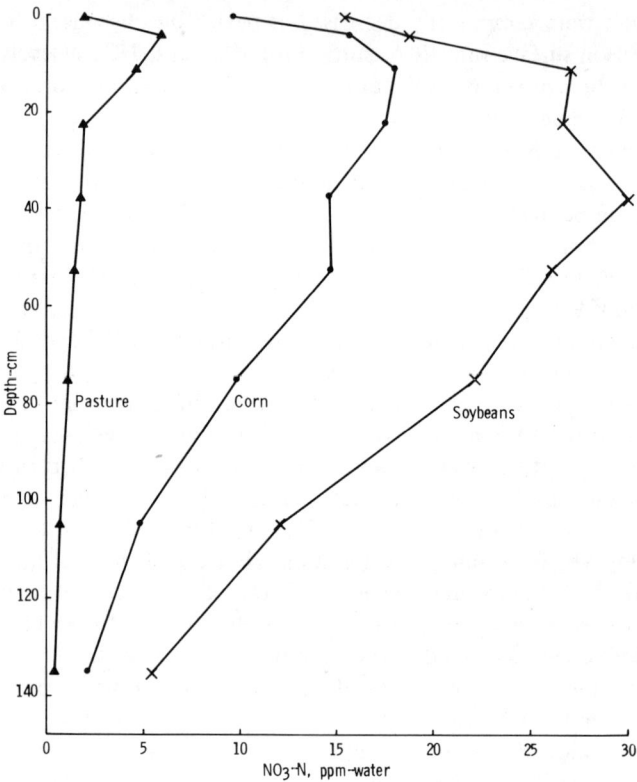

Figure 7. NO_3-N in soil water as a function of depth for corn, soybean and pasture watersheds, fall 1977.

the 0.5-ppm standard for public water supplies. Figures 3 through 6 show NH_4-N concentrations relative to flow and other measured concentrations. During low-flow periods, defined as base flow, NH_4-N concentrations ranged from zero to about 0.5 ppm. At higher flows, during which some surface runoff may occur, NH_4-N levels instream usually rise. As shown in Table III, NH_4-N levels in surface runoff are usually higher (particularly for the pasture site) than levels in tile flow. The impact on water quality of surface runoff occurring within 24 hr of application and incorporation by disking or urea and ammonium fertilizers (135 kg/ha) was evident for site 1 where NH_4-N concentrations averaged 5.8 ppm in the 4.5 mm of surface runoff from that event. For the rest of the year, NH_4-N concentrations in surface runoff averaged 0.1 ppm.

NH_4-N loss from the watershed over the 2-yr study (0.3 kg/ha) was much less than that deposited with precipitation (11.4 kg/ha). In fact, the average

NH_4-N concentration in precipitation of 0.8 ppm exceeded the 0.5-ppm public water supply standard.

Phosphate

The PO_4-P levels instream often exceeded the 0.01- to 0.03-ppm level believed to be sufficient for the growth of algae. Figures 3 through 6 show PO_4-P concentrations relative to flow and other measured concentrations. During low-flow periods, defined as base flow, PO_4-P concentrations averaged roughly 0.02 ppm (with some scatter possibly due to animal influence nearby). At higher flows, during which some surface runoff may occur, PO_4-P levels instream usually rise. As shown in Table III, PO_4-P levels in both surface runoff and tile flow are higher than 0.02 ppm, with tile flow, surprisingly, slightly higher than cropped sites (excluding runoff from site 1 the day of fertilization). Surface runoff from the pasture site, however, has very high PO_4-P levels averaging 1 ppm; presumably the result of cattle wastes, lack of incorporation of applied fertilizer, and decaying vegetation.

The impact on water quality of surface runoff occurring within 24 hr of application and incorporation by disking of 37 kg/ha of phosphorus fertilizer was obvious for site 1 where PO_4-P concentrations averaged 1.69 ppm in the 4.5 mm of surface runoff from that event. For the rest of the year PO_4-P concentrations in surface runoff averaged 0.10 ppm.

The low level of PO_4-P in base flow is believed to be caused by extraction of PO_4-P from water enroute to the stream by deep subsoils low in PO_4-P. The shallow subsoils in this watershed, however, are quite fertile with respect to available P as shown in Figure 8 for the November 1977 soil sampling. Similar data exist for five other samplings. There usually is a positive correlation between available P levels in a soil and the equilibrium-soluble PO_4-P levels in water in contact with that soil. This would explain the higher than expected PO_4-P levels in tile flow.

A close examination of the PO_4-P and sediment concentration data in Figures 3 through 6 shows that, during a surface runoff event, when sediment concentrations increase, a decrease in PO_4-P concentrations usually occurs at the point of peak sediment concentration. Also, the highest PO_4-P concentration usually succeeds the peak sediment concentration. The adsorption-desorption of PO_4-P by sediment and the sources and size (and thus chemical activity) of sediment, and therefore the erosion processes, must play a role in the observed decrease and subsequent increase.

PO_4-P loss from the watershed over the 2-yr study period (0.12 kg/ha) was less than that deposited with precipitation (0.52 kg/ha). Greater losses would be expected during a normal rainfall period.

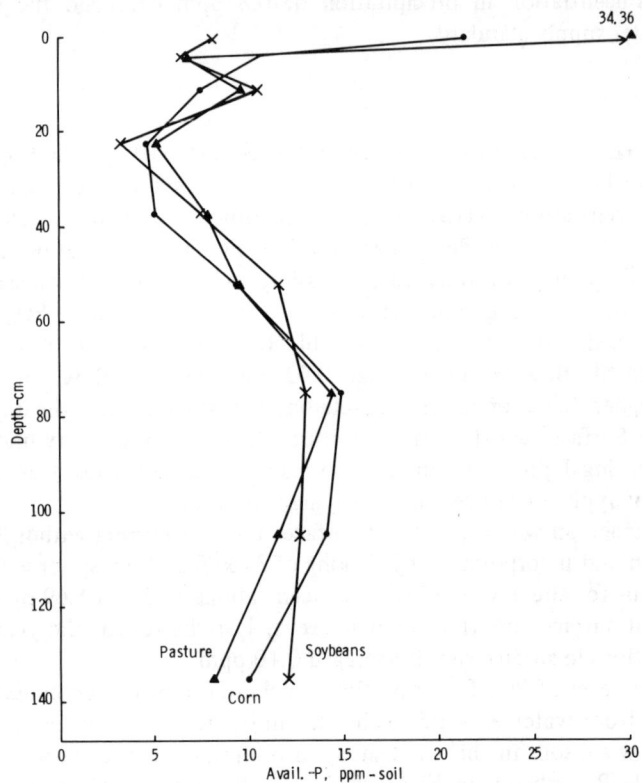

Figure 8. Available P in the soil profile for corn, soybean and pasture watersheds, fall 1977.

Chloride

Chloride in agricultural drainage is of interest, not because of a pollution potential, but because of the information it may provide. The chloride ion is considered a conservative ion (*i.e.*, it does not undergo transformations that change the amount present), and because it is very soluble and not adsorbed, it moves readily with water.

Figures 3 through 6 show Cl concentrations relative to flow and other measured concentrations. During much of the time when there was no surface runoff, Cl concentrations ranged from 15 to 30 ppm, encompassing the average Cl contents in tile flow as shown in Table III. For some reason, possibly drought-related, Cl concentrations in both tile flow and creek flow averaged about 50% higher in 1977 than in 1976. For one week of very low flow in June 1977, Cl concentrations exceeded 50 ppm continuously. High flows related to surface runoff events usually reduced instream Cl levels as would be expected from the Cl levels given for runoff in Table III.

The Cl lost from the watershed over the 2-yr period (18.6 kg/ha) is very nearly equal that deposited with rainfall (20.0 kg/ha). This resulted from about one-tenth of the precipitation lost as stream flow at about ten times the Cl concentration in precipitation (1.5 ppm, Table III). Whether the Cl ion can provide useful hydrologic information remains to be seen as more data are gathered, but the scatter in the data may preclude this.

Sediment-Transported Nutrients

Concentrations of extractable NH_4-N and available P in sediment have been measured for a small number (limited by a shortage of runoff events and low sediment concentration) of runoff samples. Extractable NH_4-N and available P are measures of absorbed N and P, defined by the chemical extraction methods used. Extractable NH_4-N in sediment from cropped sites 1 and 2 have averaged 5.2 ppm with a range of 2.2 to 8.6 ppm (excluding data for site 1, 4/19/77, the day of fertilization when extractable NH_4-N in sediment averaged 267.5 ppm). No samples for either extractable NH_4 or available P have been analyzed from site 3.

Available P in sediment from cropped sites 1 and 2 is seemingly affected by the degree of incorporation of P fertilizer. In 1976 available P in sediment from site 1, which had fertilizer plowed under, averaged 26.0 ppm; whereas sediment from site 3, which had fertilizer only disked in, averaged 40.4 ppm. In 1977 available P in sediment from site 1, which had fertilizer only disked in, averaged 42.0 ppm; whereas sediment from site 2, which had fertilizer plowed under, averaged 22.4 ppm. (The average for site 1 excluded the data for 4/19/77 when available P in sediment averaged 69.2 ppm.)

Measurements of total N, total P, organic matter content, and pH of soil samples taken to 150 cm have been made. In the 0- to 7.5-cm layer average values of 1600 ppm, 440 ppm, 2.9% and 6.9 have been obtained, respectively, for sites 1 and 2. Value for site 3 of 2800 ppm, 470 ppm, 3.9% and 6.5, respectively, have been obtained. Owing to selective erosion processes, it is probable that total N and P contents of sediments would be higher than for soil in place.

Total Dissolved Solids

Total dissolved solids (TDS) concentrations in Four Mile Creek are seldom above the 500-ppm standard set for public water supplies. During low-flow periods considered base flow and during higher flows containing shallow subsurface drainage, but not surface runoff, TDS concentrations are the highest, usually in the 300- to 400-ppm range. During surface runoff events TDS concentrations can decrease to the 100- to 200-ppm range depending on the proportion of flow made up by surface runoff. As shown in Table III for the 2 yr of record, TDS concentrations in surface runoff averaged from 60 to 172 ppm; whereas, tile flow averaged 341 ppm. During this time it is estimated that 600 kg/ha TDS were deposited with precipitation, and 420 kg/ha left the watershed with creek flow.

Pesticides

During the 2-yr duration of this study, there have been only two runoff events, both in 1976, of any consequence in the interim between herbicide application and nearly complete herbicide degradation. Figures 9 and 10 present data on amounts of herbicides remaining in the soil as a function of time for sites 1 and 2 in 1976. On the basis of calculations from spray concentrations, nozzle calibration, and ground speed, and from herbicide deposits

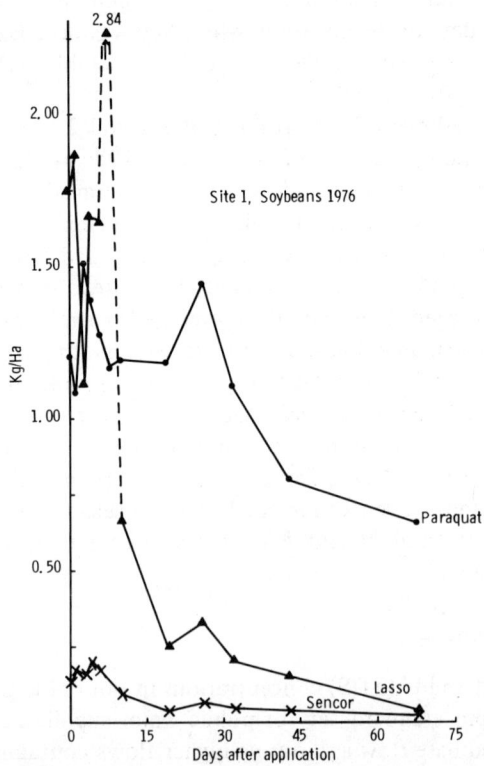

Figure 9. Herbicide amounts remaining in the soil as a function of time after application to soybean watershed.

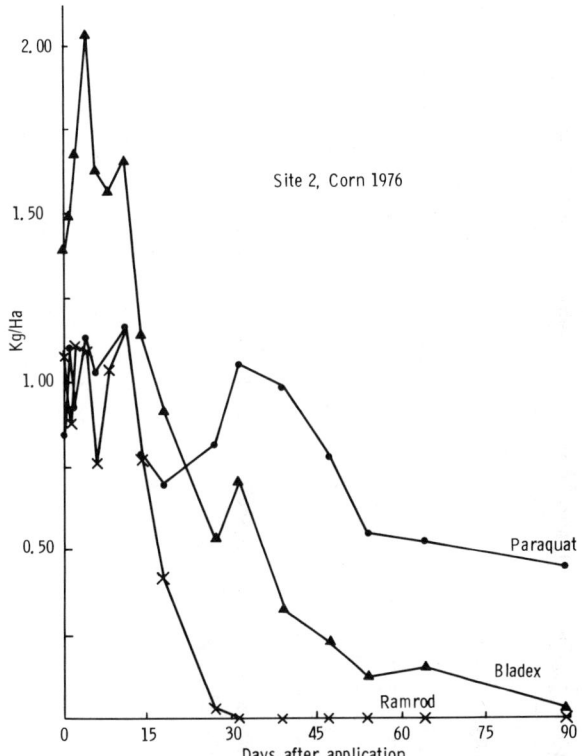

Figure 10. Herbicide amounts remaining in the soil as a function of time after application to corn watershed.

on filter paper disks, it was estimated that 2.40, 0.60 and 1.56 kg/ha of alachlor, matribuzin and paraquat were applied, respectively, to site 1 (soybeans in 1976). The data of Figures 9 and 10, although there is scatter common to most field sampling, indicate the possibility of loss immediately after application when soil samples taken within 4 hr contained less herbicide than applied, particularly for metribuzin and propachlor. Volatilization from a hot, dark-surface soil could be the explanation.

The two storms of record (30 mm of rain on 5/29/76 and 36 mm of rain on 6/13/76) came 9 and 24 days after herbicide application on site 1, and 30 and 45 days after herbicide application on site 2. Average herbicide concentrations and losses are given in Table V. Concentrations of herbicides in water of Four Mile Creek peaked on 5/29/76 at 96.0, 6.7, 10.2 and 5.8 ppb for alachlor, metribuzin, propachlor and cyanazine, respectively, in one of two successive samples taken 3 hr apart. Concentrations of herbicides in sediment

Table V. Average herbicide concentrations and losses in runoff in 1976.

Site 1 Date	Flow (mm)	Sediment (MT/ha)	Alachlor Water (ppb)	Alachlor Water (g/ha)	Alachlor Sediment (ppb)	Alachlor Sediment (g/ha)	Metribuzin Water (ppb)	Metribuzin Water (g/ha)	Metribuzin Sediment (ppb)	Metribuzin Sediment (g/ha)	Paraquat[a] Sediment (ppb)	Paraquat[a] Sediment (g/ha)
5/29/76	3.8	1.82	162	6.1	—[b]	——	75	2.8	—[b]	——	6840	12.4
6/13/76	8.1	0.75	64	5.2	183	0.1	18	1.5	0	0.0	9280	7.0
Site 2			Propachlor				Cyanazine				Paraquat[a]	
5/29/76	11.2	4.04	40	4.5	81	0.3	150	16.9	485	2.0	17200	69.5
6/13/76	1.1	0.05	1	0.0	0	0.0	37	0.4	361	0.0	4640	0.2

[a] Paraquat was never detected in water samples.
[b] Because of some unknown factor neither alachlor nor metribuzin was detected in sediment samples, although both were present in the 0 to 1-cm soil samples at 5280 and 533 ppb, respectively, taken the following day.

in Four Mile Creek peaked on 6/14/76 at 900 and 200 ppb for alachlor and cyanazine, respectively, in one sample. Neither metribuzin nor propachlor was detected in sediment at any time. Paraquat was detected in one sediment sample from Four Mile Creek but never detected in water. For both runoff events, essentially all the herbicides transported with stream flow passed through the gaging station within a 24-hr period.

A comparison is made in Table VI between the percentage of the herbicides applied to the Four Mile Creek watershed that left the watershed with sediment and water, and the percentage of herbicides applied to the small cropped watersheds (sites 1 and 2) that left the field with sediment and water. For the most widely used herbicide, alachlor, one-fifth as much was lost per hectare from the whole watershed as from the monitored field, indicating from the limited data that attenuation processes are probably occurring between the field borders and the stream. Data for the other four herbicides applied to sites 1 and 2 also indicate that attenuation processes are probably occurring for them.

The values given in Table V show that the herbicides studied, with the exception of paraquat, are lost primarily with water. Therefore, losses of these pesticides and several others in common use today will not be controlled by erosion control alone.

Watershed Inventory

Table VII presents the results of the inventory of the study watershed for 1976 and 1977 as well as for 1970 (data from an earlier project[12]). It is evident that the watershed is currently intensively farmed with 75% or more of the area in row crops, with nearly all the corn fertilized, and with nearly all the corn and soybeans treated with herbicides.

A comparison between 1970 and 1977 shows some striking changes. The amount of row crop has increased from 55 to 78% (at the expense of land in oats, hay, pasture and soil bank). In addition, the percentage of row-crop area receiving fertilizer has increased, and in the case of nitrogen on corn, the rate of application has also increased. As a result, 441 MT of N was applied to row crops in 1977; whereas, 214 MT of N was applied in 1970. Even though the rate of P application was down (possibly because of increased P levels in the soil from past applications), 73 MT was applied to row-crops in 1977 relative to 55 MT in 1970. For K, 162 MT was applied to row-crops in 1977 relative to 94 MT in 1970. The percentage of row-crops treated with herbicides also increased 20% or more from 1970 to 1977. The use of insecticides on corn also increased, although use is variable, being somewhat dependent on known infestations. No insecticides have been used on soybeans.

Additional inventory data collected show that, in 1976, 55% of the watershed area in corn, oats and soybeans was plowed. In 1977 only 42%

Table VI. Percentages of applied herbicides lost with runoff in 1976.

	Alachlor	Metribuzin	Propachlor	Cyanazine	Paraquat
Four Mile Creek	0.10	0.04	0.14	0.08	1.28
(% of watershed treated)	(28.4)	(11.0)	(1.7)	(3.8)	(0.2)
Sites 1 and 2	0.48	0.72	0.21	0.96	3.34

Table VII. Crop and ag-chemical use inventory for 1970, 1976 and 1977.

Crop-Year	Watershed Area[a] (%)	Crop Area Fertilized (%)	Average Fertilizer Application[b]			Crop Area Pesticide-Treated	
			N	P	K	Herbicides (%)	Insecticides (%)
			(kg/ha)				
Corn-70	39	88	123	31	53	71	54
Corn-76	55	96	159	22	43	99	58
Corn-77	54	97	166	25	54	98	80
Soybeans-70	16	4	17	33	64	75	0
Soybeans-76	20	12	8	20	42	94	0
Soybeans-77	24	23	7	24	70	95	0

[a] 5053 ha.
[b] Average application rate equals total kg applied divided by number of ha fertilized.

of the corn, oats and soybeans was plowed, with a corresponding increase in the amounts that were either disked, chisel plowed, or received no tillage. Only 2% of the corn, oats and soybeans was strip-cropped, and only 1% was on terraced land.

The numbers of livestock in the area are very market dependent. The number of cattle in the watershed decreased from about 3600 to 2500 from 1970 to 1977. The number of hogs decreased from about 9500 to 7200.

SUMMARY

Nonpoint pollutants from agricultural drainage are transported primarily by surface runoff, flow from drains (or interflow), and base flow. Each flow component is capable of carrying certain pollutants depending on the soil and chemical involved. In the Four Mile Creek study, measurements of flow and concentrations are made of those components from watersheds varying from 6 ha to 50 km^2 in size. Data for sediment, nutrient (N and P), and herbicide concentrations and loads have been collected for two relatively dry years.

Sediment loads from the heavily cropped dominantly loess-covered watersheds for 4 yr before this study were relatively low for Iowa, averaging about 130 MT/km^2/yr from the largest watershed. During the period of this study maximum loss from the smallest row-cropped watershed was 5380 kg/ha; maximum sediment loss was 142 mg/km^2 (1420 kg/ha) for the 50-km^2 watershed area. Loss from the small grass watershed was low (40 kg/ha) as expected.

Concentrations of sediment, averaged over a year, ranged from a maximum of 38,380 ppm (corn) to 1350 ppm for the large watershed, and 310 ppm for the grassed watershed. Average concentrations in runoff for single events varied from a maximum of 66,700 ppm for corn to a low of 1800 ppm for grass. Individual samples taken during storm runoff varied from more than 100,000 ppm from row-crop land, 17,100 ppm for the mixed cover large watershed, to 1818 ppm for grass. Sediments transported at all points in the watershed are dominantly in the silt and clay range.

Nitrate-nitrogen in agricultural drainage is one of the primary nonpoint pollution concerns with respect to both concentrations and losses, particularly if the 10-ppm NO$_3$-N standard for public water supplies is a valid value. NO$_3$-N losses, which affect downstream water quality, can be large enough to become an agricultural economic concern. The energy loss associated with NO$_3$-N loss is also a concern. The high NO$_3$-N concentrations in Four Mile Creek (average 8.6 ppm, often exceeding 10 ppm) are believed to be the result of high NO$_3$-N concentrations in shallow subsurface drainage (average of 12.1 ppm). In a study of the Skunk River in Iowa[13] the high NO$_3$-N concentrations found in the river also were believed to be the result of tile drainage. Other studies in Iowa[14-16] have found high nitrate concentrations in tile drainage. A summary of shallow subsurface drainage water quality data[17] shows that this tendency is not unique to Iowa. The potential for leaching of the very mobile NO_3-N with excess moisture, and the quick drainage of this NO_3-N laden water through tile lines, is greatest during periods of high precipitation and high NO_3-N levels in the soil. Careful management of nitrogen addition to the soil system to match plant needs is needed to decrease losses with subsurface drainage.

Ammonium-nitrogen in agricultural drainage may be toxic to fish and increases the oxygen demand during nitrification. There is a 0.5-ppm standard for public water supplies. NH_4-N losses with agricultural drainage might adversely affect downstream water quality, but are quite low and not an agricultural economic concern. Higher concentrations in Four Mile Creek (sometimes in excess of 2 ppm) were associated with surface runoff events where overland flow may contact animal wastes, decaying vegetation, and unincorporated ammonium or urea fertilizers. In a rainfall simulation experiment[18] crop residue and reduced incorporation of N fertilizer were believed responsible for higher NH_4-N concentrations in surface runoff from plots planted to conservation-tilled corn relative to concentrations in runoff from plots planted conventionally where both crop residue and fertilizer were incorporated with a plow.

Phosphate-phosphorus in agricultural drainage has often been blamed for acceleration of the eutrophication process in lakes and reservoirs. PO_4-P concentrations often exceed the 0.01 to 0.03 ppm believed to be sufficient for the growth of algae. Concentrations, however, vary considerably with the type of agricultural drainage. As with NH_4-N, higher concentrations of PO_4-P in Four Mile Creek (sometimes in excess of 0.5 ppm) were associated with surface runoff events. Concentrations of PO_4-P in runoff from the pasture averaged 1 ppm where overland flow contacted animal wastes, decaying vegetation and, in 1977, unincorporated P fertilizer. Reduced incorporation of P fertilizer (disking versus plowing) resulted in higher PO_4-P concentrations in runoff from the corn relative to the soybean watershed. Crop residue and reduced incorporation of P fertilizer were believed responsible for higher PO_4-P levels in surface runoff from areas planted to conservation-tilled corn relative to concentrations in runoff from conventionally planted areas.[18,19] The PO_4-P concentrations in tile drainage water were higher than expected,[14] probably because of high available-P levels in the subsoils in the Four Mile Creek watershed. Hanway and Laflen[15] found a positive correlation between PO_4-P concentrations in tile drainage water and available P levels in subsoils. PO_4-P losses are quite low from economic and energy viewpoints.

Concentrations of N and P in sediment are usually many times higher than concentrations of dissolved N and P in runoff water. Nutrients associated with sediment, however, either are adsorbed or are part of the organic complex and have limited immediate impact on water quality. Indeed, it is questionable whether the nutrients associated with sediment significantly affect water quality once the sediment is deposited and covered by subsequent sediment deposition. Concentrations of extractable NH_4-N and available P in sediment from the cropped sites averaged 5 and 33 ppm, respectively. Runoff closely following the application of fertilizers resulted in values of 270 and 70 ppm, respectively. Analyses of soil in place indicate that, without considering the enrichment by the erosion process, sediment from the cropped sites would have about 1600 and 440 ppm total N and P, respectively.

Pesticides in agricultural drainage are of concern because of their presence in public water supplies, their effects on aquatic life, and the potential for presently unforeseen problems. Although all the pesticides studied are herbicides, the possible reactions, interactions with soil, and modes of movement can be related to other pesticides of similar chemical and physical properties. Detectable herbicide concentrations in Four Mile Creek (maxima of 0.1 ppm in water and 1 ppm in sediment; both for alachlor, which was applied to 28% of the watershed) were associated with surface runoff events within 6 weeks after application. Losses from the cropped watersheds were from 1.5 to 18 times greater than from the watershed as a whole (on a percent-of-applied basis), indicating that attenuation processes are occurring. With the exception of strongly adsorbed paraquat, herbicide losses occurred primarily with runoff water. This also has been observed in other Iowa studies.[17,20,21] Therefore, losses of most pesticides in common use today, which are not strongly adsorbed to the soil fractions, will not be controlled by erosion-control practices alone.

ACKNOWLEDGMENTS

The authors gratefully acknowledge the support of the following: The U.S. Environmental Protection Agency for cooperation and financial support of the study; The U.S. Geological Survey, Iowa District, for cooperation with stream gaging; Dr. George Bailey, EPA, for guidance as first project director; Charles Smith, EPA, for assistance and guidance in field procedures; Nancy Rasmussen, ISU, and Jacquelyn Benner, EPA, for assistance with chemical analysis; Steve Kimes for assistance with particle-size analysis and field work; and Dr. Craig Beer and Dr. William Shrader for advice and assistance on the project.

This is Journal Paper No. J-9161 of the Iowa Agriculture and Home Economics Experiment Station, Ames, IA, Project No. 2153.

REFERENCES

1. Donigian, A. S., D. C. Beyerlein, H. H. Davis and H. N. Crawford. "Agricultural Runoff and Management (ARM) Model Version II: Refinement and Testing," Ecological Research Series EPA-600/3-77-098 (1977).
2. Negev, M. "A Sediment Model on a Digital Computer," Tech. Report No. 76, Department of Civil Engineering, Standford University, Standford, CA (1967).
3. David, W. P., and C. E. Beer. "Simulation of Soil Erosion: Development of a Mathemtical Erosion Model," *Trans. Am. Soc. Agric. Eng.* 18:126-133 (1975).
4. Onstad, C. A., and G. R. Foster. "Erosion Modeling on a Watershed," *Trans. Am. Soc. Agric. Eng.* 18:288-292 (1975).
5. Onstad, C. A., R. F. Piest and K. E. Saxton. "Watershed Erosion Model Validation for Southwest Iowa," Proc., Third Inter-Agency Sedimentation Conference. Water Resources Council (1976), pp. 1-22 to 1-34.
6. Crawford, N. H., and A. S. Donigian. "Pesticide Transport and Runoff Model for Agricultural Lands," Environmental Protection Technology Series EPA-660/2-74-013 (1973).
7. Bailey, G. W., R. R. Swank, Jr., and H. P. Nicholson. "Predicting Pesticide Runoff from Agricultural Land: A Conceptual Model," *J. Environ. Qual.* 3:95-102 (1974).
8. Donigian, A. S., and N. H. Crawford. "Modeling Pesticides and Nutrients on Agricultural Lands," Environmental Protection Technology Series EPA-600/2-76-043 (1976).
9. Frere, M. H., C. A. Onstad and N. H. Holtan. "ACTMO, An Agricultural Chemical Transport Model," U.S. Dept. Agric. Res. Serv., ARS-H-3 (1975).
10. Kunkle, G. R. "A Hydrogeologic Study of Groundwater Reservoirs Contributing Base Flow to Four Mile Creek, East-Central Iowa," U.S. Geological Survey, Water Supply Paper 1839-0 (1968).
11. Payne, W. R. Jr., J. D. Pope, Jr., and J. E. Benner. "An Integrated Method for Trifluralin, Diplenanid, and Paraquat in Soil and Runoff from Agricultural Land." *J. Agric. Food Chem.* 22:79-82 (1974).
12. Baker, J. L., and H. P. Johnson. "Impact of Subsurface Drainage on Water Quality," in Third National Drainage Symposium Proc., Am. Soc. Agric. Eng., St. Joseph, MI (1976).
13. Johnson, H. P., and J. L. Baker. "Water Quality and Cropland Nutrients," in *Physical Relationships with the Agricultural Sector,* Ames Reservoir Environmental Study, Iowa State Water Resources Research Institute, Ames (1973).
14. Baker, J. L., K. L. Campbell, H. P. Johnson and J. J. Hanway. "Nitrate, Phosphorus, and Sulfate in Subsurface Drainage Water," *J. Environ. Qual.* 4:406-412 (1975).
15. Hanway, J. J., and J. M. Laflen. "Plant Nutrient Losses from Tile-Outlet Terraces," *J. Environ. Qual.* 3:251-356 (1974).
16. Willrich, T. L. "Properties of Tile Drainage Water," Completion Report Project-A-013-IA, Iowa State Water Resources Research Institute, Ames (1969).

17. Baker, J. L., and H. P. Johnson. "Tillage System Effects on Runoff Water Quality: Pesticides," paper presented at winter meeting (#77-2504), Am. Soc. Agric. Eng. (1977).
18. Barisas, S. G., J. L. Baker, H. P. Johnson and J. M. Laflen. "Effect of Tillage Systems on Runoff Losses of Nutrients," *Trans. Am. Soc. Agric. Eng.,* in press (1978).
19. Johnson, H. P., J. L. Baker, W. D. Shrader and J. M. Laflen. "Tillage System Effects on Runoff Water Quality: Sediments and Nutrients," paper presented at the winter meeting (#77-2504), Am. Soc. Agric. Eng. (1977).
20. Baker, J. L., J. M. Laflen and H. P. Johnson. "Effect of Tillage Systems on Runoff Losses of Pesticides," *Trans. Am. Soc. Agric. Eng.,* in press (1978).
21. Ritter, W. F., H. P. Johnson, W. G. Lovely and M. Molnau. "Atrazine, Propachlor, and Diazinon Residues on Small Agricultural Watersheds," *Environ. Sci. Technol.* 8:38-42 (1974).

17

THE FATE OF NITRATE IN SMALL STREAMS AND ITS MANAGEMENT IMPLICATIONS

J. B. Robinson

    Department of Environmental Biology
    University of Guelph
    Guelph, Ontario, Canada

H. R. Whiteley, W. Stammers

    School of Engineering
    University of Guelph
    Guelph, Ontario, Canada

N. K. Kaushik, P. Sain

    Department of Environmental Biology
    University of Guelph
    Guelph, Ontario, Canada

INTRODUCTION

In the Southern Ontario Great Lakes Basin, streams draining areas of productive soils which are intensively farmed have nitrogen concentrations which reflect the proportion of cropland and the numbers of livestock in the drainage area.[1] The nitrogen in these surface waters is mainly in the form of nitrate and occasionally exceeds a concentration of 10 mg/l. Concern about eutrophication of surface waters has centered on phosphorus, and will probably continue to do so. However, it is clear that a concentration of nitrate in surface water exceeding drinking water standards must be viewed as a distinct and important problem apart from effects on productivity of lakes and reservoirs. Nitrogen is one of those elements which has a complete cycle involving permanent sink process, and thus there is potential for manipulation to prevent the buildup of nitrate in systems where its presence creates a problem.

The process by which nitrogen is returned to the atmosphere, denitrification, has been studied exhaustively in soils[2] and increasingly in lakes[4-6] and other aquatic systems, including flooded soils.[7] Recently, attention has been directed to the probability that denitrification occurs in groundwater[8] under some conditions. All of these studies are important to our understanding of nitrogen transport in the soil-water system and increase the possibility of successful management of nitrogen in the environment.

The nitrogen transport and cycling which occurs in streams has received little attention although order-of-magnitude studies were attempted by Owens et al.[9] who made N budgets for two English rivers. Inputs were derived from actual sewage loadings along with estimated loadings from land based on physiographic and land-use characteristics of the basins; outputs were measured. While the budgets tended to balance during cold months, Owens et al.[9] inferred that there was a deficiency of output during summer. Losses of nitrogen appeared to be as high as 274 g/yr for each square meter of river bed. The authors assumed that a large part of this loss resulted from bacterial denitrification in the mud. More recently, Toms et al.[10] observed nitrate loss from aerated water overlying columns of river mud, ascribed this to denitrification and studied effects on the process of varying pH, temperature and oxygen concentration in the water. Van Kessel[11,12] has similarly observed nitrate loss in water overlying columns of sediment from ditches receiving treated sewage.

For some years, our group has been monitoring a small spring-fed brook (Swifts Brook) and conducting laboratory studies on nitrogen transformations in stream-sediment-water systems. A preliminary report was made on apparent nitrogen loss during stream transport[13] and on some laboratory studies of the phenomenon.[14] A mathematical model of nitrogen transport in streams[15] recently has been described. The present paper provides further evidence for significant amounts of nitrogen loss during transport in streams and indicates the stream management options which should enhance the process.

FIELD OBSERVATIONS

The field data reported below were obtained from studies on Swifts Brook, a small tributary of Blue Springs Creek which, in turn, is tributary to the Grand River in southern Ontario. Swifts Brook watershed is about 16 km northeast of Guelph and is underlain by dolomitic limestone which, in this area, is fractured and quite porous. An extensive groundwater study,[16] conducted as part of the project reported here, showed that nitrogen from cultivated fields probably accounted for high nitrate concentrations observed in groundwater discharging at the spring which is the major source of summer stream flow for the brook. The watershed is about 2.2 km^2 in area, and the main channel length between the spring and the measuring point just above the union with Blue Springs Creek is 2 km. Surrounding the spring and upstream half of the channel is mixed deciduous and coniferous bush, while the lower reach of the brook is bordered by unused pasture and old field vegetation.

As shown in Figure 1 the watershed was instrumented to measure flow rates at various points in the main channel and in major tributaries. The principal measurement site (P9), the most downstream station, provided continuous flow rate data and automatic (timed) sampling capability. Samples could be obtained as frequently as every 6 hr. Continuous gaging was also available at the spring (P4) while staff gages were installed at P6, I1 and E5. Weirs at I2 and I3 were difficult to maintain but provided useful information on flow in these channels. Thus all significant surface flow into the main channel could be measured at least periodically. In addition to samples collected by the automatic sampler, weekly samples were collected at all the perennial sites shown in Figure 1 and at all of the flowing intermittent and ephemeral sites as well. During a number of thaw and storm events producing surface runoff, extensive sampling was conducted at all sites.

Despite the intensity of gaging and sampling, it is difficult to specify accurately the nitrogen inputs to Swifts Brook except for certain periods discussed below. It has therefore not been possible to make a precise annual nitrogen budget, but it appeared that an estimate would be useful in detecting indications of nitrogen removal in the input/output data.

As a first step in obtaining a budget estimate the daily ouflow at P9 was separated into base flow and storm flow components using traditional hydrographic techniques. Precise measurement of nitrogen discharge at P9 and of loading from the spring were available as shown in Table I. It was necessary then to estimate N contributions in the remaining base flow (after P4 was subtracted) and for storm flow. The base flow nitrogen concentrations were estimated by a subjective weighting of the means of a number of analyses of samples from I2 and I3 for all 3 nitrogen species. Storm flow concentrations of N were found to differ depending on whether their origin was cultivated land or naturally vegetated areas. Storm flow inputs were therefore calculated by weighting according to the approximate distribution of land use in the watershed.

The total inputs arrived at by the above procedure were 1500 kg for 1975 and 2050 kg for 1976. Comparing these to the outputs shown in Table I suggests losses of 250 kg N in 1975 and 450 kg N in 1976 for an average N loss rate of 350 kg/yr. We stress that this is an estimate only but calculations on a different basis, described below, lend credence to this estimate.

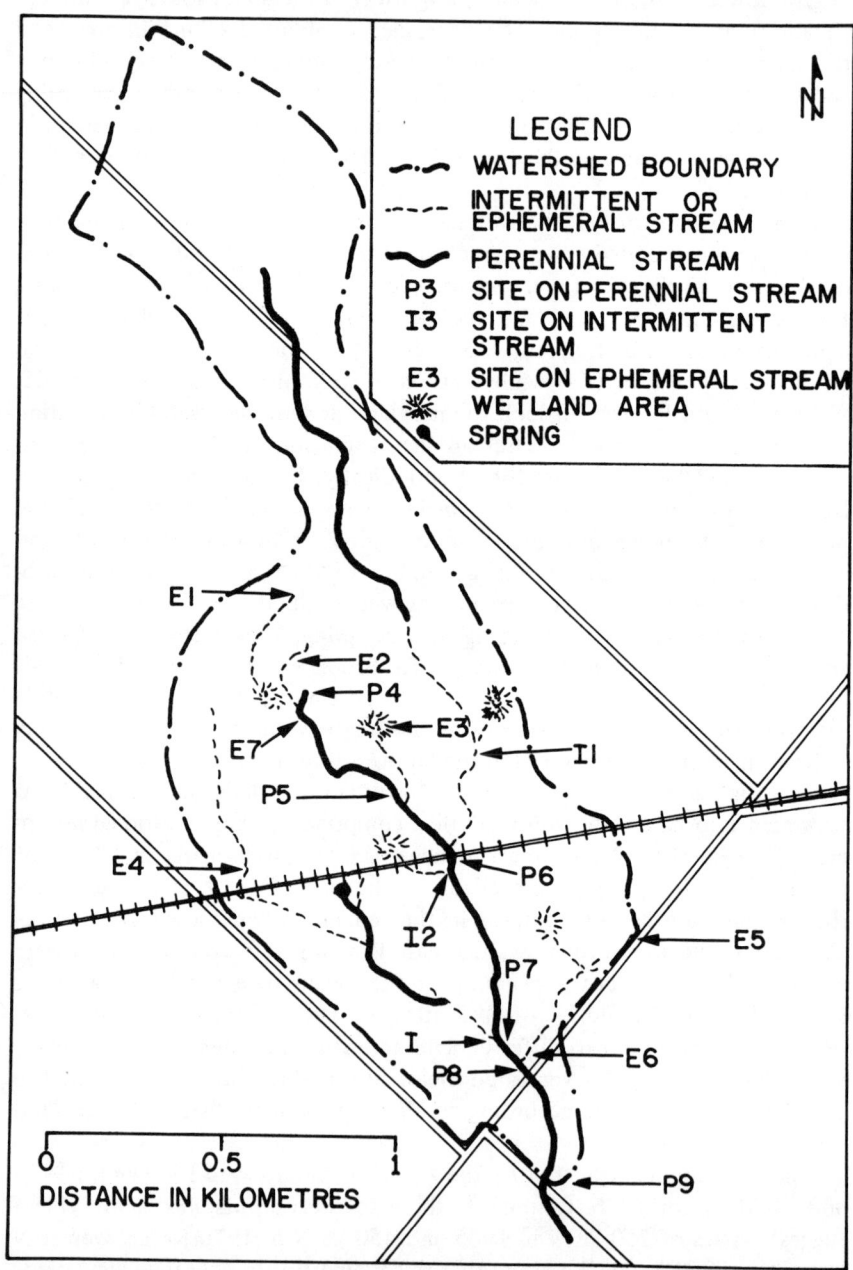

Figure 1. Swifts Brook watershed showing gaging stations and sampling points.

Table I. Weighted-by-flowrate mean concentration of nitrogen species and total N loading at stations P4 (spring) and P9 (outflow) on Swifts Brook.

	P4		P9	
	1975	1976	1975	1976
NH_4-N mg/l	0.05	0.05	0.05	0.02
Organic-N mg/l	0.17	0.21	0.56	0.61
NO_3-N mg/l	4.31	4.10	1.15	1.05
Total N Load (kg)	500	600	1250	1600

During the study period there were 37 days for which nitrogen concentration profiles of the main channel were available and during which flowrates at P9 were classified as baseflow only. However on the majority of these days there was a substantial increase in flowrate between P4 and P9 indicating inputs of groundwater to the channel. Therefore, while there were continuous declines in nitrogen concentrations down the reach on each of these days, it is impossible to state with certainty that part of this decline was not due to dilution.

On three days for which concentration data are plotted in Figures 2 and 3 the flowrate change from P4 to P9 is either negative or increases by less than 20% while the measured flowrate at P6 falls within the range of flowrates at P4 and P9.

On these days, the diminution in N concentration with distance downstream is clearly the result of instream processes acting upon the nitrate-N in transport (Figure 2). On all three days plotted, the concentration of organic N increased between stations P4 and P9. Thus the apparent loss of total N (Figure 3) is not as great as the apparent loss of nitrate N. The concentration of ammonium-N was almost negligible at all stations, ranging from 0 to 0.13 mg/l with a median value of about 0.04.

A detailed survey of the stream bed width was used to calculate stream bed area. Apparent removal rates were then calculated for the three dates shown in Figures 2 and 3 by dividing flowrate times drop in nitrate concentration for the reach by the area of the stream bed. The results are shown in Table II.

As pointed out earlier, organic-N concentration increased downstream and inclusion of organic N in the calculation to provide loss rates for total N (NH_4-N being negligible) yielded figures slightly lower than those shown in Table II. For July 17/75, June 15/76 and August 19/76 they were 2.8, 4.9 and 3.8 x 10^{-3} mg N/m^2-sec. The average of these figures (3.4 x 10^{-3} mg/m^2-sec) extrapolated to the whole stream for the whole year would

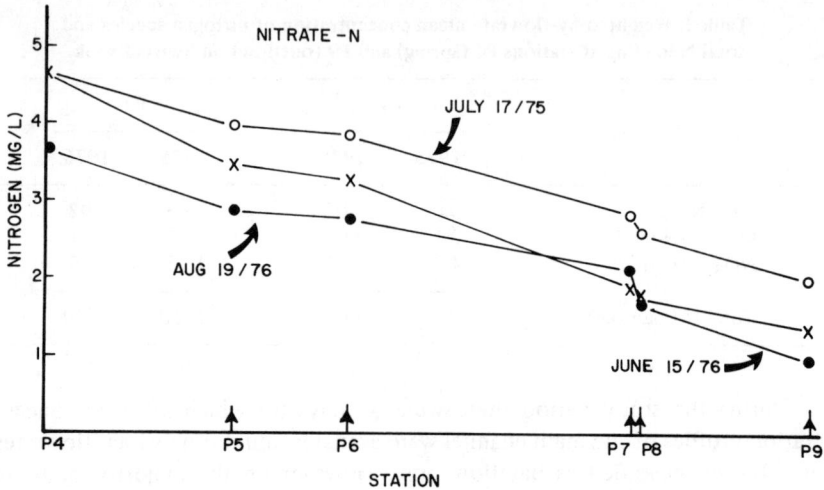

Figure 2. Concentrations of nitrate-N in water samples from Swifts Brook on three selected days of extreme low flow.

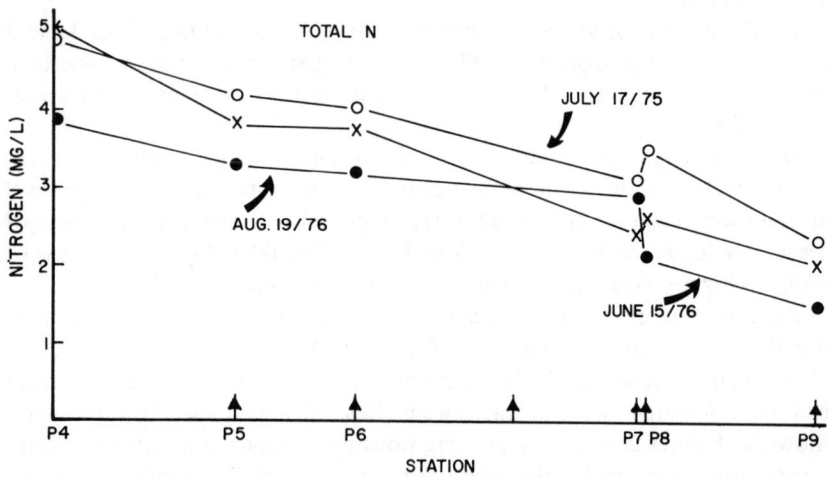

Figure 3. Concentrations of N (total) in water samples from Swifts Brook on three selected days of extreme low flow.

account for 327 kg/yr of N removal which is remarkably close to the figure (350 kg/yr) estimated earlier as a budget deficit.

The field data strongly suggest that nitrate-N is being removed continuously throughout the year from the water between P4 and P9. Processes

Table II. Removal rates in Swifts Brook during selected base flow periods.

	Apparent Nitrate-N Removal Rate (mg N/m-sec)		
	July 17/75	June 15/75	August 19/76
Range for Individual Reaches	1.1×10^{-3} to 4.3×10^{-3}	1.7×10^{-3} to 10.5×10^{-3}	1.5×10^{-3} to 5.1×10^{-3}
Average for Total Reach P4-P9	3.0×10^{-3}	5.5×10^{-3}	4.3×10^{-3}

which might be implicated are those of uptake, immobilization and denitrification. The first two are temporary sinks and, in earlier work[13,17] we showed, by measuring macrophyte standing crop (to estimate uptake) and autumn leaf-fall (to estimate carbon available for immobilization of N) that neither process could account for much more than 1% of the observed nitrogen deficit. The next section describes some results of experiments which showed that the removal process is denitrification in the stream bed sediment.

LABORATORY STUDIES

Much information was obtained from column experiments in which sediment was overlain with aerated water containing nitrate-N. These columns have been described in detail elsewhere.[14] Briefly, they were plexiglass tubes of 5.6-cm i.d., usually with about 10 cm of sediment in the bottom and with 500 ml of nitrate solution over the sediment. Aeration was provided by passing 21% oxygen in helium through a bubbling device about 1 cm over the sediment surface. CO_2 evolution from the columns was measured in traps of the type described by Sain and Broadbent.[18] The columns were equipped with redox probes permanently inserted through the column walls and into the sediment at various depths. The probes were prepared as described by Whisler et al.[19] In addition, in some columns, ports equipped with clamped tubes extending into the sediment at various depths permitted periodic sampling of interstitial water.

Figure 4 shows the changes in nitrogen content of water overlying sediments from a number of southern Ontario streams including Swifts Brook. There is remarkable similarity in apparent rate of N removal and in the shape of the curves with one exception discussed below.

It was shown earlier[14] that the rate of nitrogen removal from water in these column experiments is influenced by temperature as would be expected for biological reactions. It was also shown indirectly that the nitrogen

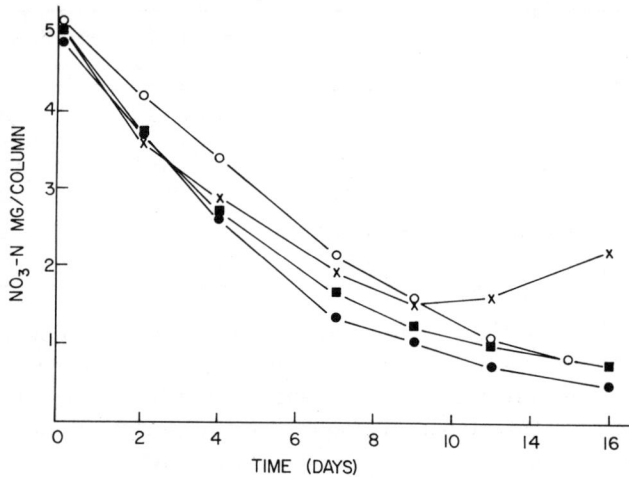

Figure 4. Changes in nitrate-N content of water overlying columns of sediment from four southern Ontario streams. Crosses: Canagagigue Creek, Open Circles: Holiday Creek, Large Closed Circles: Little Ausable Creek, Small Closed Circles: Swifts Brook.

removal from the water occurred in the upper part of the Swifts Brook sediment, probably in the top 2.5 cm. Direct evidence of this interpretation is contained in Figures 5 and 6 which show results for an experiment in which 10-cm-deep columns of Swifts Brook sediment were overlain with aerated, nitrate-containing water and the columns maintained at 22°C. After nearly all the nitrate had disappeared from the water, it was replaced with fresh nitrate-containing water and the incubation temperature was reduced to 10°C.

The concentration of nitrate-N in the supernatant and at various depths in the sediment is shown in Figure 5 along with loss of carbon as CO_2 from the columns. At 22°C supernatant nitrate-N declined very rapidly, falling below 1 mg/l in about 18 days. Concentration at the 1 cm depth in the sediment reached 2 mg/l at two days then declined rapidly to negligible values between 8 and 12 days. A trace of nitrate N was detected at 2.5- and 5-cm depth on day 2 but not thereafter. At 10°C supernatant concentration declined more slowly, and did not reach 1 mg/l until about the 34th day. The concentration at 1-cm depth was at or above 1 mg/l for the first 24 days and traces were detected at 2.5-cm depth throughout the incubation period. Carbon dioxide evolution at 10°C was about 25% that at 22°C for equivalent time periods.

The redox potentials shown in Figure 6 confirm that nitrate-N is unlikely at 22°C to diffuse much further than 1 cm into the sediment before being

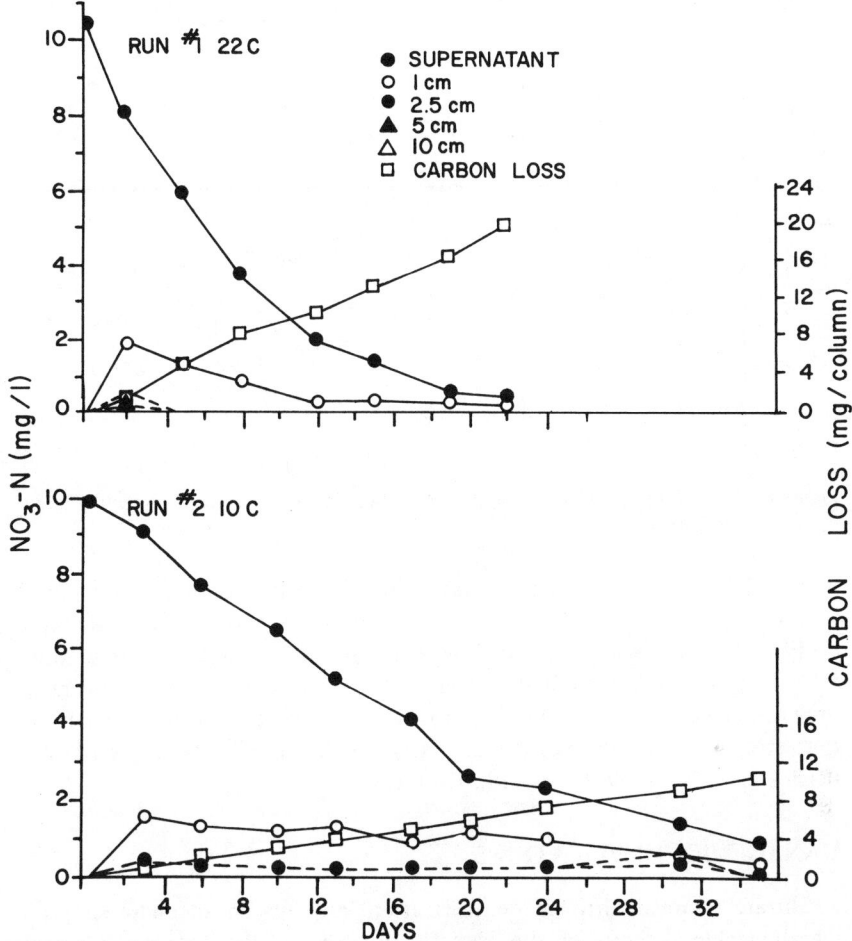

Figure 5. Supernatant and sediment concentrations of nitrate-N and CO_2 evolution from columns incubated at 22°C (top) followed by 10°C (bottom).

denitrified. However, when the temperature was lowered to 10°C, it appeared that the upper layer of the sediment became more oxidizing as might be expected if microbial activity declined while diffusion of nitrate and O_2 remained constant. The result was the higher concentration of nitrate in the sediment at this temperature.

Recently, L. Chatarpaul working in our laboratory[20] has used ^{15}N to provide direct evidence that the losses of N in the experiments reported here resulted from denitrification. In stream sediments similar to the less organic ones shown in Figure 4 he has found that only a trace of the added nitrate-N

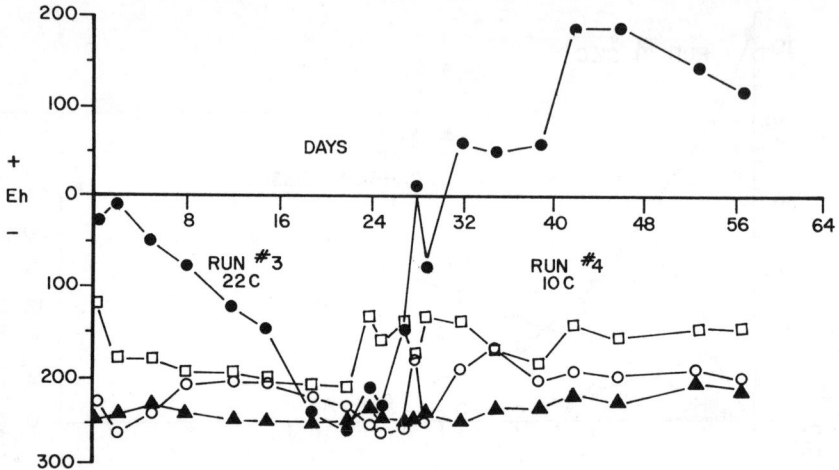

Figure 6. Eh values at various depths in sediment columns with overlying nitrate solution.

is immobilized. Furthermore, he has found that nitrification goes on actively in the upper layers of the sediment and that the anomalous curve shown in Figure 4 (Canagagigue, site 4) is the result of nitrification of sediment in NH_4-N which is manifested as an increase in concentration of nitrate-N only when denitrification begins to slow down. It seems evident, therefore, that stream sediments can act similarly to the flooded soils described by Patrick and his associates[7] as sinks for nitrogen.

MANAGEMENT IMPLICATIONS

Nitrate removal, through denitrification, requires an adequate supply of metabolizable carbon for the bacteria involved in the process. It appears from our results that stream denitrification is not very sensitive to carbon content of sediment because the sediments represented in Figure 4, although causing similar changes in nitrate concentrations of the supernatant, range in loss on ignition from 2.2 to 58%. It appears that the process is limited by the rate at which nitrate diffuses into the sediment rather than by the microbial activity. Nevertheless, in work not reported here, we have shown that the rate at which nitrate is removed from water slows down most rapidly over sediments which have little carbon. It can be assumed, therefore, that as the carbon content of the sediment falls below some critical level, the denitrification rate will begin to decline.

The very high carbon content of Swifts Brook sediment appears to be a characteristic of wooded, headwater streams which experience relatively small variation in flow, have relatively poorly defined channels and therefore retain much of the carbonaceous input from the surrounding vegetation. The protection of headwaters against agricultural and other development will help to maintain actively denitrifying sediments in the upper reaches of such streams. Similarly, the establishment of vegetated buffer strips adjacent to downstream channels will tend to increase the carbon inputs to stream sediments. However, while stream bank vegetation ensures a *source* of carbon for streams, it is likely that the hydraulic factors mentioned above are of major importance in *retention* of carbon in the stream bed.

Channelization to increase drainage capacity would tend to enhance removal of sedimented carbonaceous materials, particularly from those channels which are subject to extreme variations in flow. On the other hand, works which tend to reduce water velocity such as reduction of channel slope or construction of reservoirs by damming would tend to retain organic matter and increase retention time. We found that nitrate-N removal was always very marked over a 25-m length of Swifts Brook (between P7 and P8) where a small pond has been constructed (Figure 2).

The work reported on here provides one more reason for treating streams as living systems rather than simple conduits by which water is conveyed from upland sources to downstream reservoirs. Bank aforestation of some streams in southern Ontario would result in a number of benefits among which are flatter flow peaks, stabilized banks, reduced overland flow inputs from adjacent land, lower water temperatures, a greater diversity of fish species and increased denitrification. The cost of the reduced drainage capability which might result may in some cases be so large as to mitigate against this kind of stream management. Nevertheless, the benefits listed above, including denitrification potential, should be considered in making decisions on the management of specific stream basins.

ACKNOWLEDGMENTS

The authors are grateful to C. K. Lee, R. Mason, D. Draper and F. Powell for excellent technical assistance. Support of this work came from the following agencies: Canada Department of the Environment, Inland Waters Directorate; National Research Council of Canada; and Ontario Ministry of Agriculture and Food.

REFERENCES

1. Coote, D. R., E. M. MacDonald and W. T. Dickinson. "Agricultural Watershed Studies," Great Lakes Drainage Basin Canada, Report to PLUARG, Task Group C. Activity 1 (1978).

2. Broadbent, F. E., and F. E. Clark. "Denitrification," in *Soil Nitrogen*, W. V. Bartholomew and F. E. Clark, Eds., *Agronomy* 10:344-359 (1965).
3. Brezonik, P. L. "Denitrification in Natural Waters," *Prog. Wat. Tech.* 8:372-392 (1977).
4. Keeney, D. R. "The Nitrogen Cycle in Sediment-Water Systems," *J. Environ. Qual.* 2:15-29 (1973).
5. Larsen, V. "Nitrogen Transformations in Lakes," *Prog. Wat. Tech.* 8:419-431 (1977).
6. Wetzel, G. W. *Limnology* (Toronto: W. B. Saunders Co., 1975).
7. Patrick, W. H. Jr., R. D. Delanne, R. M. Engler and S. Gotoh. "Nitrate Removal from Water and the Water-Mud Interface in Wetlands," U.S. Environmental Protection Agency Report # EPA-600/3-76-042 (1976).
8. Gillham, R. W., and J. A. Cherry. "Field Evidence of Denitrification in Shallow Groundwater Flow Systems," 13th Can. Symp. Water Poll. Res. Hamilton (in press, 1978).
9. Owens, M., J. H. N. Garland, I. C. Hart and G. Wood. "Nutrient Budgets in Rivers," *Symp. Zool. Soc. Lond.* 29:21-40 (1972).
10. Toms, I. P., M. J. Mindenhall and M. M. I. Harman. "Factors Affecting the Removal of Nitrate by Sediments from Rivers, Lagoons and Lakes," Tech. Rep. TR14, Water Research Centre, Stevenage Heats, England (1975).
11. van Kessel, J. F. "Influence of Denitrification in Aquatic Sediment on the Nitrogen Content of Natural Waters," Agricultural Res. Rep. 852, Centre for Agricultural Publishing and Documentation, Wageningen, Holland (1976).
12. van Kessel, J. F. "Factors Affecting the Denitrification Rate in Two Water-Sediment Systems," *Wat. Res.* 4:259-267 (1977).
13. Kaushik, N. K., and J. B. Robinson. "Preliminary Observations on Nitrogen Transport during Summer in a Small Spring-Fed Ontario Stream," *Hydrobiologia* 49:59-63 (1976).
14. Sain, P., J. B. Robinson, W. N. Stammers, N. K. Kaushik and H. R. Whiteley. "A Laboratory Study on the Role of Stream Sediment in Nitrogen Loss from Water," *J. Environ. Qual.* 6:274-278 (1977).
15. Stammers, W. N., J. B. Robinson, N. K. Kaushik, H. R. Whiteley and P. Sain. "Modeling Nitrogen Transport in a Small Upland Stream," Proc. 13th Can. Symp. Water Poll. Res. Hamilton (in press, 1978).
16. Stiebel, W. H. "An Investigation of the Occurrence and Origin of Nitrate in Groundwater in the Swifts Brook Watershed," M.Sc. Thesis, University of Waterloo, Ontario (1977).
17. Kaushik, N. K., J. B. Robinson, P. Sain, H. R. Whiteley and W. Stammers. "A Quantitative Study of Nitrogen Loss from Water of a Small Spring-Fed Stream," Proc. 10th Can. Symp. Water Poll. Res., Toronto (1975), pp. 110-117.
18. Sain, P., and F. E. Broadbent. "Decomposition of Rice Straw in Soil as Affected by Some Management Factors," *J. Environ. Qual.* 6:96-100 (1977).

19. Whisler, F. D., J. C. Lance and R. S. Linebarger. "Redox Potentials in Soil Columns Intermittently Flooded with Sewage Water," *J. Environ. Qual.* 3:68-74 (1974).
20. Chatarpaul, L., and J. B. Robinson. "Nitrogen Transformations in Stream Sediments: ^{15}N Studies," ASTM D19 Symposium on Methodology for Biomass Determinations and Microbial Activities in Sediments, Fort Lauderdale, FL, January 1978 (in press).

SECTION IV

SILVICULTURE

18

BEST MANAGEMENT PRACTICES FOR SILVICULTURE

W. C. Harper
  Weyerhaeuser Company
  Tacoma, Washington

INTRODUCTION

The Federal Water Pollution Control Act, as amended in 1972 and 1977, was developed by Congress to provide a means of improving the quality of the nation's water, with the ultimate goal of making all waters fishable and swimmable. The initial thrust of regulation promulgated under that act was directed toward control of point sources of pollution discharge. Control was exercised through the NPDES permitting system and was geared around water quality standards. With control of point sources well underway, emphasis has now shifted to nonpoint source considerations. A point source of pollution, that confined within a discrete conveyance, can be controlled either through preventive technology or through "end-of-pipe" treatment. Nonpoint sources, on the other hand, can only be realistically controlled through preventive practices.

The important parameters related to nonpoint source pollution from silvicultural activity consist of four primary constituents: sediment, stream temperature, forest chemicals and dissolved oxygen. For much of the forest land in the U.S., the importance of the four pollutants given above are presented in order of priority. With the exception of forest chemicals, all parameters are constituents of a natural stream system. Herein lies another important distinction between point and nonpoint sources of pollution. Because these parameters occur naturally, it is extremely important to be able to separate that which occurs naturally from that which is caused by man.

Best management practices (BMPs) have been described as the means to control nonpoint sources of pollution. The actual definition of BMPs, means of establishment and means of evaluating effectiveness, continues to be somewhat confused. The current EPA definition addresses not only water quality considerations but also the technical and economic feasibility of a given practice. The inclusion of technical and economic considerations is important if we as a nation are to provide the food and fiber necessary for an expanding population. Because of the extreme variability of the system with which the forest manager must deal, BMPs can only be developed on a site-specific basis. Even though practices must be developed on a site-specific basis, assessment of effectiveness of those practices cannot. When first considered, this would appear to be a dilemma; however, under further consideration, it can be shown to be a completely rational approach. Because of the variability inherent in the natural system suggested above, it is inappropriate to expect compliance with water quality standards at every point on the stream. If such an attempt is made, it can be very quickly demonstrated that any management practice will be unable to meet water quality standards under all situations. Consider for example, water quality values under flood conditions, particularly as related to sediment load. Considering this variability, effectiveness of applying management practices can only be accomplished on a broad basis, including both time and space, considering the "health of a system." In this way, evaluation is not based on each individual stream segment, but rather is based on a larger land area and/or a large number of like practices.

Most states now feel that it is necessary to submit a list of BMPs to EPA in order to satisfy requirements under Section 208. I would like to offer an alternative to that concept, a program that would be easier for both the EPA and individual states to administer. The proposed alternative would be to consider the 208 plan as the documentation of a control strategy which would define the process by which a given state proposed to control nonpoint source pollution. Such a control process would likely include an explanation of the way in which BMPs are to be utilized for control. In addition, it would clearly outline methods of achieving compliance, such as cost incentives, regulation, education, etc., and would outline responsibility within the state organization of those departments charged with ensuring that the goals of the Clean Water Act are met. EPA approval of such a plan would not be for a listing of management practices, but rather it would be for approval of the control strategy envisioned by each individual state. Such a strategy would allow change in recommended forest practices as the situation and improvement in knowledge dictates. This would maintain control of actual forest practices at the state level where expertise exists which could address local and regional problems.

OBJECTIVES OF THE 208 PLANNING PROCESS

With objectives of Section 208 of the Clean Water Act in mind, and recognizing the need for continued production of wood fiber, the following is presented as a discussion of objectives that should be considered when developing the silvicultural portion of a state 208 plan.

Of primary importance is the fact that the 208 planning process should address only water quality considerations. A common trap in development of many state 208 plans is the tendency to include such things as consideration for wildlife and esthetics. This is not to suggest that such considerations are unimportant, but rather to point out that it is an inappropriate consideration under the Act, which was developed for the protection of water quality. In addition, without a full understanding of specific requirements of the Act, there seems to be a real tendency to use 208 planning as a means of land use planning. While land use control may be a viable means of controlling nonpoint sources of pollution, the political complications of such control are obvious. As specialists, we have a large enough job ahead of us in defining "how" we conduct given forest operations to concern ourselves with the politically complex problem of "where" and "when."

It is important to ensure that the 208 plan developed allows the forestland manager latitude to meet consumer demand for wood fiber at a reasonable price, within environmental constraints for water quality protection. The resulting plan should not be so cumbersome and complex that it offers disincentives to forestland owners. Current figures show that the present forestland base consists of 28% public, 59% private, and only 14% industrial. This would indicate that if we were to meet future fiber needs, production by the small private land owner is of extreme importance. The private forestland owner, both large and small, must be given incentives to increase forest management activity, not disincentives through complex regulation.

The resulting plan should ensure compatibility with long-term forest management objectives. As forestland managers, we are in a somewhat enviable position. For the most part, goals of water quality protection are generally compatible with long-term forest management objectives. Regarding long-term forest management and the forest manager, an all too often forgotten fact is that the forester was the original conservationist and/or environmentalist. His job has always been to ensure a healthy forest, and a healthy forest generally means good water quality. Considering sediment as an example, if the objective is to grow trees as quickly as possible, then it is important to protect site productivity by keeping soil out of the stream and on the side slopes. Problems arise primarily when economic demands force a land manager into a short-term management position. When adverse market conditions dictate liquidation of standing inventory in a "least-cost" way to increase capital flow, or when a land owner cannot "see" beyond his own life span and, therefore, only one rotation, sediment production, may be significant. The key then is good long-term forest management. Reduction of sediment produced by forest road systems may not have quite the same economic incentive. Roads properly located and designed, however, may result in a reduction in road miles, reduction in road width, and reduction in maintenance. Even though construction costs may be increased, a combination of these factors can have a positive economic benefit through reduction in overall road costs.

The completed 208 plan should allow for predictability. There is a tendency for often conflicting regulations, such as the Occupational Safety and Health Act (OSHA), zoning, forest practices, shoreline management, environmental impact assessment, etc., to produce a situation where the land manager following all regulations as closely as possible may still be unable to operate or may be held in violation. Confusion over the relationship between management practices and water quality standards can be cited as an example. If control strategy is not clearly defined, a land manager could conceivably be in compliance with recommended forest management practices and his operation terminated due to violation of a water quality standard, only because the particular measurement was taken during an extreme storm event. The point is, following the expenditures necessary to initiate a project, it is important to be able to proceed. Just as a regulatory agency needs to be able to plan, so does the land manager.

Somewhat related to predictability, guidelines developed in the 208 plan should not conflict with other rules and regulations. Control as related to size of landing in a forest management situation might be given as an example. The BMP from a water quality standpoint would most likely be to maintain the size of landing as small as possible. However, safety requirements related to OSHA would suggest that the size recommended by good forest practices may in fact be too small for safety purposes.

Control strategies developed in many states are leaning toward a regulatory permitting system. The creation of paperwork resulting from such a permitting system would seem to be unnecessary. The prescribed or recommended BMPs become the contract under which a forest manager operates. The implementing agency should be able to check for compliance against the recommended practices, with or without going through a formal permit process. Even under extreme situations, notification should be sufficient.

Finally, it should be recognized that the control of nonpoint sources of pollution can best be achieved through high-level performance, as measured against prescribed practices, rather than through enforcement of water quality standards. As pointed out above, the key for control of nonpoint pollution is prevention and not treatment. It is very difficult to rely on water quality standards for direct control of nonpoint sources, both due to variability of the system and due to problems inherent in defining "natural background."

In addition, it is important to understand the relationship between water quality standards and water quality goals. In my view, the overall water quality goals are those established by Congress in the Clean Water Act, *i.e.,* fishable and swimmable. Water quality standards are only a means to measure the receiving water to ensure that water quality goals are met. Inasmuch as the water quality standards adequately reflect those goals, they are appropriate. Most specialists dealing in this subject area would agree that current state water standards do not adequately reflect the goals/standards relationship for nonpoint sources.

While most forest managers would agree with the concept of utilizing management practices as opposed to water quality standards for purposes of control, it is important to maintain those standards as a yardstick for evaluating the effectiveness of forest practices. Without maintaining the water quality standards, there is a danger of moving toward a set of technology-based practices which have no basis in fact as related to the actual needs of the receiving water. This occurred during the development of regulation for control of point sources, *i.e.,* best practicable technology (BPT) and best available technology (BAT), prior to the 1977 amendments.

CONSIDERATIONS IN THE DEVELOPMENT OF SILVICULTURAL 208 PLAN

Based on the objectives for the 208 plan outlined above, it is appropriate to discuss a few points that should be considered for inclusion in that plan. Of the many points that need to be covered, this discussion will deal with only six. These points are not always adequately covered in current 208 plan proposals.

A point often overlooked in development of control strategy for nonpoint sources of pollution is that the first effort should be directed at a determination of where the nonpoint sources are significant. It makes little sense to develop control for a problem that does not exist. An early effort therefore should be directed toward identification of problem areas where significant nonpoint source pollution exists.

In most cases there will be more than one BMP for each site. The concept of "a" best management practice is incorrect. The land manager must be given the latitude to select among an appropriate range of acceptable management practices.

In addition, the selection of a BMP must allow for appropriate trade-offs between alternative practices. The "cookbook" approach often suggested can be counterproductive as far as water quality protection is concerned. For example, construction of twice the road mileage to avoid yarding through a stream course may produce more problems than it solves.

An extremely important concept to consider is that of the Streamside Management Zones (SMZ). For whatever reason, the present understanding of buffer strips is synonymous with "leave strips." Under certain circumstances, leave strips may be a proper forest management alternative, but misapplied it can create environmental problems. Unmanaged zones left for a long period of time can become overmature and begin to break up. Trees left often experience sun scald and increased disease problems, becoming more susceptible to wind-throw.

In contrast, the SMZ concept embodies the principle that streamside zones are managed, but managed recognizing the need for water quality protection. For example, the key in management for control of sediment production is in protection of stream banks and stream channels, while the key in the management for control of stream temperature is in providing the necessary shade. Regarding sediment then, overstory can often be removed as long as the streams themselves are protected and site disturbance is maintained at minimum levels. Regarding temperature, the small streams that are most sensitive often need only the understory vegetation for shade, and large streams may not need shade due to high thermal capacity. The key for temperature control in most cases may be more related to understory vegetation than to the overstory. Finally, due to the variability of stream characteristics, SMZs must be designed on a case-by-case basis. Many will recognize that this SMZ concept is entirely compatible with the original definition of buffer strips. Because of the apparent change in definition and the problems created, it is important to include the concept of management in the descriptive term.

The development of a stream typing system is an important consideration during the planning process. Because forest management practices necessarily vary depending on the size and flow characteristics of the stream involved, such a classification system is extremely important. Any system developed should be based on field-observable physical criteria and not on subjective biologic terminology that requires a biologist to interpret.

Finally, it is extremely important to carefully consider the way in which computer modeling is utilized in the 208 plan. Under current state-of-the-art, computer modeling for nonpoint loading and for nonpoint source evaluation may be inappropriate. There has been a great deal of activity to develop deterministic models for prediction and control of nonpoint sources of pollution. The level of accuracy and precision of these models is completely inadequate for such a task. This inadequacy is not a result of incompetent computer programmers, but relates to the extreme variability of the system with which they must deal. Even if such a model were possible, it would be of such an immense size that our current computer systems could not handle it. If computer models do have a role, then it is probably more in the realm of stochastic probabilistic models used for comparison of alternatives for a much larger land base. Such models would not be used to predict water quality precisely for a given management practice but would be used for assessing impact for a larger land area or range of condition. In addition, such models would provide an aid in deciding between various management alternatives.

IMPLEMENTATION OF THE 208 PLAN

In implementing the 208 plan, education is an important first step. Regulatory agencies must thoroughly understand the plan and land managers must likewise understand why they are being asked to apply certain preventive techniques. This is an area in which improved information related to biological impact is extremely valuable.

Both land managers and forest practices officers must understand exactly what the guidelines say and how they are to be used. In developing such an educational effort, training aids will necessarily need to be developed to explain the "what, why, when and where" of the guidelines. In addition, workshops can be considered as a means of technical and educational transfer.

ASSESSMENT OF EFFECTIVENESS

An assessment of the effectiveness of BMPs in meeting land management goals, including both water quality and timber production, must be conducted following implementation. Such an evaluation must include both the adequacy and excessiveness of management practices. Following such an assessment, consideration should be given to changing either the water quality goals (standards) and/or changing the recommended forest practices. The key is to adequately consider both aspects.

In conclusion it is hoped that the above discussion has stimulated some thought for those involved in the development of 208 plans. Answers to the questions posed, as well as development of a viable 208 plan, are vitally important if we are to arrive at a viable management plan for our forest lands, a plan which provides a climate for both timber production and for water quality protection. Forest management and intensive stand management is the key to those lands relegated to timber production, if we are to meet future demands for wood fiber. Increased production on these lands will reduce the pressure for timber on those forest lands set aside for recreational pursuits. It does present a real challenge for biologists concerned with timber production to work closely with biologists concerned with the aquatic ecosystem, to objectively define impacts and to work toward solution of conflicts.

19

AN APPROACH TO WATER RESOURCES EVALUATION OF NONPOINT SOURCES FROM SILVICULTURAL ACTIVITIES— A PROCEDURAL HANDBOOK

J. B. Currier, L. E. Siverts, R. C. Maloney

Watershed Systems Development Group
U.S. Department of Agriculture—Forest Service
Ft. Collins, Colorado

INTRODUCTION

The Federal Water Pollution Control Act Amendments of 1972, commonly referred to as PL 92-500, established definite goals regarding the restoration and maintenance of the physical, chemical and biological integrity of the nation's waters. The Act requires that water quality management planning, carried out under Section 208 of the Act, include a process that identifies nonpoint sources of water pollution and that establishes methods to control those sources to the extent feasible. Nonpoint sources associated with silviculture and related runoff are among several sources specifically mentioned in the Act as areas to be addressed during 208 planning and implementation.

The purpose of this paper is to provide a systematic, procedural and analytical methodology for identifying and assessing alternative technical solutions to existing or potential nonpoint source problems associated with site-specific silvicultural activities. While the specific analytical methods presented are not the only methods available, they were carefully chosen according to the capabilities of the science and the present state-of-the-art.

Nonpoint sources of pollution result from natural causes, human actions, and the interactions between natural events and conditions associated with human use of the land and its resources. To control these sources, the U.S. Environmental Protection Agency (EPA) has adopted, through federal regulation, the concept of Best Management Practices. As defined by EPA, Best Management Practices (BMP) means:

> a practice or combination of practices that are determined by a state (or designated areawide planning agency) after problem assessment, examination of alternative practices, and appropriate public participation to be the most effective, practicable (including technological, economic and institutional considerations) means of preventing or reducing the amount of pollution generated by nonpoint sources to a level compatible with water quality goals.

This paper deals specifically with the concern and requirement for control of nonpoint sources of water pollution related to silvicultural activities as expressed in the Federal Water Pollution Control Act Amendments of 1972 and the Clean Water Act of 1977. The paper covers only the technical aspects of nonpoint source water pollution control. It does not address the economic, social and institutional aspects that are also an important part of the BMP identification process. The economic considerations are described in "Silvicultural Activities and Nonpoint Pollution Abatement: A Cost-Effectiveness Analysis Procedure." The social and institutional considerations are manifested through public involvement during environmental assessment review processes.

DEFINITION OF EXISTING WATER QUALITY AND WATER QUALITY OBJECTIVES

A prerequisite for use of this technical evaluation procedure is the identification of existing water quality and water quality objectives as quantifiable numerical expressions. This type of objective provides a base against which the impacts of the proposed silvicultural activities can be compared so the degree of additional control measures necessary can be identified.

In defining water quality objectives against which analysis results will be compared, it must be noted that the present state-of-the-art is, at best, a rational estimation procedure. Comparative analysis will often fall short of predicting absolute values.

APPLICATION OF THE PROCEDURE

Silvicultural activities to which the described procedures apply include timber harvesting, transportation systems, and various cultural practices such as site preparation and timber stand improvement. These silvicultural activities are discussed in relationship to the principal potential water pollutants that may be generated and transported from the site. Such pollutants include inorganic sediment, nutrients (primarily nitrogen and phosphorus), heat, organic debris and introduced chemicals such as pesticides and fertilizers.

The technical procedures give proper recognition to space and time variations occurring in natural environments, to the pollution generation processes involved, and to defined water quality objectives. Thus, they permit evaluation of water quality management options at a level compatible with other resource evaluations. They also permit comparision of the effects of proposed management alternatives on water quality in different watersheds and on different areas within a specific watershed, given the same data base.

Application of the technical methodology generally requires a basic knowledge of hydrology plus a working knowledge of forestry, soil science, and engineering principles as they are applied in a natural environment. For all practical purposes, analysis and prediction of nonpoint sources of water pollution is a rational estimation procedure that is useful in comparative analysis of alternatives. Therefore, it is necessary for informed professionals to use local experience in applying the analysis techniques.

Although primarily a guide for the technical specialist, the handbook is also designed for water quality management planners and other land managers. The flowcharts in the "Introduction," "Procedural Summary," and "Control Opportunities" sections guide these managers in defining technical assessments needed. The analytical procedures and references in the technical chapters guide technical specialists or consultants in making those assessments. The step-by-step illustrations and the "Control Opportunities" sections guide project designers and managers in identifying appropriate practices for the particular activity and site conditions.

CHARACTERIZATION OF THE SITE

Because the character of a site largely determines the nonpoint sources that might be encountered and the effectiveness of specific control measures, good site characterization data are essential.

Soil survey reports, stream survey reports, and geologic, climatic, topographic and vegetation maps with accompanying descriptive materials all provide input for development of water quality plans and other environmental assessments. The level of detail in these documents should be compatible with the degree of reliability expected from the analysis (recognizing the sensitivity as well as the strengths and weaknesses of the analytical procedures in terms of data input).

In order to evaluate nonpoint sources on specific sites or projects, the level of information must be compatible with the map resolution used to identify the first-, second-, or third-order drainage basins as described by Strahler.[1] The analysis procedure is applicable only to these headwater areas (third-order basins or smaller).

A larger basin may be characterized from selected third-order drainages within that basin through data analysis and extrapolation based upon the similarities in site and management activities. These evaluations may be useful in identifying general types of practices which may represent BMPs and in analyzing responses for specific silvicultural activities basinwide. However, the site-specific analysis is the only option that considers site and activity variability and the identification of a site-specific BMP.

An environmental setting is a continuum which includes the hydrologic cycle, the nutrient cycle, and the erosion/sediment processes. The nature of the nonpoint process is such that the potential pollutant must be traced as thoroughly as possible through the entire system. Therefore, all major environmental factors significantly affecting its generation and transport (into the receiving waters) must be recognized. Then these factors must be related to the physical and biological processes that govern the pollutant's ultimate disposition. This process is critical in determining controls for nonpoint sources caused by silvicultural activities, because most water quality constituents identified as pollutants also occur naturally within the system. The analysis methodology is structured to differentiate natural pollution sources from those which may have resulted from human activities.

GENERAL PROCEDURAL DESCRIPTION

The procedure addresses the examination of the factors associated with generation and transport of pollutants; it discusses identification, in comparitive, numerical or qualitative terms, of the changes in pollutant output expected to follow particular silvicultural activities on a specific site.

The techniques suggested for comparing existing water quality with the water quality changes expected from proposed silvicultural activity provide a rational approach for dealing with the following facts: (1) day-to-day variations in water quality in undisturbed forest watersheds are substantial, particularly during the periods of changing flows; (2) fluctuations in undisturbed systems may be as great as those in apparently similar, but disturbed, systems.

The procedure evaluates proposed silvicultural plans to identify expected changes in water quality and to determine the type and degree of control needed, if any, to meet water quality objectives. The evaluation process continues until: (1) a combination of preventive and mitigative controls that meet the objectives have been determined, or (2) an acceptable land use alternative, which meets the objectives, has been determined. Mitigative controls may be necessary to correct existing nonpoint sources before any new activities can be made technically acceptable.

This document includes an introduction; a procedural summary; a control opportunities section; five technical sections with quantitative discussions of hydrology, surface erosion, soil mass movement, total potential sediment, and temperature; an example demonstrating the quantitative procedures; three technical sections with qualitative discussions of nutrients, dissolved oxygen and organic matter, and introduced chemicals; and a glossary of terms. The procedural summary provides a general overview and a simplified analysis methodology for each subsequent section showing the general processes and their relationships. The control opportunities and technical sections present a detailed discussion of the procedures involved and the interrelationships between processes.

The general procedure and interrelationships between the control opportunities and the technical sections, both quantitative and qualitative, are presented in Figure 1. The diagram depicts the iterative process that may be required if the proposed silvicultural activity does not meet water quality objectives. During this process, the control opportunities are evaluated and the silvicultural activity revised as needed.

CONTROL OPPORTUNITIES

Because silvicultural activities change certain landscape characteristics, primarily by disturbing the soil mantle, by altering the vegetative cover, and by changing local drainage patterns, the generation and transport of potential pollutants may be accelerated. Utilization of effective control techniques must then be considered.

In this paper, control techniques are grouped into procedural, preventive and mitigative categories. Procedural controls are those concerned with administrative actions. Preventive controls apply to the preimplementation, planning phase of a silvicultural activity. Mitigative controls are physical, chemical or vegetative measures applied to ameliorate problems that exist now, as well as those that exist after a silvicultural activity has taken place.

Procedural, preventive and/or mitigative control practices can be prescribed for various reasons, commonly including: (1) protection of water quality, (2) protection of capital investments, such as roads and buildings, and (3) protection of site productivity. It may not be necessary to specifically formulate controls for water quality because the controls imposed for site protection may be adequate to meet water quality goals. It is logical to first design a management plan to ensure protection of site productivity and capital investments. If subsequent analyses show such a plan to be inadequate to meet the water quality objectives, additional controls can be prescribed as needed.

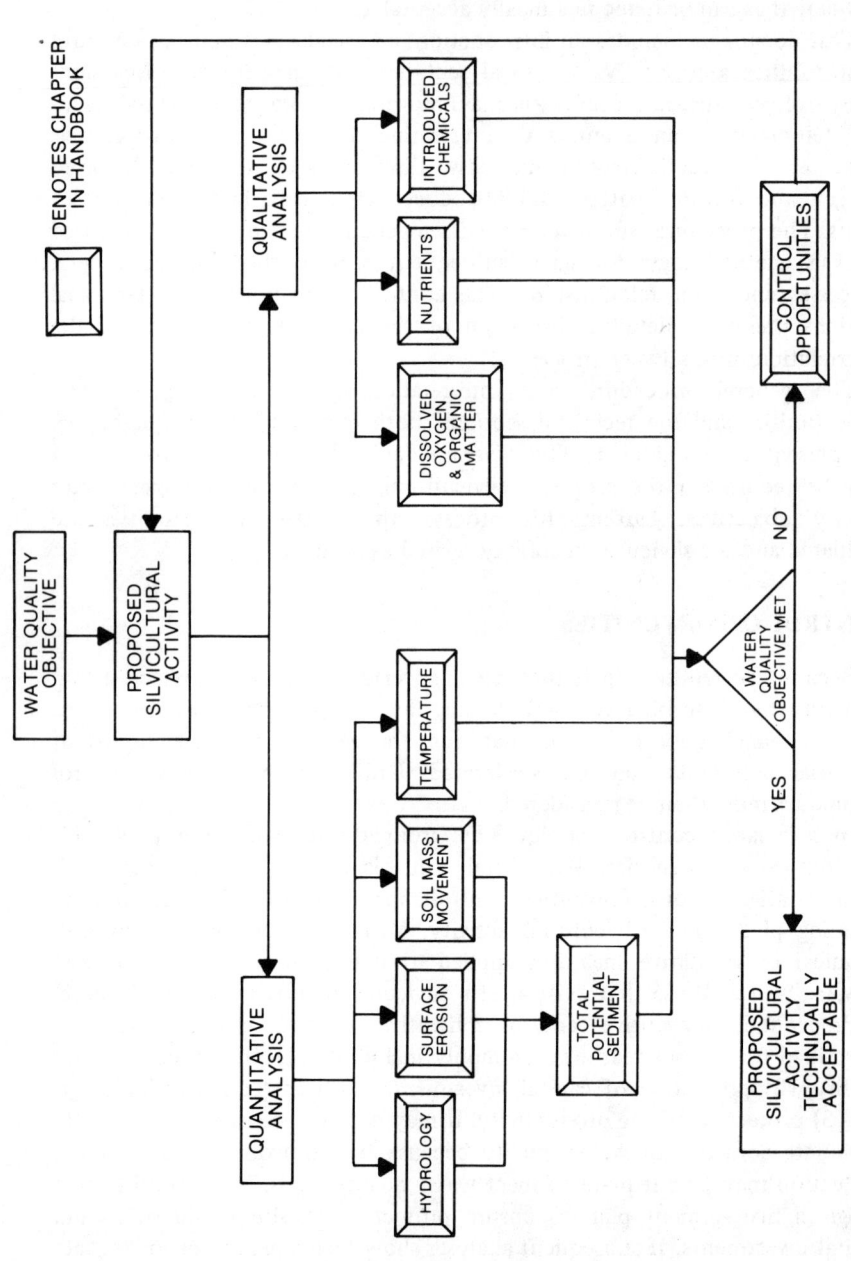

Figure 1. Interrelationships between the quantitative, qualitative and control sections and their application to a proposed silvicultural activity.

The control measures are presented in four different ways. First, there is an activity-impact list that describes each silvicultural activity and its associated resource impacts. Next there is a list of resource impacts and possible control opportunities. Then each control is presented in a series of tables that displays its relationship to the variables in each of the technical sections. Finally, there is a description of each control and whether it is procedural, preventive or mitigative.

HYDROLOGY

The hydrology section begins with a description and an analysis of the hydrologic system of the area under study. Among the many variables considered in the evaluation are precipitation, evapotranspiration, soil-water status, and stream flow, all of which influence, either directly or indirectly, the availability of energy for generation and/or transport of nonpoint source pollutants. Thus, results of the hydrologic analyses provide essential input information for subsequent analysis of nonpoint source pollution potentials using methods described in the Total Potential Sediment and Temperature sections.

Hydrologic response to silvicultural activities varies greatly from hydrologic region to hydrologic region, as well as from site to site within a hydrologic region. In those hydrographic regions where snowfall dominates the hydrologic cycle, all pertinent processes, including snow redistribution, are discussed and methods presented for their evaluation. However, in other parts of the country, some of the processes (such as snow redistribution) are not significant. Hence, guidelines are presented for modifying the basic, more comprehensive analytical framework to account for regional hydrologic differences.

SURFACE EROSION

A Modified Soil Loss Equation is presented in the Surface Erosion section as a method that may be used to estimate surface erosion from disturbed sites. Tables, graphs and equations are used for the evaluation process. To apply these tools, information on soils, topography and cover characteristics must be provided in the appropriate format.

Estimates of sediment which may be delivered to a stream system are based on seven factors which influence sediment delivery. The procedure delivers eroded material across a reference boundary, such as out of clearcut block and into an adjacent area.

SOIL MASS MOVEMENT

The chapter on soil mass movement provides a method for identifying and qualitatively assessing the site factors and management activities that increase the hazard of mass movement. Mass movements are classified into two general types: the debris-slide and debris-avalanches and the slump-earthflow types. Overall ratings can be made in terms of high, moderate or low hazard.

Only material that is delivered directly to a channel system is considered under soil mass movement. It is recognized that mass movement produces a supply of erodible material that may reach stream channels at a much later date than the actual mass movement event. It is also recognized that considerable onsite resource damage may occur; but, unless the material reaches a channel, there would not be any water quality degradation. The effect that any failure will have on water quality degradation depends primarily on the size and volume of material reaching a channel and the energy of the stream system for transport.

The information obtained from the soil mass movement evaluation is used as input to the total potential sediment production section.

TOTAL POTENTIAL SEDIMENT

The total amount of sediment present in a stream is the combination of sediment contributed from surface erosion, soil mass movement, and channel erosion. The procedure for estimating total potential sediment involves the determination of sediment supply and energy availability. Sediment supply is a function of both the size and concentration of inorganic sediment from all sources. Supply changes with variations in surface erosion, soil mass movement, and channel erosion. Energy is primarily a function of stream gradient and volume of flow. Energy changes occur with seasonal fluctuations in rainfall and stream flow, as well as changes in water flow due to reduction of vegetative cover—thus, the determination of total potential sediment requires inputs from analyses described in the Hydrology, Surface Erosion, and Soil Mass Movement sections.

TEMPERATURE

Increased water temperature can be either beneficial or detrimental to the water resource. For streams that are cooler than optimum, a moderate increase in temperature could increase productivity and have a beneficial effect on the aquatic environment. However, streams having temperatures that approach critical threshold limits during the summer months could reach lethal levels if the temperature were increased.

Removal of shading vegetation along stream channels increases the potential for a rise in water temperature if such removal increases exposure of the stream to heating from solar radiation. The magnitude of the increase is a function of the following variables: the amount of canopy removed, length of time the stream is exposed to direct solar radiation, stream bed material, stream width, stream discharge, and subsurface flow. The described procedure, based upon the use of a temperature model, provides a means of assessing the influence of these variables as they are affected by silvicultural activities and control practices. Downstream temperature changes are evaluated using a mixing ratio.

EXAMPLE

An example is provided to illustrate the analysis procedures that have been described in the preceding technical section. Two hypothetical watersheds and silvicultural activities, one in a show-dominated area and one in a rain-dominated area, are presented. Each step in the analysis procedures is described, along with the data needs and any subjective evaluation that is required. Use of the control section is also illustrated by using the control analysis procedure to select preventive and mitigative controls.

DISSOLVED OXYGEN AND ORGANIC MATTER

Silvicultural operations can potentially reduce the concentration of dissolved oxygen in the water to the lethal level for some aquatic species through introduction of organic materials and increased water temperatures. The state-of-the-art is such that it is not possible to rigorously quantify the impacts associated with the introduction of organic material to the aquatic system. The section describes in general terms the processes involved and identifies situations which may create undesirable consequences.

Water temperature, elevation, aeration potential, type of aquatic life present, and stream uses are considered in the discussion. Essential control measures can then be selected which will protect the values involved.

NUTRIENTS

Nitrogen and phosphorus are the nutrients generally cited as having the greatest potential for impacting water quality in a forest environment. Streams may show symptoms of overenrichment if there is a continuous supply of nutrients and substantial periods of low water flow, but generally there is minimal opportunity for buildup of nutrients in streams due to continual transport by water.

The discussion in this section places major emphasis on the sources of nitrogen and phosphorus in the forest environment, the intracycle processes in the forest, and the nitrogen and phosphorus outputs from the forest.

Models for predicting soluble and insoluble nutrient losses from silvicultural activities are not sufficiently developed and tested for general application. Therefore, only qualitative guidelines are given for evaluating soluble and insoluble nutrient changes within a system.

INTRODUCED CHEMICALS

Insecticides, herbicides, fungicides and fertilizers are chemicals commonly introduced into a watershed as part of silvicultural management. Introduced fertilizers enter a watercourse by direct application of fertilizer to the water surface or by leaching and subsequent subsurface flow of dissolved compounds or decomposition products. The impact of pesticides on water quality depends primarily on the following five factors: (1) toxicity to man and aquatic organisms, (2) mobility, (3) persistence, (4) accuracy of placement, and (5) orientation to streams.

This section is directed primarily to a discussion of the types of pesticides and fertilizers used and to the type of impacts that has been observed onsite and in the aquatic ecosystem. Effective controls are discussed but, due to the present state-of-the-art, no attempt has been made to quantify control effectiveness.

SUMMARY

The document is designed as a management tool to be integrated into the overall resource management process. The analysis is compatible with other resource evaluations and can be compared between watersheds and between proposed management alternatives within a specific watershed given the same data base. By following the logic and procedures outlined for the planning and implementation phases of a silvicultural activity, water quality in streams throughout the nation can be maintained or enhanced.

REFERENCES

1. Strahler, A. N. "Quantitative Analysis of Watershed Geomorphology," *Trans. Am Geophys. Union.* 38:917-920 (1957).

20

ESTIMATING IMPACTS OF SILVICULTURAL MANAGEMENT PRACTICES ON FOREST ECOSYSTEMS

F. R. Larson
   Rocky Mountain Forest and Range Experiment Station
   U.S. Department of Agriculture-Forest Service
   Flagstaff, Arizona

P. F. Ffolliott, W. O. Rasmussen
   School of Renewable Natural Resources
   University of Arizona
   Tucson, Arizona

D. R. Carder
   Rocky Mountain Forest and Range Experiment Station
   U.S. Department of Agriculture-Forest Service
   Flagstaff, Arizona

INTRODUCTION

A prototype family of computer simulation models is being developed to help forest managers and land-use planners estimate the impact of silvicultural management practices on forest ecosystems. This family, called ECOSIM (Ecosystem COmponent SImulation Models), includes three general modules: FLORA for estimating reponses of the forest overstory, herbaceous understory, and organic material; FAUNA for evaluating animal habitats, carrying capacities, and population dynamics; and WATER for assessing streamflow yield, sedimentation and chemical quality. A command system enables users to operate all modules through a common language written in straightforward user terminology. This design provides flexibility in representing management activities by users operating selected modules interactively with a computer on an appropriate data base.

ECOSIM is being designed so that resource management professionals at remote locations can readily learn and afford its use, obtaining reliable predictions while still using modest computer terminal equipment and readily accessible data. This paper highlights the development of ECOSIM to date and outlines the future directions in the development effort.

FLORA MODULE

The FLORA module will consist of computer simulators that predict the growth, yield and diversity of forest overstories, the production and, to some extent, composition of herbaceous understories, and the development and accumulation of organic material on the forest floor.

Forest Overstories

Currently, the component simulators designed to estimate the growth and yield of forest overstories generally fall into two categories: first, models that are broadly structured to represent a wide variety of tree species (or tree species groups), and second, models that are specifically structured to represent a particular tree species (or tree species group).

Within the first category, three simulators have been developed, or are being developed. These simulators, called TREE, STAND and FOREST, will estimate the growth and yield of an individual tree, a forest stand (by definition, a community of trees possessing sufficient uniformity, *e.g.*, composition, age, spatial arrangement, or condition to be distinguishable from adjacent communities), and an entire forest property, respectively.

TREE is an interactive modification of a computer model that simulates the growth of an individual tree from knowledge of diameter, height and volume.[1] While this is not a new concept in forestry, the approach exemplified by TREE differs from that of others[2-4] who have employed mathematical formulas to simulate tree growth phenomena. The primary reason for including TREE in the ECOSIM family of models is to analyze changes in tree growth as influenced by alternative silvicultural management practices.

STAND, which is presently under initial testing, is structured to estimate the growth and yield of forest stands comprised of single tree species (ponderosa pine, slash pine, aspen, etc.) or a mixture of tree species (southwestern mixed conifers, shortleaf-loblolly pine, balsam fir-black spruce, etc.). In essence, the measurable input to STAND to date has involved the simplification of prior stand projection methods applicable to uneven-edged forest stands.[5,6] Subsequent work will be directed toward considerations of growth and yield predictions in even-aged forest stands.

The simulation objective of STAND is to predict the growth (both gross and net) and yield of forest stands prior to and, if appropriate, following the implementation of various silvicultural management practices. Inputs to this modular component include a listing of trees per acre by size class, and associated diameter growth rates and volume expressions. As management is prescribed to change these inputs, posttreatment growth and yield are interactively generated. Silvicultural management practices that can be simulated within STAND represent an array of viable options for the different forest stand compositions being considered.

Outputs derived from STAND, including summaries of basal area levels before, through time and following a management redirection, are readily used by other modular components in FLORA and by other modules in ECOSIM. Since the manipulation of forest overstories is a primary management activity affecting many aspects of an ecosystem, such interfaces among modular components are critical to realistic simulation of an ecosystem's overall behavior.

FOREST is being assembled as an interactive version of other general computer models that have been structured to simulate the growth and yield of single or mixed tree species, and even-aged or uneven-aged forest properties.[7,8] This modular component addresses topics of forest growth and yield such as seed production, dispersal and germination, as well as composition, mortality, and stocking manipulation by man.

In concept, inputs to FOREST include a set of real or generated tree locations and associated tree characteristics. Each tree is then grown for a specified number of projection periods, based on potential growth functions modified by competition measures synthesized from relative tree size, crowding, and shade tolerance. Mortality is generated stochastically and depends, in part, upon the competitive status of the individual tress. Reproduction is represented by simulating seed production and germination and, if appropriate, sprout production from the forest overstory. Numerous site alternations and harvesting options can be specified as the forest develops. Outputs from FOREST will be in the form of periodic tables displaying data on stocking, mortality and yield for an array of primary wood products and total biomass.

FOREST is initially being written as a set of individual computer subroutines, each of which correspond to particular stages or processes in the development of a tree, forest stand or forest property. Each subroutine is being designed to facilitate inputs by noncomputer-oriented users not necessarily familiar with computer technology or the operational details of the original computer models.

An example of a simulator designed to estimate the growth and yield of a particular tree species (or tree species group) within the ECOSIM family of models is PIPO.[9] This modular component is an interactive management simulator for ponderosa pine stands. Users can initialize the model by selecting prestored stand tables, by entering tallies from point sample inventories, or by entering the number of trees per acre by size class. Then harvests can be specified at intervals through a sequence of questions and answers to meet a particular management objective. PIPO can be used to determine stand density levels or cutting practices for desired volume production, rotation lengths and sustained yields.

In addition to computer simulators that predict growth and yield, a modular component is under development in FLORA to estimate the diversity of (and within) forest overstories. This component, called DIVER, has two primary options with respect to manipulations of forest overstories—clearing and thinning. The clearing option derives a diversity index that represents the edge irregulatiry of a clearing (or other type of forest opening). The thinning option calculates a diversity index that represents the proportion of an area that is stocked to different forest density levels.

The diversity index derived by the clearing option is based on a previously reported analytic model that quantifies wildlife habitat.[10] The geometric shape with the greatest area and the least perimeter or edge is a circle. If the ratio of circumference to area of a circle is arbitrarily given an index of 1, a formula can be used to compute an index for comparison of any area with a circle. The higher the index value is above 1, the greater the irregularity, and by definition within DIVER, this value is an expression of diversity.

In the thinning option, the calculated diversity index is obtained through solutions of forest stocking equations that are developed for the particular forest type and size class distribution being evaluated. Stocking equations define exceedance curves which describe the proportion of a forest area (the dependent variable, expressed in percent) that is stocked to minimum basal area levels (the independent variable). Values that represent minimum basal area levels for alternative silvicultural management practices are the required inputs to the simulator. To date, the set of stocking equations in the DIVER thinning option only characterizes southwestern ponderosa pine forest conditions.[11] Work is underway to expand this option for use in other ecosystems.

Herbaceous Understories

A computer simulator has been structured within the FLORA module to estimate herbage* production from knowledge of forest overstory

---

*As used in ECOSIM, *herbage* refers to all understory species, while *forage* refers to that component of the understory that is considered palatable for a given animal species.

parameters, precipitation amount and, if appropriate, time since the implementation of a silvicultural management practice. Depending upon the particular simulation objective, a user may operate this component, called UNDER, individually or as part of an interfaced package. In the latter case, outputs from other modular components in FLORA and other modules in ECOSIM are utilized as inputs. An interactive language is used in either case.

Many of the previous attempts at developing computer simulation techniques to estimate herbage production have been dependent, primarily, on input variables depicting forest density conditions.[1,2] While this approach remains viable and has been utilized in several UNDER subroutines, the herbage production simulator eventually will also utilize knowledge of forest overstory growth. Estimates of herbage production that are based on knowledge of this variable appear consistently of higher precision than those based on knowledge of forest density alone.

Subsequent additions to UNDER will facilitate partitioning of simulated herbage production into (at least) three categories: grasses and grasslike plants, forbs and half-shrubs, and shrubs.

Organic Material

Two modular components are being assembled to describe the development, accumulation and distribution of organic material on a forest floor. One component, referred to as FLOOR, estimates the accumulation of tree needles and leaves (by layer of decomposition) on a forest floor, the rate of accumulation, and the spatial distribution. Another component, called CROWN, will predict the magnitude of tree crown and branchwood accumulation associated with alternative silvicultural management practices being simulated. These models are being designed such that they may be executed individually or as an interfaced package.

FLOOR is an interactive component which outputs parameters that describe the development, accumulation and distribution of tree needles and leaves as a function of forest density levels (usually expressed as basal area) for different management practices. The following individual layers are considered: litter, fermination, humus and total. To date, the rate of litter accumulation is the only FLOOR simulation output that provides a time dimension. Only the total forest floor (all layers) is represented in spatial distribution.

While in the early stages of development, CROWN is intended to be an interactive version of a previously documented computer program that determines tree crown volumes by layers.[13] With this model, individual trees are assigned one of 15 geometric solid forms to represent the gross volume occupied by the crown (including leaves, branchwood and air between them). Then a density variable is introduced to adjust the estimate of gross crown volume. Knowledge of adjusted crown volumes by area for a given forest stand prior to implementation of a silvicultural management practice, hopefully, will provide a reference point to assess the quantity of tree crowns and branchwood that will occur as logging residues on the forest after implementation.

FAUNA MODULE

The FAUNA module includes interactive computer simulators that describe the habitat quality for a variety of animal species, the potential animal carrying capacity of an area, and the dynamics of selected animal populations within specific ecosystem situations.

Habitat Assessment

Simulators that assess habitat quality fall into two categories—models broadly structured to represent a variety of animal species (including game, nongame and domestic) and models specifically structured to represent a particular animal species.

The primary example of a modular component in the first category is HABRAN (HABitat RANking). This component involves the synthesis of ranked response predictions with, in turn, can be summarized and arrayed as pattern recognition models. Within HABRAN, animal habitats are assigned numerical values ranging from 0 to 10, with habitat quality in an ecosystem increasing with numerical value. The specific assignment of these values is achieved through analyses of functions that relate habitat preference to readily available inventory-prediction parameters, the magnitude of which are altered by alternative silvicultural management practices. By comparing numerical habitat quality values for existing conditions with those predicted for habitats modified by management redirection, either an increase (+), a decrease (-) or no change (0) is determined. Then, a matrix of plusses, minuses and zeros arrayed for all animal habitats and management alternatives of interest (by definition, a pattern recognition model) can be displayed to provide insight as to comparative management impacts.

Initial efforts in the development of HABRAN, centered in the southwestern ponderosa pine forest ecosystem, have been directed toward: big game (specifically deer and elk); small game (including tree squirrels and cottontails); small rodents (such as mice, chipmunks, and ground squirrels); nongame bird species (as grouped by feeding categories); and domestic livestock (including cattle and sheep).

The HABRAN component of the FAUNA module is, in a sense, a first-level-of-interest assessment of the impacts of alternative silvicultural management practices. In many instances, this sort of analysis may be all that is required. However, if estimates of carrying capacities and animal distributions are needed, other modular components may be called into play.

Other habitat quality simulators broadly structured to represent a variety of animal species in ECOSIM that are in the early stages of development include: FEATUR, which is evolving from the featured species concept of timber and wildlife management[14]; and LIFE, which is based on the notion of life forms as related to community and successional stage of habitat dynamics.[15] These two components, along with HABRAN, will provide users with a choice of approaches to the simulation of habitat quality.

Within the category of FAUNA simulators designed to represent a particular animal species, perhaps the best present example is SCAB, an interactive version of a system for rating habitat quality for Abert squirrel in southwestern ponderosa pine forests.[16] This sytem brings together information on food, cover and diversity to produce a simple rating of habitat from poor to excellent. In brief, ratings are based upon: food—the occurrence of cone-producing ponderosa pine and acorn-producing Gambel oak (often found intermixed in southwestern ponderosa pine forests); cover—forest density and tree size criteria; and diversity—combinations of tree groups, dominance and spacing. Changes in food, cover and diversity resulting from the implementation of a silvicultural management practice are reflected by changes in the rating of habitat quality.

Animal Carrying Capacity

In the modular component that has been structured to predict animal carrying capacity, referred to as CARRY, herbage production (entered as a direct input by the user or obtained from the herbage production simulator) is partitioned into usable forage for deer, elk, cattle and sheep. Appropriate plant species to include in each forage component were ascertained from existing literature relevant to the preferred foods for these animals, along with information about appropriate or proper utilization percentages.

It has been assumed that the proper use factors to be applied in CARRY will be introduced by the user in an attempt to meet specific management objectives. It may be necessary, for example, to reduce a proper use factor on a particular range that has been subjected to prolonged overgrazing pressures. As baseline information relating to proper use factors for deer, elk, cattle and sheep increases, the ability to predict carrying capacities will improve accordingly.

The amount of usable forage required per animal unit month (AUM) for the animal species being considered in this prototype model was derived from a base value of 1.0 for cattle. Specifically, animal unit equivalents for deer, elk and sheep, as determined from existing literature, were: 0.19 for deer, 0.50 for elk and 0.15 for sheep. Assuming 750 lb of forage per cattle unit month (again, as determined from available information), approximately 140, 375 and 112 lb of forage are required per AUM for deer, elk and sheep, respectively.

As an alternative to the above-mentioned approach to obtaining AUM values, the user can explicitly (and directly) input AUM values, assuming that knowledge is available to do so.

With respect to the number of months that deer, elk, cattle and sheep will actually be consuming forage on any tract of rangeland, this value is quite variable depending, in part, upon weather factors that characterize the particular ecosystem and year of simulation (time of snowfall in the autumn, time of snowpack disappearance in the spring, etc.). At best, only estimates based on local knowledge of average situations in the long run can be made. However, to provide a point-of-departure in utilizing CARRY, specific forage consumption time durations have been selected. It should be emphasized that the user can readily override these default duration values to more accurately reflect local conditions if better information is available.

At this time, relatively little can be said about possible constraints that may affect the distribution of animals that are considered by CARRY. While it is known that various factors may restrict (or at least modify) animal movements, explicit identification and subsequent quantification are currently difficult. Conceivably, portions of a tract may be eliminated from use because of movement constraints (physiography, fences, etc.) which may necessitate appropriate reductions in animal stocking rates.

The effects of alternative silvicultural management practices on animal carrying capacities of a given area are primarily evaluated through predictions of changes in the level of herbage production. As forest overstories are reduced in density, a corresponding increase in herbage production commonly occurs. The increased production of herbage is then partitioned into forage which, in turn, is converted into AUM values that are distributed over the range.

Population Dynamics

Although still in the formulation stage, an interactive population dynamics model, called DYNAM, is intended to predict the impacts of silvicultural management practices on the reproduction, growth, mortality and structure of selected animal populations. More specifically, this modular component is to predict the manner by which a given population, specified by the user as reflecting existing conditions within an ecosystem, will respond to changes in food, cover and diversity that are attributed to management redirection.

The initial effort is to synthesize a branch of DYNAM to estimate the effects on mule deer populations of silvicultural management practices imposed in southwestern ponderosa pine forests. The subject forest types are essentially summer range; therefore, the model will emphasize that segment in the life of a defined mule deer population. Hopefully, once developed, the generalized structure of this model can be utilized to represent other animal populations in other ecosystems.

WATER MODULE

The WATER module is comprised of generalized components to predict stream flow yield, as well as suspended sediment and chemical quality of stream flow.

Stream Flow Yield

The requirement for a small model with simple data needs to represent stream flow yield has led to the development of YIELD. This modular component, which is a modification of a previously developed computer model that describes hydrologic behavior on forested watersheds,[17] simulates a water balance on a daily basis. It is designed such that the few data inputs required from the user are commonly available.

The primary "driving variable" within YIELD is daily precipitation. Initial values for the amount of moisture stored in the soil are also requested; the only other initialization variable is a measure of forest density conditions, expressed in terms of basal area. Outputs from the model are values representing daily runoff, change in soil moisture, evapotranspiration, and deep seepage. Linkages to other components in ECOSIM are used to obtain basal area, while the outputs of daily runoff are, in turn, inputs to the components used to predict sediment and chemical quality.

To account for snow regimes and freezing temperatures not included in the original version of the water balance model, a subroutine developed from a modification of a documented computer model of snowmelt[18,19] has been added to YIELD. This subroutine, which provides for modeling intermittent snowpacks, is dependent on four daily input variables: maximum and minimum temperatures, precipitation, and shortwave radiation. Initializing requires only limited knowledge of watershed and snowpack parameters.

Sedimentation

A prototype modular component in WATER, called SED, predicts the amount of suspended sediment in stream flow. While structured to facilitate extrapolation to other ecosystems, the present version of this interactive simulator relies on relationships developed in an assessment of suspended sediment data collected from several watersheds in the ponderosa pine forests of central Arizona.

The model is structured to offer a choice between two alternative sets of input data requirements. Input data, which either represents forest density conditions or spatial distributions or organic materials on a forest floor, are entered directly by the user or generated by STAND or FLOOR, respectively. The other data input needed, stream flow yield, may be obtained from YIELD. The program outputs the maximum concentration of suspended sediment each day, the maximum stream flow discharge, and the total weight of suspended sediment produced under alternative silvicultural management practices simulated.

Chemical Quality

A simulator has been devised to estimate maximum concentrations and daily volumes of selected dissolved chemical constituents as an initial attempt at developing a modular component to predict the chemical quality of stream flow. This component of WATER, called QUAL, is specifically aimed at providing descriptions of the chemical quality of low-volume, winter (snowmelt runoff) discharges from watersheds characterized by southwestern ponderosa pine forests and southwestern mixed conifer forests. The primary "driving variable" is streamflow quantity, the magnitude of which will often vary with alternative silvicultural management practices. This input variable can be entered directly by the user or obtained from outputs from YIELD.

Thirteen constituents are estimated within the QUAL framework: calcium, magnesium, sodium, chloride, sulfate, carbonate, bicarbonate, fluoride, nitrate, phosphate, total soluble salts, hydrogen ion (pH) and conductivity. Efforts are underway to include other water quality parameters, such as heavy metals (Zn, Fe, Cu, Pd, Cd, etc.) and dissolved oxygen.

THE COMMAND SYSTEM

The command system of ECOSIM is largely dispersed into the respective modules. In fact, there is little evidence of a main command system in the overall operation of the ECOSIM family of models. Initial selection of the modules and components to be used and subsequent assignment of default values needed in the operation is handled by the user interacting through the command system. Also, timing and sequencing of operation of individual modular components is carried out in this manner. Additionally, summary displays (tables, graphs, maps, etc.) of the simulation results are achieved through the command system.

All of the modules in ECOSIM have been structured to have three modes of operation: initialization, cycling in time, and summary. In the initialization mode, all needed input data are either introduced directly by the user or entered from stored files. The second mode of operation is a cycling in time of the processes being simulated (daily stream flow, yearly forest growth, etc.). Finally, the third mode of operation is summary and, if appropriate, other activities at the end of a simulation problem.

When a user informs the command system which modular components are to be operated, he also states when they are to be used in the simulation problem. For example, the component QUAL may be required to operate only in the fifth year of simulation, while all other components may be operated every year. The command system stores this directive and acts accordingly.

The entire system is designed to operate with minimal input data. Default values are offered with nearly all of the interactive questions posed so that, whether or not the user has the required data or reason for overriding the default values, simulation can still proceed. Similarly, if a module or component is not directly included in a simulation problem, default values are loaded into the system to provide estimates of needed parameters normally obtained as output from the unused modular components.

After the system cycles through the specified number of simulation years, individual models are entered into the summary mode of operation. Any needed computations to allow display summaries of the operation to be output are carried out at this point. Output summaries may be obtained either on a local computer terminal or at a central computer location. These summaries may be brief or detailed, depending upon the user need. In general, the parameters shown are representative of the various modules and components used in the problem. If a component is not used and default data are utilized, the parameters for the unused component will not alter the display.

ADDITIONAL COMPONENTS

As currently conceived, work in the development of ECOSIM will follow two directions: synthesis of other modules and components, and extrapolation of the interactive system into other forest and range ecosystems, as appropriate.

While in various stages of development, several other modules and components are now recognized as part of the ECOSIM family of computer models. For example, to facilitate overall planning with respect to a particular simulation problem, a module called PLAN is being structured to generate a PERT network of activities necessary to reach an objective. Another module, referred to as AREA, calculates the adjusted surface area of management units within an ecosystem, correcting for sloping or broken terrain.

As knowledge of site quality is required as input to some modular components in ECOSIM, a module called SITE is under development to generate site quality directly (through estimation of site indices) or indirectly (through analyses of plant indicators, physiography, soil surveys, etc.). Outputs from this module will describe productivity potentials for both forest overstories and herbaceous understories.

To evaluate depth and quality of view within an ecosystem in terms of current and, if appropriate, anticipated conditions, a module named SEEN is being evaluated as part of ECOSIM. Another module, called FIRE, predicts the probability occurrence of wildfires of given intensities from knowledge of fuel properties and sequencing of meteorological events; this module also estimates the impacts of fire on an ecosystem.

SNOW is a module that interactively simulates the dynamics of snowpack accumulation and melt within forests comprised of trees in varying spatial arrangements. ROAD allows for predictions of sediment loads resulting from the construction of roads with alternative design criteria.

To further aid managers and planners in analyzing land-use alternatives, a module that facilitated the development and subsequent display of basic production economics models (production functions, product-product relationships, etc.) has been synthesized. This module, referred to as ECON, also includes components that represent various linear programming and goal programming techniques. Other modules and components will be considered within ECOSIM to more completely provide socioeconomic simulation capabilities.

The primary emphasis in the initial developmental work on ECOSIM has been placed on simulation within three specific forest ecosystems: southwestern ponderosa pine, southwestern mixed conifer, and southeastern loblolly and slash pine. Currently there is interest in extending this work into other forests and into range and arid ecosystems.

Extrapolation of the ECOSIM concept into other ecosystems will be made easier by the generalized structure commonly followed in much of the initial development. In fact, many of the modular components that have been synthesized only require "localization" of coefficients for extrapolation. Others, particularly those structured to represent an explicit plant or animal species, are only appropriate for use in simulating those ecosystems in which they occur and must be replaced by other species-specific models that characterize other ecosystems. However, even here, replacement will be relatively easy within the overall structure of the command system.

ACKNOWLEDGMENTS

The research reported herein is being conducted by the Rocky Mountain Forest and Range Experiment Station, in cooperation with the School of Renewable Natural Resources, University of Arizona, and is being sponsored in part by the Environmental Research Laboratory, EPA, Athens, Georgia, in accordance with Interagency Agreement EPA-IAG-D4-0437 between EPA and the Forest Service.

REFERENCES

1. Pierce, W. R. "Simulating Tree Growth by Computer: A Direct Approach," *Western Wildlands* 3:27-32 (1976).
2. Beck, D. E. "Predicting Growth of Individual Trees in Thinned Stands of Yellow Poplar," in *Growth Models for Tree and Stand Simulation* (Stockholm, Sweden: Skegshoyekolan, Royal College of Forestry, 1974), pp. 47-55.
3. Lin, J. Y. "Stand Growth Simulation Models for Douglas-Fir and Western Hemlock in the Northwestern United States," in *Growth Models for Tree and Stand Simulation* (Stockholm, Sweden: Skegshoyekolan, Royal College of Forestry, 1974), pp. 102-118.
4. Burkhart, H. E., and M. R. Strub. "A Model for Simulation of Planted Loblolly Pine Stands," in *Growth Models for Tree and Stand Simulation* (Stockholm, Sweden: Skegshoyekolan, Royal College of Forestry, 1974), pp. 128-135.
5. Larson, R. W., and M. H. Goforth. "TRAS: A Computer Program for the Projection of Timber Volume," USDA Forest Service, Agricultural Handbook 377 (1970).
6. Hazel, A. A. "A Design for Multipurpose Surveys," USDA Forest Service, unpublished report (1961).
7. Botkin, D. B., J. F. Janak and J. R Wallis. "A Simulator for Northeastern Forest Growth: A Contribution of the Hubbard Brook Ecosystem Study and IBM Research," IBM Research, RC 3140 (1970).
8. Ek, A. R., and R. A. Monserud. "FOREST: A Computer Model for Simulating the Growth and Reproduction of Mixed Conifer Forest Stands," School of Forestry, College of Agricultural and Life Sciences, University of Wisconsin (1974).
9. Larson, F. R. "Simulating Growth and Management of Ponderosa Pine Stands," PhD Dissertation, Colorado State University (1975).
10. Patton, D. R. "A Diversity Index for Quantifying Habitat 'Edge'," *Wildlife Soc. Bull.* 3:171-173 (1975).
11. Ffolliott, P. F., and D. P. Worley. "Forest Stocking Equations: Their Development and Application," USDA Forest Service, Research Paper RM-102, Rockey Mt. For. and Range Exp. Stn., Fort Collins, CO (1973).
12. Myers, C. A., and P. O. Currie. "Simulation Techniques in Forest-Range Management. Modeling and Systems Analysis in Range Science," Range Science Department, Colorado State University (1970), pp. 17-22.
13. Mawson, J. C., J. W. Thomas and R. M. De Graaf. "Program HTVOL: The Determination of Tree Crown Volume by Layers," USDA Forest Service, Research Paper NE-354, Northeastern For. Exp. Stn., Broomall, PA (1976).
14. Holbrook, H. L. "A System for Wildlife Habitat Management on Southern National Forests," *Wildlife Soc. Bull.* 2:119-123 (1974).

15. Thomas, J. W., R. J. Miller, H. Black, J. E. Rodiek and C. Maser. "Guidelines for Maintaining and Enhancing Wildlife Habitat in Forest Management in the Blue Mountains of Oregon and Washington," *Trans. North American Wildlife and Natural Resources Conference* 41:452-476 (1976).
16. Patton, D. R. "Managing Southwestern Ponderosa Pine for the Abert Squirrel," *J. Forestry* 75:264-267 (1977).
17. Rogerson, T. L. "Simulating Hydrologic Behavior on Quachita Mountain Drainages," USDA Forest Service, Research Paper SO-119, Southern For. Exp. Stn., New Orleans, LA (1976).
18. Leaf, C. F., and G. E. Brink. "Computer Simulation of Snowmelt Within a Colorado Subalpine Watershed," USDA Forest Service Research Paper RM-99, Rockey Mt. For. and Range Exp. Stn., Fort Collins, CO (1973).
19. Solomon, R. M., P. F. Ffolliott, M. B. Baker, Jr., and J. R. Thompson. "Computer Simulation of Snowmelt," USDA Forest Service, Research Paper RM-174, Rocky Mt. For. and Range Exp. Stn., Fort Collins, CO (1976).

21

SIMULATION OF STORMWATER RUNOFF AND SEDIMENT YIELD FOR ASSESSING THE IMPACT OF SILVICULTURE PRACTICES

R. M. Li, K. G. Eggert, D. B. Simons
  Department of Civil Engineering
  Colorado State University
  Fort Collins, Colorado

INTRODUCTION

As part of the emphasis on planned utilization of natural resources, silviculture and other forest management activities, it is necessary to predict sediment yields. Watershed sediment yields reflect erosion rates and also serve as the primary source of fine sediment to rivers. In addition, many pesticides and other pollutants have an affinity for sediment and are transported through the environment bound to sediment particles. Accordingly, turbidity and high toxicity may result from sediment, thereby posing a serious threat to riverine life.

Increasing stress on the forest environment results from almost any activity in the watershed. Since there appears to be no decrease in the demand for timber, resort and recreation sites, and agricultural enterprises, the forest manager must continue to practice silviculture in a manner that minimizes the impact on the environment. A water and sediment simulation provides the manager with a method that allows quantitative comparison of management alternatives.

Previously, sediment yield was predicted using essentially empirical methods. Such techniques include the Universal Soil Loss equation and sediment rating curves. However, these methods assume the system does not change with time. If the system is changed, the historical data are invalidated.

Therefore, a simulation based on hydraulic principles offers the best means of calculating sediment yields because the parameters of the model are sensitive to change.

Simulations based on hydraulic principles have disadvantages that increase with complexity. In general, a more complex simulation requires more data. Since sediment data are often sparse, a complex simulation is often inconsistent with the available sediment data. In addition, complex simulations usually require large computation times, resulting in greater expense. Finally, there are usually too many parameters to be entered for implementation of one interactive format. Therefore, a less rigorous hydraulic simulation model that retains sensitivity to management activities, but requires less data and computation time, would provide a useful alternative. The sediment yield simulation program, SEDWAT, combines an interactive format with such an alternative method.

GENERAL DESCRIPTION

The program presented simulates the stormwater runoff and sediment yield response of a two-plane, one-channel watershed. The two-plane approximation is illustrated in Figure 1 and is essentially that proposed by Wooding.[1] A flowchart of the model appears in Figure 2.

Stormwater Runoff

The modeled stormwater runoff processes include: (1) interception, (2) infiltration, (3) overland flow, and (4) channel flow. The processes are explained below.

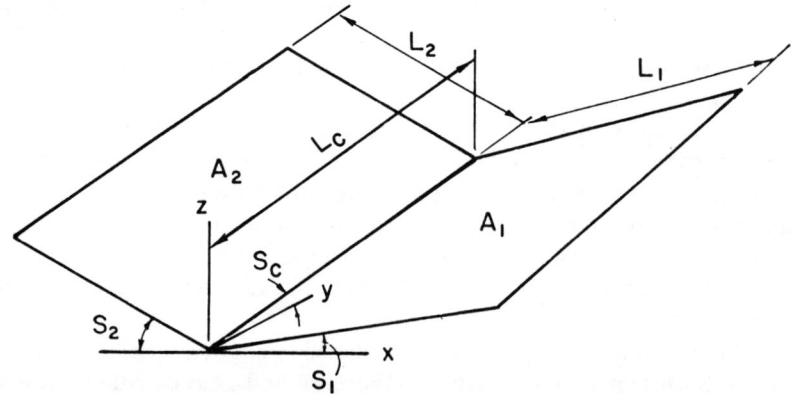

Figure 1. Geometric representation of a watershed.

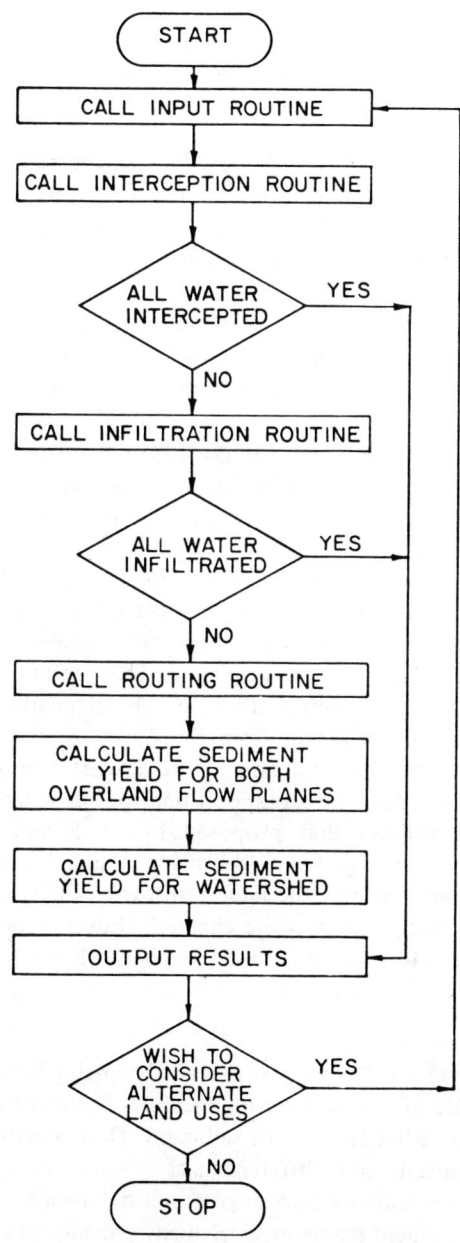

Figure 2. Flowchart of SEDWAT.

Interception. Interception is modeled by rainfall volume reduction at the beginning of the storm. This simple formulation assumes that the interception process requires that a storage volume must be filled before rainfall begins to hit the ground. The volume of water intercepted is assumed to be lost to evaporation.

Infiltration. Infiltration response is calculated using an explicit formulation of the Green-Ampt infiltration equation.[2] This formulation allows the calculation of an excess rainfall histogram resulting from a time-varying, spatially uniform rainfall. The method is described fully by Simons, Li and Eggert.[3]

Overland and Channel Flow. Excess rainfall is routed by the method of characteristics solution to the kinematic wave approximation for overland and channel flow. The method used was, in part, proposed by Harley, Perkins and Eagleson.[4] This method could be used without modification to obtain the discharge at the downstream boundary. However, times when discharge is calculated are irregular and dependent on individual plane parameters. This situation causes a computational difficulty when overland flow discharges from the two planes are added to form the lateral inflow to the channel. Therefore, to circumvent this problem, the program uses Harley's method only to calculate a minimum number of characteristics to subdivide the solution domain. Another method is used to actually calculate the discharge at the downstream boundary at a desired time. This method determines the discharge by calculating the depth along the characteristic line in the upstream direction. Moving upstream from the downstream boundary allows specification of the time that discharge at the downstream boundary is calculated, thereby eliminating the computational problem mentioned above. The approach used follows that proposed by Li, Simons and Stevens.[5]

The overland flow discharge from each plane is added and then averaged, time-step by time-step, to form the lateral inflow to the channel. The same routines are used to route water in the channel. Only the formulation of the channel routing parameters differs from the overland flow case.

Sediment Yield

Erosion and sediment yields are processes of balancing the soil erosion rate and the transport rate of the transporting medium. Sediment yield estimation is made according to different sizes of sediment. Determination of sediment yield by size is important since different sizes of sediment have different uptake rates for water pollutants such as nitrogen or phosphorus. In addition, routing by a mean sediment diameter is frequently inadequate when attempting to predict the transport capacity of overland and channel flow. The sediment transport capacity is divided into bed load and suspended load transport. The resulting sediment transport capacity is given as

$$q_t = \sum_{i=1}^{N} (q_{bi} + q_{si}) i_s \qquad (1)$$

where q_t is the total sediment transport rate
q_{bi} is the bed load transport rate for the ith size fraction
q_{si} is the suspended sediment transport rate for the ith size fraction
i_s is the percentage of sediment in the ith size factor, and
N is the number of sediment size fractions.

The potential volumetric sediment capacity is given by

$$V_t = \frac{W}{\gamma S_s} \int^{D_e - t_p} q_t \, dt \qquad (2)$$

where V_t is the total potential volumetric sediment transport capacity
W is the width of the overland flow plane or channel
γ is the specific weight of water
S_s is the specific gravity of the sediment
t is time
D_e is the duration of the rainfall event, and
t_p is the time of ponding.

Sediment yield is calculated by comparing the volume of sediment detached by rainfall energy and tractive force to the transport capacity by size faction. The lesser of these volumes is assumed to be the sediment yield. The same physical processes for sediment transport are active in the watershed and stream channels. Therefore, the theory used is the same for both cases.

Bed Load. Bed load transport for both overland and channel flow is calculated using the Meyer-Peter, Muller equation. The form used is that presented by the U.S. Bureau of Reclamation.[6]

Suspended Load. Suspended sediment transport is calculated using the Einstein[7] approximation. Concentration integrals that appear in the formulation are evaluated by a power series approximation presented by Li.[8]

Sediment Supply. Supply of sediment comes from two mechanisms: detachment by raindrop splash and detachment by overland and channel flow. Raindrop splash detachment can be formulated as a simple power function of rainfall intensity.[9]

$$V_r = a_1 i^2 \, LW \, (1 - \phi) \, A_b \qquad (3)$$

where V_r is the nonporous volume of detached material by raindrop splash
a_1 is an empirically determined constant describing erodibility of the soil, and
A_b represents the fraction of unprotected or bare soil in the area.

Detachment is determined by

$$V_f = D_f (V_t - V_r) \qquad (4)$$

where V_f is detachment by overland flow and d_f is the flow detachment coefficient. An equation similar to that used in overland flow detachment is used for channel flow detachment. For many cases, however, the flow detachment coefficient for channels is assumed as zero due to natural armoring, riprap or other forms of bed and bank protection. This is especially true for manmade channels.

If $V_t < V_r$ there is no flow detachment because the transport rate is limited by the transporting capacity. Total available sediment supply is then

$$V_a = V_r \tag{5}$$

otherwise,

$$V_a = V_r + V_f \tag{6}$$

Sediment Yield Comparison. The available supply is

$$V_{ai} = i_s V_a \tag{7}$$

where V_{ai} is the available supply for the ith particle size. Values of V_{ti} and V_{ai} can be compared. If V_{ti} is greater than V_{ai}, supply controls. If V_{ai} is greater than V_{ti}, demand controls or

$$V_{yi} = V_{ai} \text{ if } V_{ti} > V_{ai} \tag{8}$$

and

$$V_{yi} = V_{ti} \text{ if } V_{ti} < V_{ai} \tag{9}$$

where V_{yi} is the volume yield for the particle size fraction. The total yield will then be

$$Y_s = \gamma S_s \sum_{i=1}^{N} V_{yi} \tag{10}$$

where Y_s is the sediment yield by weight.

Sediment output volumes for the overland flow hydrographs are calculated on an incremental basis. At each time increment, available sediment supply is determined and compared to the overland flow transport capacity. The smaller of the two is stored in an output matrix, and the total volume is summed.

Because of timing difficulties, sediment yield calculations for channel flow are somewhat different. At each time increment, the channel transporting rate and volume capacities are calculated. Total volume of lateral sediment inflow, the sum of the two overland flow sediment volumes plus a channel detachment volume, is compared to the channel transporting capacity. The smaller of the two is the sediment yield. A time-incremented lateral inflow is not possible due to lack of knowledge of the inflow time lags along the stream. The sediment yield calculation is described in detail in Simons et al.[10]

DATA REQUIREMENTS

The data are entered in interactive format. However, for repetition of large amounts of data the user may use data files. The flexibility allows the user to easily compare alternatives. The following data are required:

Rainfall Data. A rainfall hyetograph is required. If a constant intensity storm is used, it should be subdivided into a sequence of periods with equal rainfall intensity.

Geometry Data. If a more detailed survey is not available, a topographic map can provide the necessary geometric data. Data derived from the map are slope and channel length, L_i and L_e, respectively. In addition, the overland and channel slopes, S_i and S_c, also are defined as shown in Figure 1.

Soil Data. Soil data required are the soil porosity, saturated hydraulic conductivity, average capillary suction head, and the final degree of saturation.

Canopy and Ground Cover Data. Percentages of an area covered by canopy and ground cover are required. These may be obtained from aerial photographs. In addition, the maximum volume of intercepted water per unit area for both canopy and ground cover is also necessary.

Flow Resistance Data. Flow resistance for overland flow is calculated by the program from the percentage of ground cover. Only maximum and minimum values of the resistance parameter need be entered. These values are upper and lower limits on the parameter as shown in the equation:

$$f = \frac{a}{N_r} \tag{11}$$

where f is the Darcy-Weisbach friction factor and N_r is the flow Reynold's number. The friction factor for channel flow is assumed to be constant and also must be supplied.

Sediment Parameters. A sediment size distribution representative of the watershed must be supplied. In addition, detachment coefficients, a and D_f, and bare ground area, A_b, are required.

Antecedent Conditions. The only initial condition required is the degree of saturation at the outset of the storm. This may be supplied by direct measurement or by an interstorm moisture balance model.

SIMULATIONS RESULTS

Use of the simulation to evaluate effects of various land-use alternatives is demonstrated using Watershed 17 in the Coconino National Forest near

Flagstaff, Arizona. The first example illustrates the ability of the simulation to predict runoff. Subsequent examples show the graphical output available to the user. Land-use modifications simulated are mechanical site preparation, degrees of timber harvest and effects of overgrazing.

Example 1

Watershed 17 is a 287-ac catchment in the Beaver Creek Drainage. It is a heterogeneous watershed partially covered by coniferous plants. The storm simulated was recorded on September 5, 1970. Model calibration is based on this event. As shown in Figure 3, the runoff hydrograph is very complex. Calibration for water consisted of adjusting the hydraulic conductivity to produce agreement in runoff volume. Subsequent agreement in peak runoff and hydrograph shape indicate the adequacy of the model. Calibration of the sediment parameters was limited to the detachment coefficients. Sediment yield from this storm was 157,336 lb. Simulated yield was 158,399 lb.

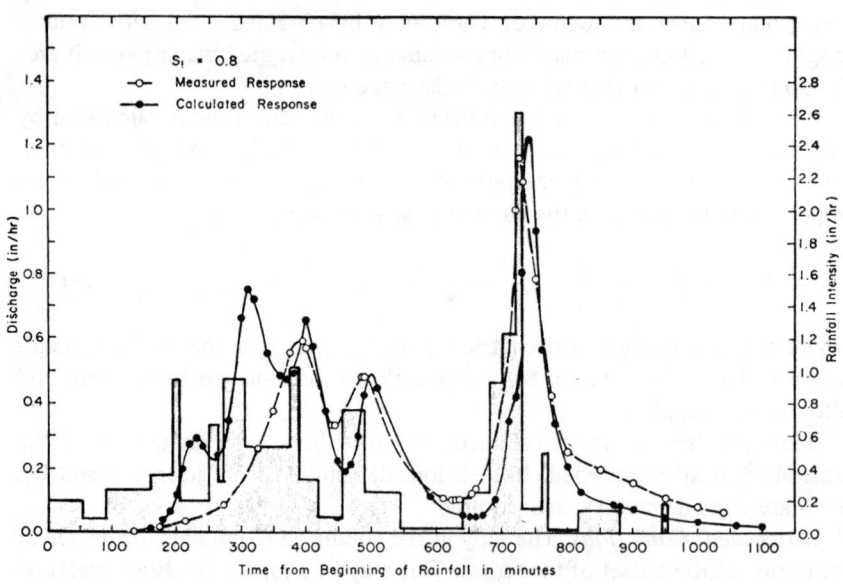

Figure 3. Rainfall-runoff response for Watershed 17, Coconino National Forest.

Example 2

Mechanical site preparation is a necessity in silviculture practice. It is simulated by a decrease in hydraulic conductivity due to compaction of the soil.

The conductivity was reduced to 25% of its original value resulting in the change of the hydrograph in Figure 4. Predicted sediment yield was 167,552 lb.

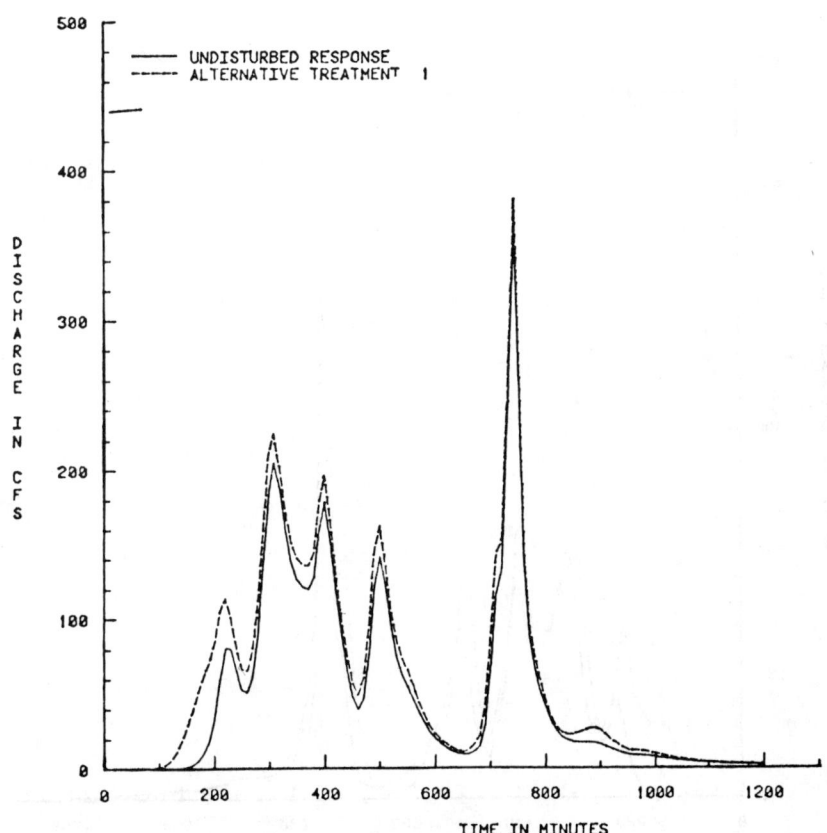

Figure 4. Effects of mechanical site preparation.

Example 3

Clear-cut timber harvesting is a common silviculture technique in the western United States. While clear-cutting is often the most economical harvesting method, it may also be the least desirable from an environmental standpoint. However, the forest manager must often evaluate the effect of timbering. This example compares the removal of canopy and ground cover with the undisturbed watershed. The operation of heavy equipment is assumed to lower the hydraulic conductivity as in Example 2. For the clear-cutting, conductivity is again lowered to one-fourth its original value. Resulting water hydrographs appear in Figure 5. The predicted sediment yield for clear-cutting was 515,468 lb.

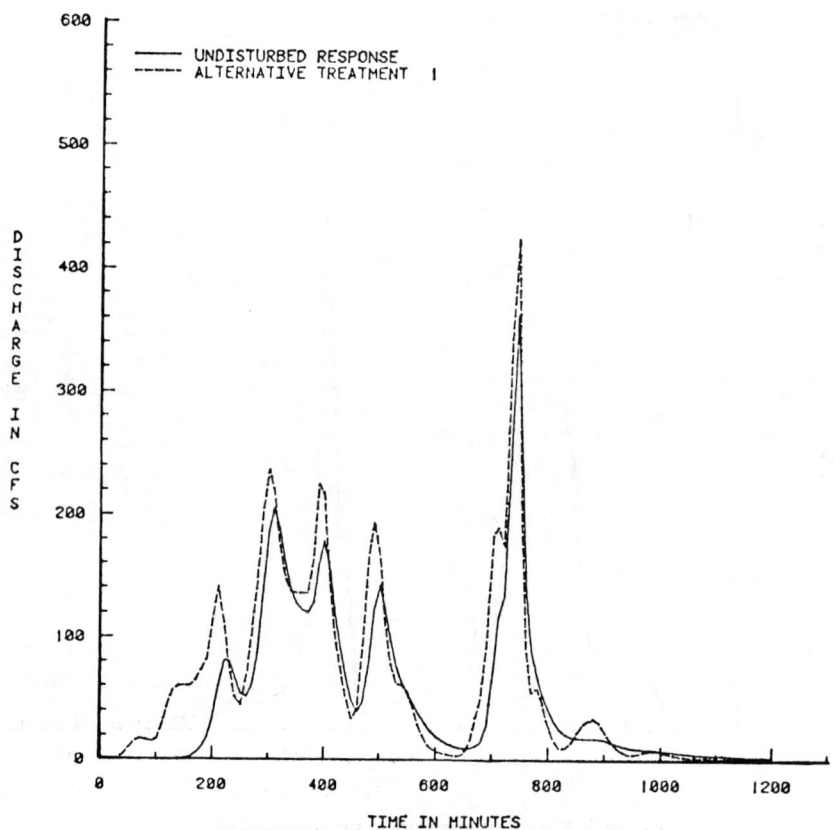

Figure 5. Effects of clear-cut timbering.

Example 4

This example illustrates the effects of overgrazing. Here, ground cover is assumed to be removed by foraging activities, and the hydraulic conductivity is lowered to 60% of its original value by trampling. The resulting simulation appears in Figure 6. Predicted sediment yield is estimated to be 441,900 lb.

Obviously, various degrees of grazing activity could be examined to select the optimum grazing level. A 50% reduction in the ground cover and a reduction in conductivity to 80% of its original value results in a sediment yield of 338,800 lb.

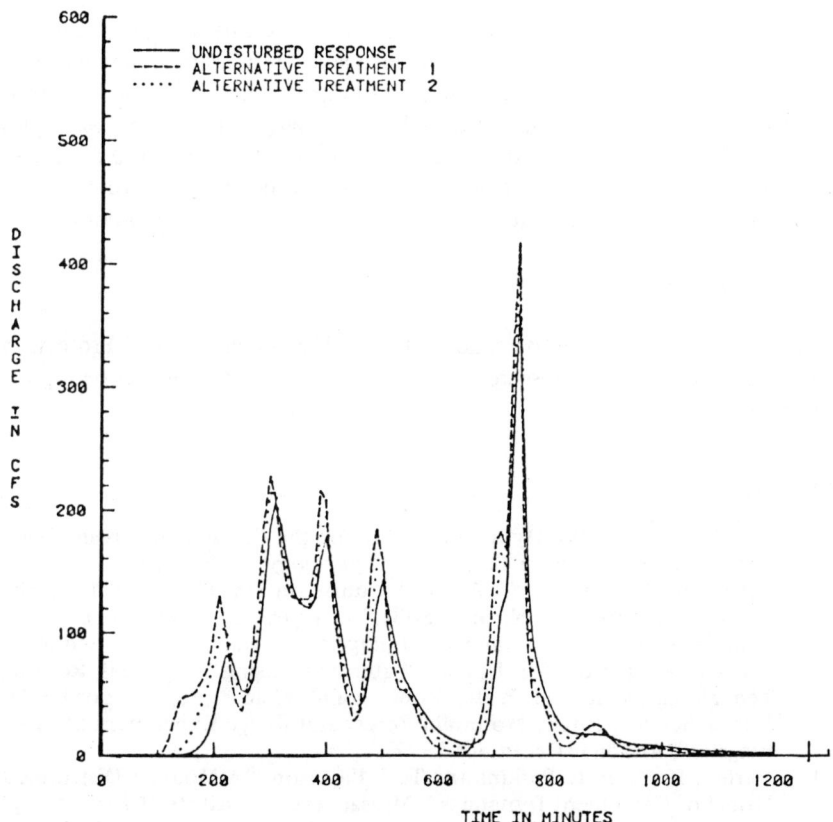

Figure 6. Effects of overgrazing.

SUMMARY

The application of the program is based on a simplified watershed geometry consisting of two planes and one channel. This single-plane model that averages the physical processes over both time and space to obtain a simple approximation of a complex model[11] is based on three assumptions:

1. The design storms are represented as spatially uniform but varying in intensity.

2. Sediment yield can be approximated by comparing the overall sediment availability during the storm and the total sediment transport capacity for the whole runoff period.
3. The protecting effect of the thin ponded water layer and the armoring effect of the loose soil is negligible.

Due to the interactive input of watershed parameters describing the physical processes of erosion and sediment yield, the simplified model can be used easily and equickly by a planner without extensive knowledge about computer programming and advanced mathematics. From the examples provided, it appears the program has great potential as a method for quantitative and qualitative prediction of water and sediment yield. In addition, its simplicity allows for a greater utility than more complex simulations.

ACKNOWLEDGMENT

The authors would like to acknowledge the U.S. Environmental Protection Agency, Environmental Research Laboratory, Athens, Georgia, for supporting this research effort.

REFERENCES

1. Wooding, R. A. "A Hydraulic Model for the Catchment-Stream Problem: 1. Kinematic Wave Theory," *J. Hydrology*, 3(3), 1965.
2. Green, W. H., and G. A. Ampt. "Studies on Soil Physics, Part I: The Flow of Air and Water through Soils," *J. Agric. Sci.* (May 1911).
3. Simons, D. B., R. M. Li and K. G. Eggert. "Storm Water and Sediment Runoff Simulation for Upland Watersheds Using Analytical Routing Techniques, Volume I, Water Routing and Yield," prepared for USDA Forest Service, Rocky Mountain Forest and Range Experiment Station, Flagstaff, Arizona (December 1977).
4. Harley, B. M., F. E. Perkins and T. S. Eagleson. "A Modular Distributed Model of Catchment Dynamics," Massachusetts Institute of Technology, Department of Civil Engineering, Hydrodynamics Laboratory Report No. 133 (June 1970).
5. Li, R. M., D. B. Simons and M. A. Stevens. "On Overland Flow Water Routing," Proceedings of the National Symposium on Urban Hydrology and Sediment Control, Lexington, Kentucky (July 1975).
6. USBR. "Investigation of Meyer-Peter, Muller Bedload Formulas," Sedimentation Section, Hydrology Branch, Division of Project Investigations, U.S. Department of the Interior, Bureau of Reclamation (June 1960).
7. Einstein, H. A. "The Bed Load Function for Sediment Transportation in Open Channel Flows," USDA Tech. Bulletin, No. 1026 (1950).
8. Li, R. M. "Mathematical Modeling of Response from Small Watersheds," Ph.D. Dissertation, Department of Civil Enginering, Colorado State University, Ft. Collins (1974).

9. Meyer, L. D. "Soil Erosion by Water on Upland Areas," in *River Mechanics,* Volume II, H. W. Shen, Ed., Ft. Collins, Colorado (1971).
10. Simons, D. B., R. M. Li, L. Y. Shiao and K. G. Eggert. "Storm Water and Sediment Runoff Simulation for Upland Watersheds Using Analytical Routing Techniques, Volume II, Sediment Yield," prepared for USDA Forest Service, Rocky Mountain Forest and Range Experiment Station, Flagstaff, Arizona (December 1977).
11. Simons, D. B., R. M. Li and M. A. Stevens. "Development of Models for Predicting Water and Sediment Routing and Yield from Storms on Small Watersheds," prepared for USDA Forest Service, Rocky Mountain Forest and Range Experiment Station, Flagstaff, Arizona (1975).

SECTION V

ECONOMIC, POLICY AND INSTITUTIONAL ASPECTS

22

NONPOINT SOURCE POLLUTION FROM AGRICULTURE: SOME SOCIOLOGICAL CONSIDERATIONS FOR IMPLEMENTING POLICY

J. C. van Es, L. C. Keasler

>Department of Agricultural Economics
>University of Illinois
>Urbana-Champaign, Illinois

Nonpoint source (NPS) pollution is one of several important issues affecting agriculture. Policies are being formulated and programs are being designed to deal with this issue. Much of the debate deals with the technical and economic aspects of various policies. The present paper deals with certain sociological aspects that may be easily overlooked but that can greatly alter the success of a program. This discussion should aid in the selection of programs, as well as help agency personnel define their roles in implementing policies.

It is important not to confuse the goal of a policy with the objective of a program. The policy goal refers to the attainment of a certain condition, while the program objective deals with the way in which the policy goal is to be attained. Thus, meeting clean water standards is a policy goal, and implementation of Best Management Practices (BMPs) is a program objective. BMP is not the only program for meeting clean water standards; public education programs and tax programs are other ways in which the policy goal of clean water may be pursued. Which programs to choose in order to attain a policy goal is frequently at the heart of the debate about public policy, although at times the goal of the policy itself is the cause of the controversy. For example, should our national energy policy be based predominantly on energy conservation with all the resulting changes in our lifestyle, or should the policy attempt to expand available energy sources to maintain our lifestyle.

It is useful to assess how closely program objectives and policy goals are related. The more remote or uncertain the relationship between program objectives and policy goal, the more likely the opportunity exists for program objectives to be attained without a comparable achievement of policy goal. Unfortunately, how closely policy and program objectives are related is frequently very difficult to determine on an *a priori* basis, an uncertainty plaguing much of the proposed NPS programs in support of the clean water policy.

Just as problem-solving can be broken down into defining the problem, defining the solution, and implementing the solution, problem-solving programs can be classified on the basis of their objectives: creating awareness of the problem, creating awareness of solution, and implementing solutions. In Figure 1, some possible NPS programs are classified according to these criteria. As Figure 1 indicates, all program objectives may contribute to attaining policy goals. However, programs that provide economic incentives for the implementation of a soil conservation plan will have a more direct relationship to achieving the policy goal of reduced water pollution than programs that direct themselves to creating awareness of the problem only.

As long as farmers are the principal decision-makers regarding on-the-farm activities, participation of farmers in the program becomes the fundamental concern. This participation either can be reached voluntarily, or can

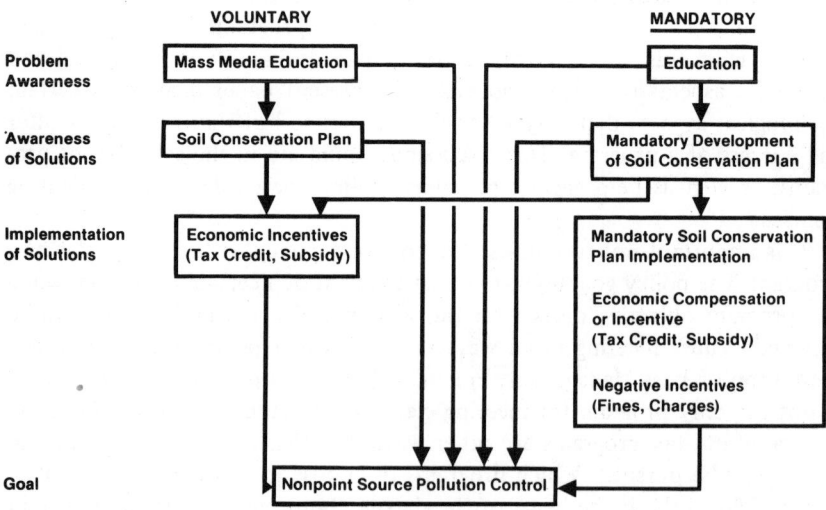

Figure 1. Interrelationships of selected approaches to the control of nonpoint source pollution.

be made mandatory. Within our political and economic framework, and supported by our value system, we prefer voluntary participation. It maintains a person's control over his or her affairs; it allows for local decisions and therefore efficient adaptations to local conditions.

Mandatory programs frequently are insensitive to local conditions and therefore inefficient and sometimes inequitable as well. However, there are many situations in which mandatory participation is required. First comes to mind the situation where the problem is so serious that reliance on voluntary participation cannot be justified. This situation is probably best exemplified by the area of public health: certain measures must be taken to prevent a contagious disease from spreading. Secondly, research indicates that even the most successful voluntary programs rarely succeed in obtaining 100% participation. Furthermore, frequently the most problematic cases are those that inspire the least voluntary participation.

In most cases mandatory participation makes it quite likely that program objectives will be attained. However, the 55-mph speed limit is one example where enforcement of mandatory participation has proven so difficult that the program's objectives could not be fully attained.

In the following section we will discuss some of the issues that arise, when we discuss simultaneously the varying program objectives and voluntary or mandatory participation strategies.

AWARENESS OF PROBLEM

Programs aimed at increasing awareness of the NPS problem are an important first step in any approach to solving the problem. Without awareness of the problem and its gravity, it is very difficult to attain participation in other programs relating to the policy goal.

A program of creating problem awareness frequently will rely heavily on use of the mass media. Past research indicates that such use can be quite successful.[1] Research also indicates, however, that to move farmers from awareness into taking some action usually requires a more complex approach, relying on sources of information other than the mass media.

Within agriculture the Cooperative Extension Service possibly has the best known record of achieving changes through educational programs. In the past, however, much of that work has focused on educational activities compatible with the profit-maximization efforts of most farmers. While much of the technology introduced to farmers in the past has helped them to increase their productivity, NPS pollution control policies have as their goal the improvement of water quality and will likely involve activities which may not be profitable to the farmer. Therefore, while the Extension Service may be an excellent organization for mounting an educational campaign, some research findings argue that we should not assume that NPS pollution control campaigns demand nothing but another application of the known strategies.[2]

While farmers are seen here as the principal on-farm decision-makers, it should not be overlooked that their actions take place within a larger social context. Society at large, as well as the members of the region and community, is an important factor in a farmer's decision-making.[3] It appears that more effort will be needed to create awareness among the general public. While no scientific polls are available, personal observations indicate that the general public has very little undersanding of NPS pollution and the more complex issue of its control. The severe apathy among the general public is not conducive to creating an environment supportive of strategies relying predominantly on educational programs among farmers. At the same time, the general low level of information among the public appears to leave the area open to those who might want to manipulate public opinion in support of special-interest positions.

PROVIDING AWARENESS OF SOLUTIONS

When awareness of the problem has been created, it is necessary to follow up with a program which provides the farmer with solutions tailored to his situation. In order to enable the farmer to take action, he will need to have specific information which allows him to make decisions pertaining to his farm. NPS pollution control is technically a very complex matter. The needs, as well as the options, differ from area to area, if not from farm to farm. Research on farmer decision-making indicates that a program aimed at making the farmer aware of solutions applicable to his farm will not be able to rely mainly on mass media. BMPs or other programs will need to be explained to the farmer in terms of the applicability of the program to his farm. Farm conservation plans constitute one approach that specifically helps the farmer determine the applicability of various options to his farm.

Farm soil conservation plans have been available to farmers for many years, but many farmers apparently have not felt the need, or possibly have not had the resources, to have a plan developed for their farm—thus raising the issue of mandatory compliance. The agricultural community, although not alone in this respect, has been an outspoken opponent of governmental regulation of its activities. Because of the fact that many pollution control measures cannot be implemented through market forces, and because of the farmer's perception of the EPA's mode of operation, the issue of mandatory participation, governmental regulations, or coercion, is always present. Requiring development of soil conservation plans for every farm actually represents a minimal interference on the farm operator's freedom to make decisions. The farmers may well perceive it, however, as a first step toward mandatory implementation.

An extensive promotional effort would be necessary, including a special effort to obtain the cooperation of leaders in the agricultural sector, to guide farmers into accepting this program. To gain farmers' acceptance of the program, it will be tempting to entice them with promises that a soil conservation plan will be a substitute for further regulation. Because it may very well be necessary to implement certain regulations at a future time, this appears to be a strategy that one ought to guard against. The government's loss of credibility among farmers in 1975, following the broken promises of unlimited grain exports, should provide an important lesson to all of us about the social and political cost of unkept promises.

IMPLEMENTATION OF SOLUTION

Instituting a program aimed at providing each farm with a soil conservation plan, or some other approach to defining the specific nature of a farmer's NPS pollution control program, would be a step in the direction of attaining the policy goal of NPS pollution control. However, program objective and policy goal would be separated from one another by a considerable margin; implementation is still within the realm of farm decision-making, where it must compete for priority with many of the other concerns farmers have. A successful program may educate every farmer in terms of his options to control NPS pollution, but at the same time show no progress toward the policy goal.

In a program with the specific objective to implement a solution, awareness of the problem and of solutions to the problem among the target population are, of course, necessary. These preconditions should be verified so that the implementation program can be adjusted to the existing levels of problem and solution awareness. It is also important to take into account whether the awareness has been created through a voluntary or mandatory program. If it has been totally voluntary, a sizable percentage of the target group can be expected to be unaware of the problem or the solution. This will require an educational effort preceding implementation, aimed at farmers not easily reached by traditional programs.

Once a program has been chosen for implementation, the manner of implementation must be considered. There are three basic categories of program implementation strategies:

1. voluntary without economic incentives,
2. voluntary with economic incentives such as tax credits or subsidies, and
3. mandatory with either positive (*e.g.,* tax deduction, subsidies) or negative (*e.g.,* fines) incentives.

Many farmers use conservation practices and work with agencies such as the Cooperative Extension Service and the Soil Conservation Service, but on a purely voluntary basis. The strength of the agencies involved in promoting such conservation programs must lie in their ability to persuade farmers to participate. Currently the agencies often deal with farmers already interested in the idea of conservation. They do not have to deal extensively with those farmers not interested in conservation.

Furthermore, as already pointed out, the control of NPS pollution frequently involves a conflict between farm profit and public welfare.[4,5] This creates further doubt about the effectiveness of an implementation strategy based solely on voluntary participation.

Voluntary implementation of conservation practices with the support of economic incentives would reduce the conflict between farm profit and public welfare considerations. These programs have been used, but in order to be truly successful the level of incentive apparently would need to be quite high. Such a program would become quite expensive to the public, and therefore very difficult to pass politically.

Making the implementation mandatory involves the greatest degree of compulsory interference with farm operations. However, both the gravity of the problem (which is not evaluated in this report) and the necessity to bring all acreage in an area under an NPS pollution control program may lead one to decide that mandatory participation is called for. The drawbacks of mandatory programs are well known. They tend to be accompanied by cumbersome administrative machinery which can be both costly and annoying to those affected by the regulations. Poor communications and misunderstandings between the regulatory agency and those regulated are a familiar part of most scenarios.

Regulations are usually created by a central authority, frequently causing inequities and inefficiencies. Soil conservation needs may be more sensitive to local conditions than almost any other sphere in which activity is regulated. There are frequent complaints that general standards are set by the political decision-making process, while bureaucratic agencies are left to decide how to implement the policies. It appears that for erosion and sedimentation control an approach of stating the policy goals while leaving the program selection to local decision-makers, including farmers, would be most appropriate.

An organizational structure which combines technical expertise with local farm decision-making participation would provide the best available guarantee that the NPS pollution control programs will be technically competent and maximally responsive to local farming needs. Effective farmer participation would aid in the efficiency of the implementation program and help guard against "over-engineering" on the part of the experts.

While we noted before that farmers place a high value on their autonomy in farm decision-making and on unrestricted property rights, they have accepted regulatory activity interfering with their decision-making autonomy in such areas as grading standards for farm products, milk marketing orders, and many public health regulations. While farmers have not necessarily cherished those regulations, there is little evidence that compliance problems have been widespread once the regulations have been introduced. However, surprisingly little research has been done on the nature of farmers' participation in mandatory programs. But it seems safe to say that without an extensive educational campaign and the active participation of farmer representatives in the decision-making process, it will be costly to overcome the expected negative reactions by farmers to any infringement on their freedom of decision-making.

In addition to the perceived threat to their autonomy, farmers will be concerned about the economic implications of the program. Under a program based on voluntary participation, a farmer may find himself at a disadvantage because his economic competitors are not participating in the program and thus not incurring similar expenditures. Under a mandatory program this problem is only partially alleviated, since the economic cost will vary depending on local conditions. To help cushion the economic impact, a mandatory program could be instituted which would provide farmers with some type of compensation which, for example, could take the form of a tax credit or a subsidization program. In general, farmers appear to favor tax credits over subsidization programs.[6] Subsidization also tends to become associated with specific structures or technological approaches, and this categorical approach may not be the most efficient one.

The voluntary or mandatory approaches have been treated here as being mutually exclusive. It is, however, possible to design policies which would incorporate a mix of voluntary and mandatory measures.[7] Farms or regions where NPS pollution poses the gravest threat to water quality may be chosen for the mandatory implementation of erosion measures, while in other regions it would be possible to rely on voluntary cooperation by farmers. This approach would place less of a burden on financial and technical resources and allow the most severe cases of NPS pollution to be treated with the urgency that is required.

CONCLUSION

While the preceding paragraphs have pointed out some of the issues involved in program selection and implementation, we have not been able to cover all the issues. Three issues come to mind which greatly affect the success of local programs:

Time Dimension

The success and acceptability of any program can be affected greatly by timing. A program may become unusually expensive or extremely threatening if it is undertaken as a "crash" effort. We realize that the gravity of the problem or political pressure may call for immediate action. Nevertheless, a well-developed timetable that indicates when various objectives need to be accomplished and that takes into account the capabilities of the organizations involved, the available financial resources, and the need to educate farmers and the general public may do much to increase the likelihood of success.

Extreme Impact on Individuals

In the present discussion we assume that farmers can afford to participate in any of the programs but need to be encouraged to do so. Or we assume that, in order to equalize the different economic impact between them, they may need to be compensated. In considering any policy, however, it should be recognized that some farmers may be forced out of agriculture if they must make heavy investments in NPS pollution control activities, must substantially change their farming operations or must take certain acreage out of row-crop production. Whatever is done, then, these individuals will understandably be resentful toward the program.

Farmer Participation in Decision-Making

Effective participation by farmers in decision-making will affect the implementation of policies at the local level. This is not the place to deal extensively with the problems involved in citizen participation in decision-making. The literature on that subject is voluminous, although few studies have examined the nature of farmer participation in the decision-making that affects their own enterprises. Research on citizen participation indicates that frequently neither the objectives of citizen participation nor the role and power of the citizen particpants have been defined well enough to allow a functional system to develop.[8] New policies which incorporate elements of farmer participation in the decision-making structure will need to specify carefully the objectives to be accomplished and the ways in which the participation is to be implemented.

We have discussed some sociological elements of the implementation of various programs, related to the goal of NPS pollution abatement. It is likely that, ultimately, elements from a number of programs will be combined in an overall NPS control program. We hope that technicians will be in a better position to administer programs if they understand the relationship of the program objectives to the policy goal and make their own actions compatible with the various program objectives. For example, programs relying on voluntary participation are not likely to reach certain farmers, and programs relying heavily on mass media communications will most likely succeed much more in creating problem awareness but have lesser results in program implementation. No solution for controlling NPS pollution will be simple; it will involve a complex approach, and at the local level different agencies will need to design complementary programs. We hope that this discussion has offered some ways to identify the elements of a complex approach to the goal of nonpoint sources of agricultural pollution control.

ACKNOWLEDGMENTS

The research was supported by a grant from the U.S. Environmental Protection Agency to the Institute for Environmental Studies, and by the Agricultural Experiment Station at the University of Illinois at Urbana-Champaign.

REFERENCES

1. Rogers, E., and P. Shoemaker. *Communication of Innovations* (New York: The Free Press, 1968).
2. van Es, J. C., and F. C. Pampel, Jr. "Environmental Practices: New Strategies Needed," *J. Extension* 16:10-15 (May/June 1976).
3. Ostrum, V. "Public Choice Theory: A New Approach to Institutional Economics," *Am. J. Agric. Econ.* 57:844-859 (December 1975).
4. Wilkening, E. A., and L. Klessig. "The Rural Environment: Quality and Conflicts in Land Use," Department of Rural Sociology, University of Wisconsin, Madison, WI (1977).
5. Pampel, F. C., Jr., and J. C. van Es. "Environmental Quality and Issues of Adoption Research," *Rural Sociology* 42:57-71 (SPring 1977).
6. Gardner, D. M., and W. D. Seitz. "Farmers' Attitudes Concerning Soil Erosion and Its Control: A Report to the Illinois EPA Agricultural Task Force," Institute for Environmental Studies with the College of Commerce, University of Illinois at Urbana-Champaign (1977).
7. Council for Agricultural Science and Technology. "Soil and Water Conservation Oversight," Report No. 60, Iowa State University, Ames, IA (1976).
8. van Es, J. C. "Citizen Participation in the Planning Process," in *Aspects of Planning for Public Service in Rural Areas,* D. L. Rogers and L. R. Whiting, Eds. (Ames, IA: Iowa State University, North Central Regional Center for Rural Development, 1976), pp. 81-105.

23

SOCIAL COSTS AND EFFECTIVENESS OF ALTERNATIVE NONPOINT POLLUTION CONTROL PRACTICES

K. F. Alt
Economics, Statistics, and Cooperatives Service
U. S. Department of Agriculture
Ames, Iowa

J. A. Miranowski
Department of Economics
Iowa State University
Ames, Iowa

E. O. Heady
Center for Agricultural and Rural Development
Iowa State University
Ames, Iowa

The purpose of this conference is a review of nonpoint pollution abatement policies, specifically a set of abatement practices referred to as "Best Management Practices" (BMPs). This paper will first discuss the concept of BMPs, point out some pitfalls in their use as a pollution abatement alternative, and then illustrate the potential reactions of farmers to such an abatement alternative. First let us present what seems to be the official definition:

> The term Best Management Practices (BMP) means a practice, or combination of practices, that is determined by a State (or designated areawide planning agency) after problem assessment, examination of alternative practices, and appropriate public participation to be the most effective, practicable (including technological, economic, and institutional considerations) means of preventing or reducing the amount of pollution generated by nonpoint sources to a level compatible with water quality goals.[1]

It is disturbing that this definition discusses only water quality goals. Nonpoint pollution involves various resources, not only an overuse of the assimilative capacity of water. In general, nonpoint pollution implies a misallocation of resources caused by a technical inability to identify the originator of certain harmful emissions and to force him to internalize the externality. The nonpoint pollution abatement policies, including BMPs, should be judged by their contribution to improving the resource allocation in the economy. If and only if the policies help to bring about such an improvement in resource allocation, hopefully even to the optimal allocation,* can the policies be judged as "good" or "best." On the other hand, a singular preoccupation with water quality goals in the determination of pollution abatement policies cannot guarantee that a socially efficient policy will be chosen.

Let us examine the soil erosion problem in more detail. The movement of topsoil has two effects, namely the onsite effect of loss of soil farmability because of gullying and decreasing soil productivity and the offsite effect caused by sediment in the streams. We may assume that each farmer is interested in maintaining the farmability of his soil, at least for the short run. However, the farmer has little economic incentive to reduce offsite damages, since almost all of the direct economic benefits of that abatement will go to others, not to him.** Consequently, the farmer will spend less of his resources on erosion abatement than would be desirable from a social standpoint.

Certain programs may be necessary to reduce the discrepancy between private and social costs which in turn would improve the resource allocation. A critical step in choosing a program and its level of operation should be the estimation of the social costs and benefits of alternative programs. Then the most efficient and equitable program can be chosen.

It is important that a comprehensive accounting of program costs be included in this evaluation. For example, if a program proposes that an area which has produced row-crops be seeded to grasses, then the cost of that program is more than the cost of the grass seed. In general, in addition to the cost of the abatement process, the program cost should include the administrative and enforcement costs borne by the governmental agency supervising the program, the net value of the crop output foregone, as well as other secondary adjustment effects.

---

*The allocation of resources is optimal if each resource is used to the point where the marginal social cost of the last unit used equals its marginal social benefit, assuming the second-order conditions are satisfied (which generally means that the marginal cost exceeds the marginal benefit for the next unit).

**In other words, the marginal private benefits of pollution abatement received by the farmer are less than the marginal social benefits realized by society.

The benefits of a program are very difficult to estimate, because the benefits may include unquantifiable environmental effects. In this context, we need to be aware of a tacit assumption. It is intuitively appealing to argue that the use of certain agricultural management practices will have a beneficial effect upon water quality. Indeed, the study on which we shall report later in this paper makes just such an assumption. However, physical processes in nature are highly variable and subject to numerous variables, parameters and random events. The identification of the physical causes and effects of nonpoint pollution and their representation in environmental models have not yet progressed to the point of high reliability. A committee of Agricultural Research Service (ARS) scientists points out that present "mathematical models to predict where a pollutant came from are probably not defensible in court."[2] In discussing onsite and offsite damages, Omstad, Young and Moldenhauer[3] point out:

> The two biggest technical constraints to implementing any program of watershed erosion and sediment control are the lack of technical knowledge about sediment movement and establishment of scientifically defensible positions regarding soil loss limits or sediment delivery limits. These are the goals of EPA for fiscal years 1976 to 1980—not only for sediment but for other water quality parameters also. The data base for defensible soil loss limits that must be exceeded before damage occurs generally is deficient.

There is a crying need for more research. However, society cannot afford to wait indefinitely. The decisions on pollution abatement cannot be postponed until more research is completed, several decades hence. The problems need a solution now.

AN EXAMPLE

Before we show the results, let us describe the objectives and assumptions of our study in some detail. We estimated the pollution abatement costs (not the total program costs) likely to be encountered by farmers faced with several possible pollution control schemes. The questions were: (1) are abatement costs negligible or overwhelming; (2) do abatement costs increase linearly with increased abatement levels; and (3) can we suggest a socially desirable level of pollution abatement?

The study area of 940,000 ac is located in a relatively hilly area in east-central Iowa. It includes the watersheds of the Iowa River between Marshalltown and the dam at the Coralville Reservoir. Almost 75% of the land area is tilled primarily for corn and soybeans. We used a linear programming model to assess the impacts of alternative nonpoint pollution policies on resource use. We constrained crop output to the historical level and assumed that farmers will minimize the cost of producing this output level subject to environmental restraints.*

Although there are other features to a farmer's decision-making process, it is difficult to include these features in a mathematical model because they may not be quantifiable. For example, a farmer who is faced with very large mortgage payments may resist any outside effort of changing his production mix to less profitable crops or more costly production processes. Similar decision variables may enter in other unquantifiable forms: as fear of the unknown or higher risk production processes (no-till), reluctance to try a process which requires a higher level of management, and unwillingness to adjust to farming techniques which will cause a loss of leisure time (if all tillage operations were forced into a short time span requiring 24-hr operations for a few days). Intangible or unquantifiable influences work on the positive side as well. Like the rest of us, farmers enjoy receiving social approval. If a farmer can garner peer approval, such as being named "Conservation Farmer of the Year," he may undertake pollution abatement efforts even at a loss of income. (This assumes that both income and approval enter his utility function.)

Our simulation of the crop-growing activities of the study area consists of 570 different production alternatives. Crop production alternatives were differentiated for each of nine soil types, six crop rotations (plus permanent pasture), and several tillage and soil conservation practices. The yields and production costs for the production activities were computed from published sources.**

The tillage practices used in the model are conventional tillage moldboard fall plowed, conventional tillage moldboard spring plowed, rotary-till plant, and no-till plant. These tillage practices vary in their contribution to erosion. In addition, several soil conservation methods could be used to reduce erosion. Terracing is the method that is possibly most effective but certainly most expensive. Contouring is another conservation method. Contouring may require the use of point rows which increase labor and machinery time requirements.

The gross erosion levels of the various crop production alternatives were computed with the Universal Soil Loss Equation.[6] The estimates of gross erosion provide the required information to make soil conservation decisions, but the level of gross erosion does not tell much about water quality. The link between water quality and gross erosion is generally specified with a

---

*Although this assumption may be subject to debate, it is impossible to predict the potential output price adjustments in the study area which may result if environmental restraints are applied at a state or national level.
**A detailed discussion of the techniques is given in Alt and Heady.[4] An example of the cost computations is also found in Alt.[5]

sediment delivery ratio. This "ratio is used as a lumped accounting for sediment load changes below the area for which gross soil loss is computed."[7] These load changes include sediment deposition at the toe of field slopes and in depressions along the runoff path to the stream.

It may be assumed that such sediment delivery ratios are related to the size of the drainage area. (Despite wide variations in topographical and other influencing factors, the sediment delivery ratio has been found to vary inversely as the 0.2 power of the size of the drainage area.[8]) This estimation process forecasts the amount of sediment delivered to the mouth of the watershed. Wischmeier[7] argues that these types of estimates are confounded by stream bank erosion and sediment accretions from nonagricultural sources and cannot provide estimates of the contribution of nonpoint cropland sources to water pollution. If the sediment delivery ratio is to be used for such estimates, it should be defined as "the ratio of sediment delivered at the place where the runoff water enters a continuous stream system to the gross erosion from the drainage area above that point."[9] The sediment-delivery ratios were developed on the basis of Wischmeier's more restrictive definition in view of the topography in each of the 18 watersheds in the study area.*

We first solved the linear programming model in the absence of environmental controls; this was our "baseline solution." Several environmental policies were then imposed. The first of these was an absolute limit on gross erosion per acre cropped. This limit was specified at three levels: 10, 5 and 3 ton/ac/yr. Another policy treated the study area as a single planning unit upon which a maximum limit on sediment delivery to the Coralville Reservoir was imposed. Such a limit simulated the effect of a water quality standard imposed upon the study area as a whole. A third policy alternative assumed payment of subsidies to farmers for construction of terraces or contouring or both.

RESULTS

The model results indicate that crop production costs will be increased if pollution control policies are imposed and enforced in this particular location. Table I shows the specific increase in production costs associated with the alternative policies. The imposition of a 10-ton/ac gross erosion limit reduces the amount of sediment from agricultural sources delivered to the Coralville Reservoir by 68% while increasing production costs by 2.5%

---

*The sediment delivery ratio estimates were developed by Ed Burr, Iowa State Geologist, and Bob Boyce, Geologist, Central Technical Unit, both Soil Conservation Service, U.S. Department of Agriculture. The estimates were tested and refined with a sediment routing model as described by Boyce.[10]

Table I. Sediment delivered to Coralville Reservoir and production costs in selected models.

Model	Sediment Delivered (1,000 ton)	Production Cost (1,000 $)
No Restraints	1,136.6	62,626
67% Subsidy of Contouring Marginal Costs	863.2	62,780
75% Sediment Delivery Restraint	852.5	62,775
100% Subsidy of Contouring Marginal Costs	780.1	62,788
50% Sediment Delivery Restraint	568.3	63,264
Erosion Limit 10 ton/ac/yr	364.5	64,212
25% Sediment Delivery Restraint	284.2	64,994
Erosion Limit 5 ton/ac/yr	193.6	67,911
Erosion Limit 3 ton/ac/yr	104.5	73,139
Erosion Limit 3 ton/ac/yr plus 67% Terrace Construction Cost Subsidy	104.3	73,207
Erosion Limit 3 ton/ac/yr plus 100% Terrace Construction Cost Subsidy	95.2	76,524

over the baseline solution values. A 5-ton/ac limit will lower the sediment delivery by 83% while raising costs by 8% over the baseline. At the 3-ton/ac limit, the biggest jump of costs (17%) is estimated; this is accompanied by a 91% decrease in sediment delivery relative to the baseline solution. The model with the lowest amount of sediment delivered (a 92% reduction from the baseline) also shows the highest cost increase (22% over the baseline value).

Earlier in this paper we posed three questions to be answered by the study. First, are the nonpoint pollution abatement costs negligible? Obviously the answer depends on the level of abatement. At high levels the costs could be overwhelming.* Second, do the abatement costs vary linearly with the level of abatement? Far from it; the costs actually increase exponentially with an increase in abatement. Third, what can we say about the optimal level of pollution abatement? In the absence of a marginal benefit function for pollution abatement we can give only a general answer. However, we can show that the marginal cost of further reductions starts to increase dramatically when gross erosion is held to less than 5 ton/ac/yr. So, we should advise to proceed carefully if the contemplated policies would require such a high level of abatement. Such policies, and their implied production cost increases, could have serious repercussions upon the viability of the

---

*For discussion of this point on a larger regional scale, see Miranowski.[11]

affected farm businesses. It is imperative that determination be made on the relationship of these costs to the environmental benefits that would accrue from a reduction of the agriculturally produced pollution. Only after such a determination can a truly optimal environmental policy be chosen.

REFERENCES

1. Code of Federal Regulations. Title 40-Protection of Environment, Chapter I, U.S. Environmental Protection Agency, Part 130.2(q) (1977).
2. Frere, M. H., D. A. Woolhiser, J. H. Caro, B. A. Stewart and W. H. Wischmeier. "Control of Nonpoint Water Pollution from Agriculture: Some Concepts," *J. Soil Water Conserv.* 32(6):260-264 (1977).
3. Onstad, C. A., R. A. Young and W. C. Moldenhauer. "Implementing Soil Loss Limits: Some Considerations," in *Soil Erosion: Prediction and Control*, Soil Conservation Society of America, Ankeny, IA (1977).
4. Alt, K. F., and E. O. Heady. "Economics and the Environment: Impacts of Erosion Restraints on Crop Production in the Iowa River Basin," Center for Agricultural and Rural Development Report No. 75, Iowa State University (1977).
5. Alt, K. F. "Economic Analysis Methodology," in *Control of Water Pollution from Cropland*, Vol. II, B. A. Stewart et al., Eds., U.S. Environmental Protection Agency 600/2-75-026b (1976).
6. Wischmeier, W. H., and D. D. Smith. "Predicting Rainfall Erosion Losses from Cropland East of the Rocky Mountains," U.S. Department of Agriculture, Agricultural Handbook 282 (1965).
7. Wischmeier, W. H. "Use and Misuse of the Universal Soil Loss Equation," *J. Soil Water Conserv.* 31(1):5-9 (1976).
8. Renfro, G. W. "Use of Erosion Equations and Sediment-Delivery Ratios for Predicting Sediment Yields," in *Present and Prospective Technology for Predicting Sediment Yields and Sources*, Proceedings Sediment Yield Workshop, November 28-30, 1972. U.S. Department of Agriculture, Agricultural Research Service, Oxford, MS, ARS-S-40 (1975).
9. Wischmeier, W. H. "Cropland and Erosion and Sedimentation," in *Control of Water Pollution from Cropland*, Vol. II, B. A. Stewart, et al., Eds. U.S. Environmental Protection Agency 600/2-75-0266 (1976).
10. Boyce, R. C. "Sediment Routing with Sediment-Delivery Ratios," in *Present and Prospective Technology for Predicting Sediment Yields and Sources*, Proceedings Sediment Yield Workshop, November 28-30, 1972, U.S. Department of Agriculture, Agricultural Research Service, Oxford, MS, ARS-S-40 (1975).
11. Miranowski, J. A. "Economic Implications of 208 Planning on the Agricultural Economy," in *The Economic Impact of Section 208 Planning on Agriculture*, Great Plains Agricultural Council Publication No. 86, University of Nebraska-Lincoln (1978).

24

MANAGEMENT AND FINANCING
OF AGRICULTURAL BMPs

J. M. Rice
URS Company
Seattle, Washington

INTRODUCTION

Water quality management planning in the past has often been an academic endeavor with few recommendations being effectively implemented. One of the major goals of the Section 208 areawide waste management planning program, mandated by PL 92-500 and administered by the U.S. Environmental Protection Agency (EPA), is to develop solutions to identified water pollution problems which can and will be put into practice. The agricultural program developed in SNOMET/King County* (Washington) represents one approach to implementing nonpoint water pollution control measures which both satisfies the technical and other requirements of EPA *vis à vis* water quality management and satisfies the more pragmatic requirements of working farmers who are concerned both about making a living and about conservation. This paper presents the results of that program with special emphasis on unique aspects of the social and political setting of the SNOMET/King County area. The implemented program is in one sense no different than those developed in many other agricultural areas of the country; yet its acceptance by the local farming community and their willingness to participate in implementation of the

---

*SNOMET is the Snohomish County Metropolitan Municipal Corporation, a single-purpose municipal government established in 1959 to perform water pollution control planning in Snohomish County. SNOMET was the recipient of the 208 grant, although a council (the Executive Panel) from SNOMET and King County directed the study.

program represent an institutional achievement in itself. Process, not product, then is a major focus of this paper.

The SNOMET/King County 208 study area consists primarily of the Snohomish River Basin in western Washington, and includes most of Snohomish County and the main agricultural portion of neighboring King County (see Figure 1). The Snohomish Basin portion of the 208 area comprises 1200 sq mi, of which about 25% is devoted to agricultural uses. Dairy farming is the predominant agricultural activity of the basin and also represents the main source of nonpoint pollution, based both on earlier water quality work[1] and on work done during the 208 (between April 1976 and January 1977). Technical work performed during the 208 included an intensive field sampling program to identify the relationship between land uses/activities and runoff water quality[2] and a modeling effort which used the field results.[3] The field work verified some but not all of the findings of previous work. The important nonpoint water pollution problems identified were:

1. bacteria, associated with runoff from barn areas and other confined animal areas;
2. bacteria, in some streams, due to animal access to streams and improper management of manure field application.

These problems did not necessarily occur simultaneously, nor were they universal to all farms in the SNOMET/King County 208 study area. Critical erosion/sedimentation and other similar runoff problems were not found in the basin, due to the fact that most farming occurs on the broad floodplain of the river, and not in the upland, sloped areas, most of which are forested. The technical portion of the 208 developed a highly usable BMP manual for use by farmers and by those providing technical assistance.[4]

(Incidentally, the modeling work performed during the 208 will be discussed in another paper to be presented at this conference: "Mathamatical Modeling of Water Quality Effects of Agricultural BMPs" by Dr. Charles Tang, URS Company, Seattle.)

The work that is described in this paper was one portion of the institutional analysis task of the 208. The institutional analysis focused on three main institutional problems (agricultural runoff, urban runoff, and areawide water quality management) and several less pressing problems. Because the Snohomish River Basin crosses jurisdictional boundaries, a host of local agencies, along with the usual federal and state agencies were included in the analysis. In addition, a major chore was to fit the agricultural program into the framework of general water quality management in the basin. The agricultural task began in June of 1976 and was completed in November 1977.

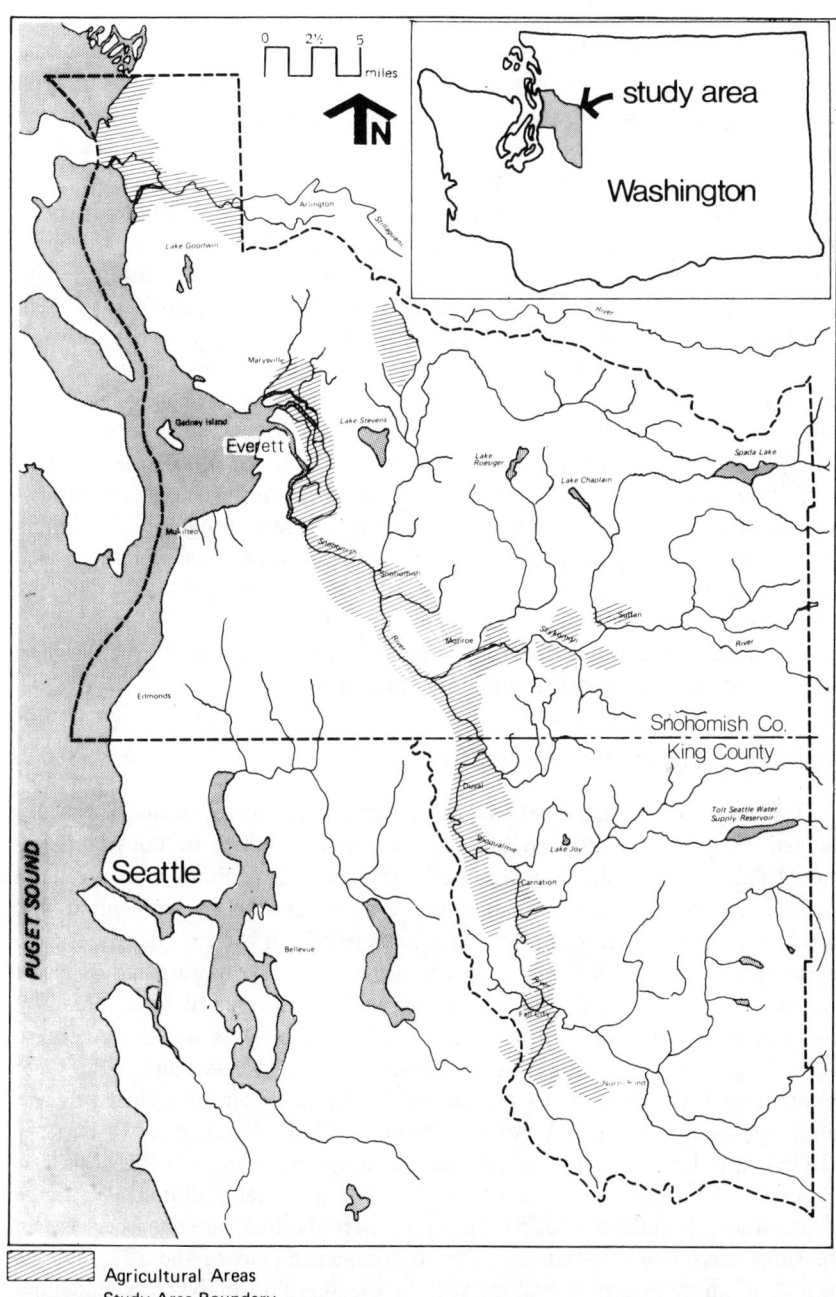

Figure 1. SNOMET/King County 208 planning area.

An institutional inventory was conducted early in the 208, and the results were used in all three of the primary analysis areas. The inventory established the institutional framework for water quality management in the 208 by identifying the jurisdictions, authorities, policies, regulations, procedures and interactions of governments and agencies in the area. This task was accomplished through a survey which included personal interviews with prospectively important agencies, such as the local conservation districts, extension agents, and Soil Conservation Service (SCS). In all (for the three institutional topic areas), over 120 agencies were contacted. Additionally, a series of public workshops was held in August and September 1976 to obtain the views of the public on various institutional approaches to water quality management. These workshops were visionary for the 208 in two ways. First, the workshops did not present alternatives for management, but rather introduced the public to the concepts, processes and possible results of the institutional analysis by allowing them to develop alternatives of their own. "Game boards" (see Figure 2) took the workshop participants through an institutional analysis including identifying goals, programs for implementation, and assignment of responsibilities to various levels of government. Second, it was the first major, substantive contact during the 208 with the public at large, and the farming community in particular. The workshops provided the basis for further development of institutional alternatives and established criteria for evaluating those alternatives.

CONSTRAINTS TO MANAGEMENT

The workshops and contacts made during the institutional inventory helped early in the 208 to identify several difficulties or constraints to establishing a workable management strategy. First, there was a general feeling within the farming community that water quality problems in the Snohomish Basin were not severe and/or were caused by activities other than agricultural. This feeling was found to be not entirely unjustified as water quality problems were generally localized and not severe. However, the total magnitude of pollutants entering the river from the farms was indeed found to be large, overshadowing other pollutant sources. This was due to the sheer number of acres devoted to farming in the basin. Balanced against this was a knowledge that both EPA and the Washington State Department of Ecology (DOE) felt that regulation of agricultural nonpoint sources in the basin was necessary to meet the goals of PL 92-500. National Pollutant Discharge Elimination System (NPDES) discharge permits had already been issued to three large dairy operations prior to commencement of the 208, and the threat of more issuances was present in the wording of NPDES guidelines for the state. (The guidelines stressed two key criteria for permitting: whether

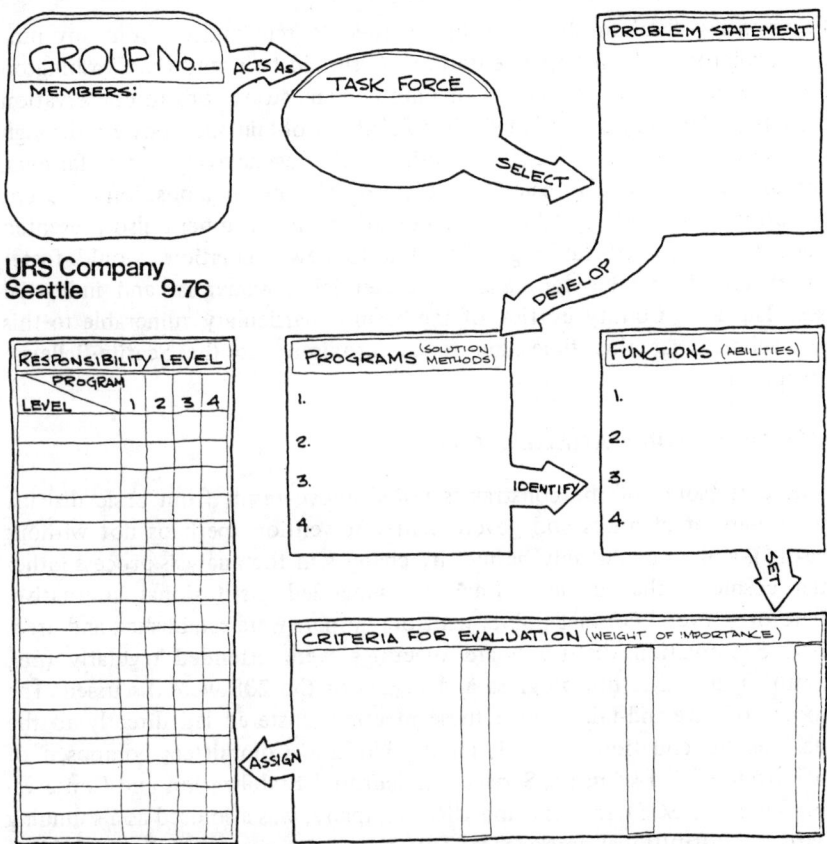

Figure 2. Gameboard used for public workshops (original size 24" x 24").

a pollutant discharge occurs due to the 25-yr, 24-hr design storm, or whether a significant pollution problem exists in receiving waters.)

Second, however, the farming community quite correctly felt that only a few farmers were causing the majority of the problems that did exist. This in itself made the use of blanket regulatory provisions and nonsite-specific technical solutions infeasible, or at least highly undesirable. Third, there was a strong desire in the farming community to remain unencumbered with bureaucratic red tape such as might be required with permitting programs. Coupled with this was an even stronger distaste for state agencies, DOE in particular, which in Washington State is powerful. This distaste was grounded in past occasional overzealousness of that agency and their general lack of "farm knowledge." Fourth, the local agency most visible to and knowledgeable of farming, the two conservation districts (one for each of the counties—Snohomish and King), wished to remain free from any hint of a regulatory role in implementation. (It should be noted that Washington State law confers no regulatory or enforcement authorities to conservation districts.) This would have made it difficult to obtain such powers through the necessary process of legislative action. Other agencies known to farmers, such as the SCS and Extension Service, were also not in a position to accept regulatory roles. Finally, the specter of urban encroachment also presented itself, *i.e.,* increased farming costs due to new regulations would foster conversion of valuable farmland to residential, commercial and industrial uses. The King County portion of the basin is particulary vulnerable to this encroachment at this time due to its proximity to the Seattle-Bellevue urban area.

SOLUTIONS FOR MANAGEMENT

In part, some of the constraints noted above result from basic distrust by farmers of planners and government intervention (perhaps not without reason!), and so could only be met by changes in the analysis process rather than cosmetic changes in a final recommended institutional alternative. Therefore, several contact points with the farmers were established and used. First, conservation district board meetings were attended regularly (*i.e.,* monthly) in which the progress and results of the 208 were discussed. The process of give-and-take which these meetings fostered led directly to the final recommendation. Second, an Ag Working Committee, composed of staff from SCS, Extension Service, Agricultural Stabilization and Conservation Service (ASCS), farmers, and URS Company, was also used as a sounding board for institutional approaches. This committee was originally established to review and comment only on the BMPs which were being developed. However, its role evolved into a much more substantial one in the 208 as it responded to and made suggestions for institutional alternatives. Finally, during the course of the project a number of general farmer meetings were held around the basin. Again, comments from individual farmers were taken seriously, resulting in a final plan which reflects the desires of the community.

Several basic alternative approaches leading to implementation and use of BMPs were formulated and evaluated during the 208 effort:

1. mandatory use of BMPs, through federal, state or local legislative action,
2. mandatory use of BMPs through federal, state or local administrative action,
3. voluntary use of BMPs supported by enforcement action where water quality problems persist, and
4. voluntary use of BMPs.

The two mandatory implementation alternatives were discarded for several reasons:

1. lack of major perceived water quality problem on the part of the farming community,
2. lack of a major areawide water quality problem as identified in the 208 study,
3. identification of only a limited number of farm operations causing localized water quality problems, and
4. probable alienation of the farming community if mandatory controls or other regulations were adopted.

EPA requirements for a nonpoint source control program are now such that a purely voluntary program is unacceptable. As a result, the entirely voluntary approach listed above was also discarded. In addition to the EPA requirements, it was also felt a purely voluntary approach would probably not reach the "problem" farms anyway, an assessment supported both by discussions with farmers and staff involved in agencies with agricultural activities and by the workshops held in fall 1976. Thus, the third alternative, in which enforcement or regulatory mechanism backs voluntary actions, was chosen. In this case, the regulatory powers were already available making implementation that much easier. The choice of the combined voluntary/ regulatory approach was, in fact, not as negative as the brief analysis would indicate. Rather the choice was made, along with the designation of participant agencies, through a long, arduous political process requiring numerous and substantive discussions with those involved, *i.e.,* the farming community.

Four programs were identified as necessary for this alternative to meet its objective of establishing use of BMPs in the 208 area. Three of the programs are centered on voluntary BMP use: (1) education and technical assistance, (2) financing and funding, (3) incentives, and (4) monitoring and enforcement.

The two conservation districts in their respective jurisdictional areas of King and Snohomish Counties were designated as lead agencies for plan implementation. The SCS, Extension Service, and the new SNOMET/King County 208 Water Quality Panel (also established as a result of the 208) will provide support to the districts. The districts will be involved chiefly with education and technical assistance, which will be the primary emphasis for the program. Specialists were initially hired by each district, using implementation money from the 208, to begin visiting all farm operations in the basin and to begin encouraging farmers to update their farm conservation plans to include BMPs. The Farm Water Quality Management Manual discussed earlier is the most important tool for this education/assistance program.

Financing and funding are necessary for the successful implementation of the programs. Individual farm use of BMPs can be financed/funded by: (1) grants from state and federal sources, (2) loans from federal sources, (3) loans from private sources, and (4) farm operator capital. The ASCS is a primary source of grants to farmers. Grants of up to $2,500 per farm per year may be obtained for cost-sharing of various capital improvements, including pollution control facilities, (*e.g.*, manure tanks), drainage control facilities, and stream bank stabilization. Unfortunately, annual funding for the ASCS varies, as does their priority schedule for eligible projects. The recent amendments to PL 92-500, the Clean Water Act of 1977 (PL 95-217), established a new, very promising source of funds. The Act added Section 208(j) to PL 92-500 which authorizes up to 50% grants to farmers to initiate activities under approved BMP guidelines. Owners taking advantage of the program will be required to enter into 5- to 10-yr contracts with EPA, SCS or other agricultural agencies (including conservation districts) verifying their compliance with requirements of the BMPs. Nationwide, $200 million is authorized for Fiscal Year 1979 and $400 million for 1980. Grants will probably not begin, however, until late next fall after administrating guidelines can be promulgated.

Traditional lending institutions, including commercial banks and member organizations such as the Federal Land Bank Association (FLBA), are the best sources of funds for most BMP activities. Federal loan sources are the Farmers Home Association (FmHA), and the Small Business Administration (SBA). FmHA loans are contingent upon the farms being unable to obtain sufficient credit elsewhere and upon the existence of a farm conservation plan. Recent legislation (PL 94-305) opened up additional loan funds for pollution control facilities to individual farmers through the SBA. These loans are taken from the Disaster Loan Fund and the amount is determined by the extent of the "injury," *i.e.*, the expenditure or loss incurred in complying with state or federal regulations, including some of the BMPs. Loan guarantees are also available for the FmHA and SBA.

Administration of the BMP programs also required financing. With the exception of the conservation districts and SNOMET, costs can be paid through existing budgets. SNOMET (*i.e.*, the Water Quality Panel) is financed by participating local governments. The conservations districts, however, are in the most difficult situation, particularly the King County Conservation District as it neither rents equipment nor provides labor to generate monies. The major push for BMP implementation will be during 1978 and 1979. The SNOMET/King County 208 provided funds for the remainder of 1977, about $5,000 each. At this writing, the funds for 1978 and 1979 had not been allocated. One promising source of funds is through the Washington DOE. Financial assistance for the abatement of water pollution stemming from agricultural practices was authorized under Referendum 26 (RCW 43.83A). DOE has not yet established the criteria for the program and its currently allocated funds ($3 million for the 1978-1979 biennium), but the program basically consists of grants to a public entity, such as a conservation district, with a 50% local match requirement which may be met with cash or in-kind services. Possible activities under this program include demonstration projects, district-owned service equipment, and community drainage projects. The EPA-USDA Model Implementation Program is also a promising source of funds.

The incentive program has two particularly attractive features which will mitigate to some extent the impact of the use of borrowed money rather than grants to implement BMPs on farms. First, the U.S. Internal Revenue Service (IRS) allows an investment tax credit for BMPs requiring capital expenditures. Secondly, IRS also allows an accelerated write-off of investments on pollution control devices approved by EPA. In both cases, certified BMPs requiring capital expenditures will qualify.

The fourth program noted previously is for monitoring and enforcement. Monitoring will fall into two categories: long-term basinwide monitoring for trend analysis, and site-specific short-term monitoring for enforcement and permitting purposes. The DOE has set up a long-term monitoring system, and the Snohomish Health District and Seattle/King County Health Department perform sampling in the Snohomish River Basin which will provide additional data. Enforcement procedures can be carried out by DOE (RCW 90.48) and the health agencies (Snohomish Health District and Seattle/King County Health Department) under existing law. Where necessary, DOE may also issue a NPDES permit to a farmer if a discharge exists. In any case, site-specific, short-term monitoring will be necessary prior to any action. Financing for participation in this program will be through existing annual budget processes. It is not believed that enforcement proceedings will be required frequently primarily due to changes in the state water quality standards first suggested as a result of the SNOMET/King County 208. The standards allow that use of approved BMPs will be sufficient for meeting water quality standards (WAC 173-201). Hence, as long as a farmer is using the BMPs, there is no fear of enforcement actions from DOE, thus circumventing one of the major constraints noted earlier.

Table I shows a more complete breakdown of the agricultural runoff program.

IMPLEMENTATION

Implementation of the agricultural nonpoint pollution control program began in late 1977 with funds provided from the initial 208 grant. The future

Table I. Assigned agency responsibilities for agricultural BMP implementation.

Participating Agency	Educational Technical Assistance	Funding/ Financing	Incentives	Monitoring/ Enforcement	Basic Responsibilities
Snohomish Conservation District	X		X		1. Lead agencies/primary contact with farm operator 2. Distribute AG BMP manuals 3. Encourage use of BMPs 4. Provide technical assistance 5. Provide construction services (at costs determined by District) (Snohomish only) 6. Make technical assistance program known to operators
King County Conservation District	X		X		
Soil Conservation Service	X				1. Assist with educational program and distribution of AG BMP manuals
Snohomish County Extension Service	X				1. Assist with educational and technical assistance program and distribution of AG BMP manuals
King County Extension Service	X				
Water Quality Panel	X	X (in certain circumstances)			1. Assist with educational program and distribution of AG BMP manuals 2. Maintain WQ data files 3. Coordinate activities

ECONOMIC, POLICY AND INSTITUTIONAL ASPECTS 339

Agency		Roles
Department of Ecology	X	1. Perform long-term monitoring 2. Bring enforcement proceedings, where necessary (RCW 90.48) 3. Notify conservation districts of impending enforcement proceedings 4. Issue NPDES permits (where necessary) 5. Approve BMPs for IRS purposes
Environmental Protection Agency	X	1. Fund BMPs and conservation district participation 2. Approve BMPs for IRS purposes
Agricultural Stabilization and Conservation Service	X	1. Matching grants up to $2,500 yr/farm
Farmers Home Administration Small Business Administration	X X	1. Loans
Internal Revenue Service	X	1. Investment tax credits 2. Accelerated write-offs for investments
Snohomish Health District Seattle/King County Health	X X	1. Bacterial monitoring 2. Enforcement (only where health hazards exist)
State Conservation Commission	X	1. Support of local conservation districts

of the program is somewhat hazy, especially with lack of funds for agency and farm operator participation. Nonetheless, sufficient inertia was established during 1977 to sustain at least low-key interest for part of 1978. Given the noncritical nature of the basin's pollution problems, implementation can be done more gradually—as funds become available—without further serious water quality degradation. In fact, the longer implementation period will undoubtedly mean even less economic and social dislocation in the community, thereby avoiding to a certain extent the problem of agricultural land conversion. The adoption of BMPs as meeting water quality standards by the state provided a step toward certainty and consistency in BMP use, again allowing BMP use to seem less repugnant. It should be noted that the approach to BMP use recommended for the SNOMET/King County 208 area was subsequently adopted for the whole state.[5]

CONCLUSION

The final recommended, and adopted, plan for control of agricultural nonpoint source pollution represents the combined visions of local farmers, 208 agency managers, and state/federal regulatory agencies. The final plan is not particularly revolutionary, but the process for developing it was substantially more satisfying than previous attempts. The commitment developed to the plan by farmers and water quality managers during the planning stage will be a key to its successful implementation, and hopefully a precursor to future environmental planning involving the farming community.

REFERENCES

1. Snohomish County, Washington. *Water Quality Management Plan*, Volume I, Water Quality Management Plan for the Snohomish River Basin (November 1974).
2. URS Company. "Water Quality Sampling," Technical Appendix III, prepared for the SNOMET/King County 208 Areawide Waste Management Planning Study (November 1977).
3. URS Company. "Water Quality Modeling," Technical Appendix III, prepared for the SNOMET/King County 208 Areawide Waste Management Planning Study (November 1977).
4. URS Company. "Farm Water Quality Management Manual," prepared for the SNOMET/King County 208 Areawide Waste Management Planning Study (September 1977).
5. Washington State Department of Ecology, Office of Water Programs. "Voluntary Program for Agriculture Announced," in *Waterline*, DOE publication (February 1978).

25

THE ECONOMIC IMPLICATIONS OF EROSION AND SEDIMENTATION CONTROL PLANS FOR SELECTED PENNSYLVANIA DAIRY FARMS

G. B. White
>Department of Agricultural Economics
>Cornell University
>Ithaca, New York

E. J. Partenheimer
>Department of Agricultural Economics
>The Pennsylvania State University
>University Park, Pennsylvania

INTRODUCTION

Amendments to the Clean Streams Law of Pennsylvania required the state's agricultural land owners to have erosion and sedimentation control plans designed and implemented by July 1, 1977. The Department of Environmental Resources (DER) of Pennsylvania is the state agency which is reponsible for implementing the guidelines for erosion and sedimentation control.[1] Procedures outlined by DER specify the following requirements for agricultural land owners:

1. All earthmoving activities, including plowing and tilling, must be conducted in a way that prevents accelerated erosion and sedimentation.
2. The methods proposed to control erosion must be set forth in a conservation plan.
3. The conservation plan must be prepared by a person trained and experienced in erosion and sedimentation control; and
4. The plan must be available at the farm headquarters.

The purpose of this paper is to highlight the results of a study dealing with the effects of implementing conservation plans on the income of commerical dairy farmers. Specific objectives were to (1) estimate the costs of control practices, (2) estimate the effects on returns to fixed resources of the adoption of practices in the plan, and (3) estimate the effects on returns to fixed resources of alternative methods of meeting soil loss constraints.

It was assumed, for the purposes of this study, that (1) most erosion and sedimentation control plans for land owners to comply with the Clean Streams Law have been, and will continue to be, prepared by the Soil Conservation Service (SCS), and (2) plans made by other agencies or individuals would incorporate many of the same practices as those recommended by SCS. These include contouring, stripcropping, terraces, diversions, waterways, cover crops, no-till or reduced tillage methods, rotation changes, and drainage systems.

THE ANALYTICAL MODEL

Twelve case-study farms were selected. The selection of case-study farms was confined to Grade A dairy farms because of the importance of dairy farming to the total agricultural economy of the state. In 1974 the sale of dairy products accounted for 39% of the total receipts from the sale of all agricultural products in Pennsylvania.[2] Case-study farms were included from Bradford, Susquehanna and Lackawanna Counties in the northeast region of Pennsylvania, Berks and Lancaster Counties in the southeast, and Cumberland and Franklin Counties in the south-central region. These counties had 33% of the state's total milk cows.

A linear programming model was used to evaluate three farm plans for each case-study farm. These three plans were as follows:

1. the "base plan," a simulation of the present farm plan before implementation of erosion control measures;
2. the "recommended plan," a simulation of the farm organization after implementing the SCS conservation plan;
3. the "alternate plan," a simulation of the farm organization considering economic as well as soil loss factors, and incorporating no-till corn into crop rotations.

Simulations for the recommended plan were run both with and without cost-sharing, in which land owners receive Agriculture Conservation Program (ACP) payments from the Agricultural Stabilization and Conservation Service (ASCS).

A highly constrained linear programming model was used to maximize returns to fixed resources subject to the relevant resource constraints. Resources which were constrained included land, labor and crop storage capacity.

Cow numbers and the number of replacement heifers were held constant for all three plans. Only existing crop activities and acreages were allowed for the base plan. Crop activities (rotations) recommended by SCS were the only cropping alternatives allowed for the recommended plan. For the alternate plan, a soil loss constraint (by field) was added, and the economically optimal mix of no-till corn grain, no-till corn silage, and hay was calculated subject to SCS soil loss limits. These limits for the average annual tons of soil loss ranged from 2 ton/ac to 4 ton/ac for the soil types represented. For most fields, the limit was 3 ton/ac. Soil losses for all plans were computed using the Universal Soil Loss Equation.

Conservation improvements were budgeted for the practices recommended by SCS for each farm for the recommended plan. Practices included in the analysis were confined to those related to erosion control. Other practices, such as wildlife management, spring development and forest management, were excluded. The annual cost of the various practices included depreciation, interest and maintenance. An interest rate of 8% was used. In the alternate plan, practices considered in addition to no-till were stripcropping and cover crops.

Management was held at a constant level of quality over all three plans. Feeding programs were not allowed to change radically from one plan to another, and the use of crop production inputs was held constant. Yields were assumed to be the same for all three plans; however, the yield increases needed to make the recommended plan's returns to fixed resources equal to returns in the base plan were computed. Thus the research emphasizes the short-term effects of the required adjustments.

Results from each individual case study were reported elsewhere.[3] In the following section of the paper, one of the 12 case studies is discussed in detail to illustrate the effects of conservation practices on returns to fixed resources. Then results from all case studies are summarized. A qualitative grouping of effects in terms of returns to fixed resources, soil loss, and total cost of conservation expenses follows.

EXAMPLE: CASE STUDY 3

In this section, results from Case Study 3 are presented. The results are not typical, but rather reflect several of the effects that can reduce returns to fixed resources when conservation practices are implemented.

This farm was located in northeastern Pennsylvania. The farm was operated by two brothers in partnership with their father. The farm operation was comprised of 365 ac of owned land and 25 ac of rented cropland. Owned land had 94 ac of cropland and 106 ac of permanent pasture.

The predominant soil type on owned land was Wellsboro channery silt loam on a C slope. The predominant soil type for the rented land was

Barbour fine sandy loam, a very productive soil on the floodplain. Soil loss data were not available for the rented land. The base plan rotation, 2 yr of corn silage and 5 yr of alfalfa hay or haylage, was retained for rented land on all plans.

Yields on the owned acreage were budgeted at 13.5 ton/ac of corn silage and 3.1 ton/ac of alfalfa hay. Since rented land had a more productive soil, higher yields of 15.8 ton/ac of corn silage and 3.3 ton/ac of alfalfa-timothy hay were estimated.

The dairy herd included 59 cows and 56 replacement heifers. Milk production for the base plan averaged 13,500 lb/yr/cow of milk sold. Labor was furnished mainly by the two brothers working full-time. Some seasonal labor was hired at $2/hr, especially during June, July and August for hay-making.

Returns to fixed resources decreased by 29.2% without cost-sharing and 27.3% with cost-sharing when compared with the base plan (Table I). Soil loss was reduced from an average of 15 ton/ac to about 1.4 ton/ac well below the 3 ton/ac limit for Wellsboro. Corn silage acreage was cut from 54.1 ac to 30.5 ac (totals for owned plus rented land). Alfalfa-timothy acreage increased by about 12 ac, and 12.3 ac were taken by diversions.

Total conservation expenditures were $9,040 without cost-sharing and $3,303 with cost-sharing. Expenditures were $1,842 for contour strip-cropping, $5,908 for diversions, $1,200 for underground drainage and $89 for fencing.

The major impact on returns to fixed resources in this case study appears to have resulted from a change from the base plan rotation on owned acreage. The rotation changed from 5 yr of corn silage and 5 yr of alfalfa-timothy hay to 2 yr of corn silage and 5 yr of hay in the recommended plan. As a result, feed purchases increased from 149 tons of a 16% concentrate ($20,413) to 188.5 tons of a 14% concentrate ($25,567). The increased cost of purchased feed was over $5,000. Other factors were the extra expense of $1,156 in annual costs for diversions without cost-sharing and $808 with cost-sharing, and the shift of 12.3 ac of land from more intensive crops to diversions. Yield increases of 37% without cost-sharing and 35% with cost-sharing would be necessary to make returns to fixed resources in the recommended plan equal to base plan returns to fixed resources.

The rotation given in the conservation plan resulted in a large decrease in returns to fixed resources, and soil losses were reduced to approximately 50% of the maximum limit. An SCS worker, after the survey, recalculated the maximum "C" value for the universal Soil Loss Equation and discovered that, with the implementation of stripcropping, diversions and the use of a cover crop, a rotation of 5 yr of corn silage and 5 yr of alfalfa-timothy hay brought soil loss just under the 3-ton/ac limit. Thus the planned rotation was unduly restrictive, given the advantage of hindsight. The base plan rotation could

Table I. Farm plans for Case Study 3.

Item	Unit	Base Plan	Recommended Plan	Alternate Plan
Returns to Fixed Resources	$/yr	26261.4	18619.1 (N)[a] 19092.9 (CS)[b]	21997.8
Soil Loss	ton	1411.9	128.3	282.0
Soil Loss per Acre	ton	15.0	1.4	3.0
Crops				
Corn Silage	ac	54.1	30.5	62.0
Ear Corn	ac	0.0	0.0	6.7
Alfalfa-Timothy	ac	64.9	76.2	50.2
Diversions	ac	--	12.3	0.0
Total Acres of Crops	ac	119.0	119.0	118.5
Labor Purchases				
March-April-May	hr	0.0	0.0	0.0
June-July-September	hr	135.0	170.7	0.0
September-October-November	hr	73.8	0.0	116.8
Feed Purchases				
14% Concentrate	ton	0.0	188.5	0.0
16% Concentrate	ton	149.0	0.0	0.0
20% Concentrate	ton	0.0	0.0	129.9
Bedding Purchases	ton	87.0	87.0	87.0
Hay Sales	ton	0.0	18.6	0.0
New Investment				
No-Till Planter	$	0.0	0.0	6310.0
Construction of Conservation Improvements	$	0.0	9040.2 (N)[a] 3302.7 (CS)[b]	1842.2
Silo Construction	$	5181.6	2438.4	9233.7
Conservation Improvements				
Contour Stripcropping	ac	--	94.0	94.0
Diversions	ft	--	8680.0	0.0
Underground Drainage	ft	--	1200.0	0.0
Fencing	rods	--	18.0	0.0
Cover Crops	ac	47.5	23.4	54.9

[a] (N) = no cost-sharing.
[b] (CS) = with cost-sharing.

have been maintained at a cost of (1) the annual cost of diversions and stripcropping, and (2) the reduced acres of intensive crops resulting from the 12.3 ac of diversions. In any case, the soil loss limit on this farm reduces returns to fixed resources considerably.

In the alternate plan, a 3-ton/ac soil loss was maintained while returns to fixed resources decreased by 12.3%. Included in the plan was stripcropping for all 94 ac, and the use of a rye cover crop on the 54.9 ac of corn silage grown on owned land. Concentrate purchases were less, but a higher protein feed (20%) was fed. This resulted from the relatively lower digestible protein (DP) and higher total digestible nutrients (TDN) which occurred with an increase in corn silage and a decrease in alfalfa. New investment of $1,842 for contour stripcropping and $6,310 for a no-till planter would be required for implementing the alternate plan.

The effects of implementing conservation practices on this farm resulted in the largest decrease in returns to fixed resources in the study. This case study is thus a caricature of potential effects. It illustrates several factors which operate to decrease returns to fixed resources when conservation measures are implemented. These effects include less intense rotations, the annual cost of conservation practices, and the removal of land from more intensive cropping to waterways and/or diversions. The results of the alternate plan also illustrate the potential of no-till corn for holding soil loss to prescribed limits while permitting more corn to be grown than could have been grown using conventional tillage.

In the following section, these effects are summarized for all farms in the study.

RETURNS TO FIXED RESOURCES

Recommended Plan

Results of the effects on returns to fixed resources are shown in Table II. Two farms had a more efficient allocation of resources with the recommended plan, which incorporated SCS recommendations. The other ten farms experienced decreases in returns to fixed resources. The greatest reduction, as shown in the preceding example, was in Case Study 3, which had a 29% decrease with no cost-sharing and a 27% decrease with cost-sharing.

The reasons for these effects were explored by assessing results of changes in farm enterprise organization and expenditures for conservation improvements. Case Studies 1 and 2 had higher returns to fixed resources under the recommended plan than under the base plan. These farms attained higher returns through intensifying the cropping program by replacing grass hay with alfalfa. Case Study 1 also brought additional cropland into production. These improvements could not be attributed to conservation planning except to the extent that some fields could be seeded with alfalfa after drainage.

A second group of farms experienced minor changes in returns to fixed resources. Included in this group are Case Studies 5, 9, 10 and 12. Characteristic of this group was the fact that there were no important changes in rotations. Small changes in returns could be attributed to the annual cost of conservation improvements and the loss of small areas of productive land to waterways or drainage ditches.

Table II. Returns to fixed resources as a percent of base plan returns and conservation practices, twelve case study farms.

Case Study	Returns to Fixed Resources			Conservation Practices[b]	
	Recommended Plan		Alternate Plan	Recommended Plan	Alternate Plan
	No Cost-Sharing	Cost-Sharing			
1	113.4	114.3	108.7	CC, CS, D, WW, UD, R	NT, CC, CS, R
2	103.5	103.7	100.1	CS, UD, R	NT, R
3	70.9	72.7	83.8	CC, CS, D, UD, R	NT, CC, CS, R
4	90.2	90.3	103.6	NT[a], CC, D, WW, R	NT, CC, R
5	97.5	97.9	97.3	CC, CS, WW, UD, R	NT, CC, CS, R
6	78.3	79.9	118.2	CS, CC, D, T, WW, OD, UD, R	NT, R
7	93.2	94.1	89.9	T, WW, R	NT, R
8	87.5	87.9	91.9	CC, CS, D, WW, R	NT, CC, R
9	97.2	98.8	118.0	NT[a], CC, CS, WW, OD, UD	NT, CC, R
10	96.7	97.8	114.9	NT[a], CS, WW, UD	NT, CC, CS, R
11	90.4	92.1	98.2	CS, D, WW, OD, UD, R	NT, R
12	99.3	99.8	98.9	NT[a], CC, CS, WW, UD, R	NT, CC, CS, R

[a] Used no-till corn in base plan as well.
[b] Code: NT - no-till corn   T - cropland terraces
          CC - cover crops   WW - waterways
          CS - contour stripcropping   OD - open-ditch drainage
          D - diversions   UD - underground drainage
                                       R - rotation changes

A third group of farms experienced rather substantial decreases in returns to fixed resources. In this group are Case Studies 3, 4, 6, 7, 8 and 11. Case Studies 3 and 4 had corn silage acreage restricted by the recommended plan. Corn silage acreage was reduced by 44% in Case Study 3 and 49% in Case Study 4. Both these farms also purchased large amounts of commercially mixed concentrate for the milking herd, and these purchases were increased as a result of the reduction in TDN produced in home-grown feeds.

Case Studies 7 and 8 produced ear corn, shelled corn and corn silage, baled corn stalks for bedding, and grew hay and small grain crops. Total corn acreage was reduced by 14% in Case Study 7 and 15% in Case Study 8. Both farms had to increase purchases of bedding, as residues from grain production were left on the ground for cover and small grain acreage was reduced.

Case Studies 6 and 11 had very high expenses for conservation improvements, but the annual cost of the improvements represented a small part of the total reduction in income. Annual costs of improvements were 11% of the total reduction for Case Study 6 and 22% of the total reduction for Case Study 11. Case Study 6 had a drastic change in rotation, with a change from a longer rotation with corn and alfalfa to a shorter rotation, with wheat and 1 yr of red clover hay. Wheat and red clover netted smaller returns than the alfalfa replaced. Alfalfa yields on Case Study 6 were very high relative to yields of other crops grown. In Case Study 8, however, red clover appeared to be a profitable alternative in comparison to alfalfa. In Case Study 11, 22 ac were placed in permanent grass hay, which reduced the acres of alfalfa. Corn acreage did not change appreciably in this case study.

These case studies showed that changing to less intensive rotations can have potentially severe impacts on income. Erosion and sedimentation control planners should be cognizant of the effects that changes in the proportion of corn, and particularly corn silage, grown on a dairy farm can have. Corn silage yields 50 to 100% more energy per acre than alfalfa on most dairy farms.[4] Although it may yield only 60 to 75% as much DP per acre as alfalfa, it still compares very favorably with alfalfa as a forage for dairy cows. Planners should approach the planning process with the objective of keeping corn acreage at or above current levels for a specific dairy farm situation if maintenance of current income is desired.

Alternate Plan

The alternate plan, incorporating no-till corn into crop rotations, resulted in higher returns to fixed resources than the base plan on six farms. This occurred even though soil loss was constrained to SCS standards for the alternate plan. Of these six case studies, 4, 9 and 10 were already planting corn using no-till methods. Thus herbicide costs were the same in the base plan and the alternate plan, and purchase of a no-till planter was not necessary in order to convert to the alternate plan. Case Studies 9 and 10 were also operating at a very low soil loss level in the base plan. Hence corn acreage was increased in the alternate plan. Case Study 6 had higher returns to fixed resources due to high alfalfa yields, not the conversion to no-till corn.

The alternate plan was also compared to the recommended plan. This comparison shows the profitability of incorporating no-till corn as a conservation method *vs* more traditional methods of erosion control, such as diversion and terraces. Higher net returns in the alternate plan in comparison with the recommended plan were obtained on seven farms. For Case Studies 3, 4, 8 and 11 soil losses were held to SCS standards, but returns were increased. These were all farms which experienced substantial decreases in net returns in changing from the base plan to the recommended plan. Case Studies 9 and 10, which were overplanned in the sense that the recommended plan had average soil losses of less than 1 ton/ac, could increase returns to fixed resources by expanding corn acreage and incurring additional soil loss.

Even though the alternative of no-till appeared profitable in Case Study 11, some difficulty in converting to this practice could occur because the farm has a high proportion of heavy, somewhat poorly drained soils. This may make no-till operation infeasible in some years with wet springs. In any given year, acreage which could be planted using no-till may be insufficient to justify purchase of a no-till planter. Operators wanting to employ no-till methods only on selected fields have an alternative. Farmers switching to no-till are advised to keep plows. Poorly drained fields on farms such as Case Study 11 could then be plowed and planted using the no-till planter. This prevents having to maintain two planters.

No-till was also a feasible alternative for adoption on Case Studies 1, 5 and 7, although it is not particularly advantageous on any one of these farms in comparison with the recommended plan. With Case Study 1, soil losses in the alternate plan met SCS standards, but losses in the recommended plan did not. In Case Study 5, the returns to fixed resources are virtually identical for the alternate plan and the recommended plan. For Case Study 7, specifying small grain as an alternative in some fields would reduce bedding purchases and result in returns to fixed resources comparable to the recommended plan.

Thus no-till probably has a place on all farms in the study with the exception of Case Study 11, which has soils which were not well adapted to no-till. Erosion and sedimentation control planners should explore the possibility of no-till corn as a way of maintaining corn acreage while reducing soil loss for those farmers who are still using conventional tillage methods. Growing no-till corn is generally thought to require a higher level of managerial skill than the use of conventional methods. The Agronomy Guide 1976[5] lists several considerations for successful no-till operations. Complete control of existing vegetation by herbicides is required. The use of no-till is advisable on shallow, droughty or erodable soils. No-till is not recommended on heavy, poorly drained soils.

SOIL LOSS

Soil loss results are shown in Table III. Maximum allowable soil loss for farms in this study, based on soil type and SCS procedures, ranged from 2 ton/ac to 4 ton/ac. Case Studies 2, 9 and 10 had soil losses in the base plan which were already considerably below the maximum allowed by SCS soil loss limits. The recommended plan reduced soil losses even further. The lower soil losses in Case Study 2 reflect the owner's desire to reduce cropping intensity. Much more corn could have been grown.

Table III. Summary of soil loss data, twelve case study farms.

Case Study Identification	Average Soil Loss, Base Plan (ton/ac)	Average Soil Loss, Recommended Plan (ton/ac)	Average Soil Loss, Alternate Plan (ton/ac)
1	10.1	3.9	3.0
2	1.0	0.6	0.9
3	15.0	1.4	3.0
4	4.6[a]	0.7	2.9
5	10.2	2.5	4.0
6	3.9	2.4	1.7
7	3.9	1.8	2.9
8	6.3[b]	2.5[b]	3.6[b]
9	1.1[a]	0.5	2.4
10	0.9[a]	0.8	1.6
11	3.9[c]	1.4[c]	1.6[c]
12	6.7[a]	1.3	2.4

[a] Used no-till in base plan.
[b] Soil loss data pertains to 164 ac of the total of 334 ac of owned cropland.
[c] Soil loss data pertains to 90 ac of the total of 190 ac of owned cropland.

In Case Studies 9 and 10, however, rotations were recommended which resulted in soil loss considerably below the maximum permitted. According to the planner for Case Study 9 and the owner for Case Study 10, this was due to the droughty soils which held down corn silage yields. In the alternate plan, corn silage acreage would increase on both these farms. These operators appeared not to be capitalizing on one of the primary advantages of no-till corn. That advantage is conservation of moisture on droughty soils.

In the recommended plan, soil losses were well below the maximum for Case Studies 2, 3, 4, 7, 9, 10, 11 and 12. Of these farms which had very low soil losses in the recommended plan, Case Studies 3, 4, 7 and 11 were among those farms which incurred large losses in returns to fixed resources in changing to the recommended plan.

In Case Study 3, the base plan rotation could have been maintained and still have met the 3 ton/ac limit. This would have greatly mitigated the reduced income which resulted on that farm due to a change in rotation. In Case Studies 4 and 11, some fields were converted to permanent hayland. There was a discrepancy between computed soil loss in this research and soil loss computed by SCS in Case Study 7.

Case Study 6 had reduced soil loss in the alternate plan compared with the recommended plan and the base plan. This was due to the optimization in the alternate plan which resulted in changing to nearly all alfalfa, and cannot be attributed to a change to no-till corn. Alfalfa yields may have been overestimated relative to other crops on this farm.

For these farms which had soil losses well below the maximum in the recommended plan, reasons for the apparent overplanning should be explored. It should be remembered that the soil loss values shown in Table III are averages for all fields on which crops (including hay) are grown. When an SCS worker encounters a specific field situation in which a year of corn in a reasonable length of rotation is estimated to result in soil losses above the maximum soil loss permitted, the land is converted to permanent hayland. Soil loss on hayland is virtually zero. Thus when significant acreage is converted to hayland, the average soil loss over the whole farm may be reduced significantly. An inspection of soil loss by individual fields, however, would reveal that fields with crops may be near the maximum limit. Hence, the conclusion that overplanning in relation to prescribed procedures was not always justified. This was true in Case Studies 4 and 11.

The preceding analysis has demonstrated that SCS soil loss limits, as applied by SCS technicians, reduced returns to fixed resources on several farms. This suggests that a flexible approach to planning, without a soil loss limit, is more appropriate than rigid adherence to absolute limits. For example, for a farm which was incurring a 12-ton/ac soil loss, planners might try to implement lower-cost practices, but avoid drastic changes in cropping intensity, to reduce soil loss to 5 tons/ac. This would preclude the large reduction in returns to fixed resources that could result if soil loss was reduced to 3 ton/ac. Since the specification of the parameters for soil loss calculation may be difficult to duplicate, some SCS planners appear to adopt a flexible approach to planning, while others do not. Modifying procedures and training of planners to emphasize soil loss goals rather than maxima should help workers in the field to use more judgment in their recommendations. Under these suggested procedures, SCS would work with any farmer who seeks to reduce soil loss without regard to a specific soil loss limit.

SCS does not currently have this flexible approach, and may not have in the near future. The responsible state agency, DER, could emphasize procedures by which farmers not wishing to implement SCS conservation plans might select less stringent erosion and sedimentation control plans. Some conservation districts currently have non-SCS employees who assist farmers in developing plans. These employees, however, typically use SCS procedures and criteria in developing plans.

DER currently has a proposal for assisting land owners of low and medium erosion hazard sites to develop their own plans. Such a procedure should be adopted into DER's guidelines for land owners, and should be highly publicized. Procedures for certifying or approving plans other than SCS plans should be established.

COST OF CONSERVATION IMPROVEMENTS

Expenditures for conservation improvements were calculated (Table IV). Costs were compared with cost-sharing at rates approved by local ASCS

Table IV. Total expenditures for conservation improvements in recommended plans, with and without cost-sharing, twelve case study farms.

Case Study	Conservation Expenditures		Amount of Cost-Sharing
	Without Cost-Sharing	With Cost-Sharing	
1	7,693.85	3,389.67	4,304.18[a]
2	1,103.80	313.60	790.20[b]
3	9,040.16	3,302.63	5,737.53[b]
4	705.48	196.36	509.12[b]
5	4,683.30	1,795.22	2,888.08[a]
6	14,128.29	5,112.62	9,015.67[a]
7	4,547.31	1,136.83	3,410.48[b]
8	6,504.27	1,490.22	5,014.05[a]
9	4,196.21	1,885.33	2,310.88[b]
10	6,881.96	3,138.51	3,743.45[a]
11	13,899.61	7,090.96	6,808.65[c]
12	3,657.13	1,412.26	2,244.87[b]
Totals for 12 Farms	77,041.37	30,264.21	46,777.16
Average per Case Study Farm	6,420.11	2,522.02	3,898.10

[a] Had LTA approved.
[b] Annual practice. Depends on adequate funding in the county and approval by the county ASCS farmer committees.
[c] Will apply for LTA.

committees. The average expenditure per farm was $6,420.11. The highest was Case Study 6 with $14,128.29 and the lowest was Case Study 4 with $705.48. If ACP payments were received at average rates for the counties in which farms were located, the average net total cost would have been $2,522.02 per farm. Five of the farms had Long-Term Agreements (LTAs), which meant that they were guaranteed payments over a 3- to 10-yr period. Six farmers indicated that they would apply for payments under the annual program.* One farmer, whose plan was being prepared at the time of the survey, indicated that he would apply for the LTA program.

A high proportion of expenditures for conservation improvements was for drainage. Nine of the twelve farms surveyed had drainage in their conservation plan (Table V). Drainage expenditures as a percentage of total expenditures ranged from 13 to 85% on these farms. For the nine farms in total, drainage accounted for 51.4% of total conservation expenditures. For the total expenditures of $77,034.97 for twelve farms, drainage accounted for 43.6%. ASCS cost-sharing for drainage was 50% for those counties included in the survey. Thus cost-sharing for drainage represented 35.9% of cost-sharing, if all farmers received payments for all eligible practices.

Table V. Estimated expenditure for drainage,[a] nine case study farms.

Case Study Number	Estimated Total Conservation Expenditures ($)	Estimated Expenditures for Drainage ($)	Percentage of Total for Drainage
1	7,693.85	2,117.60	27.5
2	1,103.80	500.00	45.3
3	9,040.16	1,200.00	13.3
5	4,683.30	2,200.00	47.4
6	14,121.89	6,886.94	48.7
9	4,196.21	3,478.13	82.9
10	6,881.96	5,827.87	84.7
11	13,899.61	9,096.89	65.4
12	3,657.13	2,274.12	62.2
Totals for 9 Farms Reporting Drainage	65,277.91	33,581.55	51.4

[a]Open drainage ditches or underground drainage.

---

*The receipt of government payments would be contingent upon approval by ASCS county committees and money available in the county. Therefore, receipt of payments by an individual farmer was uncertain.

Technical assistance and ACP payments for drainage are incentives for farmers to ask for planning assistance from conservation districts. As one farmer stated, "I had a problem with poorly drained land. I got stripping, contouring, and waterways along with drainage." Drainage also has a positive impact on yields and sometimes permits more intensive cropping.

YIELD INCREASE PARAMETERIZATION

Various procedures have been used to predict yield increases attributable to conservation measures. Swanson and Harshbarger,[6] using a procedure developed by agronomists in Illinois, predicted a yield loss of 0.47 bushels of corn per year for continuous corn on a 4% slope with up-and-down-slope cultivation. This was equivalent to less than a 1% annual reduction. Coutu et al.[7] estimated yield increases of 35 lb of tobacco for one set of conservation practices over yield without conservation plans after 15 years. This was an increase from a base yield of about 1,500 lb. Atkins[8] estimated corn yields on a deep, productive soil, at 95.3% of their initial yields after 100 years, even with 15 ton/ac soil loss. While not directly applicable to Pennsylvania conditions, these examples suggest that dramatic increases in yields may not result from implementing conservation practices.

There was no satisfactory way to predict yield increases or reduction of yield decreases from adoption of recommended conservation practices. An alternative procedure was employed to assess the likelihood of erosion and sedimentation practices being profitable for the individual farmer. The yield increases needed to bring returns to fixed resources for the recommended plan up to the base plan were computed for the ten farms which experienced income decreases (Table VI). The procedure involved computing the needed permanent yield increase such that its annual value equals the annual reduction in net income that occurred from the base plan to the recommended plan. Thus the time profile of yield increases was ignored.

Case Studies 1 and 2 were already shown to be more profitable operating under the recommended plan, and are not shown in Table VI. Less than a 5% yield increase would bring returns to fixed resources in the recommended plan up to those of the base plan for Case Studies 5, 9, 10 and 12. These farms also had substantial expenditures for drainage. There would appear to be a reasonable chance for yield increases to pay for conservation expenditures within a relatively few years on these farms. Other farms would appear to have little chance of making practices pay during the average remaining tenure of the operators. This would especially be true for Case Studies 3, 4, 6, 8 and 11. All these farms would require yield increases of 16% and higher with cost-sharing to reach the break-even point in returns to fixed resources.

Table VI. Yield increases needed to make estimated returns to fixed resources
in the recommended plan equal to returns in the base plan,
without cost-sharing and with cost-sharing.

Case Study	Percentage Yield Increase Needed (No Cost-Sharing)	Percentage Yield Increase Needed (Cost-Sharing)
3	37	35
4	19	19
5	1	1
6	28	26
7	8	7
8	23	21
9	1	1
10	4	3
11	16	12
12	1	1

ORDER OF ADOPTION OF PRACTICES

The results of this research showed that in several farm situations there was a trade-off beween income and soil loss, especially in the short run. Erosion and sedimentation control laws, if they require soil loss constraints at or near 3 ton/ac, could cause economic hardship for many farmers if the farms in this research are typical of dairy farms across Pennsylvania.

If soil loss constraints at or near the 3-ton/ac limit are to be imposed on a particular farm, careful consideration should be given to which practices should be tried first. A tentative order of adoption of practices, suggested by the results of this study, would be as follows:

1. Change to no-till corn if managerial skill and soil conditions are favorable. Cover crops may be needed in conjunction with no-till silage.
2. Implement contour stripcropping if soil loss remains above constraints. Contour stripcropping was one of the least costly structural practices to implement. Soil losses are reduced to 25% of up-and-downhill practices, and 50% of the contouring practice.
3. Implement other construction practices such as cropland terraces, diversions and waterways. These practices are more costly to construct and the latter two reduce the acreage available for intensive crops.
4. Change the intensity of rotations.

The rationale for instituting (3) before (4) is that changes in rotations were associated with significant changes in returns to fixed resources on several case study farms. Structural practices usually resulted in lower reductions consisting of the annual costs of depreciation, interest and maintenance. Furthermore, land owners are often able to get ACP cost-sharing which usually reduces the initial cost of these practices by 50 to 75%.

SUMMARY AND IMPLICATIONS

The results of this research showed that there was generally a trade-off between income and soil loss in the short run. Major findings were as follows:

1. Changing to the recommended plan resulted in increased returns to fixed resources for two case study farms. Four farms experienced reductions in returns to fixed resources of less than 5% and six farms incurred substantial reductions of 7% to 30%. For most of the farms which experienced substantial reductions, a change in rotations was the major contributing factor.
2. Yield increases of 1 to 37% would be needed to make returns to fixed resources in the recommended plan equal to returns in the base plan.
3. The alternate plan, incorporating no-till corn, stripcropping and cover crops, resulted in higher returns to fixed resources than the base plan on six farms. Seven farms obtained higher returns to fixed resources for the alternate plan than for the recommended plan. No-till appeared to be a viable alternative for all but one farm which had a significant amount of heavy, somewhat poorly drained soils.
4. Estimated expenditures for conservation improvements required in the recommended plan averaged $6,420.11 per farm with no ACP cost-sharing and $2,522.02 with cost-sharing.
5. Nine of the twelve farms had drainage planned. For these farms, drainage accounted for 51.4% of their total estimated expenditures for conservation improvements.

There are major implications for various state and federal agencies, and for land owners. Erosion and sedimentation planners should take a flexible approach toward the planning process. Helping each individual farmer reduce soil loss should be the goal of planning rather than a rigid adherence to soil loss limits for each field. A change in emphasis is needed at the national level of the SCS. SCS should be less oriented toward whole-farm planning and more oriented toward erosion control. DER, the state agency responsible for administering Pennsylvania's Clean Streams Law, should develop alternatives other than SCS conservation planning for farmers who desire less stringent erosion and sedimentation control plans.

ACKNOWLEDGMENTS

Appreciation is expressed to Nelson Bills and Roger Hexem for helpful comments on an earlier draft of this paper.

REFERENCES

1. Department of Environmental Resources (DER), Pennsylvania. "Soil Erosion and Sedimentation Control Manual for Agriculture," Harrisburg, PA (1974).
2. Pennsylvania Crop Reporting Service (CRS). "Crop and Livestock Annual Summary, 1975." Harrisburg, PA (1975).
3. White, G. B. "The Economic Implications of Clean Streams Legislation for Selected Commercial Dairy Farms in Pennsylvania," PhD dissertation, The Pennsylvania State University (1978).
4. Dum, S. A., F. A. Hughes, J. G. Cooper, B. W. Kelley and V. E. Crowley. *Farm Management Handbook,* College of Agriculture, The Pennsylvania State University, University Park, PA (1977).
5. The Pennsylvania State University. *Agronomy Guide, 1976,* College of Agriculture Extension Service, University Park, PA (1976).
6. Swanson, E. R., and C. E. Harshbarger. "An Economic Analysis of Effects of Soil Loss on Crop Yields," *J. Soil Water Conser.* (September-October 1964):183-186.
7. Coutu, A. J., W. W. McPherson and L. R. Martin. "Methods for an Economic Evaluation of Soil Conservation Practices," Technical Bulletin 137, Agricultural Experiment Station, Raleigh, NC (January 1959).
8. Atkins, S. W. "Economic Appraisal of Conservation Farming in the Grenada-Loring-Memphis Soil Area of West Tennessee," Bulletin 369, Agricultural Experiment Station, University of Tennessee (October 1963).

26

FARM-LEVEL ECONOMIC EVALUATION OF EROSION CONTROL AND REDUCED CHEMICAL USE IN IOWA

J. M. McGrann and J. Meyer
    Department of Economics
    Iowa State University
    Ames, Iowa

The impact of alternative erosion control methods and/or reduced fertilizer and pesticide use on net farm income will be a primary factor determining farmer cooperation and the public cost required to effectively reduce nonpoint pollution in Iowa. Farmers faced with the necessity to comply with both federal and state laws aimed at reducing nonpoint pollution will also need increased information on alternative ways to comply with laws in order to make a positive contribution in setting standards at the local level.

Previous studies in this area have concentrated on the macro-level impacts of soil erosion control including national studies by Nagadervara et al.,[1] a corn belt study by Taylor and Frohbert[2] and river basin studies by Alt and Heady.[3] The economics of erosion control in western Iowa was analyzed before new technologies, especially high-residue tillage technology, were fully developed.[4] Many of the economic studies have not considered the interaction between the cropping and livestock enterprises in evaluation of erosion control. Much of the information used in educational programs for farmers has concentrated on the technical aspects of how to control erosion with only limited consideration of the economic impacts and the interaction of the cropping and livestock enterprises.

This study concentrates on the farm-level economic impact of soil loss control and reduced chemical use. The farms analyzed are typical of three major soil associations in Iowa which cover a wide spectrum of farming conditions in the state. The analysis investigates the impact of restricting soil loss from the present to several reduced levels on net farm income and land rents for both specialized crop and mixed crop and livestock farms. The waste disposal problem of the limited sized livestock enterprises was not addressed in this study. A limited analysis is made of the effect that reducing fertilizer and banning pesticide use would have on farm income as an alternative to reduce nonpoint pollution. Policy alternatives to control erosion, including tax rates to achieve reduced soil loss and cost-sharing schemes for terracing, are evaluated in farm-level models. Consideration is also made as to how information generated in the study can be used for extensionist and farmer training.

The procedure used in this study to evaluate the impact of erosion control included the following steps.

1. Typical or representative farms were identified for each soil association in terms of size and composition of soil types.
2. The universal soil loss equation was used to define cropping and soil management systems that would keep soil loss within defined soil loss limits.
3. Input-output relationships were specified for each cropping and soil management system for the representative farm.
4. Enterprise budgets were developed for each crop with the use of a computerized budget generator and then weighted by cropping required to determine costs and returns by crop rotation and soil management practices.
5. Costs were estimated for the terracing erosion control option.
6. Linear programming models were formulated for each representative farm. The models also incorporated livestock enterprises and labor, machinery and capital costs.
7. Linear programming models were used to analyze impacts of reducing soil loss and other policy parameters.

FARM MODELS

Farm models were specified for the Monona-Ida-Hamburg (MIH), Clarion-Nicollet-Webster (CNW) and Tama-Muscatine (TM) soil associations.* The three soil areas are located in western, north-central and eastern Iowa, respectively, and account for approximately 12 million ac or one-third of the farmland in Iowa.[5] These areas represent a spectrum of productivity level and soil erosion problems. Productivity ranges from an estimated average corn

---

*Four soil types were specified for the soil associations. Each soil type and slope are as follows: (a) MIH, Monona 10C2, 5-9% slope, Monona 10D2, 9-14% slope, Ida 14-18% slope, and Napier 0-5% slope; (b) CNW, Clarion 2-5% slope, Nicollet, Webster and Canisteo 0-2% slope; (c) TM, Tama 2-5% slope, Dinsdale 5-0% slope, Muscatine 0-2% slope and Garwin 0-2% slope.

yield based on typical farm soil composition of 124 bu/ac in the TM association to 111 bu in the CNW to 85 bu in the MIH association.[6] The estimated annual soil loss, using the conventional moldboard plow and straight-row farming practices and based on the universal soil loss equation, is over 15 ton/ac annually in the MIH association, over 10 tons in the TM area and less than 5 tons in the CNW association.*

Typical sized farms were specified for the three areas with 320 ac of cropland in the MIH and CNW farms, and 240 ac in the TM farm. Farms were analyzed as owner-operator, specialized crop or mixed farms with farrow to finish hogs, cow-calf and cattle feeding enterprise alternatives. Cropping enterprises varied between farms and soil types within farms, but livestock enterprises were considered the same for the representative farms. Livestock enterprises were limited to the on-farm produced feed grain and forage. Operator and family labor was restricted to 3000 hr annually.

The universal soil loss equation was used to specify alternative cropping and soil management systems for a range of maximum soil loss levels for each farm.† The conventional moldboard plow-based farming system was used as a base for comparing alternative systems because, although to a lesser extent than in the past, it is dominant on Iowa farms. The conventional system also is a basis for estimating soil erosion without a soil loss restriction policy.

Enterprise budgets were developed with the Oklahoma Budget Generator and from published and unpublished information used in the Iowa State University Farm Management Extension program. Input levels correspond with recommendations by extension agronomists in the different soil associations, and yields are based on published historical data for each soil type.[6] Machinery costs were based on current replacement costs for a four-row complement in the MIH and TM associations and a six-row complement in the CNW associations.** All coefficient are for 1978 technology and input prices.

The price of corn used in the analysis was $2.25/bu, and other commodity prices correspond to the relative price of the respective commodity to corn during the 1970-77 period.†† Actual prices were soybeans $5.85/bu, hay $33.75/ton, hogs $42.00/cwt and slaughter cattle $49.00/cwt. Commodity prices *do not reflect* potential adjustments in production to comply with laws required for reducing soil erosion.

---

*The maximum level of annual soil loss allowable per acre to maintain productivity capacity (T-limit) for the TM and CNW farm is 5 tons, the MIH farm 4-5 tons.[7]

†Cropping rotations include continuous corn, corn-soybeans, corn-oats-meadow combinations and continuous meadow. Soil management alternatives include: residue and nonresidue tillage, spring or fall tillage, straight-row, contouring and terracing.

**Procedures used in calculating machinery costs with the Oklahoma Budget Generator are described in a manual prepared by Kletke.[8]

††The relative price of hay was above this level, but was reduced to reflect local market conditions if production increased due to conservation program.

RESULTS

Alternatives to Reduce Soil Erosion

Space does not permit inclusion of all the information showing the cropping sequences, estimated soil loss and enterprise budgets associated with alternative soil management practices for each soil loss level. However, some general comments can be made from this initial step of analysis.*

In order to reduce soil loss the farmer has the options of (1) reducing row crop intensity, (2) using higher residue tillage practices,† (3) moving primary tillage from fall to spring, (4) contouring, (5) terracing, and (6) or some combination of the above. Reducing row-crop intensity is very costly, especially when it means reduced supplies of corn for profitable hog production. With use of the chisel plow and less tillage, higher-residue tillage is a low-cost option in reducing soil loss. Reduced machinery cost and time of the high-residue tillage system are offset somewhat by increased pesticide requirement, especially for corn following corn, but, in general, it is less costly than conventional tillage. High-level management and timeliness are more critical with the high-residue systems. Contouring also requires a higher level of management, but farmers in Iowa do not consider the costs greater than straight-row farming. Terracing is the highest cost means to reduce soil erosion in Iowa and will be discussed in more detail later in this paper.

Economic Impact

Table I and Figure 1 show the economic impact of reducing soil erosion determined by the linear programming models of each for the three soil associations, with and without livestock production enterprises. The linear programming technique is used to select the income-maximizing alternative cropping and soil management system consistent with the maximum soil loss constraint on each soil type. The conventional nonresidue moldboard plow, straight-row system is to represent the base soil loss and farm income level.

Livestock production obviously makes a large contribution to farm income but also mitigates some of the income-reducing impact of restricting soil loss. The livestock farm uses all 3000 hr of operator and family labor except at very restricted soil loss levels on the MIH farm. Hog production is the most profitable enterprise and enters the optimal solution to levels restricted by

---

*There was a total of 48 crop rotations specified for the three alternative soil management systems and four soil loss levels for both the MIH and TM farms and 36 rotations specified for the CNW farm's three soil loss levels.

†High-residue tillage used in this study refers to tillage practices that leave between 2000-3000 lb of residue on the surface at planting time.

Table I. Economic impact of reducing soil loss in different soil associations.

Soil Association[a]	Maximum Soil Loss Restriction	Weighted Avg Soil Loss (ton/ac)	Weighted Avg Soil Loss (index)	Net Farm Income[b] (value)	Net Farm Income[b] (index)	Row-Crop (ac)
MIH	With livestock					
	Conventional	16.71	100	$33,267	100	207
	15 tons	11.62	70	31,744	95	193
	10 tons	7.98	48	29,454	89	169
	5 tons	4.34	26	23,199	70	119
	No livestock					
	Conventional	16.59	100	15,706	100	207
	15 tons	11.24	68	13,405	85	193
	10 tons	7.98	48	11,212	71	164
	5 tons	4.41	27	9,507	61	104
TM	With livestock					
	Conventional	12.29	100	55,678	100	240
	10 tons	4.27	35	56,501	101	240
	5 tons	3.23	26	54,136	97	240
	2 tons	1.31	11	48,448	87	204
	No livestock					
	Conventional	12.29	100	35,874	100	240
	10 tons	4.42	36	36,147	101	240
	5 tons	3.37	27	33,783	94	240
	2 tons	1.84	15	30,292	84	222
CNW	With livestock					
	Conventional	4.54	100	59,031	100	320
	5 tons	2.65	58	59,031	100	320
	2 tons	1.76	39	50,474	86	243
	No livestock					
	Conventional	4.54	100	39,064	100	320
	5 tons	2.65	58	39,064	100	320
	2 tons	1.76	39	30,577	78	243

[a] Monona-Ida-Hamburg (MIH), Tama-Muscatine (TM) and Clarion-Nicollet-Webster (CNW) soil associations.
[b] Income to land, management, risk and profit.

labor or availability of on-farm produced feed grain. The feed grain restriction limits hog production on the MIH farm at the 5-ton-maximum soil loss level. It should be noted, however, that if off-farm employment is available the crop farmer could offset part of the income difference.

As can be observed in Table I and Figure 1, the income-reducing impact of restricting soil loss from the conventional to the first restricted loss level

364 BMPs FOR AGRICULTURE AND SILVICULTURE

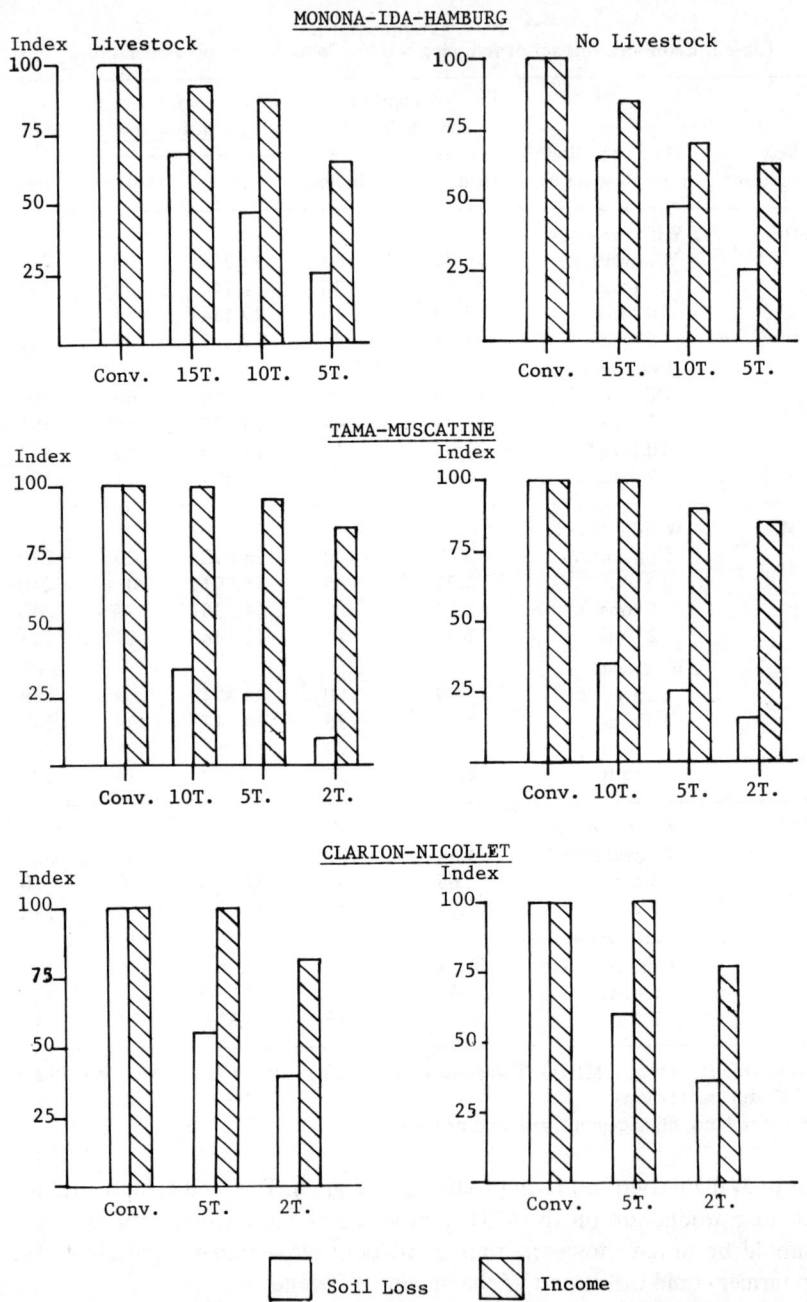

Figure 1. Indices of soil loss and income for different levels of soil loss restrictions by soil association.

considered is small or, as in the case of the TM and CNW farms, there is no income-reducing impact. This first level of soil loss reduction is accomplished with increased use of residue tillage and contouring. This result is consistent with data that show a rapid increase in farmers' use of conservation tillage in Iowa.*

The results of the analysis also indicate that soil erosion can be reduced below the 5-ton level or the T-limit in the TM and CNW soil associations with little or no sacrifice of income. However, income is reduced in the MIH soils area where productivity is lower and erosion greater due to the soil types and topography. Row-crop intensity can be maintained as soil loss restrictions are lowered, except at the very restrictive 2-ton level for the TM and CNW farms through residue tillage and contouring. Reducing soil erosion on the MIH farm requires not only reducing row-cropping intensity but also substantial investment with respect to income in high-cost terracing.

Machinery Cost

Increased machinery investment to change tillage practices is a concern of farmers. In the three soil association areas, adoption of increased residue tillage practices requires the substitution of the moldboard plow for a chisel plow. Since horsepower requirements are lower and field operation time is reduced with residue tillage, there is no necessity to increase the investment in tractor size. Depending on farm size, adding the chisel plow increases machinery investment from $2.75 to $3.67 and annual ownership cost $0.50 to $0.65/ac. This suggests increased machinery investment and cost are not a primary limitation on adoption of residue tillage.†

Impact on Land Rent

The linear programming models can also be used to evaluate the impact that restricting soil loss has on land values through the inputted land rents.** For the MIH farm where livestock production is limited by feed grain supply as increasing restrictive soil loss levels sharply increase the value of additional

---

*Nearly 45% of the land in Iowa in 1977 was tilled with some form of conservation tillage rather than the conventional moldboard plow, disking and harrowing practice. In 1968 0.5 million ac used nonconventional tillage. In 1977, the estimate was over 9 million ac.[9,10,11]

†The low cost associated with the adoption of this implement explains in part the doubling of crop acres between 1976 and 1977 to over 4 million ac tilled with the chisel plow in Iowa.[11]

**Shadow prices or marginal value product values are used to estimate the value of the last acre of land available. One should recall commodity prices were not changed as production adjustments were made to meet soil loss restrictions.

land that can produce corn, land forced to permanent pasture because of high erodibility rent is reduced sharply. Without livestock, and in the cases of the TM and CNW farms with adequate feed grain to fully employ labor, the inputted rents of land are reduced as soil loss is restricted. Very restrictive soil loss limits reduce row-crop intensity and/or require the more costly terracing alternative to control soil loss.

From this analysis the effect restricting soil loss has on land rents varies between soil associations and the resources available to complement livestock production. One general conclusion which can be made is that poor-quality, highly erodible land will decrease sharply in value relative to the highly productive, low-erodible land if soil loss restrictions are enforced. This factor alone should have a major impact on land values.

The Terracing Alternative

Terracing as a means to control soil erosion has received considerable support in Iowa, including a state cost subsidy or cost-sharing program. This study evaluates the economic impact of terracing as a means to control erosion as opposed to the other alternatives included in the farm models and alternative subsidy programs for the Ida and Tama farms. Two cost-sharing programs were considered, a 75% subsidy on investment cost and a 75% subsidy on the investment interest rate.

The analysis of the subsidy alternatives is made for the 5-ton soil loss level on the MIH farm where a limited amount of terracing becomes competitive with contouring and residue tillage as the least-cost alternative to control soil loss with no cost-sharing on some of the soil types. Without livestock and at a 5-ton soil loss level, the maximum income level of terracing is 58 ac (21% of the crop acres) with or without the subsidy program. With the inclusion of livestock in the analysis and, in turn, higher value of the corn, the number of acres profitable to terrace remains at 58 ac until the subsidy reduces cost and terracing increases to 187 ac (67% of the crop acres). The farm income is $700 greater (2.5%) with the investment subsidy than with the interest subsidy, or either alternative would give similar income and land use results are the same.

Results of the analysis on the effect of cost and interest cost-sharing for the more productive TM farm differ from the MIH in that as either form of cost-sharing increased, the amount of land terraced increased from 36 to 132 ac for both the specialized crop as well as the crop-livestock farm. The net farm income difference between the cost-sharing alternatives is, however, less than $1000 more than the nonsubsidized option that meets minimum soil loss limits by spring residue tillage.

The conclusion one can derive from the analysis is that the economic impacts of cost-sharing vary by type of farm and soil associations. Terracing is a high-cost option for farmers and/or the public to control soil erosion, and other alternatives should be considered before a cost-sharing program is implemented. Alternative cost-sharing strategies can give similar results. This should be recognized in choice of implementation procedures. The alternative cost-sharing schemes considered here are only two of a wide variety of options that could be evaluated using the linear programming farm models.

Soil Loss Tax

One of the considerations in terms of implementation of a soil loss control program is using a soil loss tax per ton of erosion rather than setting maximum soil loss limitations on a soil type as was used to evaluate the impact of reducing soil erosion in this study. The two alternative methods were evaluated on the MIH farm where the cost of reducing erosion was most severe. The economic impact of the two methods *with livestock* are quite similar. Taxing soil loss at a marginal rate of $8.33/ton would give a farm income of $425 more than restricting soil loss below five ton/ac on all soil types. The land use, however, was quite distinct for the two alternatives. The profit maximizing, cropping and soil management system did not include the low-producing, highly erodible land. Under the taxing alternative more of the productive land was terraced, but practices were used that would give an average of 8.5 ton/ac of soil loss on over 30% of the land. This would compare to all alternatives being under the maximum level of less than 5 tons if a maximum rate was placed on each soil type.

Analysis of the MIH farm without livestock gave almost identical results with either the soil loss tax (a marginal rate of $5.56/ton) or the maximum soil loss restriction. Again this would indicate the importance of including livestock enterprises when analyzing policy questions.

Conclusions as to the most appropriate method to restrict soil loss in terms of water quality cannot be made until information is available on actual sediment delivery and the associated impact on water quality. When this information is made available, the farm models can be broadened to analyze this aspect.

Reducing Chemical Use

Since the use of fertilizer and chemicals accounts for a significant portion of nonpoint pollution, one alternative means to reducing this source of pollution is limiting chemical use.* Table II shows the impact of limiting fertilizer and banning pesticide use on one soil type on the MIH farm at the 5-ton soil loss level.

---

*Data on estimated yield reduction can be found in the Iowa State University 208 Technical Assessment Report.[11]

Table II. Net returns per acre for different fertilizer, insecticide and herbicide levels for the Monona C2 soil and 5-ton soil loss

Level of Input Use			Net Income Per Acre for Alternatives[a]		
Fertilizer	Herbicide	Insecticide[b]	(1) COMM, NR[b]	(2) CCOM, R[c]	(3) Cont. C, R[d]
Full	Full	Full	48.58	57.25	68.21
Full	Full	Zero	48.58	55.48	61.13
Full	Zero	Zero	43.73	45.70	41.22
3/4	Full	Full	45.26	49.04	59.76
3/4	Full	Zero	46.05	47.70	51.25
3/4	Zero	Zero	40.11	36.71	31.07
1/2	Full	Full	33.54	35.97	42.11
1/2	Full	Zero	34.28	34.25	33.20
1/2	Zero	Zero	28.12	23.29	13.08

[a] Net return to land, management, risk and profit. Alternative (1) is straight row, (2) is contoured and (3) is with terraces. C = corn, O = Oats, M = meadow, NR = conventional tillage and R = residue tillage.
[b] No rootworm insecticide is required for corn following meadow.
[c] This was the maximum income alternative in the whole farm analysis.
[d] Does not show income reduction to offset terracing cost.

As can be observed in Table II, the impact of banning the use of insecticide (corn rootworm treatment) varies because it is only required where corn follows corn in a rotation.

In continuous corn eliminating the insecticide would reduce income an estimated $7/ac while eliminating herbicide use would reduce income approximately $13/ac.

Reducing fertilizer use has the greatest income-reducing impact for a continuous corn rotation. Income is reduced by 12 and 39%, yields by 10 and 25 as fertilizer use is reduced to three-quarters and one-half the recommended level, respectively. As fertilizer use is reduced the income sacrificed is less with the oats-meadow rotations requiring less nitrogen fertilizer.

Elimination of pesticides and reduction in fertilizer use to one-half the recommended level would reduce yields by one-half and income to land, management, and risk to the point it would be infeasible to use the MIH farm to produce row-crops without an increase in commodity prices.

A similar analysis of limiting fertilizer and pesticide use was done on the Nicollet soil of the CNW farm. A corn-soybean rotation can be used on this soil to stay under the 5-ton soil loss restriction. Rootworm insecticide is not required in the rotation.

As can be observed in Table III, the net income is less at full chemical use level with the residue tillage. This is due to an assumed small increase in herbicide use of high-residue tillage compared to the nonresidue tillage alternative.

The estimated economic impact of eliminating the use of herbicide is to severely reduce income, about 35% for the corn-soybean rotation. Reducing fertilizer use by one-half actually has a less income-reducing impact than

Table III. Net returns per acre for different fertilizer and herbicide use levels for the Nicollet Soil and 5-ton soil loss.

Level of Input Use[a]		Net Income Per Acre for Alternatives[b]	
Fertilizer	Herbicide	(1) CSb, R	(2) CSb, NR
Full	Full	155.91	158.74
Full	Zero	103.27	103.27
1/2	Full	138.17	141.00
1/2	Zero	85.54	85.54

[a] No insecticide is required on corn following soybeans.
[b] Net return to land, management, risk and profit. Alternative (1) is straight-row with residue tillage, Alternative (2) is straight-row conventional tillage. CSb = corn-soybean rotation.

reduction of herbicide because of less nitrogen needed by corn following soybeans but would still reduce income over 10%. Elimination of herbicide and limiting fertilizer to one-half the recommended level would cut income nearly in half again making it economically infeasible to farm the land without a substantial increase in commodity prices.

Price Changes

Changes in the price levels in general will not alter the optimum farm crop and livestock enterprise combination, providing the price changes are proportional for all factors and commodity prices. However, net profits will be directly affected by changes in the price level. Increased prices of feed grains for the crop farms or alternatively increased hog or cattle prices will increase the economic impact of reducing row-crop intensity and thus favor soil erosion control alternatives that allow for more feed grain production, *i.e.,* contouring, high residue tillage and terracing. Without soil erosion limitations, increased grain prices encourage intensive row-crop farming and, in turn, the high levels of soil erosion that have been experienced since the sharp increase in commodity prices in 1973.

Throughout the analysis commodity prices were not changed as production adjustments were made to meet soil loss restrictions. This is a correct assumption for the individual farmer but, if all farmers in the corn belt had to meet the same restrictions adjustment in land use, commodity prices would increase offsetting at least a portion of the economic impact of restricting soil loss.

EDUCATIONAL MATERIAL

The necessity of public involvement in nonpoint pollution control planning and identifying "best management practices" with consideration of technical, economic and socially feasible control alternatives has been well communicated. Extension personnel face a formidable task in their educational responsibility to provide assistance to county and district committees and to the general public to help increase understanding of nonpoint pollution and its control.

One of the most important sources of educational materials available from the procedure used in this study is the information developed from defining the cropping rotations for each soil management system and the associated enterprise budgets. Farmers and extensionists are accustomed to working with enterprise budgets, and these can be used as a simple straightforward way to illustrate economic consequences of alternative strategies to reduce soil loss.

Developing farm models that closely represent local conditions is essential to communicate with farm people. This is one reason it is important to include both the crop and livestock enterprises in the analysis.

Much of the information developed in Iowa was at the county level involving extension and soil conservation personnel. By getting their involvement they not only can make an important contribution in selection of alternatives to be evaluated but can more effectively articulate the results of the analysis. Capability to use and interact with the models during local meetings is not possible in Iowa but many states that have terminals could provide this educational experience.

The procedure used in this study can generate the type of information necessary to train extensionists and farmers. Evaluation of alternatives in the economic framework has made programs more effective than the traditional approach depending almost exclusively on the technical evaluation of alternative erosion control practices.

CONCLUSIONS

The procedure used in this study can be used to evaluate the economic impact of erosion control or reduced chemical use at the farm level. The approach can also be an important means to generate agricultural extension information to give a clearer understanding of the economic consequences in dealing with nonpoint pollution at the local level.

The analysis shows that soil erosion can be reduced below 5 ton/ac, or to a level where productive capacity can be maintained with little or no sacrifice of farm income in two of the soil associations studied. These areas account for over 10 million ac or 28% of the states' land. In the third association, subject to greater erosion problems, reducing soil erosion below 5 ton/ac annually would reduce farm income 30 to 40% resulting in a high farmer and/or public cost. In all soil associations studied, inclusion of livestock enterprises helped to mitigate some of the income-reducing impact of restricting soil loss.

While the analysis shows some very positive results in terms of limited income reduction from restricting soil loss, the economic impact of reducing fertilizer use and banning pesticides would be most severe.

A number of factors restrict the extent that nonpoint pollution problems can be analyzed at the farm level in Iowa including: (1) limited information on relationships between erosion control, sediment delivery and water quality is available; (2) data is lacking as to expected long-term consequences of excessive erosion on crop yields; (3) more information is needed on the external or off-farm cost of nonpoint pollution and offsite sediment damage; and (4) data are needed on the cost of alternative off-farm means to improve water quality as opposed to the controlling of soil erosion and reducing farm chemical use.

As the information is made available, analysis and education programs can be more complete and nonpoint pollution programs can become effective and responsive to local needs.

REFERENCES

1. Nagadevara, V. S., E. O. Heady and K. J. Nicol. "Implications of Application of Soil Conservancy and Environmental Regulations in Iowa Within a National Framework," Iowa State University Center for Agricultural and Rural Development, CARD Report No. 57 (June 1975).
2. Taylor, R. C., and K. K. Frohberg. "The Welfare Effects of Erosion Controls Banning Pesticides and Limiting Fertilizer Application in the Corn Belt," *Amer. J. Agr. Econ.* 59:24-35 (1977).
3. Alt, K. F., and E. O. Heady. "Economics and the Environment: Impacts of Erosion Restraints on Crop Production in the Iowa River Basin," Iowa State University Center for Agriculture and Rural Development, CARD Report 75 (December 1977).
4. Dean, G. W., E. O. Heady, S. M. A. Husain and E. R. Duncan. "Economic Optima in Soil Conservation Farming and Fertilizer Use for Farms in the Ida-Monona Soil Area of Western Iowa," Research Bulletin 455, Iowa Agr. and Home Econ. Exp. Station, Ames (1958).
5. Oschwald, W. R., F. F. Riechen, R. I. Dideriksen, W. H. Scholtes and F. W. Schaller. "Principal Soils of Iowa," Special Report No. 42, Iowa State University of Science and Technology, Cooperative Extension Service, Ames (1965).
6. Fenton, T. E., E. R. Duncan, W. D. Shrader and L. C. Dumenil. "Productivity Levels of Some Iowa Soils," Special Report No. 66, Iowa Agr. and Home Econ. Exp. Station, Ames.
7. USDA-SCS-Iowa. "Field Office Technical Guide," Section III-1 (June 1977).
8. Kletke, D. D. "Operation Manual for the Oklahoma State University Enterprise Budget Generator," Research Report P-919, Agriculture Experiment Station, Oklahoma State University, Stillwater (1975).
9. Amemiya, M. "Conservation Tillage in the Western Corn Belt," *J. Soil Water Conserv.* (January-February 1977).
10. Department of Agronomy, Iowa State University. "Conservation Tillage in Iowa, 1976-1977," unpublished extension information developed from SCS data, Iowa State University, Cooperative Extension Service, Ames (1978).
11. Iowa State University. 208 Technical Assessment Report, unpublished draft, Iowa Agr. and Home Econ. Exp. Station, Ames (January 1978).

27

ECONOMIC IMPACTS OF POLICIES
TO CONTROL EROSION AND SEDIMENTATION
IN ILLINOIS AND OTHER CORN-BELT STATES

W. D. Seitz, C. Osteen
    Institute for Environmental Studies
    University of Illinois
    Urbana, Illinois

M. C. Nelson
    Department of Agricultural Economics
    Prairie View A&M
    Prairie View, Texas

Erosion and resulting sedimentation has been identified as an important agricultural nonpoint source of pollution to be controlled by the states under PL 92-500. A number of important economic issues relating to this process are examined in this paper.

Achieving the objectives of this legislation will generate costs that will be borne by society. In this paper the results of several studies on the magnitude and the distribution of these costs will be reported. These studies indicated that the aggregate cost will not be as great as indicated by some watershed level studies. This is due in large part to the response of farm operators to the price changes induced by the soil loss control policy. Another important factor is the significant decreases in soil loss that can be accomplished by shifting to conservation tillage, generally without an adverse impact on farm income. These results indicate that application of varying degrees of control on soil erosion among states will probably have little impact on the distribution of farm income among states. The heaviest adverse impact, however, will be borne by the owners of erodible land. Although erosion controls will bring about substantial reductions in erosion over time which will maintain high soil productivity, the economic incentives to individual farmers are not great enough to induce them to institute the changes without incentives to do so.

METHODS

One of the basic tools used was a linear programming model of the cornbelt region of the United States. The marketing and production of corn, soybeans, wheat, oats, hay and pasture in a single year were modeled. The 17 land resource areas (LRAs) which define the corn belt produce 70% of U.S. corn production and 60% of U.S. soybean production.[1] The model estimated market equilibrium prices and production for corn and soybeans under different environmental constraints by including demand functions estimated by Taylor and Frohberg.[2] Fixed requirements of hay and pasture were set for each LRA, while small grain requirements were set for the entire corn belt.

Crop production relationships considered costs and yields on different soil types with different rotations and conservation and tillage practices. Each of the 17 LRAs was divided into 11 land capability units (LCUs) to account for differences in productivity and erodability of the soil. Each crop production activity is a crop rotation combined with a tillage practice and a conservation practice which is applied to an LCU within an LRA. Conservation practices included straight-row planting, spring plowing, and chisel plowing. Rotations accounted for the influence of previous crops on fertilizer and pesticide requirements for the current crop. Costs of labor, fertilizers and pesticides were included with the annual variable costs of production. Soil losses per acre were estimated using the Universal Soil Loss Equation for each crop production activity.

For the regional analysis, this model was modified by incorporating a different method of including demand functions,[3] defining the boundaries of Illinois which created new production regions, aggregating some LRAs to reduce computational costs, and revising soil loss restrictions based on new Soil Conservation Service (SCS) information. It was also assumed that any land removed from production by the model would be allocated to hay production to provide revenue for the land owner.

A watershed analysis was conducted on the Big Blue Creek watershed in western Illinois to look at long-term impacts of soil erosion controls.[4] Nine hypothetical farms representative of the physical and topographical characteristics of the watershed were developed to generate information at the farm level. A linear programming model of a 10-year period was used to determine the optimum combination of crop rotation, conservation and tillage practices, and nitrogen application. Yields and net revenues were based on the depth of remaining topsoil and discount rate. This procedure was repeated in 10-year intervals over a 100-year period accounting for soil losses.

IMPACTS ON THE CORN BELT

An important implication of this work is that prices and production of crops will change in response to erosion controls to improve water quality. It is likely that some crop prices will increase (and production decrease) when they are removed from production to remove soil loss, while the prices of substitute crops will decrease (and production increase). These changes in prices and production, as well as the costs of erosion controls, will have an impact on farm income and food costs. Society will bear a cost to reduce erosion, and hopefully improve water quality, due to costs of erosion control practices applied and changes in crop production associated with rotation changes. The price impacts of these constraints will affect how the cost is distributed among producers and between producers and consumers. Important factors in determining price impacts and social costs and their distribution are the type of erosion control policy applied (soil loss tax, soil loss restriction, or cost-sharing) and the distribution of crops over different soil types.

Seitz et al.,[4] Taylor et al.[5] and Osteen and Seitz[6] indicate that significant increases in conservation tillage may occur without government regulations. This is due to the lower labor and energy costs associated with these practices.[7] For example, Osteen and Seitz[6] predicted that 74% of Illinois cropland would be allocated to conservation tillage when no erosion control policies are applied. Presently, only about 28% of Illinois cropland is tilled in this manner.[8] Education programs emphasizing economic advantages of conservation proactices should help to speed the adoption of these practices.

In Seitz et al.,[4] Taylor et al.[5] and in Taylor and Frohberg[2] it was reported that a soil loss tax is more efficient that a soil loss restriction which, in turn, is more efficient than a soil loss restriction combined with a terracing subsidy. A $2.00/ton soil loss tax caused a reduction of soil loss of 337 million tons at a social cost of $192 million, averaging $0.57/ton. A 3-ton/ac/yr soil loss restriction (TAY) caused a reduction of soil loss of 353 million tons at a social cost of $480 million, averaging $1.36/ton.* Under a 3-TAY restriction combined with cost-sharing of terracing at 50%, soil loss was reduced by 360 million tons at a social cost of $495 million, averaging $1.38/ton.

---

*Estimates were also generated with a revised (lower) set of soil loss coefficients.[4] With the 3-ton limit the social cost is $151 million in order to accomplish a 160-million-ton reduction in soil loss.

In each case, the reduction in soil loss is in addition to the reduction that could be achieved by switching to conservation practices on approximately 70% of the land.

A soil loss tax is more efficient than the other two approaches because it encourages the cheapest allocaton of soil erosion control practices and crop rotations to different soil types in terms of cost per ton of soil loss reduction. The approach, however, will be difficult to implement. A soil loss restriction requires that soil loss on all acreage be reduced to the level established regardless of the cost per ton, while a terracing subsidy encourages terracing over cheaper methods of reducing soil loss such as conservation tillage or contouring. For example, the $2.00/ton tax resulted in 7.7 million acres of terracing and 79.8 million acres of chisel plowing, while the 3-TAY restriction resulted in 9.2 million acres of terracing and 79.1 million acres of chisel plowing. The 3-TAY restriction with a 50% cost-share caused 12.4 million acres to be terraced while 78.2 million acres were chisel-plowed. Soil loss restrictions and terracing subsidies concentrate efforts on areas with the highest rates of erosion, while taxes concentrate on these areas where erosion can be reduced most cheaply. The major problem with cost-sharing is that less cost-efficient practices would be applied if they were cost-shared. If all appropriate practices were cost-shared at the same percentage rate under a soil loss restriction policy, this approach would be as efficient as restrictions alone, although the distribution of the burdens would be affected.

Seitz et al.[4] and Taylor and Frohberg[2] show that different programs have different impacts on crop prices, food costs and farm income. Under the $2.00 tax, the $192 million social cost consists of a decrease in farm income of $960 million, of which $515 million is paid as taxes, and a decrease of $252 million food costs. Soybean prices increased ($0.18/bu) and production decreased, while corn price decreased ($0.02/bu) and production increased. Prices of the other crops also decreased. However, under the 3-TAY restriction, both farm income and food costs *increased*. Farm income increased by $527 million and food costs increased by $1,007 million.* Both corn and soybean prices increased (by $0.12/bu and $0.50/bu) and production decreased. Other crop prices increased slightly. The 3-TAY restriction combined with a 50% terracing subsidy produces results similar to the 3-TAY restriction.

A comparison of the Osteen and Seitz[6] and Seitz et al.[4] results indicates the importance of the distribution of crops over different soil types on prices, food costs and farm income. More soybeans were allocated to erodible land than under Seitz et al.[4] and Taylor and Frohberg[2] in the unconstrained

---

*With the revised set of soil loss coefficients[4] the $151 million is comprised of a cost to consumers of $197 million and a gain to farmers of $46 million.

model runs. This difference in crop allocations results in different price impacts when constraints are imposed which, in turn, causes a *decrease* in both farm income and food costs when a 3-TAY restriction is applied in Osteen and Seitz.[6]* Soil loss was reduced by 384 million tons, in addition to the initial reduction due to the shift to conservation tillage, at a social cost of $64 million, averaging $0.17/ton. The social cost is lower in this analysis than in Seitz *et al.*[4] because of the use of revised soil loss coefficients (lower) and the assumption that hay would be produced on land removed from production by the model. Food costs decreased by $258 million, while farm income decreased by $322 million. In this analysis soybean price increased ($0.42/bu) and corn price decreased ($0.09/bu) while both decreased in the Seitz and Taylor studies. In this study, corn, which causes less erosion than soybeans but more than small grains and hay, was substituted for soybeans to reduce erosion causing corn production to increase and price to decrease.

These analyses suggest that the aggregate impacts of soil erosion control programs on farm income probably would not be severe. Taylor and Frohberg[2] predicted an increase in farm income of $4.71/ac under a 3-TAY restriction,† while Osteen and Seitz[6] predict a loss of $2.88/ac. In both cases there would be a gain associated with the conversion to conservation tillage practices. In comparison, Kasal[9] implied a much higher cost in a watershed study in which he estimated a cost of $10.77/ac with a less restrictive 5-TAY limit. This comparison indicated the need for aggregate models for determination of the impacts of policies applied on a wide area.

The estimated impacts on farm income and food costs are not consistent among our several models. A key factor appears to be the impact on the price of corn. If corn price decreases, as it does with the soil loss tax and the 3-TAY restriction under Osteen and Seitz,[6] farm income and food costs decrease. If corn price increases, as it does under the 3-TAY restriction under Seitz *et al.*,[4] the opposite occurs. The direction of change in corn prices will depend on a complex interaction of the initial allocation of crops and the shifts among the several crops when restrictions are applied. The price impacts on farm income and food costs are due to the revenue effects of inelastic demand for food. The price of soybeans consistently increases (and production decreases) because it is the most erosive of the major crops grown in the corn belt.

---

*The method of including demand curves used in Osteen and Seitz[6] might not be exactly equivalent to that used in Taylor and Frohberg.[2] Also, some of the net return coefficients have been changed somewhat in the process of revising coefficients for the regional model. These two factors may have brought about the difference in the allocation of crops over soil types.

†The increase is only $0.41/ac with the lower soil loss coefficients.

REGIONAL IMPACTS

Since the planning for erosion control is being conducted at the state level, the question of whether imposing varying constraints will result in significant regional impacts is of interest. Osteen and Seitz[6] utilized the modified corn-belt model to address this question under several alternative policies. The policies analyzed are 3-TAY restrictions (1) imposed throughout the corn belt (policy CB), (2) imposed throughout the corn belt except in Illinois (NI), (3) imposed in Illinois but not in the remainder of the corn belt (I), and (4) imposed in Wisconsin but not in the remainder of the corn belt (W). Illinois' share of farm income decreases less than 1% under policies I, NI and CB, and remains unchanged under policy W as compared to unrestricted conditions. Wisconsin's share of farm income remains unchanged under policy I, increases slightly under policies NI and CB, and decreases slightly under policy W.

The impacts of soil loss restrictions on farm income shares are the result of price changes, erosion control costs, and production shifts among regions. Price changes have the same impact on returns in areas with erosion constraints as they do in other areas. Thus farm income shares among states are not changed significantly by the price changes. There are shifts of crops grown among regions in response to soil loss constraints imposed in parts of the corn belt. For example, soybeans are shifted to those areas without controls, and to areas where they can be produced without erosion problems, while less erosive crops are shifted to the constrained areas. These shifts occur because it is more efficient to meet the constraints in this manner than by adopting more expensive soil erosion control structures.

The combination of rotation shifts and price changes results in the small decrease in Illinois' share of farm income under policies NI and CB. This negative impact is greater on Illinois because it has a relatively high proportion of land in corn production, the price of which decreases, while it has a lower proportion of land in soybean production, the price of which increases. A very small shift of soybeans into and corn out of Illinois under policies NI and CB moderates the decrease in farm income in Illinois.

Erosion control costs incurred are the prime reason that Illinois' share of farm income is less under policy I and Wisconsin's share is less under policy W than under unrestricted conditions. They also cause Illinois' share of farm income under policy CB to be less than under policy NI.

Shifts in crop production brought about by the application of soil loss restrictions in large areas outside of Wisconsin's borders are the important reason for the increases in Wisconsin's share of farm income under policies NI and CB. The production of corn and soybeans, crops with a high return, increases in Wisconsin, while production of small grains, crops with a low return, decreases. Small grains are substituted for corn and soybeans on class 3E soils in eastern Kansas and Nebraska and central Missouri when soil loss restrictions are applied in those states. The relatively low soil losses in Wisconsin, primarily the result of less intense rainstorms, encouraged the shifts.

The results show the expected soil loss decrease in areas applying restrictions and increase in areas not applying restrictions. The decreases are greater than the increases, and thus total soil loss decreases as the area of application of the restriction increases. Soil loss in the corn belt was 17 million tons less under policy W, 33 million tons less under policy I, 348 million tons less under policy NI, and 385 million tons less under policy CB than under unrestricted conditions. Soil loss in Illinois increased by 10.5 million tons under policy NI.

SUBSTATE IMPACTS

Osteen and Seitz[6] showed that variation in impacts occur over soil types and regions of a state. In Illinois the burden of controlling erosion falls more heavily on the individuals on the more erodible land. When a 3-TAY restriction is applied in Illinois, income reductions on erosive soils averaged $1.60/ac more than other soils. The greatest reductions occurred in class 3E and 4E soils in southern Illinois. Income reductions were greater in southern Illinois because there are more soils with erosion problems in that part of the state.

Variations in farm income on different soil types reflect erosion control practices and changes in crop prices brought about by those practices. Erosion control costs under a 3-TAY restriction in Illinois resulted from 1.1 million acres of terracing and 6 million acres of contouring. Less than 1% of Illinois land was affected by rotation changes. Price impacts actually caused returns on Illinois land not requiring erosion controls to increase under policy I when corn and soybean prices increased slightly. Interestingly, the individuals on this less erosive land in Illinois are somewhat more adversely affected under policies NI and CB than are individuals on more erosive land. This is due to the shifts of corn out and soybeans into the state noted above.

LONG-TERM IMPACTS

Another important issue is the impact of erosion controls on long-term productivity and economic returns. The watershed level analysis in Seitz et al.[4] indicates that erosion controls would bring about substantial reductions in erosion over time but that the private economic incentives to individual farmers are not great enough to stimulate adoption.

They found that 57% of the land in the Big Blue Creek watershed, which was selected because it is somewhat representative of the corn belt and surrounding areas, may lose all of its topsoil in 100 years if current practices are continued. If soil loss was restricted to SCS t-values, only 9.3% of the acreage would lose all of its topsoil.

A comparison of returns per acre under current practices and under soil loss constraints at t-values indicates that the returns are higher under current practices for the first 25 to 30 years, after which they fall off drastically to levels much below the level under constraints. The application of soil loss constraints on a large scale would not generate the same results due to the interaction of price and production effects as indicated in the corn-belt analysis.

Long planning horizons and low interest rates are required for an individual, behaving in an economically rational manner, to apply erosion controls. The planning horizon must be at least 40 years before the deferred benefits of erosion control justify the initial cost and lost income in early years when the returns are not discounted. When the discount rate is 5%, the planning period must be 60 years long before an individual would choose to adopt conservation practices.

Since it is unlikely that many people will have such long planning horizons or low discount rates, it seems likely that few people will apply erosion controls without financial incentives or regulations.

The policy interest in nonpoint source control may be viewed as an implicit recognition that offsite damages from erosion may be considerably greater than onsite damages and that the land owner cannot be expected to account for those costs in this decision-making.[10] Increasing cost-sharing programs or stronger programs such as regulations or taxation will be necessary to bring about increased application of erosion control practices.

CONCLUSIONS

When considering the imposition of a 3-TAY soil loss limit, which is stricter than the SCS recommended limits on most Illinois land, the economic impacts would not be great. Some relocation of crop production would occur, considerably more conservation tillage would be used, some adjustments in crop prices would occur but, in aggregate, the impact on farm income and food costs would not be as severe as may have been expected. When the prospect of varying rates of soil loss control among states is considered, the impacts on state shares of farm income also appear small, although there would be shifts of crop production among states, and soil loss rates may be increased on some land if controls are not adopted throughout the production area. We do find that there will be greater impacts on the owner and operator on more erosive than less erosive land.

We find that there may well be substantial negative productivity consequences in the long run if current practices are continued. Given expected planning horizons and discount rates, farmers cannot be expected to adopt conservation practices without increased incentives from the public sector. If the potential for reduced water-based damages was added to this analysis, it appears that the increased concern for soil erosion control would be justified.

ACKNOWLEDGMENTS

Research leading to this paper was supported by USEPA Contract No. 68-01-3584 and Illinois Insititute for Environmental Quality Contract No. 20.079. We gratefully acknowledge others on the research team, especially C. R. Taylor, now Assistant Professor, Texas A and M University.

REFERENCES

1. Swanson, E. R., and C. R. Taylor. "Potential Impact of Increased Energy Costs on the Location of Crop Production in the Corn Belt," presented at the 31st Annual Meeting of the Soil Conservation Society of America, Minneapolis, Minnesota, August 1-3, 1976.
2. Taylor, C. R., and K. K. Frohberg. "The Welfare Effects of Erosion Controls, Banning Pesticides, and Limiting Fertilizer Application in the Corn Belt," *Am. J. Agric. Econ.* 59:25-35 (1977).
3. Duloy, J. H., and R. D. Norton. "Prices and Incomes in Linear Programming Models," *Am. J. Agric. Econ.* 57:591-600 (1975).
4. Seitz, W. D., D. M. Gardner, S. K. Gove, K. L. Guntermann, J. R. Karr, R. G. F. Spitze, E. R. Swanson, C. R. Taylor, D. L. Uchtmann and J. C. van Es. "Alternative Policies for Control of Nonpoint Sources of Water Pollution from Agriculture," USEPA Contract No. 68-01-3584, Institute for Environmental Studies, University of Illinois at Urbana-Champaign (1978).
5. Taylor, C. R., K. K. Frohberg and W. D. Seitz. "An Aggregate Economic Analysis of Potential Erosion and Plant Nutrient Controls in the Corn Belt," accepted for publication in a forthcoming volume of the *J. Soil Water Conserv.* (1978).
6. Osteen, C. and W. D. Seitz. "Regional Economic Impacts of Policies to Control Erosion and Sedimentation in Illinois and Other Corn-Belt States," accepted for publication in a forthcoming volume of the *Am. J. Agric. Econ.* (1978).
7. Moore, S. Extension Entomologist, Illinois Natural History Survey, Champaign, Illinois, personal correspondence (1977).
8. Lessitor, F. "No-Till Still Increasing," *No-Till Farmer* (March 1976), pp. 4-5.
9. Kasal, J. "Trade-Offs Between Farm Income and Selected Environmental Indicators: A Case Study of Soil Loss, Fertilizer, and Land Use

Constraints," Technical Bulletin No. 1550, ERS, USDA, Washington, DC (August 1976).
10. Guntermann, K. L., M. T. Lee and E. R. Swanson. "The Offsite Sediment Damage Function in Selected Illinois Watersheds," *J. Soil Water Conserv.* 30:219-224 (1975).

28

PROCEDURE FOR ECONOMIC EVALUATION OF BEST MANAGEMENT PRACTICES

T. H. Dempster, J. H. Stierna
  U.S. Department of Agriculture
  Soil Conservation Service
  Washington, D.C.

INTRODUCTION

Recent estimates of soil erosion on the nation's cropland indicate that soil losses occur at an average rate of 9 ton/ac/yr—nearly twice the rate considered acceptable by soil conservationists. M. Rupert Cutler, Assistant Secretary for Conservation, Research and Education, U.S. Department of Agriculture (USDA), stated in a news release earlier this year, "Soil losses could be reduced from the present average of 9 ton/ac/yr without adversely affecting our ability to produce more farm commodities."

Conservation treatment for erosion control should be site-specific, and a single practice rarely is sufficient. A combination of practices is generally necessary. The Soil Conservation Service (SCS) defines a resource management system (RMS) as that combination of conservation practices needed to protect the resource base, sustain an acceptable standard of living, and maintain the quality of the environment. A somewhat similar term, best management practices (BMPs), was coined by the Environmental Protection Agency to refer to those practices needed (singly or in combination) to reduce nonpoint source water pollution to meet water quality goals. This paper discusses the ongoing attempts of an SCS work group to assess the adequacy of SCS procedures currently used to evaluate the expected impacts of alternative RMSs and BMPs. This work group will identify areas of concern where new evaluation procedures should be developed or current ones improved.

The approach used for estimating these impacts is comparable to that used to evaluate water resource projects—the with-and-without approach. The impacts of each alternative RMS or set of BMPs are the differences between the measured variables with and without the practice or combination of practices installed. In this way the approach allows for a comparison of alternatives.

Evaluating BMPs is an integral part of resource program planning. These programs range from site-specific measures installed on the lands of individual cooperators or groups of cooperators to project-type programs such as PL 566 (small watershed program) and river basin studies. Evaluation activities need to be expanded because of new programs, such as the Rural Clean Water Program provided for in the 1977 amendments to the Clean Water Act (PL 92-500) and the national inventory and appraisal mandated by the Soil and Water Resources Conservation Act of 1977 (PL 95-192).

PHYSICAL IMPACTS

Soil depletion and its long-term effects on the land resource are a prime concern of soil and water conservationists. To examine fully the implications of soil depletion, it is necessary to recognize the potential long-term decline in soil productivity under different conditions and with different levels of specified production and capital inputs. Often, however, the long-range influence on crop yields is masked by increases in inputs such as fertilizer or improved technology. Another onsite impact is the change in farmability caused by gully erosin. One primary offsite impact occurs when sediment carried from eroding farmlands causes progressive siltation of streams, lakes and reservoirs. Stream bed siltation frequently results in increased flooding. The ability of an area to sustain fish and wildlife populations is impaired as the habitat is damaged by high sediment concentrations. All of these are detrimental to water quality.

To estimate how much sediment reaches a watercourse, both gross erosion (the amount of soil moved within a field) and the sediment delivery ratio (the percentage of the eroded soil that is actually carried to some downstream location) must be determined.

The accepted method for estimating gross erosion is the Universal Soil Loss Equation (USLE).[1] This equation predicts the amount of soil that is moved within the field by the erosive action of rainfall and surface runoff. Factors considered in the formula include rainfall and runoff erosivity index, soil erodibility, slope length, slope steepness, cover and management, and farming practices. The USLE predicts the long-term average gross soil loss from sheet and rill erosion but does not predict that from gully and stream bank erosion. The gross soil loss value must be reduced to account for the percentage of eroded soil deposited before the runoff water enters a stream system. At this point, offsite impacts begin.

Models have been developed to estimate the effects of BMPs on the economy and the stream environment. Results from these techniques must be interpreted with care, because the results are no better than the accuracy of the physical data inputs, for example, production input changes and crop yield responses. One evaluation tool used by SCS is the technical guide kept at each SCS field office. Sections of the guide covering soil and site information, resource use and management systems, and cost-return information can be used for an integrated evaluation of BMPs. The interrelationship of soil, treatment and use, yield responses, and monetary estimates of costs and returns can be incorporated from material in the guide. Where complete, these data provide much of the information needed to determine the expected impacts of resource deterioration and describe the long-range treatment needed for maintenance or improvement.

Except for isolated areas, there are very few statistically verified data for determining the impact of erosion on crop yields. This major deficiency limits the ability of analysts to evaluate the economic impact of installing conservation measures, including BMPs. Field observations have not been consistently made and recorded either over a large enough area or over enough time to be statistically reliable. Literature reviews of the available research have not been made to sort out findings that might be of practical use. To overcome this problem there is a need to establish a nationwide data collection and monitoring system. This will require direct and continuous participation of soil scientists, agronomists, biologists, engineers and other specialists. Crop response data should be assembled at the field level by soil type, water and capital input levels, and levels of management. These data should be collected in a systematic way so that the statistical significance of the different variables can be tested, thus making the data more useful.

BENEFICIAL IMPACTS OF BMPs

Monetary Effects

Monetary effects are evaluated within the framework of the with-and-without analysis. Onsite monetary effects are the differences in net returns to land, overhead, risk, and management to the farm operator for the alternative under consideration.

Offsite monetary evaluations are in early stages of development. They depend on the impacts identified by physical scientists and include many variables for which market prices are not available. These variables include such items as enrichment of water by nutrients, quantity of suspended solids, and chemical content. Although many of society's goals for water quality are noneconomic, the questions of who will benefit, who will pay, and how much they will pay for water quality improvement will continue to be important.

386 BMPs FOR AGRICULTURE AND SILVICULTURE

Many significant challenges are still ahead for those who evaluate offsite monetary impacts.

The SCS work group, which consists mainly of regional and field-level economists, is developing procedures for evaluating RMSs. These procedures will apply to a broad range of resource concerns, such as water quality, water conservation and production efficiency. Procedures will be kept flexible enough for easy adaptation by field personnel, but consistent with established economic principles.

The work group's first activity was to prepare an inventory of current SCS procedures. The inventory includes information on identifying onsite and offsite impacts, basic data needed and their availability, status of evaluation methods, and display of impacts in the format of the "Principles and Standards for Planning Water and Related Land Resources."[2] An additional activity is the preparation of flowcharts describing recommended techniques for evaluation purposes (see Figure 1).

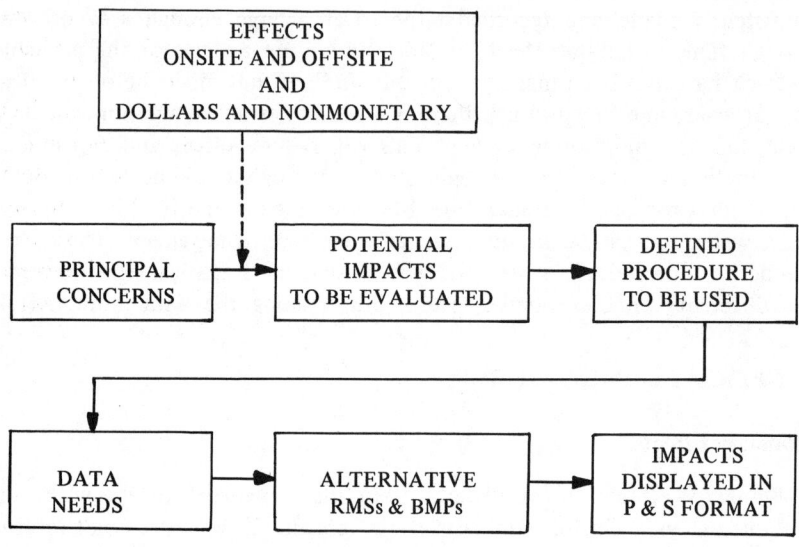

Figure 1. Example flowchart.

The procedure will consider (1) changes in net returns due to more efficient crop production or higher yields, (2) least costly alternative systems, and (3) changes in environmental variables. In addition, a rating system for estimating the qualitative variables will be studied.

Several other technical tools are available to SCS personnel for evaluation of BMPs. The SCS Crop Budget System, linear programming, and the Value of Agricultural Production Program are three examples.

The SCS Crop Budget System[3] has been designed to prepare and display alternative crop budgets. Adapted by SCS from a system developed by the Economics, Statistics and Cooperatives Service (ESCS) of USDA, this system is designed to reduce the need for continually updating economic data and to incorporate the costs and returns of conservation measures. Although not an optimizing system, it can combine various resource management systems with enterprise operations to produce a set of alternative crop budgets. Capabilities of the system include:

1. regular updating of crop budgets from the ESCS system;
2. readily modifying budgets to meet local soil and climatic conditions;
3. modification of budgets for individual farms;
4. production of data applicable to areawide analyses, such as in Resource Conservation and Development (RC&D), river basin, or watershed studies;
5. consistency in concepts and methods from area to area and state to state; and
6. flexibility to permit results to range from summary level to extensive detail, depending on the user's preference.

One advantage of the Crop Budget System is that it can readily incorporate changes in farm operations associated with a specified BMP. The budget then becomes a documented display of each alternative.

This tool can assist field level evaluation of BMPs by storing many existing cost-return data within computer facilities. These data can be readily retrieved and modified to meet local conditions. The data include average yields, current product prices, quantities and prices of operating inputs, machinery requirements, related engineering and economic coefficients and selected data on conservation practices. Individual users must determine to what extent these data apply to their problems and what changes may be needed in the data. The Crop Budget System can serve as one medium for storage, retrieval and updating of needed data.

Linear programming has been used to determine the effects of land treatment on soil losses and net income. Information required for this methodology includes land use, cropping patterns, crop yields, production costs, and erosion data by soils. From a specified number of alternative activities, this optimizing technique can select the set that produces the greatest economic return or the greatest erosion reduction depending on which variable the user chooses to optimize.

The Value of Agricultural Production Program (VAGPR) is an SCS computer program that replaces manual computations for calculating future without-treatment returns by various crop rotations. It compares these returns with alternate conditions of land treatment.

Environmental Effects

The effects of alternative systems on environmental as well as economic concerns need to be evaluated. Environmental effects may be expressed in either quantitative physical and ecological terms or appropriate qualitative terms. The evaluation may include effects on (1) areas of natural beauty; (2) water, land and air quality; (3) biological resources and selected ecosystems; and (4) irreversible or irretrievable commitments of resources.

Index systems may help quantify value judgments concerning existing and probable future environmental conditions. Their advantage is the ease of displaying effects and conveying significance. Their principal weakness is the risk of oversimplification and the substitution of professional for societal judgments.

Current ecological conditions can be described and compared with conditions that would exist without a system or practice. Similar descriptions can be prepared by time frames of those conditions to be expected with alternative BMPs installed. The with-and-without approach thus applies to environmental as well as monetary considerations.

ADVERSE IMPACTS OF BMP

Typically, the major adverse impact of BMPs is the cost of materials, labor and equipment needed to install and maintain the BMP or RMS. Flat-rate cost schedules are available at most SCS state and field offices and can be used for evaluating alternative systems (see Table I). These schedules are not site-specific since they are averages. The land user can obtain the most accurate site-specific cost information when he solicits bids from contractors and vendors.

Flat-rate installation costs are based on quoted price lists, actual costs experienced by soil conservation district cooperators, and research data published by state universities and federal agencies. They include taxes, insurance, safety equipment and other items ordinarily included as overhead. Rates generally do not include land charges.

Several factors must be considered. They include the length of time the practice will be effective, considering reasonable operation and maintenance (O&M) costs. Annual O&M costs are generally estimated as a percentage of the installation costs.

Costs and expected returns must be compared for an equivalent time period. Generally, annual equivalent costs are computed by amortizing

Table I. Example of flat-rate schedule for conservation practice cost.

Conservation Practice	Indicator Unit	Flat-Rate Installation Cost ($)	Life Span (Years)	Annual O&M Cost (%)	Total Annual Cost ($)
Dam, Diversion	Job. est.	1,000	25	3	110
Dam, Multiple-Purpose	Job. est.	1,000	25	3	110
Debris Basin	Job. est.	1,000	25	5	130
Deferred Grazing	Job est.	100	Annual		100
Dike					
Class I	1,000 yd^3	1,000	50	1	75
Class II	1,000 yd^3	750	25	1	70
Class III	1,000 yd^3	500	10	1	75
Diversion	1,000 ft	400	10	5	75

installation costs over the expected economic life of the practice at the interest rate established for evaluation purposes. The land owner or operator may wish to amortize the investment in the shortest practicable time consistent with the benefits.

Work is now being completed by SCS to incorporate practice cost data into the Crop Budget System. The installation costs are amortized over a specified life span at the desired interest rate; O&M costs are incorporated as a separate item.

Income foregone because of applied measures should be recognized. It may either be included as part of the installation costs of the BMP or be incorporated into the basic with-and-without evaluation. An example is a grassed waterway. The land in the waterway is either taken out of production or changed from cropland to a use with lower returns such as pasture.

Currently, the methodology often used for estimating adverse environmental impacts is keyed to a variety of rating systems. Several methods have been used, and others are being studied in more detail.

DEVELOPMENT AND SELECTION OF BMP

The BMP selected for installation should ordinarily be no more costly than alternative means of accomplishing the same objective. The effects of the alternatives must be nearly identical for a cost comparison to be valid.

From an economic viewpoint, the optimum level of treatment occurs at the point where the return added by the last increment of input is equal to the cost of adding that increment. However, it is essential to identify noneconomic impacts for the decision-makers. Until a uniform and workable rating system is developed and widely accepted, evaluating the total impact of conservation activities will continue to be difficult. In 1974 USDA developed procedures to help implement the Water Resources Council's Principles and Standards (P&S). They include a system of accounts to display beneficial and adverse effects, both monetary and nonmonetary.

Similarly, the beneficial and adverse effects of BMPs can be displayed for economic development and environmental quality accounts:

	Economic Development		Environmental Quality	
	Beneficial Effects	Adverse Effects	Beneficial Effects	Adverse Effects
BMP	XX	XX	XX	XX

Additional displays can be made for each alternative system.

Although the evaluation of BMPs may not lend itself to a complete adoption of the P&S format, displaying beneficial and adverse impacts to a system of accounts can be a valuable asset to the evaluation and selection of alternative BMPs.

SUMMARY AND CONCLUSIONS

BMPs may have a variety of beneficial and adverse effects on economic development and environmental quality. Environmental effects are generally characterized by their nonmarket nature. They are important criteria for judging proposed practices or systems. Many environmental impacts cannot be readily labeled beneficial or adverse, since judgment varies among individuals. However, the effect itself should be identified, measured and displayed for purposes of decision-making.

The lack of physical data on yield response, changes in water quality, and the like is a major problem in evaluation. In identifying and quantifying effects, an interdisciplinary approach is needed.

If sufficient data on crop responses and impacts were available, the Crop Budget System would be a versatile tool in estimating changes in net returns to land owners or operators who install BMPs. Economic models can be used to evaluate both onsite and offsite impacts.

Offsite impacts are of major importance in BMP evaluations, but they are difficult to identify and relate to nonpoint problem sources and alternative treatment. Diligent efforts are needed to develop and improve techniques for quantifying and/or qualifying these impacts.

Techniques for evaluating BMPs must not be limited to economic justification. The evaluation of BMPs must include both the monetary and nonmonetary impacts.

REFERENCES

1. Wischmeier, W. H., and D. D. Smith. "Predicting Rainfall-Erosion Losses from Cropland East of the Rocky Mountains," *USDA Agric. Handbook* 282 (1965).
2. U.S. Water Resources Council. "Principles and Standards for Planning Water and Related Land Resources," (1973).
3. Soil Conservation Service, USDA. "Crop Budget System Users Guide," (1977).

29

AN ECONOMIC ANALYSIS OF EROSION CONTROL OPTIONS IN TEXAS

D. R. Reneau, C. R. Taylor
> Department of Agricultural Economics
> Texas A&M University
> College Station, Texas

INTRODUCTION

The Federal Water Pollution Control Act Amendments, PL 92-500, established a national goal of eliminating the discharge of pollutants into the nation's waterways by 1985.[1] As a step toward this goal, Section 208 of that law requires states to develop plans which: (1) identify nonpoint sources of pollution; (2) recommend "Best Management Practices" to control the identified pollutant; and (3) set forth procedures and regulatory controls to reduce the pollution to the extent feasible.

The Texas Soil and Water Conservation Board was given the overall responsibility to coordinate the state planning process and develop the standards and methodology necessary to meet the federal requirements for agricultural and silvicultural pollutants. In order to be able to assess the situation in those areas of Texas which are primarily agricultural, the Board contracted with the Texas Water Resources Institute at Texas A&M University to do a study of the economics underlying agricultural nonpoint source pollution in various Texas watersheds. The resulting work was designed to: (1) estimate the present offsite damages resulting from sediment leaving agricultural land and how these damages might change with changes in the level of sediment loss; (2) estimate the expected soil loss from agricultural land under specified conservation practices; and (3) estimate the economic consequences of various possible options for controlling agricultural soil loss.

Six watersheds from different regions of Texas were chosen as test areas on which the model to be developed could be tried. This paper will focus on one, and that only so far as is necessary to elucidate the mechanics of the model developed. Nonetheless, it is typical of the size and type of watershed that the model is designed to simulate and, hence, is a good example of where the model might be used.

DESCRIPTION OF THE WATERSHED

The Lavon reservoir watershed (Figure 1) covers an area of 477,613 acres, which is primarily in Collin county, but also includes part of Grayson, Fannin and Hunt counties, Texas.[2,3,4] The watershed lies entirely in the Blackland Prairie Land Resource Area. Soils in this nearly level to rolling prairie can be divided into three principal soil groups: (1) bottomlands of alluvial soils that are highly productive; (2) black, waxy upland soils that are used primarily for the production of small grains and pasture, and (3) light-colored, deep and shallow upland soils over limestone and marble. The model uses individual soil mapping units as its basic land division. This particular watershed is composed of 21 mapping units.

Over the period 1972-1975, approximately 9% of the land in the watershed has been planted to cotton, 12% to small grains, 14% to feed grains, and the remainder to hay, pasture and minor crops.[5] An extremely small amount of the cropland is irrigated.

Lavon dam, which is located about 25 m northwest of Dallas, was constructed for water supply, flood control and recreational purposes in 1953. The dam was modified in 1974 to increase its capacity. Since the reservoir was designed with a large sediment pool, silt has not diminished the water supply and flood control capacity of the reservoir, although it could in the future.

Lavon watershed is comprised of three PL 566 watershed protection project areas.[6] Construction of 191 flood control structures has been approved for these watersheds with 147 of the structures in place as of October 1976. These flood control structures along with land treatment have reduced the siltation of Lavon reservoir.

AN OVERVIEW OF THE MODEL

The general model (Figure 2) incorporates data on each soil series and its possible crop rotations and conservation practices* to estimate: (1) the

---

*Reduced tillage methods are not feasible in the area because of the high clay content of the soils.

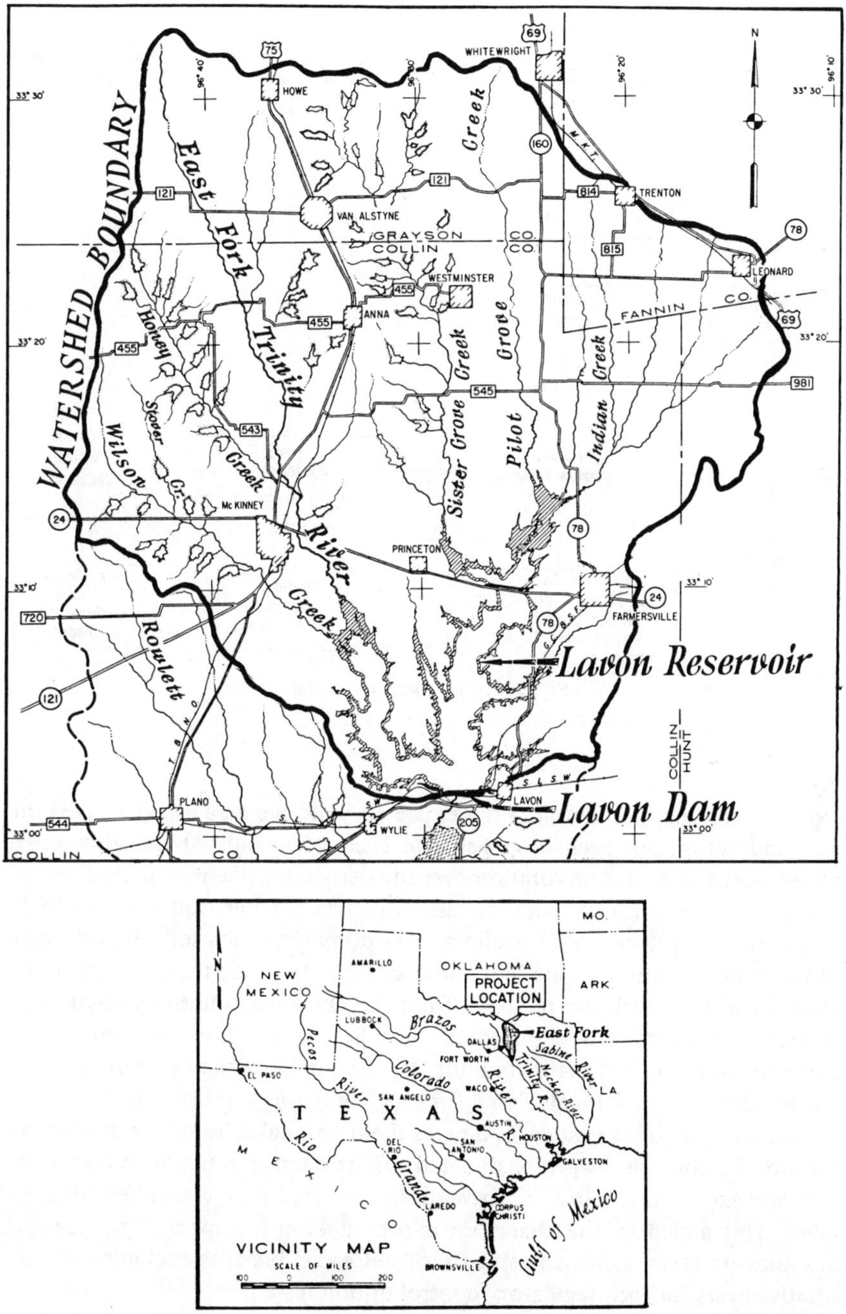

ECONOMIC, POLICY AND INSTITUTIONAL ASPECTS 395

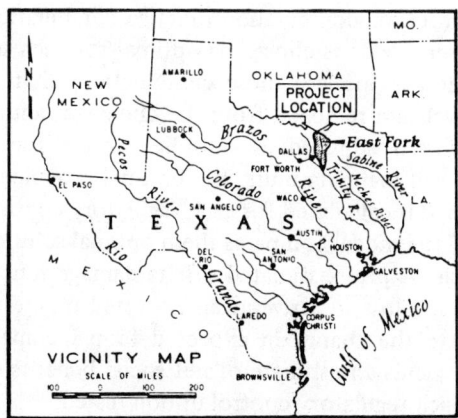

Figure 1. Lavon watershed.

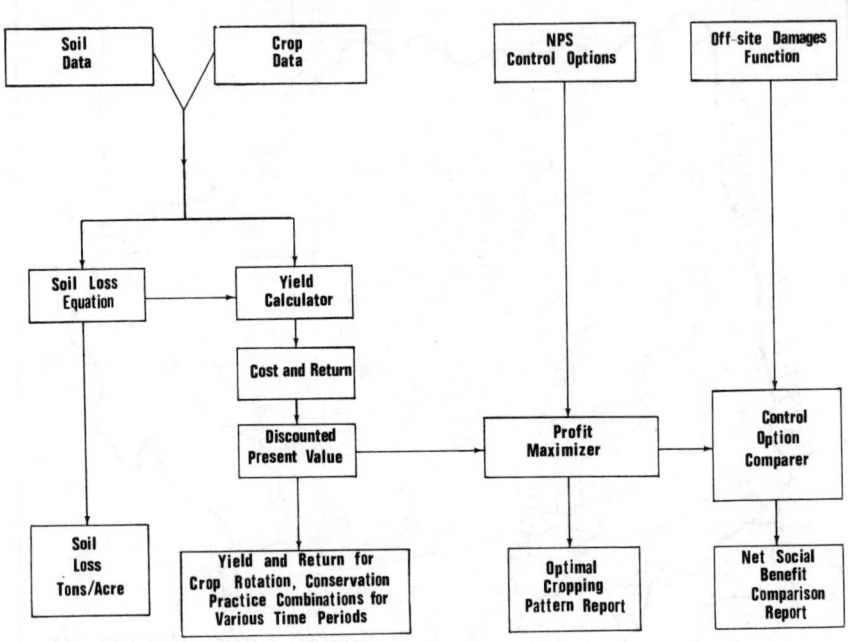

Figure 2. Flowchart of model.

expected soil loss per acre; (2) the expected yield per acre each year; (3) the cost and return for each crop per acre each year; and (4) the discounted present value of that crop rotation over the designated planning period.

Next, the crop rotation-conservation practice combination with the highest discounted present value is chosen as optimal for each soil. As each regulatory control option is applied, those combinations that cannot meet the control soil loss levels are removed from the choice set and the present value of each combination is adjusted as subsidies or taxes are introduced. By summing over the soil mapping units, the optimal cropping pattern and soil loss for the watershed is calculated for each regulatory control tested.

The last part of the model compares these optimal solutions with a benchmark of no controls. A prior-estimated offsite damage function is combined with the expected soil loss to arrive at an expected offsite sediment damage value. This added to the change in expected farm income and government subsidies or taxes yields an estimate of net social benefits excluding administrative costs for each regulatory control option tested.

SOIL AND CROP DATA

A considerable amount of technical data, both physical and economic, is required for each watershed. The Universal Soil Loss Equation (USLE) developed by Wischmeier and Smith was used to estimate the expected soil loss.[7] This requires information for each factor in the equation for each soil in the watershed. Also required are: (1) the acreage in each soil mapping unit in the watershed; (2) the cost of terracing the different soils; (3) expected crop yields for each soil; (4) the depth of the topsoil and which of three estimated productivity loss functions is applicable; and (5) what crop combinations or rotations could be found on a particular soil.

Besides the soil data, information was developed on the expected production costs and returns for each of the major crops in the watershed. Preharvest variable cost was subdivided into fertilizer costs, insecticide and herbicide expenses and machinery and labor costs. This division allowed the flexibility of restricting fertilizer use if a control on nitrates or phosphates was contemplated or varying the insecticide and herbicide use if certain chemicals were banned. It also allowed the machinery and labor cost to be increased proportionally to reflect the added field time required with contouring and terracing conservation practices. Harvesting costs were calculated on a per yield basis to allow yield changes to effect production costs. A fixed cost for machinery depreciation was added but no land or management charges.

Expected prices were defined as the average price received by Texas farmers for the specified crop between 1958 and 1975 adjusted to 1977 dollars by the index of prices paid for production items.[7]

Crop rotations rather than individual crops were considered the basic unit of production in the model. This was done for two reasons. First, the previous crop influences the amount of erosion from the current crop. The average erosion for a rotation is not a simple average of the erosion for each crop grown continually. Second, the expected yield of some crops will be proportionally higher (or lower) when grown in rotation due to fertility or moisture carryover, weed control or other considerations.

ON-FARM ECONOMICS OF SOIL LOSS

Soil Loss Calculation

The average expected soil loss for each crop rotation-conservation practice combination is calculated for each soil in the watershed using the USLE. A sample listing of the resulting values is shown in Table I for three of the soils in Lavon reservoir watershed.

Table I. Expected soil loss (ton/ac/yr) for each crop rotation, soil type and conservation practice.

Crop Rotation	Soil Mapping Units					
	Austin Silty Clay 3-5% Slopes, Eroded			Houston Black Clay 2-4% Slopes, Eroded		
	Straight-Row	Contour	Terrace	Straight-Row	Contour	Terrace
Cotton	44.51	22.25	10.83	20.50	10.25	8.49
Grain Sorghum	29.67	14.84	7.22	13.66	6.83	5.66
Wheat	14.84	7.42	3.61	6.83	3.42	2.83
Hay	0.74	0.37	0.18	0.34	0.17	0.14
Common Bermuda Pasture	1.48	0.74	0.36	0.68	0.34	0.28
Improved Bermuda	0.74	0.37	0.18	0.34	0.17	0.14
Native Pasture	2.97	1.48	0.72	1.37	0.68	0.57
Cotton/Cotton/Sorghum	39.31	19.66	9.57	18.10	9.05	7.50
Cotton/Sorghum/Sorghum	33.38	16.69	8.13	15.37	7.69	6.37
Cotton/Wheat/Wheat	22.25	11.13	5.42	10.25	5.12	4.25
Sorghum/Wheat/Wheat	17.06	8.53	4.15	7.86	3.93	3.25
Cotton/Sorghum	36.35	18.17	8.85	16.74	8.37	6.93
Cotton/Sorghum/Wheat	25.96	12.98	6.32	11.96	5.98	4.95

Yield Loss Calculation

In a long-run analysis of soil conservation the relationship between erosion and future crop yield is critical. This is because the on-farm benefits from conservation practices arise mainly from the relatively higher future crop yield resulting from that conservation practice. Unfortunately, very little experimental or field data on this important relationship are available. Consequently, for purposes of this study it was necessary to develop estimates of this relationship for each soil mapping unit.

Yield loss attributal to topsoil loss depends to a certain extent on the suitability of the subsoil for crop production. Soils in each watershed were classified into one of three groups. Group A consists of soils that have subsoil that is unsuitable for field crop production. For this group, crop yield was assumed to be zero after all topsoil was eroded. Group B consists of soils with subsoils that are slightly suitable for field crop production. It was assumed that crop yield on Group B soils would be 25% of the currently attainable yield after all the topsoil was eroded away. Group C consists of those soils with subsoils that are somewhat more suitable for crop production. After the loss of all topsoil, yield in this group was assumed to be 50% of present yield. The group to which each soil belongs and the initial average topsoil depth is part of the soil data required for each watershed.

Due to paucity of experimental or field data on the relationship between topsoil thickness and yield, it was necessary to subjectively specify this relationship for each soil group. After considerable discussion with Soil Conservation Service (SCS) and Texas A&M University scientists, the three relationships shown in Figure 3 were specified. The functions in Figure 3 have two important characteristics. One is that each function is expressed in terms of *percent* of topsoil lost and *percent* of initial yield attainable after erosion. This reflects the fact that the loss of 1 in. on an initially shallow soil will decrease yield more than the loss of 1 in. of an initially deep soil. For example, the loss of 1 in. of a soil in Group A with an initial depth of 20 in. will reduce yield by about 2%, while the loss of 1 in. on a soil with an initial depth of 5 in. will decrease yield by about 8%.

The second important characteristic of the Figure 3 function is that the loss of the last remaining topsoil will reduce yield by more than the loss of the upper portions. For instance, the loss of the first 20% of topsoil in Group A reduces yield by about 8%, while the loss of the last 20% of topsoil reduces yield nearly 46%. Because of the critical nature of the relationships shown in Figure 3, additional experimental and field research appears warranted.

Cost and Return to Crop Production

The production cost information was developed for each watershed from a set of 1977 crop budgets prepared by the Texas Agricultural Extension

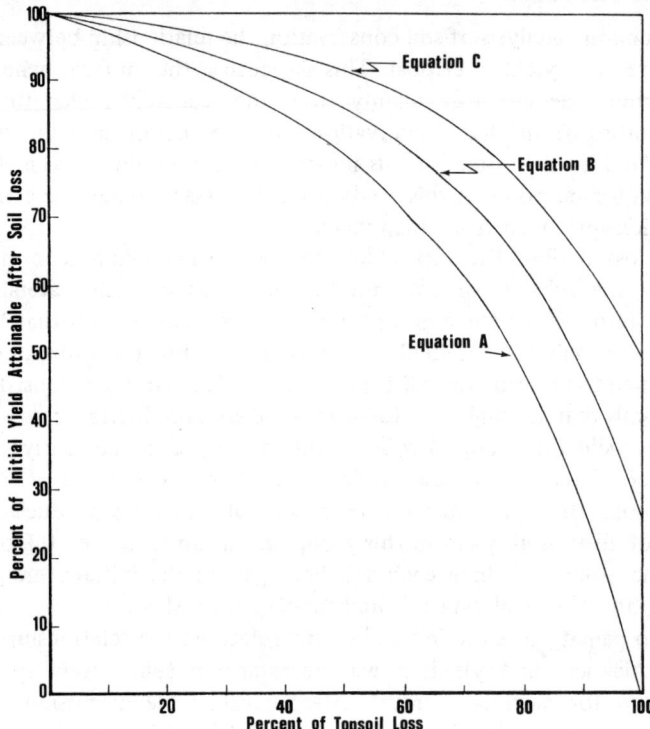

Figure 3. Relationships between yield and topsoil eroded.

Service for the appropriate Land Resource Area.[9] The program modifies the basis cost data to fit each soil by changing the harvest costs proportional to the yield for that crop in each rotation and by adding the additional costs incurred with the specified conservation practice. For the conservation practice of contouring, preharvest machinery and labor costs were increased a certain percentage. In the case of Lavon watershed this was 5%. Terracing was assessed the added cost of building and maintaining terraces besides the increase in machinery and labor costs. This cost was the discounted sum of: (1) the initial construction cost, (2) an annual maintenance charge equal to 5% of the construction cost, and (3) the cost of rebuilding the terraces every 10 years, assumed to be equal to one-third of the construction cost.

Discounting of Future Benefits and Costs

As a point of reference from which to calculate the present value of future benefits and costs, 1977 was designated the base year.

All future benefits and costs were discounted to 1977 dollars using standard discounting techniques and a real interest rate of 1.5%. The 1.5% rate was arrived at by subtracting the average inflation rate of the last 10 years which is 5.8% from the 7.3% average private interest rate charged by banks over the same 10 year period.

The present values of net returns associated with particular crop production activities are calculated at this point. The present value of net returns is computed as:

$$PV = \sum_{t=1}^{T} [B_t (\frac{1}{1+i})^t - C_t (\frac{1}{1+i})^t]$$

where Σ = summation of discounted benefits and costs over time
t = time, in years
B_t = gross benefits in year t
C_t = gross costs in year t
i = interest rate minus inflation rate
$(\frac{1}{1+i})$ = discount rate
T = length of planning horizon

Yield and Return Over Various Planning Periods

The effect of soil conservation and erosion control on the agricultural economy is felt over a period of years as the mix of inputs changes for a given output. Erosion carries away the topsoil reducing soil fertility, thus reducing crop yields. If erosion is slowed, future crop yields will be higher than they would otherwise have been, given the same level of management.

Farmers make many short-run decisions because they are concerned with next year's income. On the surface this suggests that farmers would use a short time horizon for planning conservation practices. However, most farmers are concerned about the future value of their land in addition to income flow. Inasmuch as the agricultural component of land values is the capitalized value (present value) of the highest and best use profit stream into perpetuity, and given the limited alternate uses for agricultural land in many parts of Texas, the value of the land is tied closely to its future agricultural productivity. Thus, it was important that this study consider not only present productivity but also the effect on future productivity and, hence, land values of cropping and conservation practices. Therefore, a long planning horizon was the only appropriate time period for determining what is the appropriate combination of crop rotation-conservation practices a land owner should employ. In order to emphasize this point and to demonstrate the importance of the length of the planning period, a series of tables (one for each soil in the watershed) was created showing the yield and expected return for crop rotation—conservation practice combinations for time horizons of 10, 100 and 200 years. For example, Table II shows the report generated for Houstin, black clay, 2-3% slopes.

PUBLIC POLICY OPTIONS FOR NPS CONTROL

In designing a nonpoint source (NPS) control plan, it is necessary to define the feasible control methods from a technical perspective. For control of sheet and rill erosion and the sediment resulting therefrom, the control methods considered are the consevation practices of contouring and terracing, and changes in land use such as shifting to a crop which causes less erosion.

Once these technical alternatives are specified, it is necessary to determine a way of implementing a pollution control method. The standard policy options for implementing a pollution control method. The standard policy options for implementing a control include regulation, provision of economic incentives, education and public investment. For point sources of pollutants, regulations are typically directed toward the pollutant at the point of emission into waterways. However, this is not possible with NPS pollutants. Hence, regulations must be directed toward the agricultural practices that cause or influence the NPS pollutants.

The economic incentive option includes alternatives such as federal or state cost-sharing arrangements for conservation practices, and excise taxes on inputs such as fertilizers and pesticides or even on soil loss. Education is a viable policy option in situations where producers or others are misusing inputs that cause pollution, or are not adopting conservation practices that would be profitable. In these situations a successful education program would increase producer's income as well as reduce the environmental damages caused by misuse of agricultural chemicals and production practices. Public investment is appropriate for controls that are not appropriate for individuals, but that can be justified by governmental units. An example would be the construction of municipal wastewater treatment plants. In any particular NPS situation, a combination of the above policy options may provide the best solution to the problem.

The specific erosion-sedimentation control options considered for Lavon watershed were:

1. restricting soil loss to be no greater than the SCS tolerance or "T" limits;
2. restricting soil loss to be no greater than 2, 5 or 10 ton/ac;
3. terracing subsidies or cost-sharing arrangements for 50 and 100% of the annual costs;

Table II. Yield loss and per acre return to land and management for Houston black clay 2-3% slopes.

ROT	CP	%TOPSOIL LOST/YR	REMAINING YR 1	YIELD YR 10	YIELD (AS A % OF YEAR 1) AND PROFITS AT					P.V. OF PRT STREAM TO		
					YEAR 10	YEAR 100		YEAR 200		10	100	200
C	SR	1.208	62.67	60.07	98.4	50.0	−18.65	50.0	−18.65	568.	2126.	1909.
	CU	0.604	57.09	56.14	99.4	87.8	37.26	50.0	−24.23	524.	2662.	2769.
	CT	0.384	54.56	50.10	99.8	94.6	45.60	77.5	17.95	464.	2649.	3101.
S	SR	0.805	66.00	64.62	99.1	74.4	28.72	50.0	−16.67	504.	2987.	3002.
	CU	0.403	63.15	62.86	99.8	94.2	54.76	74.4	25.87	582.	3109.	3655.
	CT	0.256	58.71	55.51	100.0	96.6	54.57	91.4	47.17	512.	2984.	3603.
W	SR	0.403	47.11	46.89	99.8	94.2	40.89	74.4	19.49	434.	2320.	2728.
	CU	0.201	44.76	44.76	100.0	97.3	41.85	94.2	38.54	413.	2261.	2735.
	CT	0.128	41.88	38.38	100.0	98.3	39.91	96.6	38.20	354.	2129.	2594.
H	SR	0.020	54.58	54.58	100.0	100.0	54.58	99.8	54.35	503.	2818.	3453.
D	SR	0.040	37.31	37.31	100.0	99.8	37.16	99.1	36.61	344.	1926.	2356.
CP	SR	0.020	53.26	53.26	100.0	100.0	53.26	99.8	53.05	491.	2750.	3370.
ND	SR	0.081	33.26	33.26	100.0	99.1	32.92	97.8	32.47	307.	1714.	2095.
C/C/S	SR	1.067	74.80	72.49	98.6	50.7	−19.18	50.0	−13.86	681.	2954.	2847.
	CU	0.533	70.12	69.36	99.5	90.7	54.44	50.0	−13.86	645.	3373.	3749.
	CT	0.339	67.29	62.97	99.9	95.4	59.36	83.8	40.10	582.	3322.	3960.
C/S/S	SR	0.906	76.10	74.29	98.7	63.8	17.43	50.0	−15.03	695.	3327.	3294.
	CU	0.453	72.35	71.86	99.7	93.1	61.12	63.8	13.68	666.	3542.	4104.
	CT	0.288	69.21	65.01	100.0	95.1	62.79	89.0	51.38	600.	3450.	4151.
C/W/W	SR	0.604	50.44	49.71	99.4	87.8	35.32	50.0	−11.57	463.	2387.	2549.
	CU	0.302	47.07	47.02	100.0	96.0	42.00	87.8	31.90	434.	2335.	2790.
	CT	0.192	43.77	39.57	100.0	97.4	40.40	94.6	37.06	365.	2198.	2664.
S/W/W	SR	0.463	51.54	51.16	99.7	92.8	43.05	61.4	5.85	475.	2516.	2900.
	CU	0.231	49.02	49.02	100.0	96.9	45.36	92.8	40.53	452.	2466.	2976.
	CT	0.147	45.47	41.27	100.0	98.0	42.96	96.1	40.80	381.	2304.	2804.
C/S	SR	0.986	81.05	78.92	98.7	53.2	1.15	50.0	−14.06	740.	3414.	3365.
	CU	0.493	76.83	76.19	99.6	92.8	63.13	53.2	−13.00	707.	3740.	4269.
	CT	0.314	73.85	69.62	100.0	95.8	66.48	86.7	51.07	642.	3671.	4403.
C/S/W	SR	0.705	67.26	66.13	99.2	82.2	41.18	50.0	−5.95	617.	3143.	3278.
	CU	0.352	63.70	63.56	99.8	95.2	56.63	82.2	37.63	587.	3159.	3756.
	CT	0.224	60.50	56.30	100.0	97.0	55.94	93.2	50.53	519.	3038.	3678.

4. contouring subsidies or cost-sharing arrangements for 50 and 100% of the additional cost for contouring;
5. subsidies of 50 and 100% on the initial cost of constructing terraces;
6. restricting soil loss to be no greater than the SCS limit or a specific limit of 5 ton/ac combined with contouring or terracing subsidies;
7. restricting soil loss to less than the SCS limit or a specific limit of 5 ton/ac combined with a 50% subsidy toward initial construction costs of terraces;
8. taxes on soil loss of 8, 10, 12, 14, 16, 18 and 20 cents per ton;
9. taxes on soil loss of 8, 10, 12, 14, 16, 18 and 20 cents per ton combined with a subsidy of 50% of the cost of terracing or contouring.

Table III lists the options considered in Lavon watershed individually with the abbreviation used for each.

These policy options were chosen to cover a wide range of available alternatives. Section 208 of the PL 92-500 does not specify the type of regulation or incentive that must be used, so decision-makers may choose from the above set of options or use the model to test others which experience or experiment may suggest.

Profit Maximizer

The crop rotation-conservation practice combination with the highest discounted present value that does not violate the control being tested is chosen for each soil. This optimal combination with its associated present value, soil loss, and terraced or contoured acreage is reported along with the totals for the watershed. Table IV is a sample of this output for three of the soils in the Lavon watershed. The complete cropping pattern for the watershed is also reported on a yearly basis so that the expected acreage and yield of each major crop and the associated fertilizer and herbicide-insecticide usage can be noted. Table V demonstrates the crop report returned using the 200-year time horizon benchmark run in the Lavon watershed.

Comparison of the Control Options Considered

One of the objectives when designing the model was to be able to compare the results of the suggested soil loss control options. It was desired to be able to compare the various options as to how much soil loss was reduced in the watershed and most importantly how economically efficient the reduction was.

To decide whether erosion-sedimentation control is justified on economic grounds and to identify the economically most efficient policy option, the following types of information are needed:

Table III. Alternate control options modeled.

Control	Table Abbreviation
Annual soil loss less than SCS tolerance limit (T)	SL <T
Annual soil loss less than 2 tons per acre	SL <2
Annual soil loss less than 5 tons per acre	SL <5
Annual soil loss less than 10 tons per acre	SL <10
Subsidy equal to 50% of annual terracing costs	TR 50
Subsidy equal to 100% of annual terracing costs	TR 100
Subsidy equal to 50% of annual contouring costs	C 50
Subsidy equal to 100% of annual contouring costs	C 100
Subsidy equal to 50% of the initial cost of constructing terraces	ITR 50
Subsidy equal to 100% of the initial cost of constructing terraces	ITR 100
Soil loss <T, 50% terracing costs subsidy	SL <T, TR 50
Soil loss <T, 50% contouring costs subsidy	SL <T, C 50
Soil loss <T, 50% initial terrace construction costs subsidy	SL <T, TI 50
Soil loss <5, 50% terracing costs subsidy	SL <5, TR 50
Soil loss <5, 50% contouring costs subsidy	SL <5, C 50
Soil loss <5, 50% initial terrace construction costs subsidy	SL <5, TI 50
A tax on annual soil loss of 8 cents per ton	TX 8
A tax on annual soil loss of 10 cents per ton	TX 10
A tax on annual soil loss of 12 cents per ton	TX 12
A tax on annual soil loss of 14 cents per ton	TX 14
A tax on annual soil loss of 16 cents per ton	TX 16
A tax on annual soil loss of 18 cents per ton	TX 18
A tax on annual soil loss of 20 cents per ton	TX 20
A 8-cent tax on soil loss with a 50% subsidy on terracing or contouring costs	TX 8, 50 TSC
A 10-cent tax on soil loss with a 50% subsidy on terracing or contouring costs	TX 10, 50 TSC
A 12-cent tax on soil loss with a 50% subsidy on terracing or contouring costs	TX 12, 50 TSC
A 14-cent tax on soil loss with a 50% subsidy on terracing or contouring costs	TX 14, 50 TSC
A 16-cent tax on soil loss with a 50% subsidy on terracing or contouring costs	TX 16, 50 TSC
A 18-cent tax on soil loss with a 50% subsidy on terracing or contouring costs	TX 18, 50 TSC
A 20-cent tax on soil loss with a 50% subsidy on terracing or contouring costs	TX 20, 50 TSC

Table IV. Optimal soil use for Lavon watershed soils with 200-year planning horizon.

	Soil Mapping Units			Totals for Lavon Watershed
	Austin Silty Clay 3-5% Slopes, Eroded	Houston Black Clay 2-4% Slopes, Eroded	Houston Black Clay 2-3% Slopes	
Crop Rotation	Wheat	Improved Bermuda Pasture	Cotton/Sorghum	
Conservation Practice	Terrace	Straight-Row	Contour	
Present Value				
Annualized ($1,000)	1,819	451	9,243	23,942
Soil Loss (1,000 Tons)	31	7	1,081	1,811
Terracing:				
Acres (1,000)	43	0	0	43
Cost ($1,000)	303	0	0	303
Contouring:				
Acres (1,000)	0	0	137	168
Cost ($1,000)	0	0	578	650

Table V. Optimal crop production in Lavon watershed with 200-year planning horizon.

Crops	Acres (1,000)	Quantity (1,000)	Yield/Acre	% of Total Acres	Fertilizer (1,000 lb.)		Chemicals ($1,000)	
					Nitrogen	Phosphorus	Insecticide	Herbicide
Cotton	90	35,880 lb	399 lb	20	5,397	5,397	1,198	641
Grain Sorghum	90	8,399 bu	93 bu	20	8,995	5,397	230	345
Wheat	80	2,456 bu	31 bu	18	6,368	3,184	193	0
Hay	74	342 ton	5 ton	17	13,284	2,952	179	163
Common Bermuda Pasture	54	259 AUM	5 AUM	12	5,400	2,160	131	119
Improved Bermuda Pasture	19	95 AUM	5 AUM	4	1,900	760	46	42
Native Pasture	42	51 AUM	1 AUM	9	0	0	0	0
Totals	449			100	41,344	19,850	1,977	1,310

1. the off-site environmental damages that would be abated by the policy;
2. the private and social costs incurred by farmers and society when alternative policy options are implemented at various levels of control; and
3. the implementation, administrative and enforcement costs associated with each policy.

These benefit and cost components, once combined, indicate whether a particular policy at a specific level of control is justified on economic efficiency grounds. Of course, in deciding between policies, the distributional or equity aspects and political acceptability must also be considered.

OFFSITE SEDIMENT DAMAGES

A procedure for estimating offsite damages resulting from sediment in a watershed was developed by Lee and Guntermann.[10] This procedure attributes damages to the following factors: (1) an increase in annual cost for a reservoir resulting from a shortened economic life; (2) an increase in the annual cost for flood control structures caused by sediment reducing their economic life; (3) the sediment component of flood damages and damages associated with sediment that remains in the watershed; (4) the increase in sediment damage that occurs after the end of a reservoir's economic life or after the end of a flood control structure's economic life; (5) the loss of recreational benefits resulting from the siltation of a reservoir; and (6) the loss of water supply benefits resulting from sediment displacing the water supply pool in a reservoir.

The Lee and Guntermann procedure implicitly assumes that sediment will not be dredged from a reservoir or removed from a flood control structure. Also implicitly assumed was that a new reservoir or a new flood control structure would not be built to replace an existing one once it is completely filled with silt. These do not appear to be realistic assumptions for many reservoirs or the flood control structures in Texas watersheds. Consequently, the Lee and Guntermann procedure was modified for estimating offsite sediment damages in the study watersheds. Sediment damages were attributed to the following factors: (1) the cost of removing sediment from flood control structures in the watershed by draining them and then removing the accumulated sediment; (2) the cost of dredging sediment from the reservoir; and (3) the sediment component of flood damages and damages associated with sediment that remains in the watershed. Computational formula and damage estimates for each of these components, using the Lavon reservoir watershed as an example, follow.

Cost of Removing Sediment from Flood Control Structures

For this component of damages, it was assumed that the sediment pool in a flood control structure would be allowed to completely fill. Then, before sediment reduced the flood control capacity of the structure, the structure would be drained in a dry period and the sediment removed by bulldozing or a similar operation. SCS engineers estimate that this type of operation would cost about $0.29 per ton of sediment removed. With N as the life of the sediment pool it was assumed that a structure would be cleaned every N years. N was computed by the following formula:

$$N = \frac{K\ C_{RS}}{G_e A_N D_R T_E}$$

where N is the life of the sediment pool in years
  C_{RS} is the capacity of the sediment pool in ac-ft
  G_e is the gross erosion based on a particular crop rotations, tillage system, conservation practice, and management level for the watershed in ton/ac/yr
  A_N is the net drainage area in acres
  D_R is the delivery ratio used to convert gross erosion to sediment delivered
  T_E is trap efficiency of the structure
  K is the conversion constant from ac-ft to tons

Values for C_{RS}, A_N and D_R were obtained from the PL-566 watershed work plans.[2-4] K was assumed to equal 1,920 ton/ac-ft, and T_E equal to 0.95.

The present value cost of removing sediment from flood control structures in the watershed into perpetuity is given by the formula:

$$PV = \sum_{S=1}^{191} \left[\sum_{t=1}^{\infty} (\frac{1}{1+i})^{N_s t}\ C_r C_{RS,S} K \right]$$

$$= \sum_{S=1}^{191} \left[\frac{(\frac{1}{1+i})^{N_s}}{1-(\frac{1}{1+i})^{N_s}}\ C_r C_{RS,S} K \right]$$

where PV = present value cost
  C_r = per ton cost of removing sediment from a flood control structure (= $0.29)
  N_s = life of the sediment pool of the Sth structure
  i = interest rate
  $C_{RS,S}$ = capacity of the sediment pool in the Sth structure in ac-ft

The annualized cost of removing sediment from flood control structures is:

$$D_{FS} = i \cdot PV = i \sum_{S=1}^{191} \left[\frac{(\frac{1}{1+i})^{N_S}}{1-(\frac{1}{1+i})^{N_S}} \right] C_r C_{RS,S} K$$

where D_{FS} = annualized cost of removing sediment from all flood control structures in Lavon watershed.

Cost of Dredging Lavon Reservoir

Annualized offsite sediment damages attributal to the siltation of Lavon reservoir were based on the cost of dredging the sediment pool each time the pool filled. Computation of the time required for the sediment pool to fill is more complicated than for a flood control structure because the calculation of sediment input is more complicated. Sediment input into the reservoir can be conceptualized as the sum of two components. One component is sediment originating in subwatersheds that drain into flood control structures, while the other component is that originating in subwatersheds not protected by flood control structures. Other things equal, sediment input into a reservoir from a subwatershed protected by a flood control structure is much lower than for the other subwatersheds. This is because the flood control structure is functioning as a sediment trap. Assuming that the trap efficiency of these structures is 0.95 and that the gross erosion rate is the same for all subwatersheds, the total annual sediment input into Lavon reservoir was computed as:

$$S = 0.05 \, D_R A_F G_E + D_R A_{NF} G_E$$

where S = annual sediment input into Lavon reservoir
D_R = delivery ratio
G_E = gross erosion rate in ton/ac
A_F = acreage in Lavon watershed protected by flood control structures other than Lavon reservoir
A_{NF} = acreage not protected by flood control structures.

Based on this, the time required for the sediment pool in Lavon reservoir to fill was calculated as:

$$N = \frac{C_{RS} K}{S}$$

where N = years required for the sediment pool to fill, with average gross erosion in the watershed equal to G_E

C_{RS} = capacity of the sediment pool in ac-ft
K = constant for converting ac-ft to tons.

The following values were assumed for the Lavon reservoir:

K = 1,920 ton/ac-ft
C_{RS} = 35,650 ac-ft
D_R = 0.3
A_F = 203,077 ac
A_{NF} = 274,536 ac

To compute the cost of dredging Lavon, it was assumed that a small portable dredge with a 10-in. line would be used. Operating costs for this type of dredge are about \$240/hr, with 200 yd^3/hr pumped.* Assuming that the average density of sediment is 1.19 ton/yd^3 the cost per ton of sediment dredge is \$1.01.

The present value cost of dredging Lavon every N years into perpetuity is given by the formula:

$$PV = \sum_{t=1}^{\infty} (\frac{1}{1+i})^N C_d C_{RS} K_1 = \frac{(\frac{1}{1+i})^N}{1 - (\frac{1}{1+i})^N} C_d C_{RS} K$$

where PV = present value cost
C_d = per ton cost of dredging sediment
i = interest rate
C_{RS}, and K as previously defined

The annualized cost of dredging sediment from Lavon reservoir is:

$$D_L = i \cdot PV = iC_d C_{RS} K_i \frac{(\frac{1}{1+i})^N}{1 - (\frac{1}{1+i})^N}$$

Sediment Component of Flood Damages and Damages Associated with Sediment that Remains in the Watershed

Estimates of this component of damages (D_S) were obtained directly from the PL 566 watershed work plans.[2-4] In 1976 dollars the damages totaled \$44,462 for a gross erosion rate of 12.8 ton/ac. For other erosion rates these damages were assumed proportional to total erosion.

---

*The assistance of the Galveston and Ft. Worth branches of the U.S. Army Corps of Engineers in obtaining this cost estimate is gratefully acknowledged.

The total offsite damages in each watershed were calculated at several assumed gross erosion rates. From these sets of points relating gross erosion with offsite damages, an offsite damage function was estimated. Figure 4 is the offsite damages function estimated for Lavon watershed.

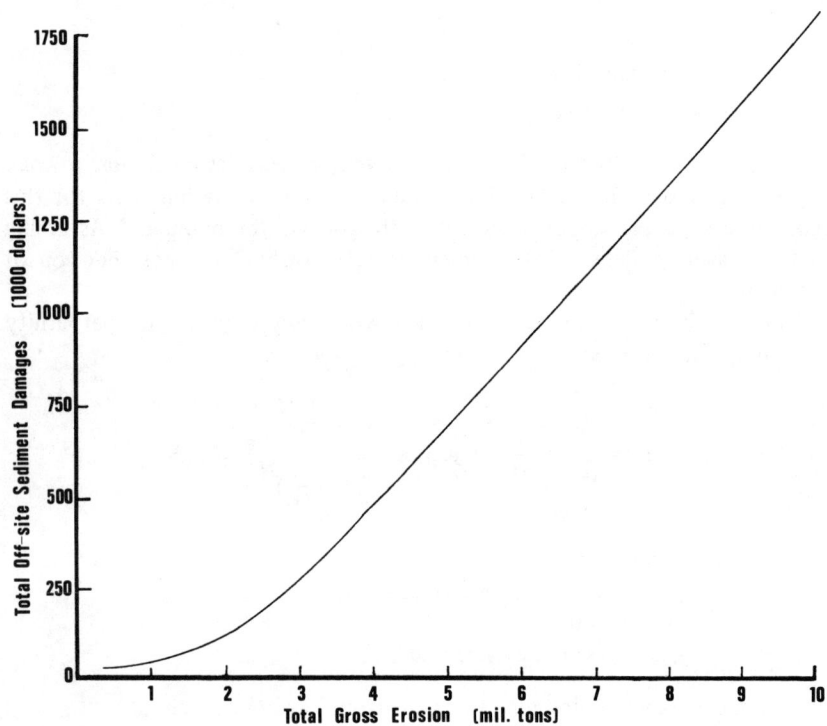

Figure 4. Total offsite sediment damages in Lavon watershed.

Administrative and Enforcement Costs

The cost of administering and enforcing any of the NPS controls considered here has been estimated to be at least $0.50/ac of land in Lavon watershed.[11] For the watershed as a whole, these costs will thus be at least $224,125 annually for the agricultural land in the watershed. The largest component of this cost estimate is based on the amount of technical assistance that would be required to implement the policies. While there will be cost differences between policies, this figure gives a rough floor to the administration and enforcement costs. This cost figure should be kept in mind when considering the benefit and cost figures given in succeeding tables.

Net Social Benefit Tables Created

Because the benefits of soil conservation accrue over time, rather than immediately, the length of a farmer's planning horizon influences the crops that will be grown and the conservation practices employed. This, in turn, influences the estimated economic impact of NPS control options. Due to uncertainty about the length of farmers' planning horizon, estimated effects were calculated for three time horizons. These are 10 years, 100 years and 200 years. Results based on these planning horizons were assumed likely to bracket the actual economic impacts of the erosion controls considered. To demonstrate the information generated the tables created for the Lavon watershed assuming a 100-year planning horizon are included.

Estimated effects of various erosion-sedimentation control policies on farm income, government cost or revenue, soil loss, offsite sediment damages abated, and net social benefits are shown in Table VI. Table VII gives the associated acreage distribution, while Table VIII shows the extent and cost of terracing and contouring by control option.

SUMMARY AND CONCLUSION

The model is designed to examine both the on-farm economics of soil conservation and the economic consequences of various NPS pollution control options. These topics are joined in this study because they deal with different facets of the same problem. Unlike some pollutants, the sediment that washes off farmers' fields to become a problem downstream is a valuable resource, not a waste product. Because the soil is valuable itself, some level of soil conservation practices is going to be economically desirable even if the downstream pollution damages are not considered by the farmer. The results have consistently shown that soil conservation does indeed pay and that its value is greater the longer the planning horizon of the decision-maker. This suggests that an educational program in this area may reduce sediment damage while increasing farm income at the same time.

The second part of the model deals with the total economic impact of various soil loss control options. Options based on regulation, taxation, economic incentive and combinations thereof are examined. Given the estimate of offsite sediment damages and the assumptions of the model, the analysis suggests that soil loss controls or subsidies are not presently warranted from a social welfare viewpoint in the Texas watersheds modeled. However, it should be noted that the estimation of offsite damages is imprecise at best. Many types of environmental damage are intangible and others are caused indirectly. Future research could be directed profitably toward calculating more precise and complete estimates of environmental damage.

Table VI. Major economic consequences of NPS control options in Lavon watershed assuming farmers have a 100-year planning horizon.

Control Option	Change in Annualized Farm Income ($1,000)	Government Cost (-) or Revenue (+) ($1,000)	Change in Gross Soil Loss (1,000 T)	Offsite Sediment Damages Abated ($1,000)	Net Social Benefits Excluding Administrative Costs ($1,000)
SL <T	-2,343.33	0.0	1,842.55	224.89	-2,118.43
SL <2	-2,449.60	0.0	1,875.04	226.91	-2,222.69
SL <5	-2,013.31	0.0	1,681.59	213.71	-1,799.60
SL <10	-0.93	0.0	6.52	1.10	0.17
TR 50	0.0	0.0	0.0	0.0	0.0
TR 100	376.88	-1,321.04	1,231.37	172.52	-771.64
C 50	339.25	-339.25	0.0	0.0	0.0
C 100	724.16	-809.02	37.44	6.28	-78.58
IT 50	0.0	0.0	0.0	0.0	0.0
IT 100	0.0	0.0	0.0	0.0	0.0
SL <T, TR 50	-1,636.91	-743.59	1,842.55	224.89	-2,155.61
SL <T, C 50	-2,336.75	-6.57	1,842.55	224.89	-2,118.43
SL <T, IT 50	-2,253.04	-4,906.40	1,842.55	224.89	-6,934.55
SL <5, TR 50	-1,385.82	-660.52	1,681.59	213.71	-1,832.63
SL <5, C 50	-1,970.35	-42.96	1,681.59	213.71	-1,799.60
SL <5, IT 50	-1,933.79	-4,321.28	1,681.59	213.71	-6,041.36
TX 8	-175.01	174.99	0.0	0.0	-0.02
TX 10	-218.75	-218.74	0.0	0.0	-0.01
TX 12	-262.48	262.49	0.0	0.0	-0.01
TX 14	-306.24	306.23	0.0	0.0	-0.01
TX 16	-349.86	348.94	6.52	1.10	0.18
TX 18	-393.49	392.55	6.52	1.10	0.16
TX 20	-437.10	436.17	6.52	1.10	0.17
TX 8, 50 T&C	169.90	-207.23	82.30	13.73	-23.59
TX 10, 50 T&C	127.81	-165.13	82.30	13.73	-23.59
TX 12, 50 T&C	85.71	123.02	82.30	13.73	-23.58
TX 14, 50 T&C	43.60	-80.92	82.30	13.73	-23.59
TX 16, 50 T&C	1.63	-39.86	88.82	14.81	-23.43
TX 18, 50 T&C	-40.35	2.11	88.82	14.81	-23.44
TX 20, 50 T&C	-82.32	44.08	88.82	14.81	-23.44

To calculate the economic consequences of the chosen control options, it was necessary to make certain basic assumptions. These assumptions can be critical to the results of the study and must be kept in mind if the model is to be interpreted correctly. These assumptions include: (1) relative expected prices will remain constant, (2) expected present value of profit is a good indicator of farmers' decision criteria, (3) farm profits, government cost or revenue and sediment damage abatement have the same value weights, and (4) farmers know and will act in their own long-run self-interest.

Table VII. Percent of acreage in each crop by control option for Lavon watershed assuming farmers have a 100-year planning horizon.

Control Option	Cotton	Grain Sorghum	Wheat	Hay and Pasture
Benchmark	20.14	20.14	17.76	41.96
SL <T	15.28	15.28	17.76	51.68
SL <2	15.28	15.28	17.76	51.68
SL <5	15.28	15.28	17.76	51.68
SL <10	20.07	20.07	17.76	42.11
TR 50	20.14	20.14	17.76	41.96
TR 100	20.14	20.14	17.76	41.96
C 50	20.14	20.14	17.76	41.96
C 100	20.14	20.14	22.00	37.72
IT 50	20.14	20.14	17.76	41.96
IT 100	20.14	20.14	17.76	41.96
SL <T, TR 50	15.28	15.28	17.76	51.68
SL <T, C 50	15.28	15.28	17.76	51.68
SL <T, IT 50	15.28	15.28	17.76	51.68
SL <5, TR 50	15.28	15.28	17.76	51.68
SL <5, C 50	15.28	15.28	17.76	51.68
SL <5, IT 50	15.28	15.28	17.76	51.68
TX 8	20.14	20.14	17.76	41.96
TX 10	20.14	20.14	17.76	41.96
TX 12	20.14	20.14	17.76	41.96
TX 16	20.07	20.07	17.76	42.11
TX 18	20.07	20.07	17.76	42.11
TX 20	20.07	20.07	17.76	42.11
TX 8, 50 T&C	20.14	20.14	17.76	41.96
TX 10, 50 T&C	20.14	20.14	17.76	41.96
TX 12, 50 T&C	20.14	20.14	17.76	41.96
TX 14, 50 T&C	20.14	20.14	17.76	41.96
TX 16, 50 T&C	20.07	20.07	17.76	42.11
TX 18, 50 T&C	20.07	20.07	17.76	42.11
TX 20, 50 T&C	20.07	20.07	17.76	42.11

Assumption one rules out any large technological breakthroughs that would drastically change production costs or yield of one crop in relation to the others. It also rules out the discovery of presently unknown ways to cheaply restore the soil fertility of eroded soils or to remove sediment from waterways at little or no cost. Furthermore, major changes in crop prices relative to the general price structure would invalidate the conclusions of this study. If crop prices fell relative to other prices, offsite damages would carry significantly more weight, and greater erosion control would be socially beneficial. On the other hand, if relative crop prices rose, offsite damages

Table VIII. Extent and cost of terracing and contouring by control option for Lavon watershed assuming farmers have a 100-year planning horizon.

Control Option	Terracing		Contouring	
	Acres (1,000)	Cost ($1,000)	Acres (1,000)	Cost ($1,000)
Benchmark	0.0	0.0	180.00	678.50
SL <T	211.00	1,487.18	5.60	13.15
SL <2	216.60	1,517.20	0.0	0.0
SL <5	180.00	1,321.04	36.60	85.92
SL <10	0.0	0.0	180.00	678.50
TR 50	0.0	0.0	180.00	678.50
TR 100	180.00	1,321.04	0.0	0.0
C 50	0.0	0.0	180.00	678.50
C 100	0.0	0.0	235.60	809.02
IT 50	0.0	0.0	180.00	678.50
IT 100	0.0	0.0	180.00	678.50
SL <T, TR 50	211.00	1,487.18	5.60	13.15
SL <T, C 50	211.00	1,487.18	5.60	13.15
SL <T, IT 50	211.00	1,487.18	5.60	13.15
SL <5, TR 50	180.00	1,321.04	36.60	85.92
SL <5, C 50	180.00	1,321.04	36.60	85.92
SL <5, IT 50	180.00	1,321.04	36.60	85.92
TX 8	0.0	0.0	180.00	678.50
TX 10	0.0	0.0	180.00	678.50
TX 12	0.0	0.0	180.00	678.50
TX 14	0.0	0.0	180.00	678.50
TX 16	0.0	0.0	180.00	678.50
TX 18	0.0	0.0	180.00	678.50
TX 20	0.0	0.0	189.00	678.50
TX 8, 50 T&C	0.0	0.0	211.00	751.27
TX 10, 50 T&C	0.0	0.0	211.00	751.27
TX 12, 50 T&C	0.0	0.0	211.00	751.27
TX 14, 50 T&C	0.0	0.0	211.00	751.27
TX 16, 50 T&C	0.0	0.0	211.00	751.27
TX 18, 50 T&C	0.0	0.0	211.00	751.27
TX 20, 50 T&C	0.0	0.0	211.00	751.27

would become less important and the optimal erosion control would depend on the on-farm trade-off between present production and future production.

The second assumption asserts that the shifts in cropping patterns will take place, as this is the decision criterion built into the model. Farmers have other criteria besides profit on which they base their decisions. These other criteria might include: personal preference for one crop over another, preference for leisure rather than more profit, varying estimates of risk and uncertainty, and others. While these other criteria play a part in farmers' decisions, it is a general assumption of economics that expected profit is the most important consideration, and focusing on it alone will yield generally accurate results.

The third assumption is the rationale behind the net social benefit calculation. It indicates that for the purposes of this study "government" is considered only as a point of accounting, *i.e.,* a frictionless point of transfer for part of the jointly held social wealth. Net social benefit does not change if money transfers from farm income to government or vice versa. Also it implies that farm income is equal in social desirability to a similar dollar amount of offsite sediment damage abatement. This can be defended by noting that if the dollar value of the offsite damages has been correctly estimated, then it would be better for farmers as a group to pay for the damages directly rather than lose an amount of profits greater than the value of the damages abated.

The last assumption rules out ignorance of, or uncertainty about, the most profitable cropping system—conservation practice. It also implies that financing will be available for any necessary equipment shifts or terrace construction. Neither of these conditions will be met always, and failure will reduce the actual change caused by implementation of any of the control options specified.

The estimated farm income consequences of NPS control options that are presented in this report were also based on the assumption that crop prices would not change in response to the implementation of a particular policy. This is a reasonable assumption as long as the policy is imposed only in a small area with no changes in outside areas. However, if a pollution control policy is imposed in a large area or for the whole nation, it is expected that crop prices will change in response to implementing a policy that significantly affects cropping patterns, yield or production costs. Thus, the results presented in this study apply only if NPS controls are imposed in small areas or in ways that do not affect comparative crop prices.

ACKNOWLEDGMENTS

This paper is the result of a series of watershed studies funded by the Texas Soil and Water Conservation Board and the Texas Department of Water Resources. The research was conducted under the auspices of the Texas Water Resources Institute, the Texas Agricultural Experiment Station and the Texas Agricultural Extension Service. The authors would like to express their appreciation to Dr. Jack Runkles, Director of the Water Resource Institute for assistance in organizing and carrying out the research. Dr. William L. Harris, a Soil and Water Specialist with the Texas Agricultural Extension Service, coauthored some of the individual watershed studies and was deeply involved in gathering the necessary soils data. He was also principally responsible for the development of the "yield loss vs topsoil loss" functions. Many other Soil Conservation Service and university scientists, too numerous to mention, made contributions to this work. Their assistance is greatly appreciated.

The responsibility for the accuracy and completeness of this paper is borne, as usual, solely by the authors.

REFERENCES

1. U.S. Congress Public Law 92-500. "Water Quality Control Act Amendment" (1972).
2. USDA—Soil Conservation Service. "Supplemental Work Plan for East Ford above Lavon Watershed of the Trinity River Watershed, Collin and Grayson Counties, Texas," Temple, TX (September 1963).
3. USDA—Soil Conservation Service. "Work Plan for Pilot Grove Creek Watershed of the Trinity River Watershed (Lavon) Collins Fannin, Grayson and Hunt Counties, Texas," Temple, TX (January 1959).
4. USDA—Soil Conservation Service. "Work Plan for Sister Grove Creek Watershed of the Trinity River Watershed (Lavon) Collins and Grayson Counties, Texas," Temple, TX (June 1956).
5. Texas Department of Agriculture and USDA Statistical Reporting Service. "Texas County Statistics," compiled by Texas Crop and Livestock Reporting Service, Austin, TX (1970-1975).
6. U.S. Congress Public Law 566. "Watershed Protection and Flood Prevention Act," 83rd Congress 68 Stat. 666.
7. Wischmeier, W. H., and D. D. Smith. "Predicting Rainfall—Erosion Losses from Cropland East of the Rocky Mountains—Guide for Selection of Practices for Soil and Water Conservation," *Agricultural Handbook* No. 282 (Washington, DC: U.S. Government Printing Office, 1965).
8. Texas Department of Agriculture and USDA Statistical Reporting Service. "Texas Prices Received and Paid by Farmers," compiled by Texas Crop and Livestock Reporting Service, Austin, TX (1958-1975).
9. Parker, C. A., and R. W. Sammons. "Texas Crop Budgets," Texas Agricultural Extension Service, College Station, TX (1977).
10. Lee, M. T., and K. Guntermann. "A Procedure for Estimating Offsite Sediment Damage Costs and an Empirical Test," *Water Resources Bull.* 12(3) (June 1976).
11. Kretschemar, G. E., Jr. Texas Soil and Water Conservation Board, personal communication.

30

THE POLICY RELEVANCE OF ALTERNATIVE INSTITUTIONAL APPROACHES TO 208 PLANNING

A. Hamilton, L. W. Libby
  Department of Agricultural Economics
  Michigan State University
  East Lansing, Michigan

INTRODUCTION

We are about midstream in the water quality planning process mandated under Section 208 of the Federal Water Pollution Control Act Amendments of 1972. Planners in various agencies throughout the country are preparing documents that will describe the steps necessary to achieve the goals of "fishable and swimmable water by 1983." Perhaps the major technical challenge facing planners is the design of "best management practices" (BMPs) which will control nonpoint pollution. It is quite clear that none of the nation's water quality goals can be met without the control of these sources of effluents.

Some of these plans are completed and approved, others are just getting started. Thus, one might expect that the control of nonpoint sources of pollution ought to be underway. Unfortunately, no plan or planning process by itself has ever produced a drop of clean water. It can be effective only if the measures the plan calls for are, in fact, put in place. The implementation of the plan is necessary if the whole exercise is to have any separable impact on the problem of water pollution.

It is almost axiomatic that nonpoint sources of pollution arise from the way people use land. It follows that any change in the amount of nonpoint pollution must come from a change in the way that people use land. For the most part, for various political, historical and legal reasons, the power to regulate land use lies with local governments.* These include counties, cities, villages and townships. It would seem, at least for the moment, that if nonpoint pollution is going to be controlled, a fair part of the burden for doing so will fall on local governments. At the same time 208 planning is being done by a multitude of organizations. The two principal types are state departments (environment, conservation, public health, or natural resources) and multicounty regions. There is a clear difference in the jurisdictional boundaries of these two organizations with the state level agencies being larger and further removed from local governments than the regional agencies.

We would like to suggest that it makes a difference as to which type of organization prepares a Section 208 plan, not because the plans are "better" in some absolute sense at one level or the other, but because the relative probabilities of implementation are different. Because the jurisdictional boundaries of the regional planning agencies are closer to local governments than state agencies, their opportunity sets† should be more closely related. The regional agencies are more closely aligned with local governments; thus there is a possibility that they will be better able to choose a set of means to achieve water quality goals that will be acceptable to local governments. The question is, how may we measure this difference in the relative probability of implementation? If we do not wish to wait until 1983 or 1985 when, presumably, we will have clean water policy in action, we must look for some intermediary proxy variables which may provide an indication.

Bartlett's model of information subsidy[1] may be used to assist us in specifying the appropriate proxy variables. Bartlett suggests that politicians (decision-makers) seek to maximize votes and operate under uncertainty with respect to the impact on their constituencies. Consequently, various groups may subsidize information to the decision-maker in hopes of influencing his/her decision because he/she operates under uncertainty. Bartlett also describes bureaucrats (planners) as security maximizers. One of the means they use to increase security is to subsidize information to decision-makers.

There are several ways in which this information subsidy process may be measured. Section 208 planning has as part of its process public participation requirements. If formal public participation meetings are viewed as iterative, then they may be viewed as an information subsidy activity.

---

*There are, of course, exceptions to this statement. Hawaii is a notable example, with comprehensive land use controls. Other states have also adopted special land use controls with respect to specific resources such as rivers, agricultural lands and coastal zones.

†An opportunity set is the perceived matrix of benefits and costs that an individual faces with respect to a contemplated action.

That is, a public hearing allows the planner to extoll the virtues of a given set or sets of agendas of means, and in so doing sells planning to the public and decision-makers, and gathers community inputs to the planning process. Budgets for public participation should also serve as a proxy for information subsidy. Other measures should include special presentations to decision-makers (*e.g.*, whether the agency writes press releases and, if so, how many, and size of public participation staff). All of these variables appear to gauge some aspect of the information subsidy process and are quantifiable. If this is the case then these variables may be used to measure the differences in the performance of alternative 208 water quality planning organizations.

The hypothesis we wish to investigate is that the regional planning agencies, because they are closer to local governments, will do a more comprehensive job of subsidizing information to local decision-makers.

SURVEY AND FINDINGS

A survey was sent to all agencies with Section 208 responsibilities in the 50 states (excluded were the District of Columbia, Guam and Puerto Rico). Of these 217 surveys, 82% were returned and 75% proved usable. Nonresponse bias was, for the most part, negligible.[2]

In trying to measure information subsidy the variables may be broken into two general classes, those that aim at widespread audiences and those targeted at specific audiences. Those variables which address general audiences include number of hearings, public participation budgets, and news releases. The variables measuring information subsidy to specific audiences attempt to gauge the information subsidy process to decision-makers at various levels of government. This process is measured by looking at the number of special presentations to decision-makers. In each of these tests a one-tailed hypothesis is specified because it is expected that each measure of information subsidy will be weighted in favor of the regions.

Tables I and II depict the general information subsidy variables. All of the parametric tests suggest that the regional agencies responding were more active in the information subsidy process as measured by the included variables. The variables included were significant at the 0.05 level or better.

One nonparametric test was done in this section. This test was the Chi-squared procedure in Table IV which tests whether there was a difference in the number of agencies that prepared news releases on Section 208 planning at the state and regional levels. The analysis found that 97.5% of the state organizations and 95.5% of the regional organizations prepared their own news releases. This difference is significant to the 0.2581 level. It would be difficult to suggest on the basis of this test that there was a difference. However, if one examines the number of news releases prepared per county

Table I. Public participation hearings, public participation budgets and public participation staff per county and per 1,000 population.

	N	Mean	Standard Error	T Value	DF	Significance (one-tailed)
Public Participation Hearings Per County						
States	28	0.67	0.711	-2.05	111	0.021
Regions	85	15.46	38.022			
Public Participation Hearings Per 1,000 Population						
States	28	0.0251	0.005	-1.66	109	0.050
Regions	83	0.0250	0.063			
Public Participation Budget Per County						
States	32	$129.00	9311	-1.73	112	0.048
Regions	83	$338.28	6645			
Public Participation Budgets Per 1,000 Population						
States	32	$ 95.00	17.67	-3.36	109	0.001
Regions	79	$215.89	21.69			
Public Participation Staff Per County						
States	31	0.1836	0.114	-2.47	112	0.008
Regions	79	0.7196	0.125			
Public Participation Staff Per 1,000 Population						
States	31	0.0018	0.000	-2.28	109	0.012
Regions	80	0.0065	0.001			

Table II. News releases per county and numbers and percentages of agencies preparing news releases.

	N	Mean	Standard Error	T Value	DF	Significance (one-tailed)
Number of News Releases Per County						
States	30	2.00	1.49	-2.48	100	0.008
Regions	72	12.48	2.64			
		Yes (%)	No (%)	N	X^{2a}	Significance
Agencies Preparing News Releases on Section 208 Planning						
States	28	(87.5)	4 (12.5)	32	1.27	0.2581
Regions	84	(95.5)	4 (4.5)	88		

[a] Chi-squared statistic is adjusted with the Yeat's correction for continuity.

(Table II) it is easy to see that on the average the regional agencies prepare far more news releases per county (12.48) than the state agencies (2.00).

Tables III and IV list the results of the analysis done on the variables which attempt to depict the information subsidy process to specific groups of decision-makers. Those decision-makers are the governor or the governor's staff, state legislators, county legislators and city and township legislators. Again one-tailed tests are used. It is hypothesized that the regions will do a better job subsidizing information to county, city and township legislators than the state agencies will. On the other hand, the jurisdictional boundaries of concern to the governors and state legislatures are probably closer to the jurisdictional boundaries of the state agencies. Thus, it is expected that state agencies will do a better job of subsidizing information to these groups.

The results in Table III show that these hypotheses are upheld. The state agencies made presentations to the governor, or the governor's staff, and state legislators (8.26 to 54.4% and 76.2 to 33.9%, respectively) more frequently than regional agencies. At the same time the regional agencies made presentations to county and city and township legislators with a greater frequency than state agencies (86.2 to 68.2% and 92.8 to 71.4%, respectively). All of these results are significant to the 0.10 level or better. Table IV describes the tests of the number of special presentations made to each specific group. In general the statistics behave as predicted, although the tests involving county, city and township legislators are much more significant than those involving the governor, the governor's staff and state legislators. Unfortunately, the authors believe that in these tests nonresponse problems have biased the results upward. This set of questions was the least answered on the entire questionnaire. Of the state agencies, 36-42% filled in the item in question. The regional agencies responded at a slightly higher rate, 43-50%, but again nonresponse bias is probably present. The authors suspect that the values excluded were probably zero more often than not, although they have no way to prove this assertion. If this assertion is correct, then there is reason to suspect that the estimates of the population means given in Table IV are biased upward. This does not mean that information cannot be gleaned from this table. While little faith can be attached to the absolute differences in the estimates of the population parameters, other information is available. If it is assumed that the direction and magnitude of the nonresponse bias in each case is about the same, then the means in Table IV may still tell us something about the relative differences in the information subsidy practices of each group.

This case proves to be just the opposite of the first set of variables. Here nonparametric tests have probably yielded more information than the

Table III. Agencies making special presentations to: the governor or governor's staff, state legislators, county legislators and city and township legislators.

	Yes (%)	No (%)	N	X^{2a}	Significance (one-tailed)
Special Presentations Made to Governor or Governor's Staff					
States	19(82.6)	4(17.4)	23	4.64	0.0312
Regions	37(54.4)	31(45.6)	68		
Special Presentations Made to State Legislators					
States	16(76.2)	5(23.8)	21	9.54	0.0020
Regions	20(33.9)	39(66.1)	59		
Special Presentations Made to County Legislators					
States	15(68.2)	7(31.8)	22	2.73	0.0983
Regions	69(86.2)	11(13.7)	80		
Special Presentations Made to City and Township Legislators					
States	15(71.4)	6(28.6)	21	5.52	0.0186
Regions	77(92.8)	6(7.2)	83		

[a] Chi-squared statistic is adjusted with the Yeat's correction for continuity.

Table IV. Number of special presentations made to the governor or governor's staff, state legislators, county legislators, and city and township legislators.

	N	Mean	Standard Error	T Value	DF	Significance (one-tailed)
Number of Special Presentations to the Governor or Governor's Staff						
States	14	2.57	0.716	1.14	64	0.130
Regions	52	1.57	0.410			
Number of Special Presentations to State Legislators						
States	12	7.08	5.74	0.69	59	0.247
Regions	49	3.51	2.16			
Number of Special Presentations to County Legislators per County						
States	13	0.252	0.091	-2.06	66	0.022
Regions	55	3.015	0.649			
Number of Special Presentations to City and Township Legislators per County						
States	12	0.943	0.482	-2.23	63	0.015
Regions	53	10.260	1.970			

parametric ones because the questions involved were simpler and consequently were answered with a higher frequency.

On the basis of the evidence presented it is possible to draw four conclusions about the differences between the information subsidy processes of the two organizations. First, the regional agencies appear to provide more information to general audiences. The analysis also suggests that the regional agencies subsidize information on a larger scale to county, city and township legislators. Since these levels of government are generally the ones that are perceived as the legitimate controllers of land use, the plans to control nonpoint pollution drawn up by the regions should have a higher probability of implementation than those drawn up by state-level agencies. Finally, the state agencies appear to do a better job of subsidizing information to state-level decision-makers.

LIMITATIONS OF THE ANALYSIS

Clearly, this analysis is not without its limitations. Information subsidy by planners is only one of many factors that affects the decisions of elected officials. The information provided by other groups, the decision-maker's personal philosophy, the existing political climate and a multitude of other factors all enter into the calculus that will determine whether a plan will be implemented. It should also be clear that this analysis has only measured some of the more formal systems of information subsidy. Anyone who has ever worked in any decision-making process can appreciate the impact that the informal system of information subsidy has on the decision-making process. Unfortunately, almost by definition this process is not quantifiable.

Finally, this analysis assumes that the power to regulate land use and hence, most nonpoint pollution, lies at the local level. In the short run this assumption can probably stand on its own merits. In the longer run this assumption must be examined. The distribution of discretion and power among levels of government is a function of the current political and economic circumstances. It is altogether possible that the power to regulate land use may shift to a higher level—perhaps to the state. There are at least two possible scenarios under which this might come about. First, local government's failure to take the necessary steps to control nonpoint pollution would in effect create a regulatory vacuum which state governments might fill either voluntarily or possibly by being forced to do so by the federal government. Secondly, we should remember that state-level planners face the same or greater advantages in subsidizing information to state decision-makers that regional planners have with local decision-makers. It is perfectly reasonable to expect that they are lobbying for state-level control of nonpoint pollution. We must remember that the planning process is not merely a reflector of status quo relationships but is intimately involved in the redefinition of those relationships.[3]

POLICY PRESCRIPTIONS

If this analysis is to some extent limited, it is nonetheless useful. Our first general observation is that implementation is not separate from planning. Too often we are lulled into a false sense of accomplishment when a plan is produced. But a plan is only an intermediate product. Unfortunately, measuring the success of Section 208 and PL 92-500 on the number of plans drawn up or the detail of those plans is like measuring the flow of electricity from a hydroelectric project by counting the number of tons of concrete in the dam. The point is that a plan is nothing more than a pile of papers unless some set of decision-makers turns it into action. Thus the planning process, if it is to succeed, must be sensitive to the political realities that face those who must ratify those plans.

What does this imply for the next round of resource planning? We would like to suggest that it is impossible to answer the question, "What is the optimum or best planning boundary for a given resource?" independently of the question, "Who must make the hard decisions that will implement that policy?" We need to be continually conscious of these parameters in the design of public planning institutions. To fail to do so is to place unnecessary barriers in the path of prudent and effective control of our natural resources.

REFERENCES

1. Bartlett, R. *Economic Foundations of Political Power* (New York: The Free Press, Macmillan Pub. Co., 1973).
2. Moser, C. A., and G. Kalton. *Survey Methods in Social Investigations* (New York: Basic Books, 1972), pp. 166-168.
3. Lieontief, W. "What an Economic Planning Board Should Do," *Challenge* (July-August 1974).

31

ECONOMIC, INSTITUTIONAL AND WATER QUALITY CONSIDERATIONS IN THE ANALYSIS OF SEDIMENT CONTROL ALTERNATIVES: A CASE STUDY

B. M. H. Sharp
   Department of Agricultural Economics
   University of Wisconsin
   Madison, Wisconsin

S. J. Berkowitz
   Water Resources Center
   University of Wisconsin
   Madison, Wisconsin

INTRODUCTION

It is becoming increasingly evident that the water quality goals of PL 92-500 will not be achieved by regulating point sources of pollution alone. Moreover, in many areas nonpoint sources (NPS) are the major contributors to regional water quality problems. The design of NPS programs is currently one of the most complex and pressing issues facing society.

Sediment is the most ubiquitous NPS pollutant and by volume is the largest pollutant of the nation's water resources.[1] Sediment restricts drainage ways and reservoirs, impacts aquatic life and is the transport vehicle for other pollutants such as pathogens and nutrients. It is also an extremely variable pollutant with mass loadings varying from one watershed to the next—even from one storm to the next.

This study has three objectives. The first is to analyze the likely on-farm impacts of establishing three alternative tolerable soil loss limits: 2 ton/ac,

3 ton/ac, and 5 ton/ac. The second objective is to estimate the change in sediment loadings to water bodies from an agricultural watershed due to the adoption of different conservation practices. The third objective is to analyze and evaluate the potential role that cost-sharing, low-interest loans and educational and extension programs can play in inducing the use of soil-conserving technology.

This research has been undertaken at the University of Wisconsin-Madison, in conjunction with the EPA-sponsored Washington County Project. The overall objective of this project is to demonstrate the effectiveness of land treatment measures in improving water quality, and to help devise the necessary institutional arrangements for the acceptance and implementation of a countywide sediment control program.[2]

DESCRIPTION OF THE STUDY AREA AND PROCEDURES

Kewaskum watershed, located in Washington County, was selected for intensive research as it was considered representative of agricultural areas in southeastern Wisconsin. The watershed, which is illustrated in Figure 1, comprises 7,936 ac, about 40% of which is prime agricultural land devoted predominantly to dairy farming with cropping playing a vital role in feed supply.

The dominant soils of the watershed are loams and silt loams in the Hochheim-Theresa association which covers nearly 50% of Washington County. The Soil Conservation Service (SCS) designates these soils in land capability Class I and II, and Hydrologic Class B, with only limited restrictions due to water and erosion hazards. These soils, from an agricultural standpoint, are potentially the most productive in the county. Lesser acreages are occupied by soils in the Casco-Fox-Rodman association which are somewhat shallower and steeper and have critical management requirements. Finally, the organic soils of the Houghton-Palms-Adrian association make up a small, but hydrologically important (Hydrologic Class D) part of the watershed. Drainage is a necessary precursor to cropping, and the predominant land use is either pasture or marshes and woodlands.

Monitoring stations have been installed in two intensively farmed upland subwatersheds of the Kewaskum Creek watershed to demonstrate the impact of various management practices on water quality. This study focuses on detailed information derived from the Kewaskum-North subwatershed (425 ac). The study site is purposely small so that both the site-specific characteristics influencing sediment loadings could be adequately assessed, and the detailed economic implications, at the farm level, of alternative sediment control strategies could be evaluated.

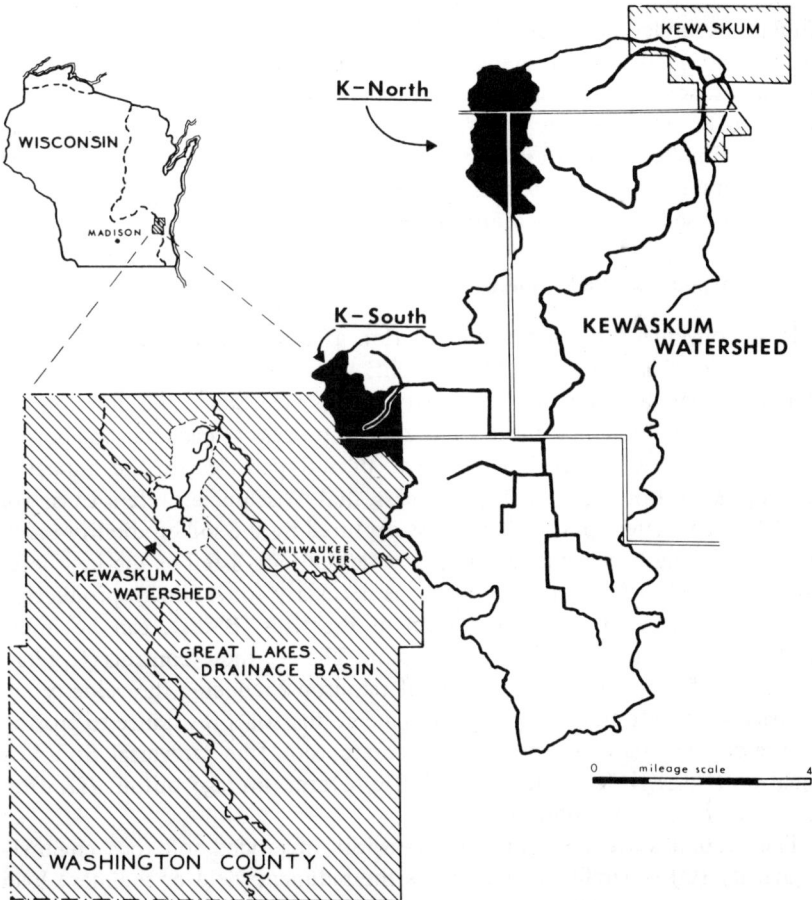

Figure 1. Location of Kewaskum watershed within Washington County, Wisconsin

A revenue-maximizing linear programming (LP) model was developed to capture the essential features of the on-farm decision-making process in the Kewaskum-North watershed. Data from field interviews were incorporated into this model wherever possible in an attempt to capture the critical activities, resources and managerial concerns of a mixed crop-livestock operation. We consider this to be an important element of this study, especially since NPS control programs will be concerned with modifying the *existing* set of farming practices.

A hydrology-based simulation model was modified and calibrated with monitored data, in order to predict the watershed sediment loadings associated with each land use configuration generated by the LP model. The simulation model considers both hydrologic and soil loss variations that result from land use differences in deriving watershed yield predictions. Sediment loadings were estimated using four 6-month sequences of historical rainfall data. This enables the issue of seasonal variability to be addressed directly. Finally, two alternative cost-sharing programs and a low-interest loan program are analyzed with the objective of identifying the land use, revenue and sediment implications of each.

Physical Aspects of the Problem

It is convenient to view sedimentation as a process involving the three stages depicted in Figure 2. This simplification facilitates an identification of the critical determinants (N_j) involved at each stage and the controls (C_j) which can influence the output of the process at each stage. The first stage involves the physical detachment of soil particles by the impact of raindrops and flowing runoff, a process that Wischmeier refers to as soil loss.[3] Soil characteristics, such as those captured by the soil erodibility index, slope, field length, vegetative cover and rainfall, are listed as the key determinants of soil erosion—a process that is necessary for sediment production. Controls, reflected by Best Management Practices (BMPs), can be implemented at this stage to reduce soil erosion. That is, these controls (C_1) have the effect of dissipating the impact of raindrops, reducing the velocity and quantity of runoff and, in general, enhancing the soil's resistance to erosion. At present we have a reasonably accurate tool [the Universal Soil Loss Equation (USLE)] to characterize the long-term, average output (G) from this stage of the sedimentation process.

The second stage concerns the transport of eroded soil to a channel— a quantity (D) generally referred to as the sediment yield or sediment load. In practice, a sediment delivery ratio provides the vehicle for transforming gross soil erosion into the amount of sediment delivered. It is possible to intervene at this stage of the process also. For example, grassed waterways, the preservation of wildlife habitats, and grassed borders are illustrations of how the sediment delivery ratio may be modified. However, there are no dependable equations capable of capturing the transport mechanism. Given this element of uncertainty, along with the functional relationship that exists with the previous stage, it is clear that the control process becomes more complex and, concomitantly, the system response is more uncertain.

Once the sediment has entered the waterway, the final stage involves in-channel transport and deposition. Again, controls (C_3) can be introduced here to reduce in-stream loadings, such as sediment traps, stream bank protection, and flow modifications. Sediment is, therefore, the end product of a highly complex and partially understood process originating at the field

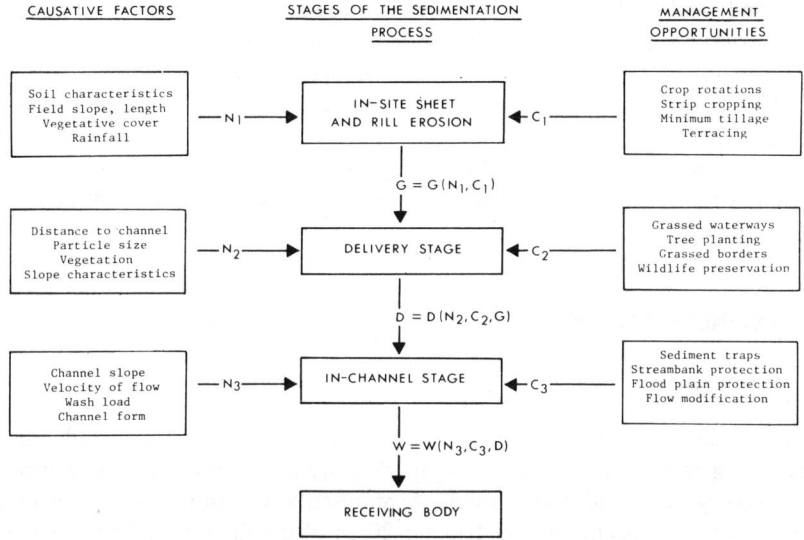

Figure 2. The interaction of causative factors and management opportunities within the sedimentation process.

level. Moreover, the complexity of the delivery and the in-channel phases makes prediction of the water quality effects of land management alterations extremely tenuous.

Natural variability is perhaps the major complicating factor. In fact, climatic influences on the number and magnitude of sediment-generating events have been shown to explain as much as an eightfold variation in observed annual sediment yields.[4] Man-induced variability can have either a compounding or ameliorating effect. Watershed response to heavy June rains on cornland will differ markedly from what its response would be to the same storm occurring in September. Thus, both natural and man-induced sources of variability must be simultaneously considered.

From an economic viewpoint it is necessary to view sedimentation and its inherent variability as a process when designing sediment control policies. This is important for two reasons. First, the commitment of resources in an economically efficient manner will depend upon which stage of the process is chosen for intervention. A cost-effective approach would demand explicit recognition of control costs at *each* level. This will in general imply multiobjective functions because the notion of water quality spans the last two stages—a cost-minimizing program based upon sediment concentrations at the in-channel stage may not necessarily correspond to a minimizing program based upon sediment concentrations in the receiving body.

Second, and perhaps more relevant to program design, are the distributional issues associated with each option within and between stages. Clearly the financial burden will depend upon the stage at which controls are invoked. For example, the establishment of tolerable soil loss limits places the share of the financial burden more directly upon the agricultural community whereas the costs of sediment traps and grade controls are shouldered by taxpayers in general. Obviously, the incidence of these costs will depend on the specific program.

THE ECONOMIC MODEL

Observed Activity

Agricultural activity in the study area typically involved a mixed crop-livestock operation with cropping supplying the major portion of a balanced diet to dairy cows and young stock. A representative farming system would be characterized as comprising 100 ac: 20 ac of cornland (C), 20 ac of oatland (O), 40 ac of alfalfa (H) and 20 ac in pasture and wasteland. This unit would carry approximately 25 dairy cows with a per cow production of 10-12,000 lb of milk, and about 10 replacement heifers would be raised. The feeding system typically involved some combination of baled hay, chopped alfalfa hay, and green chop alfalfa as a forage source, with oats, shelled corn, corn cob meal and purchased protein providing the necessary concentrates. These are family-operated units and, generally, labor is hired only during the harvest period. Other resources such as barn space and feed storage facilities were invariably used to capacity.

The harmonizing of feed supply with a relatively fixed feed demand profile is of pivotal concern to this type of farming operation. Other managerial considerations are related to the seasonal demands placed on labor, the maintenance of soil fertility and the level and distribution of revenue within the planning period. The above information, in addition to the more detailed management practices obtained by interviewing farmers in the watershed, provided the basic data source for the LP model. This information was supplemented with technical data where necessary.

The Linear Programming Model

An optimization model based upon a mix of empirical and engineering data was synthesized in order to capture the basic characteristics of the decision-making process at the farm level. Here the watershed was treated as one farming unit, the assumption being that all resources are mobile across farm boundaries, in addition to each farm production process being characterized by constant returns to scale. Any loss of information resulting from this aggregation process was considered minimal owing to the small area involved (425 ac) and to the homogeneity of the farming systems.

The LP model may be viewed as an ensemble of three interacting subsystems: a set of cropping activities, a dairying submodel and a financial submodel. Each of these subsystems was the focus of research with the overriding concern being to understand how the integrity-preserving characteristics of the model influenced system performance under different policy alternatives. The format utilized is illustrated in Figure 3 where a 6-yr rotation (O-HHH-CC), defined on field 1, is identified along with an outline of the milk-producing submodel. The financial submodel is captured by a set of investment, buying and selling activities which are shown to be coupled with the net revenue function.

Briefly, the 6-hr rotation is shown to utilize inputs such as energy, labor and machinery to produce acres of cornland, oatland and hayland. Two alternative tillage systems, conventional plowing and chisel plowing, were considered for cornland, each requiring different inputs of energy, labor, herbicides, seed, etc. For example, the use of chisel plowing increases herbicide usage by 15% and seeding rates by 10%, while on the other hand it reduces fuel usage and labor consumption by approximately 20%/ac. Five alternative rotations were considered representative of what was observed in the study area. They are listed in Table I along with the average soil loss in tons per acre.

We can imagine the first level of the production process to yield a snapshot of land use in any particular year. Additional inputs such as pounds of nitrogen (N), phosphorus (P), potassium (K) and machinery then account for the productivity of these crops and their processing before sale or use as an intermediate product. Take cornland, for example; this crop may be harvested for grain and sold, or picked and ground into corn meal for livestock production. These two options are shown to supply feed for milk production. A similar set of alternatives exists for oatland and hayland.

Gross margins for the activities which produce directly marketable outputs are listed in Table II. The long-run planning prices developed by Hughes and Weigle[5] were utilized in this study, and the machinery costs were based upon the analysis of Willett et al.[6]

The dairy subsystem provides the alternative of using crops such as corn and alfalfa as inputs to milk production. Since ration composition is an important consideration in dairy farming, only balanced rations were included in the model. Three alternative levels of milk production were defined along with four alternative combinations of forage and concentrates supplying the necessary total digestable nutrients (TDN) for each production level. Two of these rations are listed in Table III. Most of the ration components

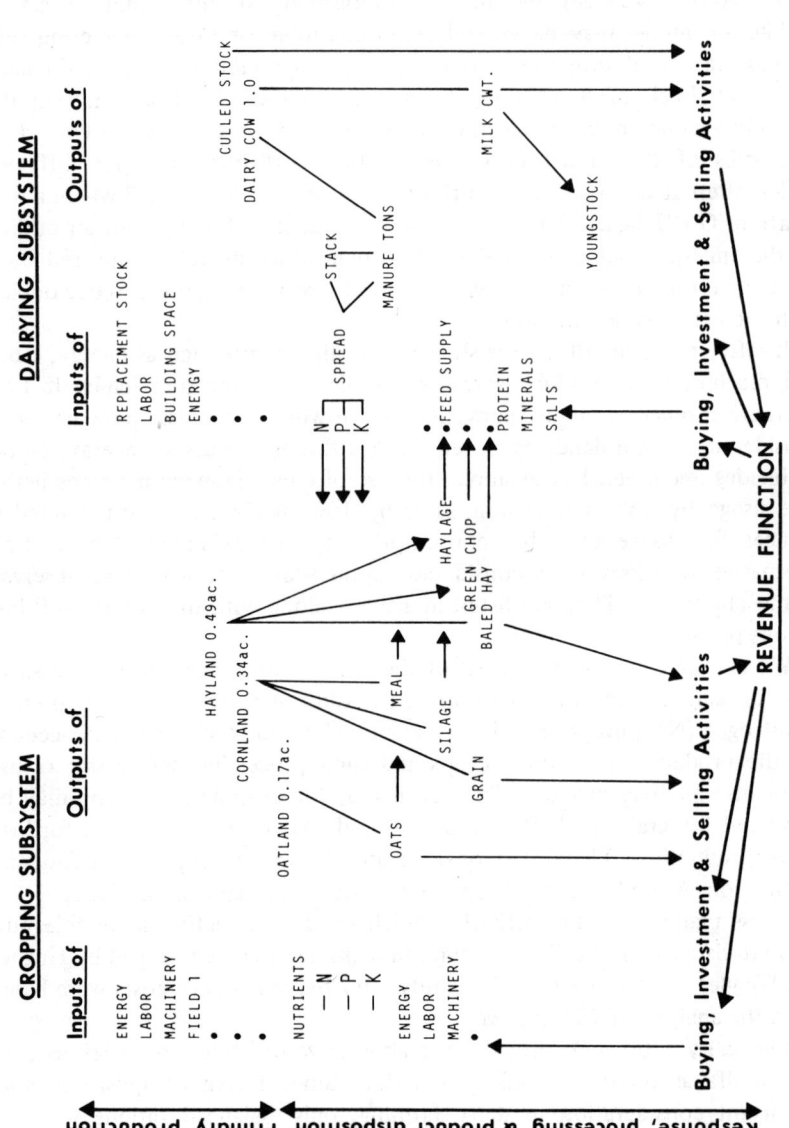

Figure 3. Production interdependencies in a mixed crop-livestock system.

Table I. Average soil loss in tons per acre for cropped fields under alternative rotations: conventional tillage system[a].

Field	Area (ac)	Continuous Corn	O-HHH-C	OO-HHH-C	O-HHH-CC	OO-HHH-CC
1	81	16.68	2.08	3.34	4.17	5.00
2	24	46.45	5.81	9.29	11.61	13.94
3	58	4.55	0.41	0.66	0.83	0.99
4	44	8.34	1.04	1.67	2.08	2.50
5	22	5.74	0.72	1.15	1.43	1.72
6	95	10.63	1.32	2.12	2.66	3.19
7	20	5.67	0.71	1.33	1.42	1.70
Cropped Area (ac)	344					

[a]For minimum tillage multiply the above numbers by 0.70.

could also be purchased at market prices. Again, the objective was to span the feeding rations observed in the watershed.

A necessary side-product to recognize is the production of manure, which is shown in Table III to vary commensurate with the feeding levels. However, the composition of the manure was assumed to remain constant within a given level and between levels. The model was constrained to dispose of all manure produced in any one year. Furthermore, it was assumed that the manure could be spread at any time of the year, but the N-P-K values of the manure would vary according to the time of application. For example, a winter-spreading activity would supply less N-P-K per ton of manure applied than autumn-spread manure. Once applied, this source of nutrients was not differentiated from commercial sources, thereby enabling the value of endogenously supplied N-P-K to be traded off against market supplied sources. That is, the nutrient value of manure at a point in time is weighted by the opportunity costs of labor and machinery for comparison with commercially supplied sources. In addition, since the daily hauling of manure was an observed practice, it was necessary to invest in a stacking facility in order to transfer manure between periods within the year.

The financial submodel plays a pivotal role in this study. Intraannual cash flow considerations and long-term investment opportunities are recognized. Each activity, if it is coupled with the revenue function, will demand or supply funds from four liquidity rows, each representing a period of 3 months. These are cash-balancing rows which may be supplemented by short-term loans at 9%.

Long-term investment activities can draw upon funds from three sources: a 3-yr loan at 9.25%, a 5-yr loan at 9.25% and a 10-yr loan at 9.50%. These loans are amortized and the repayment annuities are incorporated into the

Table II. Activities producing directly marketable outputs from the cropping submodel.

	Corn		Oats	Oats Undersown with Alfalfa	Baled Alfalfa
	Plow	Chisel Plow			
Receipts					
Yield (bu/ac)	110.00	99.00	90.00	90.00	3.00
Price ($/bu)	2.40	2.40	1.40	1.40	35.00
Gross Revenue ($/ac)	264.00	237.60	126.00	126.00	105.00
Expenses ($/ac)					
Fertilizer	50.18	50.18	16.88	16.88	16.98
Seed	10.16	11.06	6.80	29.19	–
Insecticide	6.88	6.88	–	–	–
Herbicide	12.88	14.81	–	–	–
Machinery	32.25	23.55	27.09	27.09	21.32
Total	112.35	106.48	50.77	73.16	38.30
Net Revenue[a] ($/ac)	151.65	131.12	75.23	52.84	66.70
Labor (hr)	3.94	3.13	2.30	2.30	5.61

[a] Net revenue before charges for land, labor and management, and interest on operating capital.

Table III. Alternative balanced rations for the dairy submodel.

Ration	Forage Source[a]					Concentrate Source[b]				Milk[c]	Manure[d]
	Haylage	Silage	Green Chop	Hay	Forage TDN	Corn Meal	Oats	Protein	Concentrate TDN		
A1	0.70	1.40	0.70	0.46	4695	18.90	44.32	4.50	2345	120.00	14.81
A2	0.00	1.96	0.00	1.47	4913	15.42	43.05	4.40	2127	do	do
A3	1.49	0.00	1.04	0.95	4533	19.25	45.15	6.04	2507	do	do
A4	1.49	1.03	0.00	0.00	4681	19.24	49.49	4.26	2359	do	do
B1	0.63	1.27	0.63	0.42	4235	17.04	39.97	4.06	2115	100.00	13.64
B2	0.00	1.77	0.00	1.32	4432	13.90	38.82	3.97	1918	do	do
B3	1.34	0.00	0.94	0.85	4089	17.36	40.72	5.44	2261	do	do
B4	1.35	0.92	0.00	0.00	4222	17.36	40.66	3.85	2128	do	do

[a] All forages in tons, TDN in lb.
[b] Corn meal and oats in bushels, protein in cwt.
[c] Milk production in cwt, price $8.25 per cwt.
[d] Manure production in tons.

revenue function. Each investment activity draws upon a row which can be constrained by the amount of debt a farmer is willing to incur to implement more conservation-oriented systems. In addition, these loans may be financed in part (up to 50%) from the revenue function.

Therefore, in terms of this study, the financial submodel plays two roles. First, it enables the existing resource base to be adjusted in accordance with the optimal plan. This results in a new level and configuration of resources which serves as a baseline for comparison purposes. Second, this submodel is the vehicle through which the farming system adjusts, in an optimal manner, to the imposition of tolerable soil loss limits.

THE SEDIMENT YIELD MODEL

In order to evaluate the sediment yield implications of alternative land management strategies, the simulation model "LANDRUN" was modified for use in the Kewaskum watershed.[7] LANDRUN was chosen from an array of hydrologic and sediment yield models because it appears to adequately characterize critical hydrologic and sedimentation processes, it provides reasonable predictions of watershed output, it is sensitive to variations in land use and rainfall characteristics, and it also meets moderate financial and personnel constraints.[8]

LANDRUN is similar to many other computer models in that it is a collection of independently derived, interconnected submodels made compatible and linked together to be run sequentially. It has two major classes of submodels: one to predict watershed runoff, and the other to predict watershed sediment yield. These key submodels will be briefly described in order to show the manner in which sediment loading predictions are sensitive to land use and management decisions.

Hydrology Submodels

Watershed hydrologic response has a dominant influence on sediment yield. LANDRUN calculates the following hydrologic budget for each time period simulated:

$$Q = P - I - DS \quad (1)$$

where Q = runoff
P = rainfall
I = infiltration
DS = available depression storage

Infiltration is the most complex, yet most important, variable calculated by the program and the following derivative of Holtan's empirical formula is utilized[9,10]:

$$f = a(s-x)^b + f_c \tag{2}$$

where f = infiltration rate
 s = maximum available storage
 x = soil moisture
 b = constant, found to be 1.4 by Holton
 a = coefficient related to how fast available storage can be filled
 f_c = limiting infiltration rate

The "a" factor varies with land use and season, being proportional to the percent of ground surface occupied by roots and stems.[9] The values we used for this factor are delineated in Table IV.

The predictions of this model are highly sensitive to the input values of f_c, the limiting infiltration rate. Ranges for this factor were developed by Musgrave[11] and have been made applicable to soils throughout the country by their relationship to the soil hydrologic group classification system by the SCS.[12] More theoretically precise models to describe infiltration have been proposed.[13] These, however, require the input of a saturation permeability value for at least one soil horizon, and an extensive review of the literature confirmed our belief that while this factor is known to vary widely between different soil types and with land use, its variability cannot yet be adequately quantified.

LANDRUN predictions were also found to be sensitive to the depression storage values used. During dry periods depression storage capacity is renewed by evaporation and infiltration. This is the factor most responsible for differences in runoff estimates from different land uses. Again, the literature provides little precise information on what the values of depression storage should be. The base values we have used and the sources from which they have been derived are listed in Table V.

Sediment Yield Submodels

LANDRUN generates watershed soil loss estimates through a modified use of the USLE[17]:

$$SL = E(LS)KCP(DR) \tag{3}$$

where SL = soil loss (in ton/ac) per simulation period
 E = rainfall/runoff energy factor
 LS = length/slope factor
 K = soil erodibility factor
 C = cropping management factor
 P = conservation practice factor
 DR = delivery ratio factors

Through the rainfall/runoff energy factor (E), the program addresses the fact that both rainfall and runoff can initiate erosion. That is:

Table IV. Seasonally varying "a" factor for Holtan's infiltration model. ($hr^{-1}\ cm^{-1}$)

Crop/Land Type	May	June	July	August	September	October
1st-yr Corn	0.08	0.08	0.08	0.08	0.08	0.08
2nd-yr Corn and Continuous Corn	0.03	0.03	0.04	0.04	0.05	0.05
Oats, Alfalfa Undersown	0.04	0.06	0.08	0.10	0.11	0.12
Oats	0.04	0.06	0.08	0.04	0.04	0.04
Alfalfa	0.13	0.14	0.14	0.15	0.15	0.15
1st-yr Corn (Minimum Tillage)	0.11	0.11	0.11	0.12	0.12	0.12
2nd-yr Corn (Minimum Tillage)	0.08	0.09	0.09	0.10	0.10	0.10
Pasture, Roads and Buildings	0.20	0.20	0.20	0.20	0.20	0.20
Woodlands	0.27	0.27	0.27	0.27	0.27	0.27
Feedlots	0.03	0.03	0.03	0.03	0.03	0.03

Table V. Selected input parameters for hydrology submodels in LANDRUN.

Field	Limiting Permeability[a] (cm/hr)	Depression Storage[b] (cm)			
		Other	Corn	Oats	Alfalfa
1	0.6		1.24	0.43	0.16
2	0.6		0.10	0.06	0.03
3	0.6		2.36	1.36	0.44
4	0.6		2.01	1.01	0.26
5	0.6		2.31	1.31	0.41
6	0.6		2.09	1.09	0.30
7	0.2		2.07	1.07	0.29
Upland Pasture	0.6	0.19			
Lowland Pasture	0.1	0.32			
Woodlands	0.6	1.16			
Roads and Buildings	0.6	0.25			
Feedlots	0.6	1.36			

[a]Based on soil hydrologic class.
[b]Slope-dependent values adapted from Musgrave,[14] Musgrave and Norton,[15] Doty and Wiersma.[16]

$$E = RA + RU \quad (4)$$

where E = rainfall/runoff energy factor
 RA = rainfall energy factor
 RU = runoff energy factor

The rainfall energy factor (RA) is derived directly from Wischmeier,[18] who found an empirical relationship between rainfall erosivity during a storm and the product of the total kinetic energy of the storm and the maximum 30-min rainfall intensity. The runoff energy factor (RU) ties the sediment yield submodels of LANDRUN to the hydrology submodels. The factor utilized is derived from Foster's[19] modification of William's[20] original formulation. Runoff-generated erosion is considered proportional to the product of storm runoff volume and peak runoff rate raised to the one-third power.

The length/slope (LS) and soil erodibility (K) factors were derived from the most sensitive factors in the soil loss equation. The cropping management (C) and conservation practice (P) factors are where sensitivity to land use is established; this provides the basic linkage between the economic and sediment yield models. To enhance the reliability of our simulations, seasonally varying, crop-specific C factors were incorporated. These are illustrated in Table VI.

Two delivery ratio factors are considered in the simulation model.

$$DR = DR_a DR_b \quad (5)$$

Table VI. Seasonally varying "C" factors for sediment submodel in LANDRUN[a].

Crop/Land Type	May	June	July	August	September	October
1st-yr Corn	0.15	0.27	0.15	0.09	0.12	0.14
2nd-yr Corn and Continuous Corn	0.34	0.47	0.33	0.18	0.20	0.24
Oats, Alfalfa Undersown	0.50	0.30	0.06	0.03	0.03	0.02
Oats	0.45	0.28	0.11	0.17	0.17	0.17
Hay and Woodlands	0.02	0.02	0.02	0.02	0.02	0.02
1st-yr Corn (Minimum Tillage)	0.11	0.07	0.07	0.06	0.10	0.14
2nd-yr Corn (Minimum Tillage)	0.22	0.23	0.23	0.11	0.17	0.22
Pasture, Roads and Buildings	0.03	0.02	0.03	0.03	0.03	0.03
Feedlots	1.00	1.00	1.00	1.00	1.00	1.00

[a] Adapted from Wischmeier and Smith.[17]

The first (DR_a) assumes that the amount of gross soil loss delivered is related to the fraction of rainfall that infiltrates as follows:

$$DR_a = 1 - \frac{I}{P} \qquad (6)$$

where I = infiltration and P = rainfall. The second factor (DR_b) must be input, and should be derived from monitoring data. This attempts to account for the other "unknowns" in the process.

The LANDRUN model was used to predict total sediment loads during the growing season (May through October) from the Kewaskum-North watershed under the land use and management configurations generated by the economic model. To predict sediment yield via LANDRUN, the watershed must be aggregated into physiographically similar areas based on land use, soil and slope characteristics. The subdivision format utilized is illustrated in Figure 4. The breakdown of cropland acreage by specific crop types within each field was based either on actual land use data or on values provided by the economic model.

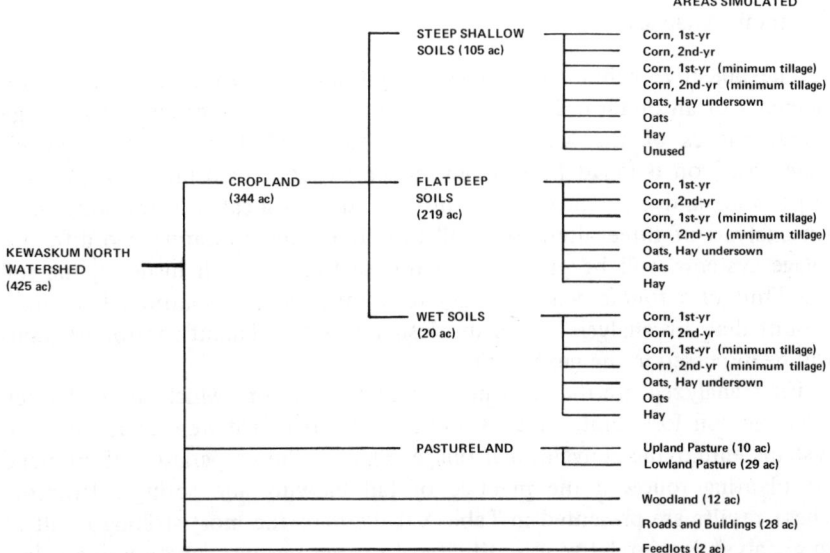

Figure 4. Aggregation of Kewaskum-North watershed for LANDRUN simulations.

Monitoring data from 1977 storm events and detailed land use information for that year were available for calibrating the model. Good agreement with the hydrology measurements was achieved by increasing the limiting soil permeabilities uniformly by a factor of 1.3 and by adjusting depression storage values from 1.4 to 0.6 times from May through October. The sediment delivery ratio factor (DR_b) of 10% yielded best correspondence between predicted and measured values, once the hydrology calibration was established.

To capture the variability in sediment yield expected from natural fluctuations in rainfall, while remaining within computer budget constraints, four periods with differeng levels of total rainfall erosivity were selected for simulation. Figure 5 shows a plot of the probability distribution of the May through October rainfall energy factors calculated from 27 years of hourly precipitation data from Hartford, Wisconsin (approximately 10 miles south of the Kewaskum-North watershed). The years encircled indicate the 4 representative years chosen for simulation, the results of which will be presented in a similar fashion, with the yield prediction on the abscissa and its probability distribution on the ordinate. The latter is assumed to be the same as the probability distribution of each simulated year's rainfall energy.

THE ANALYSIS

Economic Analysis

The objective behind the foregoing synthesis was to capture the critical elements of an on-farm decision-making process and the essential hydrology characteristics of the Kewaskum-North watershed. Both of these models were based on detailed information collected from field interviews, on site inspections and on data from monitoring stations located in the watershed. An analysis of three alternative soil loss limits, incorporating two different tillage systems, will be presented in this section. In each instance, the soil loss limit of 5 ton/ac will be considered as the baseline solution. The implications that this analysis has for the design of NPS sediment control programs will be discussed in the next section.

First analyzed are the farm management conditions which resulted when tolerable soil loss limits of 5, 3 and 2 ton/ac per field were established for systems employing conventional tillage. That is, the preparation of cropland for planting followed the practice of fall plowing and spring cultivation. These results are presented in Table VII. Perhaps the most striking result of this analysis is the relative insensitivity of the net revenue function to a reduction of allowable soil loss limits. This stems from the opportunity to substitute soil-conserving rotations and pasture while at the same time maintaining approximately the same number of milking cows.

This process of rotation adjustment results in the decline of two activities. Hay purchases decline as the tolerable soil loss decreases, the result of a move to more alfalfa-intensive rotations. Corn sales also decline, which is a result

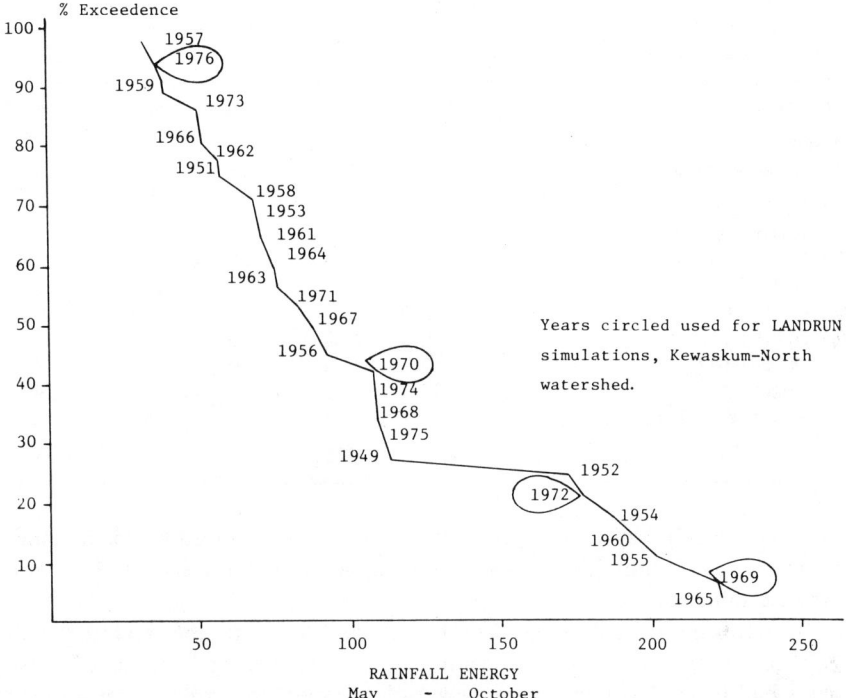

Figure 5. Probability distribution of rainfall energy at Hartford, Wisconsin, May through October period, 1949-1976. Rainfall energy = (EI)/100.[17] Exceedence: % of May-October period rainfall energy equal to or greater than the value shown.

of the high soil loss coefficients associated with the more corn-intensive rotations. This is a predictable trend given the average soil loss figure presented in Table I.

The second model assumed the availability of chisel plowing as an alternative for preparing cornland. As a tillage technique it dominated the existing set of conventional practices as evidenced by a higher value for the objective function for each level of the soil loss constraint. The fundamental reason behind this clear preference for minimum tillage is that it expands the set of feasible rotations by reducing soil loss. That is, more farming options become possible while still meeting the tolerable soil loss constraints. This is clear when comparing the acreages devoted to corn-intensive rotations under conventional tillage with those under minimum tillage. It was also possible

Table VII. Revenue, livestock and land use implications under conventional tillage and minimum tillage technology.

Model Soil Loss Constraint (ton/ac/field)	Conventional Tillage			Minimum Tillage		
	2	3	5	2	3	5
Net Revenue	89,512	90,209	90,772	90,913	91,392	91,904
Livestock:						
Number of Cows	121	120	119	125	124	124
Ration Fed	A2	A2	A2	A2	A2	A2
Land Utilization in Acres:						
Continuous Corn	21	21	23	57	63	70
O-HHH-C	148	58	21	274	274	274
O-HHH-CC	156	253	297	0	0	0
OO-HHH-C	0	0	0	0	0	0
OO-HHH-CC	0	0	0	0	0	0
Pasture	58	51	42	52	46	39
Roads, Woodlands, etc.	42	42	42	42	42	42
Watershed Total	425	425	425	425	425	425

to meet the soil loss constraints with minimal acres diverted to pastureland. At the 5-ton/ac limit under chisel plowing no cropland needed to be taken out of production.

By referring to Table I it is clear that fields 1 and 2 are the greatest potential contributors to sediment loadings in the watershed. Portions of these fields had to be devoted to pasture in order to meet the soil loss constraints. Even at the 5-ton/ac limit, 3 ac of pasture were necessary under conventional tillage. In view of the potentially high contributions from these sources, a terracing alternative was introduced into the conventional tillage system. The terraces were priced at $260.00/ac.

Two financial scenarios were utilized to analyze the terracing option. The first assumed that the terraces would be fully financed by a 5-yr loan at an interest rate of 9 1/2%. At a soil loss limit of 2 ton/ac the loan annuity (approximately $67.00/yr) far exceeded the shadow prices associated with the soil loss constraint for both fields. Consequently, terracing did not enter the LP solution.

The second alternative assumed a 75% cost-sharing level in conjunction with a 5-yr loan at a subsidized interest rate of 6 5/8%. The loan annuity in this situation is approximately $16.00/yr. A program of this nature enabled both fields to be fully cropped at the 3-ton and 5-ton limits, with 23 ac and 6 ac of terracing being constructed, respectively. Furthermore, the value of the objective function exceeded that for the conventional tillage system without the terracing option. However, it was not as remunerative as the minimum tillage system.

Sediment Loading Analysis

The numerical results of the May to October LANDRUN simulations are shown in Table VIII. The simulations of watershed loadings under conventional and minimum tillage systems are depicted in Figure 6. The most striking observation is the wide variability in predicted loadings due to natural differences in rainfall characteristics. There is a difference of at least two orders of magnitude in growing-season sediment yields under all systems analyzed. These results indicate that about 75% of the long-term watershed sediment yield is attributable to less than 20% of the seasons.

Table VIII. Predicted sediment yield from Kewaskum-North watershed.

System	Period	Sediment Yield Soil Loss Constraint (ton/ac)		
		2	3	5
Conventional	May-October 1969	53.9[a]	68.5	83.0
	May-October 1972	11.9	14.7	17.3
	May-October 1970	3.0	3.7	4.8
	May-October 1976	0.2	0.2	0.2
Minimum Tillage	May-October 1969	41.9	45.4	48.9
	May-October 1972	9.0	9.7	10.3
	May-October 1970	2.8	2.9	3.1
	May-October 1976	0.2	0.2	0.2
Conventional with Terraces	May-October 1969	50.5	68.3	83.2
	May-October 1972	10.8	14.2	17.5
	May-October 1970	1.8	2.6	4.7
	May-October 1976	0.2	0.2	0.2

[a]Tons per May-October period.

Management variances do, however, result in significantly different predicted loadings. The loading rate curve under the 5-ton/ac constraint with conventional tillage (curve C5 in Figure 6) is considered as the basis for comparison. Except for the extremely low rainfall periods (such as 1976), sediment yield reductions of approximately 30 to 60% were achieved, depending on the land use configuration, the management practice, and the period's rainfall energy characteristics.

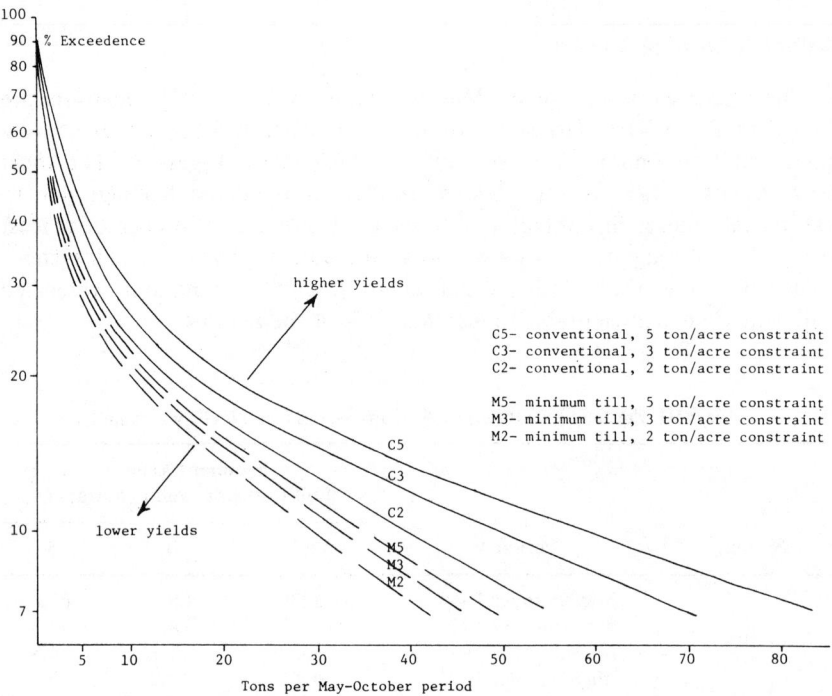

Figure 6. Predicted sediment yields from Kewaskum-North watershed. (Exceedence: % of May-October periods expected to have sediment yields equal to or greater than indicated value.

The economic impact of constraining three management systems (conventional tillage, minimum tillage and conventional tillage with a terracing option) was presented above. Now we turn to the predicted differences in the watershed sediment yield response corresponding to each of these policies.

Lowering the soil loss constraint from 5 to 2 ton/ac under a conventional tillage system results in a 30-40% loading reduction. This effects both low and high rainfall periods in a similar fashion. Meeting the 2-ton/ac constraint utilizing the minimum tillage option results in sediment yield reductions of 40-50%. In this instance, the percentage reduction is greater during the more intense, high-sediment-generating periods (50% reduction in loadings for 1969, vs 42% reduction for 1970). This contrasts with the results obtained when the 2-ton/ac constraint is met through the use of terraces in conjunction with conventional tillage. Under these circumstances, the yield reduction during the high rainfall years is about 40% (1969 and 1972 simulations), while for the moderate rainfall years (1970) simulation) loadings drop by over 60%. These response differences between the minimum tillage and terrace systems are considered to be due to differences in their relative impacts on runoff and sediment generation. The crop management factor (C) is the most significant parameter changed when minimum tillage technology is adopted. Since soil generation increases exponentially with discharge, it is reasonable that the percent sediment yield reduction is higher during the higher rainfall periods. Depression storage capacity, on the other hand, appears to be the most significant parameter changed under the terrace system. The effect of increased depression storage capacity on reducing runoff, and thus sediment generation and delivery, is of more consequence during low to moderate rainfall periods.

CONCLUSIONS AND POLICY IMPLICATIONS

In this study we have analyzed the impact of three alternative soil loss limits on agricultural activity and sediment loadings. The watershed utilized for this research was small. However, the implications for policy-making becomes more apparent when it is recognized that 13 million acres of the agricultural land in the U.S. portion of the Great Lakes Drainage Basin are devoted to dairying and livestock enterprises.

The methodology employed in this study of the control of sediment loadings from agricultural land has a dual interpretation. First, the soil loss limits may be considered typical of a regulatory approach. On the other hand, a set of BMPs associated with each soil loss limit is also implied. That is, this research highlights the tillage alternatives, the rotation options, and the terracing options available to reduce on-farm soil loss.

The major policy implication is that the NPS planning process should adopt an inward-looking approach to management. By focusing on small, yet representative watersheds, the complexities of the adjustment process will be more fully appreciated. This perspective will enable agencies to tailor management plans according to site-specific needs.

Cost-sharing and low-interest loans were shown to dampen the financial burdens of a regulatory approach. However, if a voluntary approach is relied on to induce the use of conservation technology, program emphasis must include more intensive educational and technical assistance. Furthermore, these efforts should focus on conservation techniques which offer private benefits to the farmer in addition to reducing sediment.

This study has demonstrated the superiority of a minimum tillage technique from both an economic and a sediment loading perspective. It also has the added attraction, over the construction of terraces, for example, of offering farmers more flexibility to deal with seasonal price variations since more land can be diverted to cropping. Furthermore, it is a conservation technique that does not depend upon the commitment of large sums of money.

The frequency distribution of mass watershed sediment loadings has been proposed as an indicator of the water quality implications of alternative cropping and management systems. To achieve water quality objectives in a given watershed, both the expected variability in rainfall characteristics of the area and the specific needs of the watershed's drainageways and its receiving waterbodies must be considered when selecting management strategies. Streams, for example, might best be protected with terraces, which would provide excellent loading reductions during regularly occurring, low to moderate rainfall years. Lakes, due to their large storage capacity, might be expected to be better protected by the adoption of minimum tillage technology due to its effectiveness in reducing loadings during the less frequent, high rainfall years, which have the greatest long-term sediment transport potential.

In conclusion, we have highlighted a number of relevant technical and economic aspects of the NPS sediment problem which are pertinent to agriculture. It is an extremely complex and partially understood problem. Furthermore, the response of environmental systems to control measures will be uncertain, especially in light of the natural variability involved. We would like to conclude by recommending a flexible approach, one that does not involve the commitment of large financial resources and one that forecloses as few options as possible at the farm level.

ACKNOWLEDGMENT

Funding for this study was provided in part by the U.S. Environmental Protection Agency (Grant # G005139-01) and by the University of Wisconsin-Madison. Our thanks are extended to Thomas Whalen for his invaluable assistance on the computer.

REFERENCES

1. General Accounting Office. "Comptroller General's Report to the Congress by the Comptroller General of the United States," December 20, 1977.
2. Washington County Project Work Plan. EPA Report No. 905/9-77-001. Environmental Protection Agency, Region V, Chicago, IL (June 1975).
3. Wischmeier, W. H. "Use and Misuse of the Universal Soil Loss Equation." *J. Soil Water Conserv.* 31:5-9 (January 1976).
4. Knox, J. C., P. J. Bartlein, K. K. Hirshboeck and R. J. Muckenhirn. "The Response of Floods and Sediment Yields to Coimatic Variation and Land Use in the Upper Mississippi Valley," Institute of Environmental Studies, Report 52, University of Wisconsin-Madison (1975), p. 72.

5. Hughes, H. G., and R. N. Weigle. "Long-Range Planning Prices—1975 to 1980," University of Wisconsin-Extension, Madison (May 1976).
6. Willett, G. S., O. I. Berge and R. N. Weigle. "Cost of Farm Machinery," University of Wisconsin-Extension, Madison (August 1972).
7. Novotny, V., and M. A. Chin. "Nonpoint Overland Pollution Transport Model, "LANDRUN" Users Manual," Department of Civil Engineering, Marquette University, Milwaukee, WI (1976).
8. Skopp, J., and T. C. Daniel. "A Review of Sediment Predictive Techniques as Viewed from the Perspective of Nonpoint Pollution Management," *Environ. Management* 2(1):39-53 (1977).
9. Holtan, H. N. "A Concept for Infiltration Estimates in Watershed Engineering," USDA—ARS 41-51, Washington, DC (1961).
10. Holtan, H. N., C. B. England and W. H. Allen, Jr. "Hydrologic Capacities of Soils in Watershed Engineering," in: *Proceedings, International Hydrology Symposium*, Volume 1, Fort Collins, CO (1967).
11. Musgrave, G. W. "How Much of the Rain Enters the Soil?" in *Water USDA Yearbook* (1955), pp. 151-159.
12. Soil Conservation Service. *Hydrology,* USDA-SCS National Enginering Handbook, Section 4 (1964), pp. 7.1-7.28.
13. Philips, J. R. "An Infiltration Equation with Physical Significance," *Soil Sci.* 77:1553-157 (1954).
14. Musgrave, G. W., and H. N. Holtan. "Infiltration," in *Handbook of Applied Hydrology*, V. T. Chow, Ed., Section 12, II ed. (New York: McGraw-Hill, Inc., 1977), pp. 12-25.
15. Musgrave, G. W., and R. A. Norton. "Soil and Water Conservation Investigations at. the Soil Conservation Experiment Station, Missouri Valley Loess Region," Sta. Prog. Report 1931-1935 and USDA Tech. Bulletin 558 (1937), pp. 58-60.
16. Doty, C. W., and J. L. Wiersma. "Geometric Shaping and Contouring of Land as Related to Potential for Surface-Water Storage," *Trans. ASAE* 12(3):322-328 (1969).
17. Wischmeier, W. H., and D. D. Smith. "Predicting Rainfall-Erosion Losses from Cropland East of the Rocky Mountains," ARS-Ag Handbook No. 282, USDA (1965).
18. Wischmeier, W. H. "Rainfall Energy and Its Relationship to Soil Loss," *Amer. Geophys. Union Trans.* 39:285-291 (1958).
19. Foster, G. R., L. D. Meyer and C. A. Onstad. "Erosion Equations Derived from Modeling Principles," Paper Number 73-2550, presented at winter ASAE meeting, Chicago, IL, December 11-14, 1973.
20. Williams, J. R. "Sediment-Yield Prediction with Universal Equation Using Runoff Energy Factor," Poceedings, Sediment-Yield Workshop, USDA Sed. Lab., Oxford, MI, November 28-30, 1972; and in "Present and Prospective Technology for Predicting Sediment Yields and Sources," ARS-S-40 (1972).

32

INSTITUTIONAL AND TECHNICAL ASPECTS OF THE DEVELOPMENT OF AGRICULTURAL BMPs IN A FIVE-COUNTY RURAL/URBAN MICHIGAN REGION

J. P. Jones, J. C. Sutherland
Williams & Works, Inc.
Grand Rapids, Michigan

Some factors which led to the completed Areawide Waste Management Plan for the Southcentral Michigan Planning Area, and especially those factors which led to development of BMPs for rural and agricultural runoff control, will be presented in this paper. Comments will also be included on the overall 208 approach to water quality management.

Michigan Region 3 includes five counties: Kalamazoo, Barry, Calhoun, Branch and St. Joseph. The larger cities are Kalamazoo, Battle Creek and Portage. There are 15 to 20 smaller municipalities in the region (Figure 1). Three major drainage basins—of the Kalamazoo, St. Joseph and Grand Rivers—are included (Figure 2).

With approval of an $800,000 grant to the Southcentral Michigan Planning Council in June 1975, the planning pattern was drawn up for the following 2.5-year period. Extensive public involvement was projected, and public involvement became a strong input and supportive factor in developing the plan. The major groups involved besides the policy boards are politicians and other elected officials; technicians such as engineers, planners, geologists, limnologists and biologists; and educators, lake association members, students, farmers, and citizens-at-large. These groups are supported by official agencies close to the area such as the Michigan Department of Natural Resources and the Soil Conservation Districts.

Different main objectives were held by the different participants. Some were principally interested in water quality data, while others were more

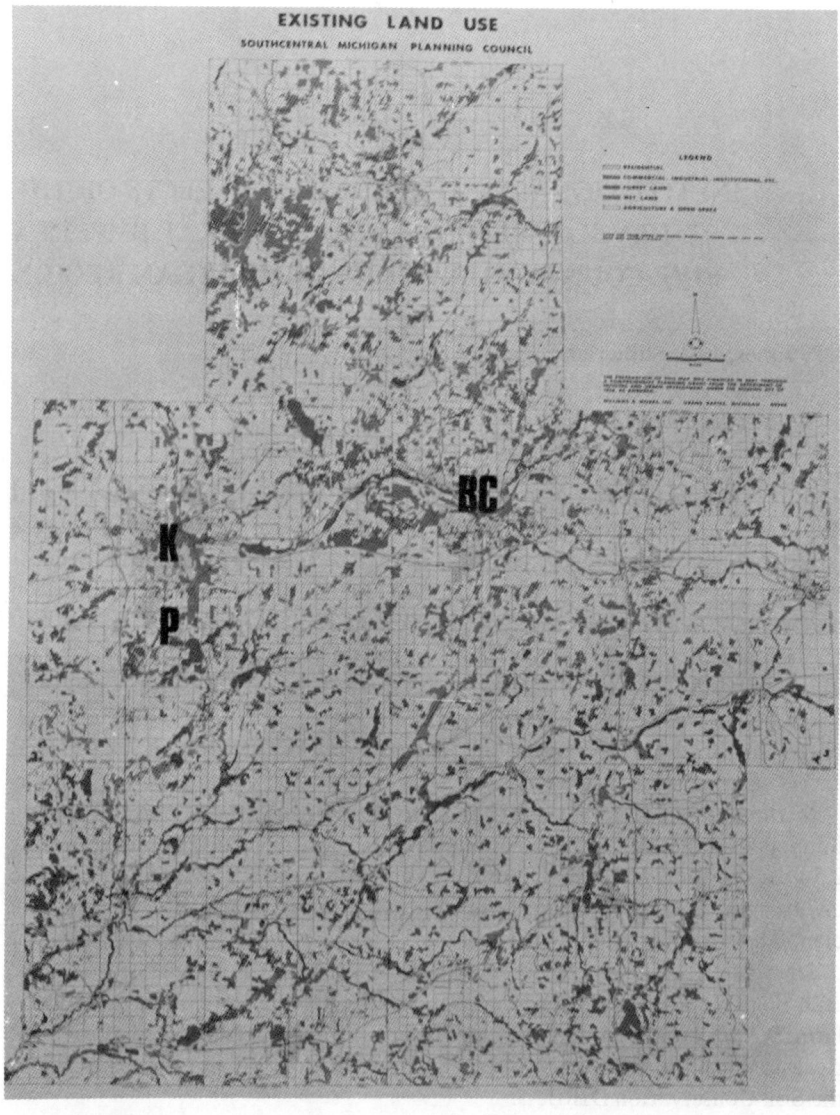

Figure 1. Existing land use; southcentral Michigan planning council.

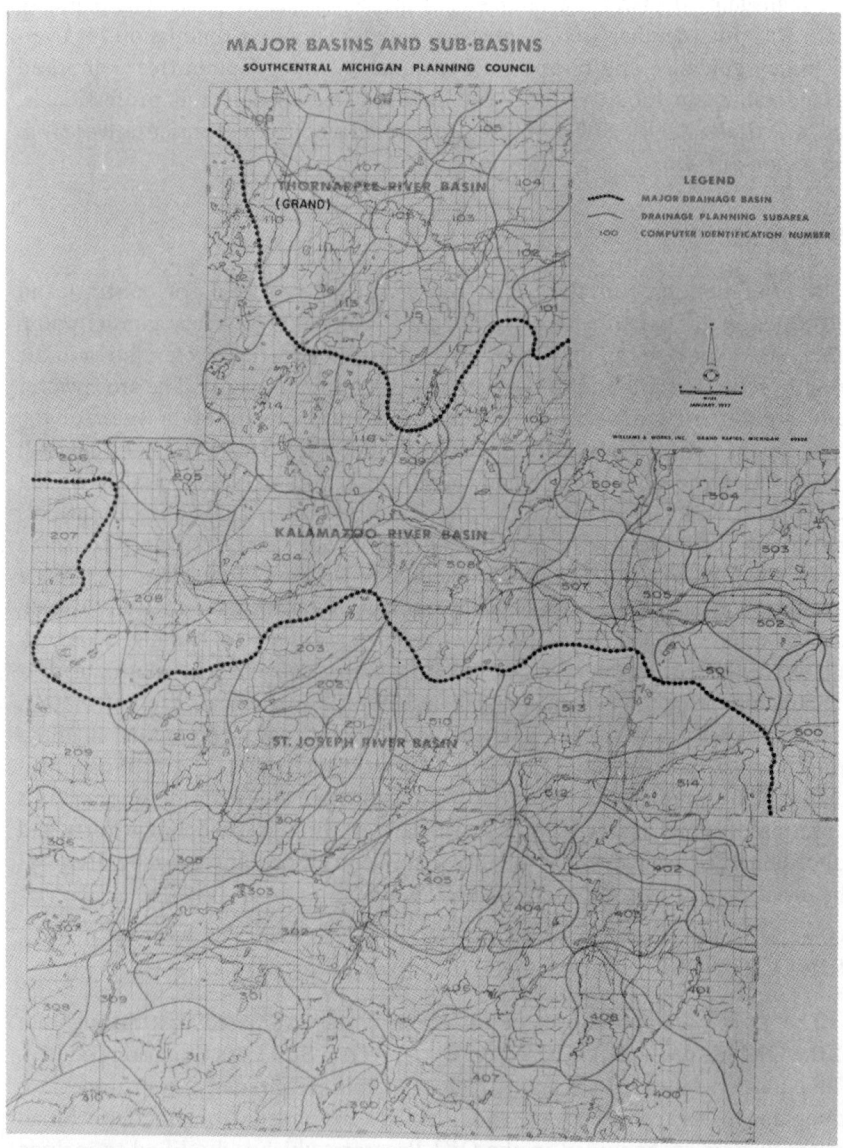

Figure 2. Major basins and sub-basins; southcentral Michigan planning council.

458 BMPs FOR AGRICULTURE AND SILVICULTURE

interested in political impact. A small core was concerned about the true management aspects of the program. As consultants, we recognized early the tendency to overemphasize the water quality data aspects, and it became our task to weld the many groups into a single-purpose body to develop an effective management plan for attainment of the water quality goals. The planning team was put together first by creating a Water Quality Commission for overall policy guidance and control, a Technical Advisory Committee comprised of federal, state, county, local and industry technicians and professionals, and a Citizens Advisory Committee composed of citizen representatives from the region at large.

URBAN RUNOFF SUMMARY

A brief summary of the urban runoff setting is useful for contrast and introduction to the subject of rural runoff control. The urban runoff pollutants are expected to be significant contributors to accelerated aquatic growth in the streams and in the downstream impoundments. The stormwater runoff pollutant loads expected from an average precipitation year for the three larger cities, Kalamazoo, Portage and Battle Creek, were calculated. The Kalamazoo urban watershed significantly affects water quality in the Kalamazoo River. Urban runoff from the Portage area also directly affects the quality of water in the Kalamazoo River because of the relatively short travel time in the Portage Creek tributary. Urban runoff from the Battle Creek area affects the Battle Creek stream and the Kalamazoo River. Overall, suspended solids are high, with values ranging from 44 to 132 mg/l. The calculated values of total nitrogen are not high enough to be alarming. But biochemical oxygen demand (BOD) and phosphorus are a different matter. Compared with existing point discharge pollutant loads, the annual loadings from urban runoff are equal to about 2.3 yr of BOD and 7 yr of phosphorus loading. The predicted urban runoff contribution to stream deterioration is a matter of some concern, and it was largely through modeling efforts and subordinately through water quality data acquisition that the magnitude of the urban runoff factor came to light.

RURAL RUNOFF

The magnitude and impact of rural runoff was estimated first through use of the OKI Model which is based on the Universal Soil Loss Equation (USLE). The equation takes into account factors of rainfall intensity, soil properties, topography, vegetative cover, and present erosion control practices. Small urban clusters are included in the OKI Program which is modified to include them. In the analysis, three rural land uses were considered: agricultural, pastureland and forests. Modern mapping of the agricultural soils had not been done to completion over any large contiguous areas, but good use was made of generalized soils maps. Sediment yield and pollutant loads of course vary with the type of land use. Landsat satellite maps provided information on the spatial distribution of land uses within the area. The percentages of each land use and each soil association were estimated. The land use/soil association matrix was further subdivided according to soil types. Soil erodibility values were obtained for each soil type, and topographic factors were developed for each. As an example, for a watershed with five soil associations and three soil types in each, 15 soil erodibility values and topographic factor values were used for each kind of land use in the analysis. These factors were used in the USLE to determine soil erosion rates for each type of soil, and these results were accumulated for each drainage subbasin. Sediment delivery ratios were developed for each watershed based on the soil textures and the size of the drainage areas. These have been applied to the erosion rates to estimate sediment flows and pollutant flows for each watershed.

Overall, the model predicts no present large-scale regional pollution problem from rural nonpoint sources, under the 1977 state water quality standards. But if these standards are upgraded as is now being considered, there will be general contraventions rather than the current situation of a few local problems.

In working toward an overall management system, the Southcentral Michigan Planning Council kept the objective of maximum local control in the forefront. The agencies selected to manage the urban nonpoint source situation are the present operators of municipal sanitary wastewater treatment facilities, because these agencies presently have the needed know-how. The agencies selected to manage rural nonpoint discharge conditions are the counties, mainly because they now administer the Michigan Soil Erosion and Sedimentation Control Act which was passed in 1972.

Perhaps the most significant rural management aspect of the 208 program is the effort to include BMPs for agriculture and construction. It was in this particular effort that the Soil Conservation Districts (SCD) and the Soil Conservation Service (SCS) played a significant contributing role. At first, due to funding limitations in the separate districts, there was no mechanism to defray costs in assigning SCD personnel to the region. After much negotiation, the regional staff together with the consultants arranged for contracts with the five SCDs for technical services in the development of BMPs and to help implement the resulting plan.

Some details about development of the rural runoff management plan are worth sharing. A uniform method of assessment of the need for BMPs was developed with the help of the SCD people on an area by area, farm by farm basis. Briefly, the method consists of evaluating the kind and magnitude of runoff from each farm within 2 miles of major watercourses. The prime contribution to the data was the personal knowledge of each soil conservation officer. The data were presented in a way that concealed the identity of each farm, yet noted three important things: the type of BMP required, the cost of implementing each BMP on a unit basis, and the types of construction necessary to control agricultural runoff entering surface waters. The total predicted cost was summarized subbasin by subbasin and for each county. The projected costs range from $6 to $10 million per county.

The input of the SCDs, on a contract basis to provide this kind of technical service, is unique so far in Michigan, but it probably will be adopted in other Michigan regions until such time as SCS and the SCDs are profitably funded to carry out this cooperative function with the regional governments. We think that the personal input to the program by the people most affected—farmers and their conservation district agents—produces a realistic program of nonpoint source control as part of the regional management process.

A list of 29 BMPs was composed by the SCS at the state level for selected use in each individual farm case by the SCD people:

- pesticide application
- contour farming
- conservation cropping system
- agricultural waste management system
- contouring orchard and other fruit areas
- cover and green manure crop
- crop residue use
- critical area planting, seed
- streambank protecting, structural, vegetation
- debris basin
- diversion
- filter strips
- grade stabilization structure
- field windbreak
- grassed waterway or outlet
- minimum tillage
- livestock exclusion
- mulching highway banks
- permanent vegetative cover
- tree planting
- irrigation water management
- contour stripcropping
- stripcropping, wind
- parallel terrace
- field stripcropping
- soil testing fertility management
- drainage
- animal waste management
- critical area sodding
- no-till
- woodland management
- critical area trees and shrubs
- grass water maximum shaping

Unit costs for the BMPs range from $10 for a pesticide application, to $20,000 for a 1.6-ha (4-ac) agricultural waste management system. The U.S. Congress has recently passed legislation to provide needed additional funding to assist individual farmers in implementing the BMPs of the types proposed within the plan.

The management plan has been approved at the local level. It is before the State Environmental Review Board and is being reviewed by the Governor's Office for certification. The plan already has been used as replacement for incomplete or additional basin studies. The program and plan were completed at least 2 months ahead of schedule at a cost of nearly $200,000 below the allocated amount of $800,000.

Some observations related to the need for general improvement in the Water Quality Management Program are offered next. We are sure that none of these is unique to this region, and that most of them have been discussed by the participants at this conference. However, we think reiteration might help them not to be forgotten in the rush to get on with other things. These observations concern impediments to effective planning and implementation:

1. Planning periods established by the Environmental Protection Agency (EPA), based upon congressional statutory time limits, are unrealistic and have done considerable harm to some regional programs that were not developed within the context of those time limits. An example is the extensive sampling and monitoring undertaken in some regions in attempting to measure water quality definitively. There is generally neither time nor money to carry out such programs meaningfully.

2. Undue emphasis has been placed by federal and state technical personnel on the water quality acquisition aspects of the 208 program, as distinct from the management aspects. Reams of material are gathered on water quality, but management for the most part has been given short shrift in plans that we have reviewed. In most instances, the management plans are a listing of the status quo and some lip service to nonpoint source controls.

3. Congress and the EPA in earlier days had looked upon the 208 program as a means to encourage, if not actually to force, federal land use planning and development controls. Local officials were defensive about their perceived right to obtain control over development at the local level. Effective land use controls are not being developed in the face of this conflict. There were some gestures made toward developing appropriate controls for runoff prevention, but these were mostly left to local implementation on a voluntary basis. As a policy, this kind of process is at best a paper exercise and at worst merely grantsmanship to meet the contract requirements.

4. Disharmony between state and federal officials and technicians reviewing and guiding the development of the management plans has resulted in deterioration of interest at the regional and local levels and erosion of credibility within the technical framework of things.

5. Inexperienced personnel at the state and federal levels are forced to tackle very complex problems without adequate training and often as a result of transfer from another established state or federal agency. Many months were invested by federal and state staffing to being inexperienced and

462 BMPs FOR AGRICULTURE AND SILVICULTURE

unknowledgeable technicians to the level of expertise of regional and consultant personnel.

As a suggestion for improvement, perhaps a 4-year period should be allocated for preparation of the management plan to allow time for meaningful water quality monitoring and sampling. If the 4-year period is impossible, the state might assume the water quality data acquisition burden after the management plan is written.

SECTION VI

STATE AND WATERSHED APPROACHES

33

THE EVALUATION OF BEST MANAGEMENT PRACTICES FOR THE REDUCTION OF DIFFUSE POLLUTANTS IN AN AGRICULTURAL WATERSHED

T. H. Cahill, R. W. Pierson, Jr., B. Cohen
Resource Management Associates
West Chester, Pennsylvania

INTRODUCTION

The magnitude of diffuse pollutants from agricultural land runoff in the Great Lakes, especially Lake Erie, has been well documented in several recent studies.[1,2] With the particular problems associated with accelerated eutrophication in Lake Erie during the past 30 years, phosphorus has received the focus of attention as the most important pollutant. Based on this conclusion, a massive program has been underway since the early 1970s to reduce significantly the input of phosphorus from wastewater treatment facilities, especially the large cities bordering the lake (Figure 1), such as Detroit, Cleveland and Toledo. This investment, which will ultimately total 0.8 billion federal dollars in waste treatment facilities, will make a major dent in the problem. However, it is estimated that it will also be necessary to reduce the phosphorus inputs from tributary drainage basins (Figure 2) by 37% in order to achieve the desired water quality levels in the lake.[1] This tributary input of phosphorus from diffuse sources was estimated at 6600 MT during 1975, with most of the phosphate mass transport occurring in association with soil particles (Figure 3). It is also estimated that most of this material originates from agricultural land (Figure 4) within the Lake Erie basin, much of which is intensively cultivated, especially in the western portion.

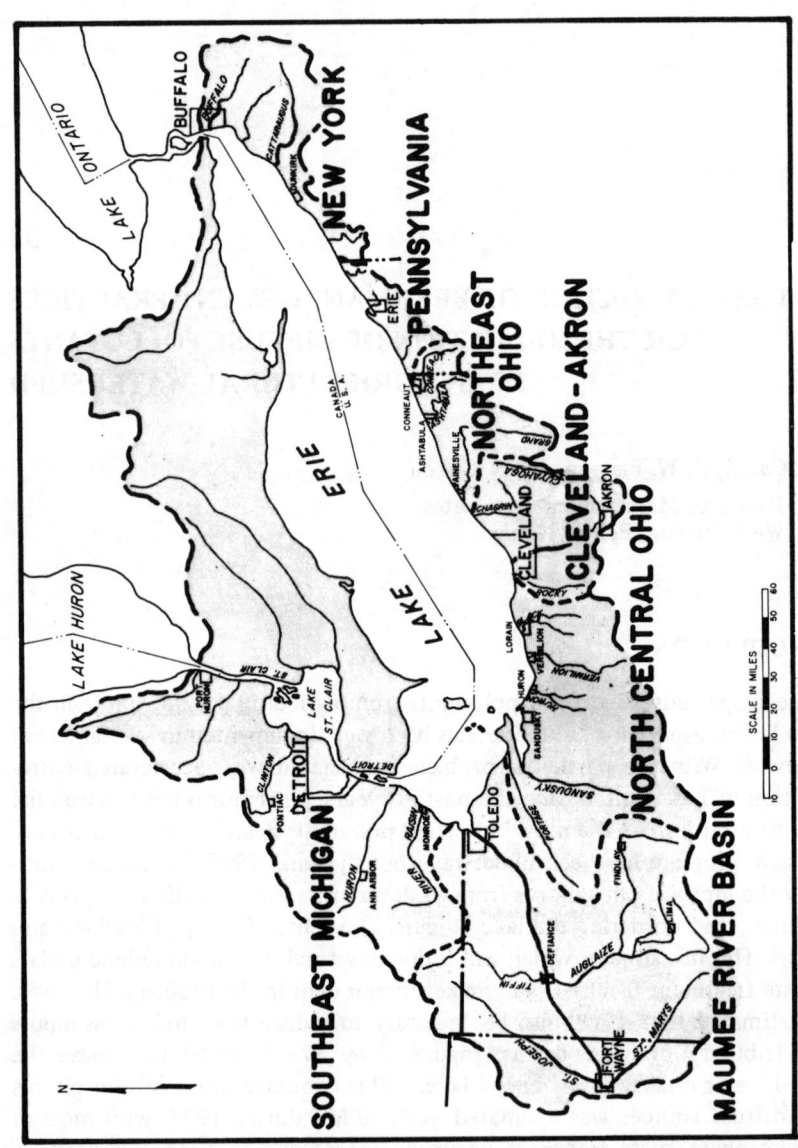

Figure 1. Lake Erie basin showing major cities and tributaries (FWPCA, 1968).

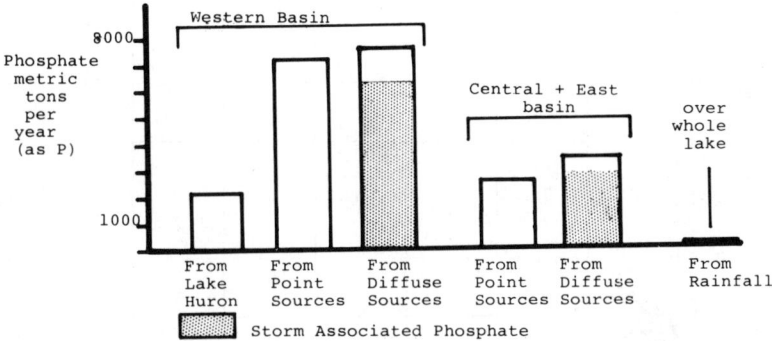
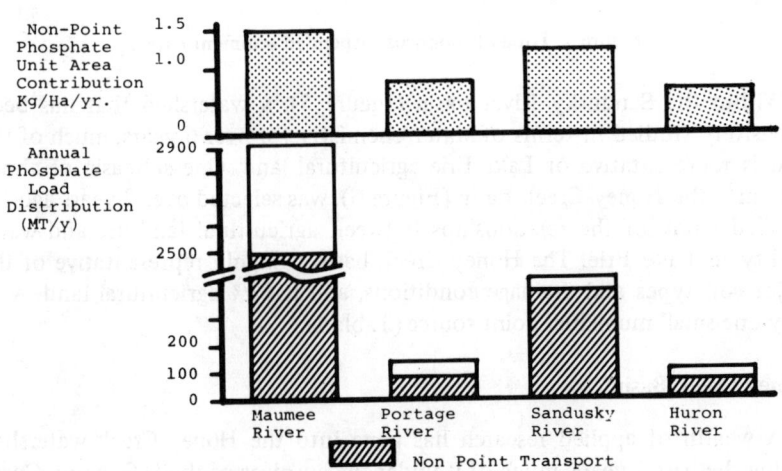

Figure 2. Phosphorus sources in Lake Erie.

While the combinations of soil characteristics, landform and cover, topography and weathering vary greatly over this 23,000 mi^2 drainage area, it can be divided into four general physiographic regions, the origins of which are found in its glacial lake history. Because of physiographic features much of the agricultural land in the western Ohio portion of the basin is flat and poorly drained but extremely productive under the necessary condition of artificial drainage.

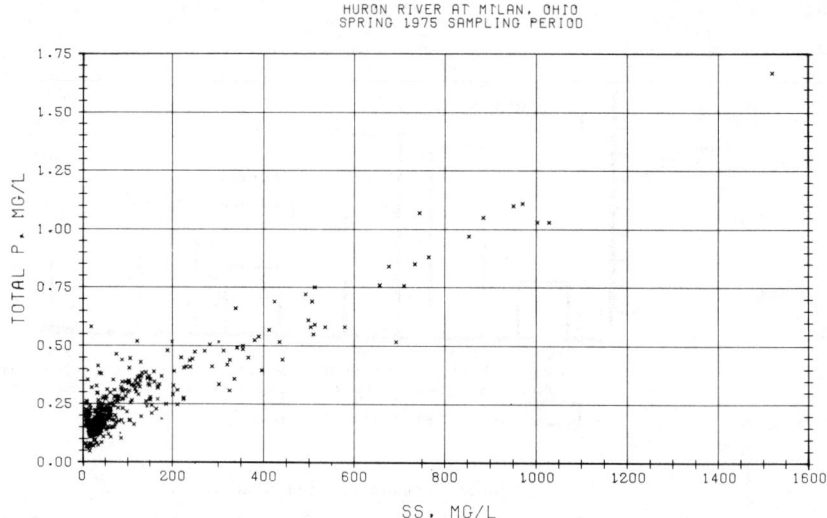

Figure 3. Total phosphorus-suspended sediment curve.

Within the Sandusky River basin (Figure 5), a watershed that has been intensively studied in terms of water chemistry for over 6 years, much of the land is representative of Lake Erie agricultural land. One subbasin of about 180 mi^2, the Honey Creek basin (Figure 6), was selected over 2 years ago for detailed study of the relationships between agricultural land use and water quality in Lake Erie. The Honey Creek basin is highly representative of the major soil types and drainage conditions, and is 83% agricultural land, with only one small municipal point source (Table I).

Honey Creek Basin

A wealth of applied research has gone into the Honey Creek watershed during the past 2 years, much of it under the auspices of the U.S. Army Corps of Engineers through the Lake Erie Wastewater Management Study, and more recently by the U.S. EPA, Athens Research Laboratory. A number of chemical sampling stations have been established in the watershed (Figure 7), but one key station at Melmore has been collecting water chemistry samples and flow measurement at 4- to 24-hr intervals since February 1976, providing an excellent continuous record of mass transport from the basin. This record shows that the 151-mi^2 drainage area above the gage discharges 38 tonne/yr (0.98 kg/ha) of total phosphorus, a value consistent with the major agricultural basins in Lake Erie, which range from 1.1 to 1.4 kg/ha/yr output. A

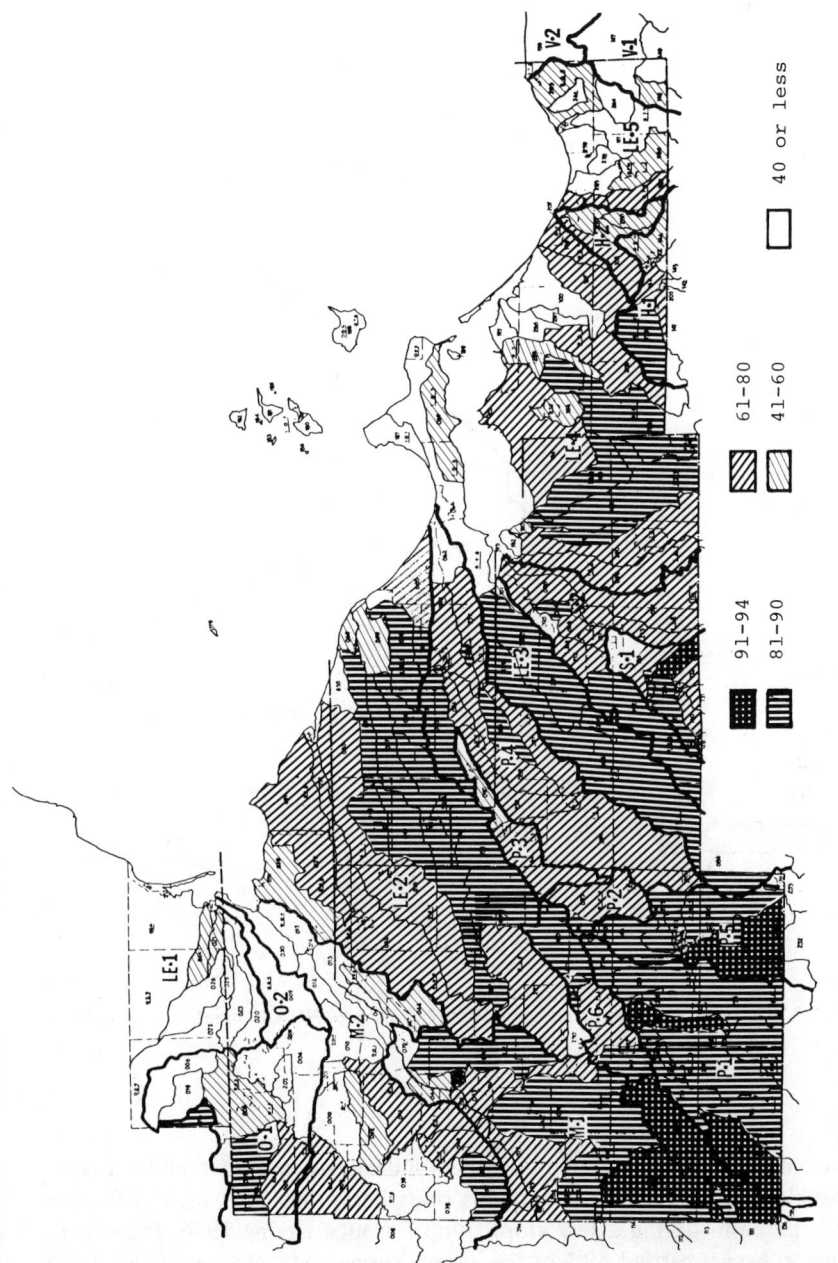

Figure 4. Agricultural land in western basin. The percent of each subbasin in cropland and disturbed cropland (LRIS) is shown.

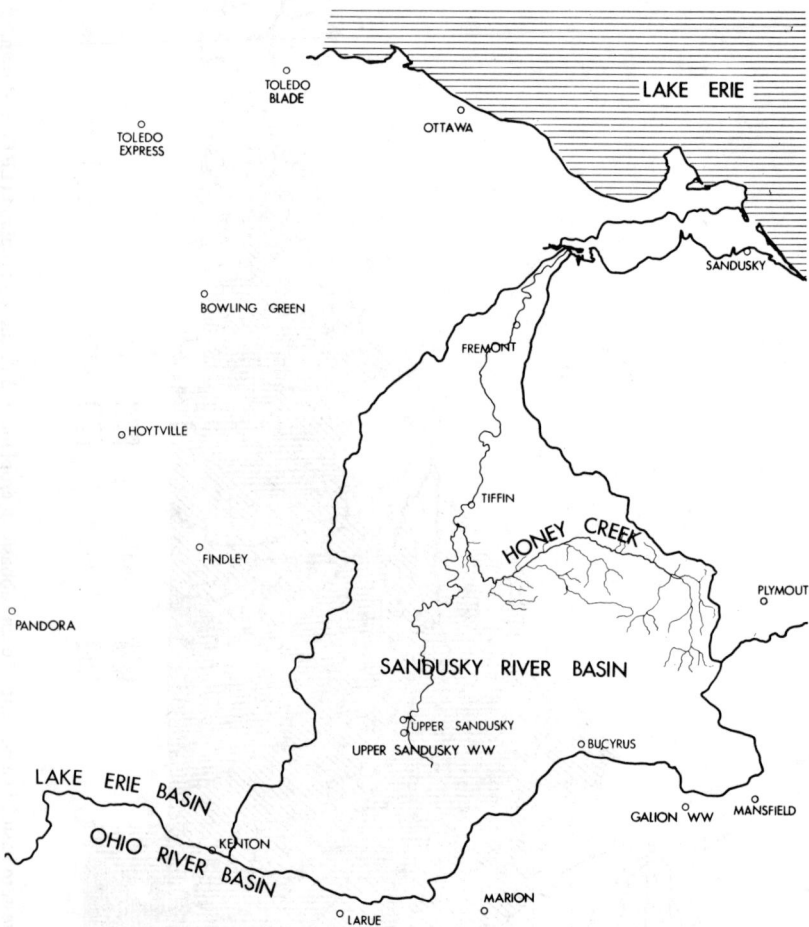

Figure 5. Sandusky River basin showing Honey Creek and NOAA climatological stations in the region.

mass balance estimate for the watershed indicates that cultivated land is responsible for a major portion of this phosphate, most of which is flushed from the basin during a few storm runoff events. During 1976, four storm transport events carried 89% of the annual suspended solids and 86% of the total phosphorus (Table II).

STATE AND WATERSHED APPROACHES 471

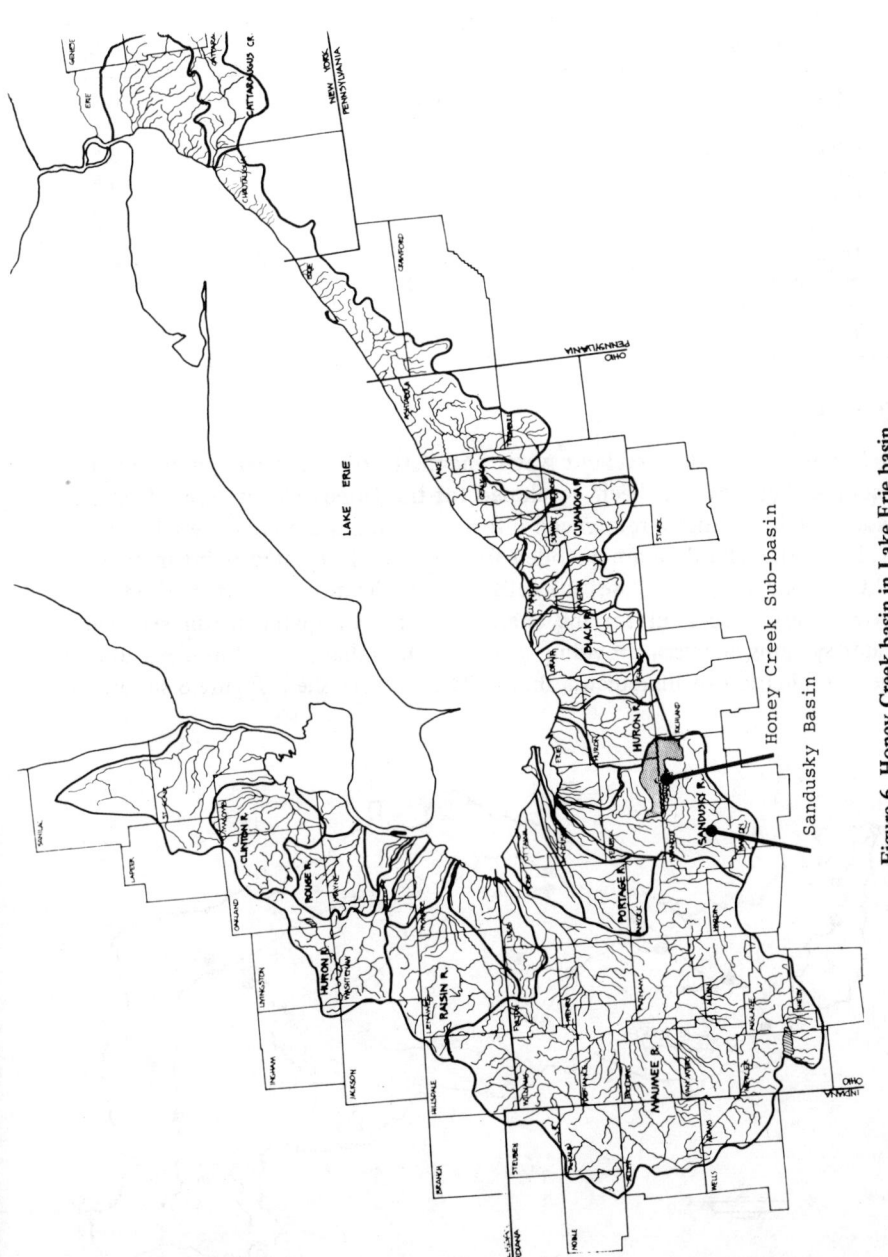

Figure 6. Honey Creek basin in Lake Erie basin.

Table I. Honey Creek basin land use above Melmore.

Land Use Category	Hectares	% of Basin
Row-Crop	19,820	50.5
Field Crop	11,992	30.5
Other Agriculture	240	0.6
Woodlot	4,096	10.5
Wetlands	840	2.2
Water	392	1.0
Urban	1,848	4.7
Missing Data	0	
Total	39,228	

Resource Data Base

In addition to the excellent water chemistry record, a computerized land resource data base has been developed for the Honey Creek basin. The data base contains spatially referenced information for those characteristics listed in Table III. The data are encoded for a randomly located point in each of 12,000 4-ha (approx. 9.88 ac) cells. The application of this method[3] has proven very successful for the characterization of spatial features in land-water systems analysis, providing a statistically reliable set of information on the combination of ingredients that exist in a watershed. Figure 8 shows the

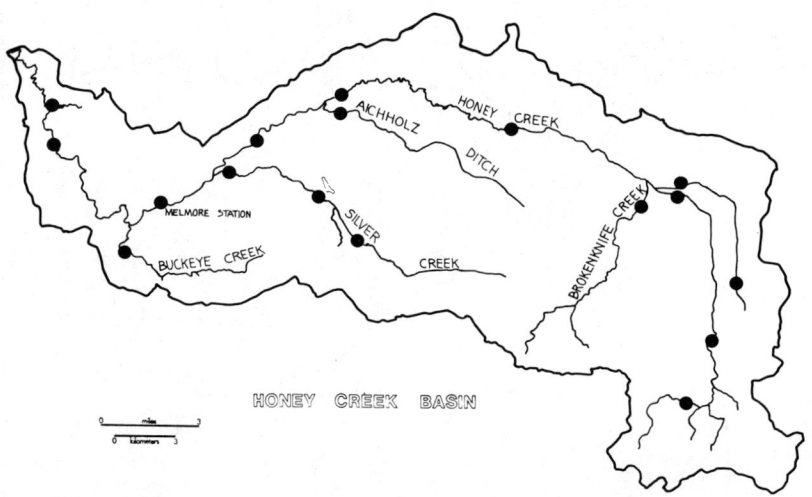

Figure 7. Chemical sampling stations in Honey Creek.

Table II. Honey Creek at Melmore comparison of mass transport for 1976.

	Late Winter Storm Period February 10 to March 10	Annual Total	%
Total Phosphorus (MT)	32.7	37.8	86%
Suspended Solids (MT)	19,233.0	20,822.0	92%

Table III. Honey Creek watershed land resource information system data files.

Land Use by both point and cell
Soil Phase by soil series and slope category
Watershed in 26 subareas
Slope Magnitude in tenths of a percent
Slope Direction by azimuth from $0°$ to $360°$
Elevation of point in tenths of foot
Physiographic Region by soil series groupings
Political Subdivision

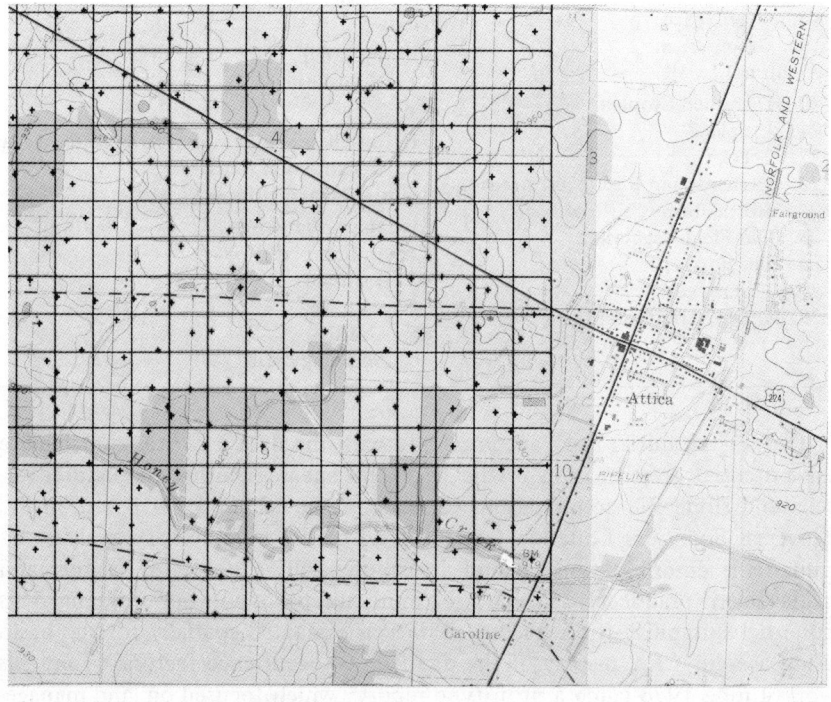

Figure 8. Example of cell-point data coding system for Honey Creek basin.

location of points and cells, as encoded, in a portion of the Honey Creek basin near Attica, superimposed on a 1:24,000 quad sheet base. For each point, some 12,000 covering the basin, the data listed in Table III are stored in files, which vary in size and complexity depending on the parameter. The soil file, for example, is taken directly from the SCS-5 data and includes complete information on the physical and chemical properties of each soil horizon, or layer. Other files, such as land cover, consist of code numbers for a particular coding scheme, as listed in Table IV. A more detailed description of the Land Resource Information System (LRIS) as it exists in the Honey Creek basin is given elsewhere.[4]

Table IV. TMACOG land cover analysis—final list of classes.

Urban
- 11. Single-Family Residential
- 12. Multiple-Family Residential
- 13. Mobile Home
- 14. Commercial and Services
- 15. Industrial
- 16. Institutional
- 17. Extractive
- 18. Open Space
- 19. Other Urban

Agricultural Land
- 20. Disrupted Cropland
- 21. Cropland
- 22. Truck Crops
- 23. Orchards and Bush-Fruit Areas
- 24. Horticulture
- 25. Old Field Vegetation
- 26. Feedlots
- 27. Farmsteads
- 28. Other Agricultural Land

Forestland
- 41. Deciduous Forest
- 42. Coniferous Forest
- 43. Mixed Forest

Water
- 51. Rivers and Streams
- 52. Lakes
- 53. Reservoirs

Wetlands
- 61. Forested
- 62. Nonforested

Barren
- 71. Beaches, Mudflats and Unvegetated Areas
- 72. Construction Activity

Transportation and Utilities
- 81. Improved Roads
- 82. Unimproved Roads
- 83. Railroads
- 84. Airports
- 85. Utilities
- 86. Shipping Ports

One of the initial tasks accomplished with the data base was to calculate the potential gross erosion in the basin, by subareas, using the land cover, soils and slope file combinations for each of the 12,000 cells, based on the Universal Soil Loss Equation (see Table IX). Other preliminary analyses included the determination of land areas potentially suitable for various land management practices, such as no-till, minimum-till or contour-cropping. This potential suitability for a selected BMP was evaluated spatially for the basin using a color CRT display of the digitized co-occurrence factors. From this work during 1976 came a prototype report[5] which focused on land management techniques and sediment runoff from cultivated lands. It was apparent that, in order to influence the local farmers in implementing these practices, several steps were necessary. First, the actual source areas, transport and impact of these pollutants had to be better defined and measured; second, the degree of success in improving water quality (while maintaining productivity) has to be proven, either on an estimated basis by modeling studies or in actual practice by demonstration.

Research Issues

There are several fundamental research questions with respect to diffuse sources that are only partially answered, with tremendous implications for the success of land management techniques to reduce soil-associated pollutants. First and foremost is the question of source areas: classic soil erosion theory as well as test plot data tell us that, all other things being equal, the more steeply sloped areas, usually in the headwaters of a basin, will contribute more erosion. On the other hand, partial area hydrologic theory[6] suggests that overland flow occurs on only a limited portion of the basin (Figure 9),

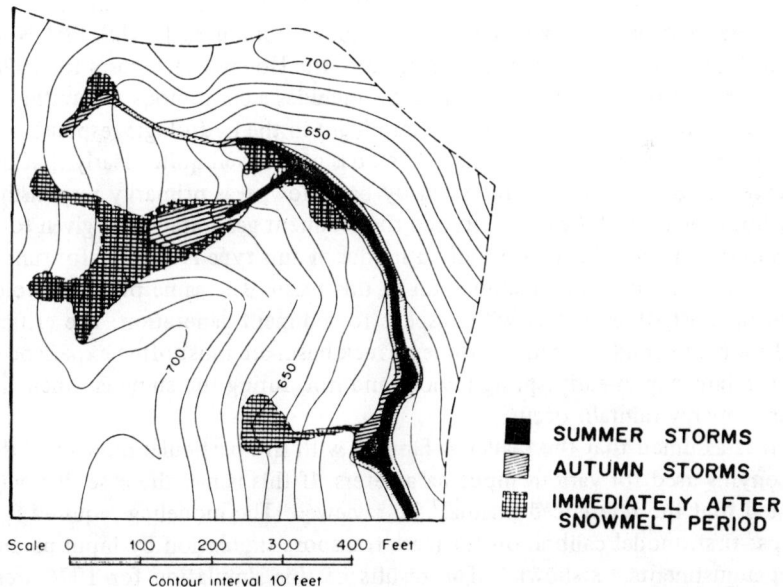

Figure 9. Partial area example. Seasonal extent of the saturated zone before summer storms, autumn storms, and immediately after the snowmelt season in a catchment with steep well-drained slopes and a narrow valley bottom. Basin WC-4, Sleepers River Watershed, Danville, Vermont.[6]

namely on those low-lying, high-water table soils contiguous to the stream channel. Alternative conclusions will lend to very different management strategies. If most of the sediment being carried from an agricultural watershed is coming from only 15 to 20% of the basin land area, and that area is the less permeable, poorly drained soil, then improved drainage and/or reduced cultivation might be more successful than those strategies which attempt to minimize rainfall erosive energy by improved cover conditions, or reduced velocity of overland flow by slope reduction.

The sediment transport question is also important. Does material take many years to move out of a watershed, with subsequent scouring and deposition in a number of temporary sinks, such as swales, ditches and stream bottoms; or does the majority of material observed in passage from a basin during a storm event originate on the land surface during that storm? This question also weighs heavily on the selection of suitable land management techniques. The time required to measure any success achieved by such alternatives as fertilizer reduction is very much dependent on the answer to this question.

NONPOINT SOURCE MODELING

In order to focus on the questions of pollutant generation and evaluate the potential for erosion reduction by various techniques, the U.S. EPA Nonpoint Source (NPS) Model[7] was calibrated to the Honey Creek basin for 1976. The major feature of this model is that it includes a continuous accounting of soil moisture (Figure 10), the key determinant in the hydrologic response of a basin. Prior research in the Honey Creek basin[8] showed quite clearly that the degree of soil moisture, as reflected by base flow, was primarily responsible for both the magnitude of runoff and the sediment generated for a given total precipitation. For this reason only a model of this type is suitable for runoff and pollutant production simulation in this basin. The same probably holds true in most other basins with respect to sediment generation. The critical soil moisture conditions in the Honey Creek basin are most often experienced in the late winter-early spring period, and not during the summer when the higher energy rainfalls occur.

It is assumed that the reader is familiar with this particular model and the acronyms used for various input parameters. If this is not the case, it is suggested that the referenced manual[7] be reviewed. The modeling required two steps: first, model calibration for mass transport simulation by input parameter adjustments, as shown.[4] The results of this simulation for 1976 were fairly good (Table V).

In order to obtain a good fit to actual suspended solids data as well as to use the model to evaluate a potential best management practice (BMP), it was found that very careful specification of land cover, fines accumulation,

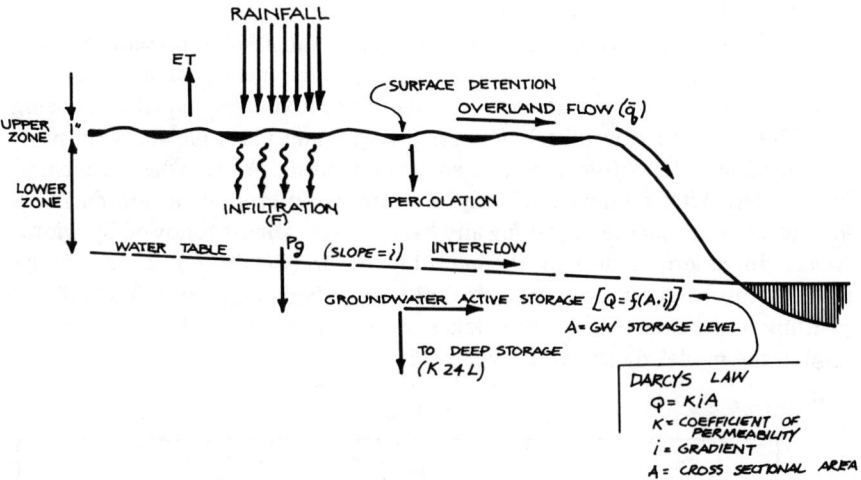

Figure 10. NPS flow schematic.

Table V. Calibration summary.

	1976 Actual Data		1976 (Run 3311) Simulated Data	
	Flow (in.)	Suspended Solids (tons)	Flow (in.)	Suspended Solids (tons)
January	--	--	--	--
February	4.39	16,965.0	4.05	16,806.0
March	1.69	3,115.0	2.19	3,009.0
April	0.29	55.0	0.42	428.0
May	0.18	155.0	0.57	499.0
June	0.18	194.0	0.39	106.0
July	0.12	166.0	0.28	58.3
August	0.08	62.0	0.23	11.8 (urban imp.)
September	0.05	30.0	0.20	11.6 (urban imp.)
October	0.04	4.0	0.12	13.6 (urban imp.)
November	0.04	2.0	0.05	8.2 (urban imp.)
December	0.02	1.0	0.18	12.2 (urban imp.)
Total	7.08	20,749.0	8.68	20,963.7

fines removal and initial fines pool were critical. As calibrated, the output is supply-limited during the high-water table period of the year and overland flow limited during most of the growing season.

The NPS model calibrated to Honey Creek for 1976 represents what is hereafter referred to as conventional (cropland) management. In Honey Creek, conventional management consists of approximately equal areas being fall (October and November) and spring (April and May) plowed with prior crop residues left, before corn and soybean planting. Winter wheat is planted in October. After its harvest in July, a winter cover crop of clover or alalfa is grown. Oats are planted in spring and harvested in August followed by winter wheat. In determination of the monthly cover (COVVEC) values for the model, crop rotation, crop sequence, plowing schedules, crop residue effects, planting and harvest dates were taken into consideration. The cover values used in the model are shown in Figure 11.

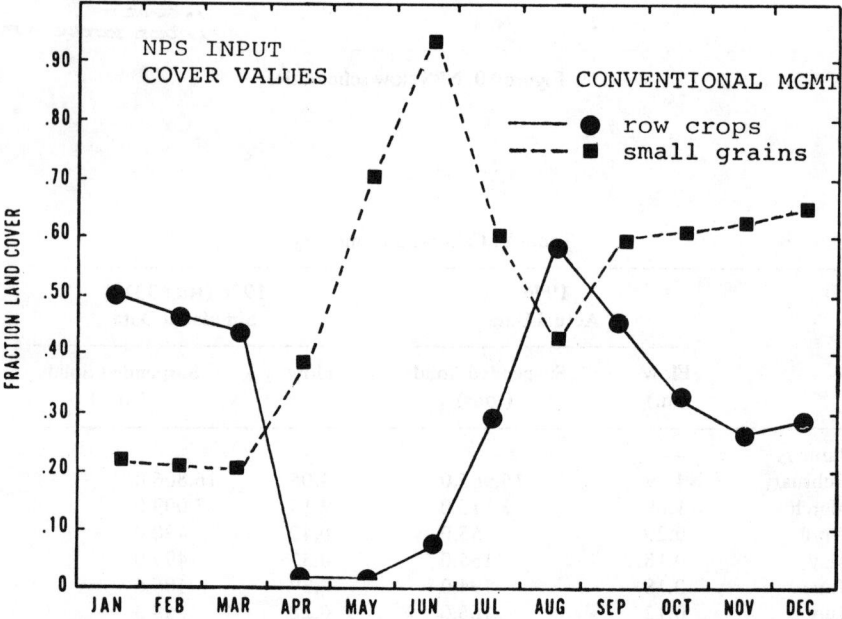

Figure 11. NPS input cover (COVVEC) values for conventional management of row-crops and small grains in Honey Creek watershed.

The second step was to overcome the "scale" problem associated with the model, which was originally designed for and calibrated on small catchments (< 10 ha). The application of such a model to a basin the size of Honey Creek required the development of a routing subroutine. Such a program, based on channel geometry and basin topography, routed simulated hydrographs by kinematic wave velocity and chemographs by fluid velocity for each of the 14 subbasins (Figure 12).

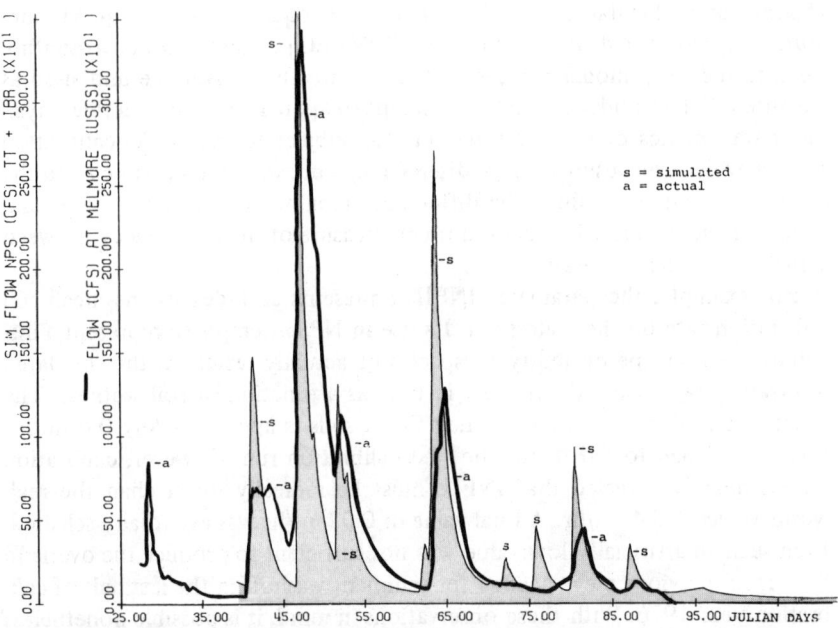

Figure 12. Simulated flow of February-March 1976 (Julian day 32-60) routed by time of travel plus intrabasin routing.

Management Alternatives Evaluation

With the model calibrated for the basin, it is possible to evaluate what specific effect various land management alternatives will have on the model parameters, and subsequently estimate their effect in erosion reduction. These management alternatives have evolved through and have been developed by the agricultural technical community, including the Soil Conservation Service, the Agricultural Research Service, and their sister agencies at the state and local level, and are generally described as BMPs. Not all of these practices will necessarily reduce sediment transport from a basin since their original intent was to optimize productivity, often by improvement of drainage from cultivated lands. However, for the most part, their effect on water quality should be positive.

Use of NPS Model to Predict Erosion Control by BMPs

The NPS model as presently formulated does not contain explicit provisions for testing alternative farm management scenarios. When calibrated to a watershed as large as Honey Creek, the model represents at a point the variety of natural processes and management practices that are occurring over a 150-mi^2 area. Insofar as the fines generation equation of the QUAL subroutine represents a distinct reality, the COVVEC parameter, set on a monthly basis, is the only model parameter that is directly measurable and used as measured. Other model parameters attempt to approximate measurable physical characteristics or process rates but are subject to the model calibration process which can result in large discrepancies between the measured and the calibrated parameter value. The difference in real-world values for the parameters and the calibrated values is a rough measure of the discrepancy between actual and model processes.

For example, the parameter INFIL represents an index of the mean soil infiltration rate on the watershed. Its use in NPS attempts to represent a distribution of soil permeability in space but actually results in the simulated infiltration rate being distributed in time as a function of soil wetness. The mean soil infiltration rate of Honey Creek soils is approximately 0.4 in./hr, a value assigned to INFIL for the first calibration run. However, calibration efforts quickly revealed that INFIL must be radically lower than the real-world value of 0.4 in./hr. A final value of 0.02 in./hr was eventually selected. Even such an artificially low value was not sufficient to produce the overland flow from pervious areas needed for sediment washoff in the last half of calibration year 1976. With these observations in mind, it is possible nonetheless to estimate conceptually the direction of change to various NPS model parameters that can be anticipated if a change in cropland management occurs. Table VI presents a preliminary attempt to do this. Across the top are listed the NPS model parameters in three groups: LANDS (hydrology), QUAL (sediment production), and land use variables. Candidate BMPs are shown down the right. The estimated direction of change in a model parameter due to implementation of a practice is indicated by an arrow. No-till compared to conventional tillage, for example, may increase the upper soil zone storage capacity (UZSN) by increasing the organic content of the surface horizon. Infiltration capacity (INFIL) may improve due to minimum disturbance of soil aggregation processes. Surface roughness (NN) will increase due to accumulation of crop residues. The index to actual evapotranspiration (K3) may decrease due to light-colored crop and weed residues retarding soil drying. The same residues increase interception storage (EXPM). Depending on the time of year the NPS model simulation is begun, initial upper zone storage may be slightly higher. No-till may have an influence on intrinsic soil erodibility (KRER), but its effect is likely to be complex. No-till may lower

Table VI. Agricultural land management practices vs nonpoint modeling parameters.

Practice	Hydrologic (LANDS)													Sediment (QUAL)						
	UZSN	LZSN	INFIL	INTER	IRC	NN	SS	L	K3	EXPN	KK24	UZS	LZS	JSER	KSER	CONEC	PMPMAX	ACUPY	REPERV	SRERI
On-Field																				
Intensity of Ground Disturbance	←		←			←			→	←		←			→	←	←	→	←	→
No-Till	←		←			←			→	←		←			→	←		→		
Minimum Till	←														↔					
Ridge Plant								→							→					
Direction of Ground Disturbance																				
Contour							→	→							→					
Contour Strip							→								↔					
Terrace							←			←						←	←			
Crop Effects															→	←				
Winter Cover																←				
Fertilizer Application			←													←				
Crop Rotation																				
Mixed (Field/Channel)																				
In-Field Berms		→		→	→						→		→							
Tile Drains																				
Diversions																				
Channel																				
Grassed Waterways																←				→
Outlet Protection																←				→
Stream Bank Protection																			←	
Sediment Basins																←				→

the basic capacity of overland flow to carry sediment (KSER) due to its effects on sediment particle size and on surface roughness. Its most obvious and measurable effect is on cover (COVVEC) which is anticipated to increase over conventional tillage every month. Potency factor (PMPMAT) for phosphate is likely to increase due to the limited fertilizer placement options of no-till. Since no-till minimizes soil disutrbance, accumulation of fines (ACUPV) during normal plowing seasons will be nearly zero. Undisturbed soil aggregation processes may increase the natural rate of fines removal (REPERV) during nonstorm periods. Finally, the initial supply of fines for no-till is likely to be a small fraction of that present under conventional farm management (SRERI).

NPS model parameters can be similarly evaluated for all on-field practices. However, applicability of the NPS model to simulate BMPs becomes tenuous for mixed field/channel effects such as tile drains or permanent diversions. The NPS model was not designed to simulate BMPs located in the stream channel. Grassed waterways, stream bank and outlet protection could be treated as practices on critical areas instead of channel practices, but this would require separate NPS model calibration of these partial areas, separate flow and water quality data which are not likely to be available.

Although it is a fairly straightforward exercise to anticipate the direction of change of a model parameter due to implementation of a BMP, it is entirely another matter to select a reasonable value. The first reason for this difficulty is that very few specific field tests have been made of these practices with the NPS model parameters in mind. Second, even if such measurements were made and significant changes noted between conventional practice and BMPs, one would be reluctant to use them as such since most model parameters are "calibrated" and as such bear a fragile and tenuous connection to field measurements. The only notable exception to this is the cover parameter COVVEC which, as mentioned above, is directly measurable in real space and used in the model without modification during calibration.

**NPS Model Simulation of Erosion
Control by No-Till Planting**

No-till crop production is selected to demonstrate the application of the NPS model to estimation of sediment load reduction by implementation of a potential BMP. No-till is selected for several reasons:

1. No-till is a field practice (see Table VII, BMP categories).
2. No-till significantly increases cover (COVVEC).
3. No-till is suitable on significant areas of Honey Creek.

No-till crop production in this discussion is limited to row crops, corn and soybeans, which for Honey Creek in 1976 accounted for 52% of the watershed area. COVVEC, ACUPV and SRERI values for no-till corn and soybeans were estimated. The steps used in the development of no-till scenarios are to determine new model parameter values for no-till, then to run the model for all land uses except row-crops, for row-crops under conventional management, and finally for row-crops under no-till management. The monthly suspended solids loads for three no-till scenarios are then calculated by an area-weighted addition of loads from the three above NPS model runs.

Suitability for No-Till Crop Production: Three Management Scenarios

The Ohio Agricultural Research and Development Center has classified major Ohio soil series according to their suitability for no-till corn production.[9] Economic criteria are considered in this classification. Corn yields of no-till suitable groups are better, the same, or only slightly lower than yields from conventionally cultivated corn. A computerized count of Honey Creek soils in cropland as they fall into the no-till suitability groups is shown in Table VII. Less than 10% of the watershed is suitable without drainage improvements. Another 25% is suitable if drainage is improved. About 15% of the watershed that is cropland is either unsuitable (10.5%) or insufficiently studied to classify (4.4%). Nineteen percent of the watershed is not cropland. The remaining third of the watershed (34.4%) is cropland that can be planted using no-till methods for row-crops if excessive soil moisture delays conventional plowing until May 10th or later.

Table VII. Suitability of Honey Creek cropland for no-till row-crop production.

	Percent of Honey Creek Watershed Area	
No-Till Suitability Class	Independent of Crop Rotation Schedule	Restriction by Crop Rotation Schedule
Cropland		
Suitable without drainage improvements	6.7	4.2
Suitable only with drainage improvements	25.0	15.6
Suitable only if planting delayed beyond May 10	34.4	21.5
Not suitable or not yet classified	14.9	--
Other Land Uses		
Urban, woodland, water or wetland	19.0	--

However, due to crop rotation schedules, a field suitable for no-till in any one of the three suitable categories is not always available for no-till row-crop production, being committed to wheat or oats during some seasons. Thus the percentage of the watershed potentially available to no-till management in these categories is reduced by approximately three-eighths which is the proportion of cropland in 1976 devoted to small grains. These no-till suitability figures are summarized in Table VII.

The three no-till scenarios for Honey Creek to be examined in detail are derived from Table VII, with the crop rotation restriction, as follows:

No-Till Scenario 1: no-till row-crop on 4.2% of watershed
No-Till Scenario 2: no-till row-crop on 19.8% (4.2% + 15.6%)
No-Till Scenario 3: no-till row-crop on 41.3% (4.2% + 15.6% + 21.5%)

Estimation of COVVEC, ACUPV and SRERI for No-Till Corn and Soybeans

No-till COVVEC values were estimated by the same bookkeeping procedure used for the calibration of the NPS model to all of Honey Creek. Crop rotation effects, prior crop residues, crop area ratios, and conventional practices for small grains were taken into account. Crop residue carryover from previous years and lack of ground disturbance in April and May were the main determinants in the increased COVVEC values estimated for corn and soybeans as shown in Figure 13. The COVVEC curve for no-till corn is compared to COVVEC values for conventionally tilled row-crops (corn and soybeans) in Figure 14. Although there is some difference in COVVEC between no-till corn and soybeans in the last half of the year (Figure 13), the no-till corn COVVEC values were those used in the no-till NPS scenarios. This choice will not affect the results obtained, since there is no overland flow and thus no suspended solids loads simulated from cropland during the last half of the year.

The NPA model permits specification of monthly accumulation rates for soil fines. The ACUPV parameter allows the user to represent soil disturbance, such as plowing, as a daily increase in the pool of soil fines available for washoff. Calibration of the NPS model for Honey Creek in 1976 set this rate at 10 lb/ac/day for row-crops during April and May and during October and November to represent pool increases due to spring and fall plowing activities. Since no-till crop production does not significantly disturb the soil, ACUPV values were reduced to zero for the no-till run of the NPS model.

Similarly, the size of the fines pool (SRERI) in February was reduced from 250 to 25 lb/ac for the no-till simulation. (Had this reduction not been made, the best no-till scenario would have been nearly indistinguishable from the 1976 conventional simulation.)

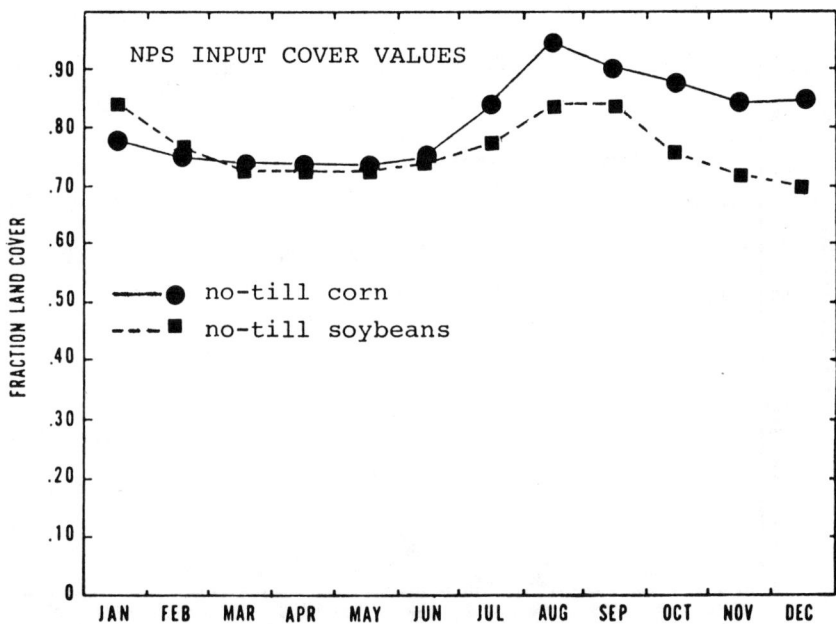

Figure 13. NPS input cover (COVVEC) values for no-till corn and soybeans. Cover values for no-till corn were used to simulate suspended solids loads in the three no-till management scenarios described in the text (RMA, 1978).

Simulated Erosion Control by No-Till Row-Crop Production

In order to estimate suspended solids loads for the three no-till scenarios, the NPS model was run separately for three situations. First, four land uses (small grains, woodland, water/wetland, and urban) with calibration parameter settings was run. Second, a 1-mi^2 area of row-crop, also with calibration parameter settings, was run. Finally, a 1-mi^2 area of row-crop with cover (COVVEC), fines accumulation rates (ACUPV), and initial fines (SRERI) modified for no-till (see above) was run. The monthly and annual loads for each scenario were calculated from these three NPS model runs by adding together appropriate proportions of the separate simulations as required by each scenario. The NPS model simulated results, summarized in Table VIII, predict that no-till row-crop production on well-drained cropland in 1976 could have reduced suspended solids loads by 3%; by 15% if artificial drainage had been improved to the extent possible; and by 33% if row-crop planting had been delayed beyond May 10th.

In considering these results, it must be remembered that they apply only to the hydrologic situation of 1976, with its unusual snowmelt-rain event in

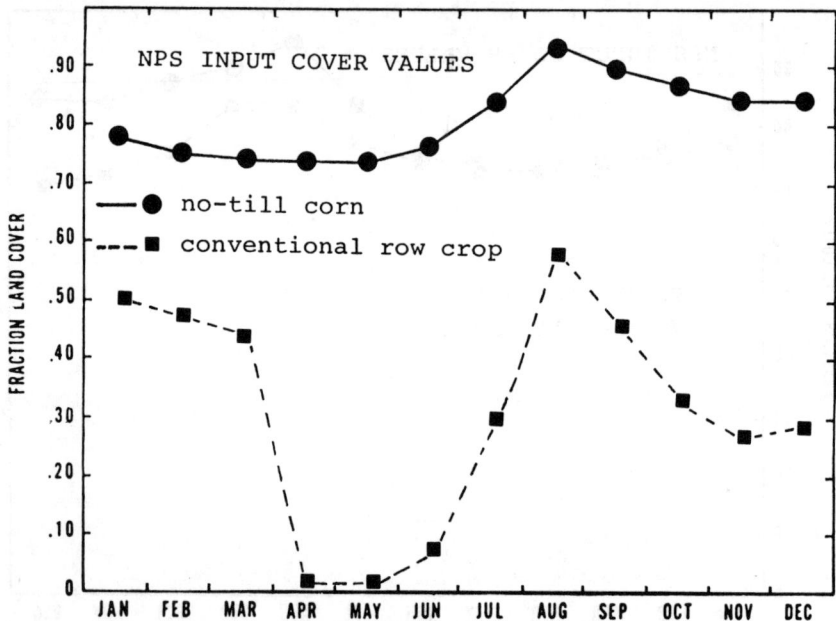

Figure 14. NPS input cover (COVVEC) values for no-till corn and conventional row-crop management (RMA, 1978).

Table VIII. Erosion control by no-till crop production in Honey Creek.

Month	Conventional Tillage (1976)	NPS Model Simulation (tons suspended solids) Crop Rotation Restriction		
		No-Till on 4.2%	No-Till on 20%	No-Till on 41%
February	16,800	16,200	14,100	11,300
March	3,000	2,940	2,680	2,330
April	430	420	380	330
May	500	500	500	500
June	110	110	110	110
July	60	60	60	60
•	•	•	•	•
•	•	•	•	•
•	•	•	•	•
Annual Total	21,000	20,300	17,900	14,080
Percent Reduction in Sediment Load		3%	15%	33%

mid-February. Examination of the monthly results in Table VIII reveals that the three scenarios differ for February, March and April only. Since there was overland flow (simulated) from May through July, the identical results in these months mean that the loads generated were not limited by the supply of fines available for washoff but by the depth of overland flow, which was the same in all scenarios. This result is reasonable in light of the relatively mild rainfall experienced during this period. There was no severe late spring or early summer storm to deplete the no-till available fines and thus distinguish it from the conventional management results. The months from August to December were lacking any NPS model overland flow component from pervious surfaces and thus could not deliver sediment to the streams from cropland. The amount of suspended solids simulated during these months is solely from impervious urban surfaces.

Implementation of scenario 3 is dependent upon a wet spring to delay planting to May 10th. Since this condition does not occur every year in Honey Creek, the long-term potential reduction in sediment loads, as predicted by the NPS model for no-till crop production, probably lies somewhere between 15 and 33%.

Although the 15% reduction in sediment load from Honey Creek estimated by the NPS model for no-till row-crop production under scenario 2 appears modest, it must be remembered that only 20% of the land area is involved. If a unit area of row-crops under conventional management is compared to the same unit area under no-till row-crop production, it is seen that the simuated delivery of suspended solids from the no-till area is 30% of that from conventionally grown row-crops or a 70% reduction. Other BMP candidates to complement no-till should be evaluated for cropland not suitable for no-till or not in a row-crop rotation, possibly by application of the NPS model as illustrated here.

Evaluation of BMPs by the Universal Soil Loss Equation

The Universal Soil Loss Equation (USLE) is another planning tool used to predict soil loss. "C" factors were estimated for conventional and no-till crop production for Honey Creek[10] and applied to the calculation of potential gross erosion rates. While the gross erosion estimates are substantially greater than the actual net transport of sediment from the basin, the relative reduction in the USLE and the NPS model as results of no-till cro production can be compared (see Table IX).

Each of the 11,500 4-ha cells were calculated using the USLE, based on a "C" factor of 0.245, "L" of 300 feet, "R" of 125 and using the "K" and slope from the computer data base. The no-till improvements to "C" were applied only on that watershed land area where no-till is suitable, (see Table

VII). Assuming that the relationship between estimated gross erosion and actual net erosion (*i.e.*, the delivery ratio) holds constant with spatially variable BMPs such as no-till, the percentage reduction comparison shows that the two methodologies of evaluating BMPs would give very different answers. The NPS model indicates far less of a reduction in sediment delivery from the basin due to no-till application on suitable lands than does the USLE. It will be an interesting future research effort to compare relative reductions in sediment yields estimated by the NPS model and the USLE for other BMP scenarios.

Table IX. Effect of no-till as estimated by the USLE.

	Total Soil Loss (ton)	Soil Loss (ton/ac)	Reduction (%)	Total Soil Loss (w/delivery ratio 0.0823)
Present Conditions C = 0.245	243,029	3.5		20,800
No-Till Class I	207,899	3.0	14.5%	17,109
No-Till Class I, II	111,006	1.6	54.4%	9,135
No-Till Class I, II, IV	63,678	0.9	74.8	5,240

Note: Analysis of cropland above Melmore (72%), using L = 300 ft.

CONCLUSIONS

The NPS model has been calibrated to the Honey Creek watershed, an agricultural area in north-central Ohio. The model was used to estimate the water quality improvement in Honey Creek obtained by the use of no-till row-crop production, a possible BMP. This application demonstrates the potential role the NPS model could play in evaluation of cropland management alternatives for federal and state planning agencies concerned with water pollution from nonpoint sources. However, much work remains to be done to describe other possible BMPs in terms of NPS model parameters before the model can become a flexible planning tool.

Regarding the development and evaluation of management plans, it is possible to view nonpoint source control as an optimization problem, with the nonpoint loading model acting as the central operator in the solution procedure. The most serious difficulty with this approach may not be the complexity of the system *per se* so much as the fact that the constraints which establish the set of feasible solutions tend to be very ill-defined. Perhaps the most convenient format is one in which water quality is optimized subject to constraints upon the use of specific management practices on given types of land. Unless there are important interactive effects among agricultural practices at different locations (which does not appear likely except at the cultural level), an "optimal" solution can often be found by disaggregation, *i.e.*, by classification of land areas according to natural characteristics and feasible management options. RMA has explored this general approach in its "opportunity mapping" for the Lake Erie project. The basic difficulty is that it is often unclear which management options are potentially feasible at which types of locations. This difficulty can be overcome by a land data base with "co-occurrence" capabilities, such as was developed in Honey Creek. Some investigators have proposed subanalysis in which the set of feasible measures in each case is established given the constraint that net farm income must be maintained. Such studies can become very complex, however, and tend to omit important cultural and political factors.

Management analysis under Section 208 is frequently described as a straightforward process of evaluating the water quality benefits of a given set of BMPs. However, we feel strongly that nonpoint modeling should contribute directly to the formulation of BMPs, since this process must involve implicit or explicit tradeoffs between social costs and water quality benefits. BMPs need not be determined through an optimization process, but feedback between water quality effects and control alternatives is obviously desirable. Given the above-mentioned difficulty of maximizing water quality subject to management constraints, a reasonable procedure may be to establish water quality constraints and then to explore possible management strategies which will achieve the chosen goals. This might be done by developing a series of scenarios, each consisting of a spatially detailed set of management practices which are shown by the model to yield acceptable pollutant loadings. Rather than choosing a "best" plan according to formal critieria, the scenarios could be submitted to policymakers who could then choose on the basis of numerous factors including intuitive judgment.

REFERENCES

1. U.S. Army Corps of Engineers. "Lake Erie Wastewater Management Study: Preliminary Feasibility Report," Buffalo District, Buffalo, NY (1975).
2. Great Lakes Basin Commission. "Maumee River Basin Level B Study, Analysis of Non-Point Sources," by Research Management Associates, (April 1976).
3. Bliss, N., T. H. Cahill, E. B. MacDougall and C. A. Staub. "Land Resource Measurement for Water Quality Analysis," Tri-County Conservancy, Chadds Ford, PA (1974).

4. Resource Management Associates. "Nonpoint Source Model Calibration, Honey Creek Basin," U.S. EPA, under Contract No. R-805421-01 (in preparation, 1978).
5. Resource Management Associates. "Honey Creek Report," for U.S. Army Corps of Engineers, Buffalo, NY (1977).
6. Dunne, T. "Recognition and Prediction of Runoff Producing Zones in Humid Regions," *Hydrol. Sci. Bull.* 20:3, 9 (1975).
7. U.S. Environmental Protection Agency. "Modeling Nonpoint Pollution from the Land Surface," EPA-600/3-76-083 (1976).
8. Cahill, T. H., and T. R. Hammer. "Phosphate Transport in River Basins," I.J.C. Fluvial Transport Workshop, Kitchener, Ontario (1976).
9. Ohio Agricultural Research and Development Center (DARDC). "An Evaluation of Ohio Soils in Relation to No-Tillage Corn Production," Research Bulletin 1068, Wooster, OH (1973).
10. Urban, D. USDA, Medina, Ohio, personal communication (1978).

34

SEDIMENT AND NUTRIENT CONTRIBUTIONS TO THE MAUMEE RIVER FROM AN AGRICULTURAL WATERSHED

D. W. Nelson
   Department of Agronomy
E. J. Monke, A. D. Bottcher
   Agricultural Engineering Department
L. E. Sommers
   Department of Agronomy
   Purdue University
   West Lafayette, Indiana

The Maumee River is usually a gently flowing river, and even in flood periods the velocities are relatively low because the gradient is not steep. However, the waters in the Maumee River never appear clear because of the suspended sediment load. For a 10-year period of record, 1961-71, the sediment discharge from the Maumee River into Lake Erie averages about 500 kg/ha annually.[1,2] However, sediment yields even of this magnitude may be important in lowering the water quality of Lake Erie because they are composed mostly of fine clay-sized particles and can carry a high nutrient load, particularly phosphorus (P). Unknown at this time, however, is the contribution which P attached to sediments actually makes to the eutrophication rate of Lake Erie. It is implicitly assumed in this study that some of the attached P becomes available and contributes to the eutrophication process occurring in the lake.

This paper reports on the sediment and nutrient yields from the Black Creek watershed into the Maumee River. In particular, it compares the data collected from two major drainage areas within the watershed. However, the reported results are based only on a 2-year period of record, 1975 and 1976, which is a rather short hydrologic period on which to make definite conclusions. As a consequence, the conclusions reached are subject to modification with succeeding years of record.

MATERIALS AND METHODS

The Black Creek watershed was chosen as being fairly representative of the soils and agricultural practices of the Maumee basin, although it is only 4950 ha in size compared to 1,711,500 ha for the Maumee basin. A map of the experimental watershed is shown in Figure 1.

The soils in the Black Creek watershed can be divided into two categories, those soils which were formed entirely in glacial till and those soils which were also influenced at various stages in their formation by shallow water cover or by wave action. These soil categories are subsequently referred to as glacial till soils and lake plain and beach ridge soils, respectively. Indiana Highway 37, as shown in Figure 1, divides these soils fairly well with the glacial till soils to the north and the lake plain and beach ridge soils to the south. The glacial till soils are gently rolling while the lake plain soils are nearly level.

Two of the major drainage areas which empty into Black Creek, the Dreisbach drain and Smith-Fry drain, were studied intensively. The Dreisbach drain is located along the western boundary of the watershed and the Smith-Fry drain is located along the eastern boundary. Their drainage areas, of comparable size, represent the greatest contrast in soils and land use within the watershed. The drainage area of the Dreisbach drain contains 74% rolling and 26% nearly level topography while that of the Smith-Fry drain contains only 29% rolling and 71% nearly level topography. The land use is also quite different with 35% of the drainage area of the Dreisbach drain in row-crops as compared to 63% for the drainage area of the Smith-Fry drain. The drainage area of the Dreisbach drain also contains the town of Harlan which has effects on water quality in that stream. Characteristics of these two drainage areas as well as those of the Black Creek watershed are given in Table I.

Sediment and nutrient yields from the Black Creek watershed and from the drainage areas for the Dreisbach drain and Smith-Fry drain were determined by integrating sediment and nutrient concentrations with flow rates. Stage-discharge relationships were developed for the outlets of these study areas to give flow rates. Water stages were recorded continuously at these locations with a presssure-type stage recorder (Model 12 Flow Recorder, Foxboro).

Water samples were collected either manually or with automated samplers. Grab samples were collected each week and also during storm events. The atuomated samplers were triggered at a set minimum stage and then continued to operate automatically until the sample storage was exhausted or the

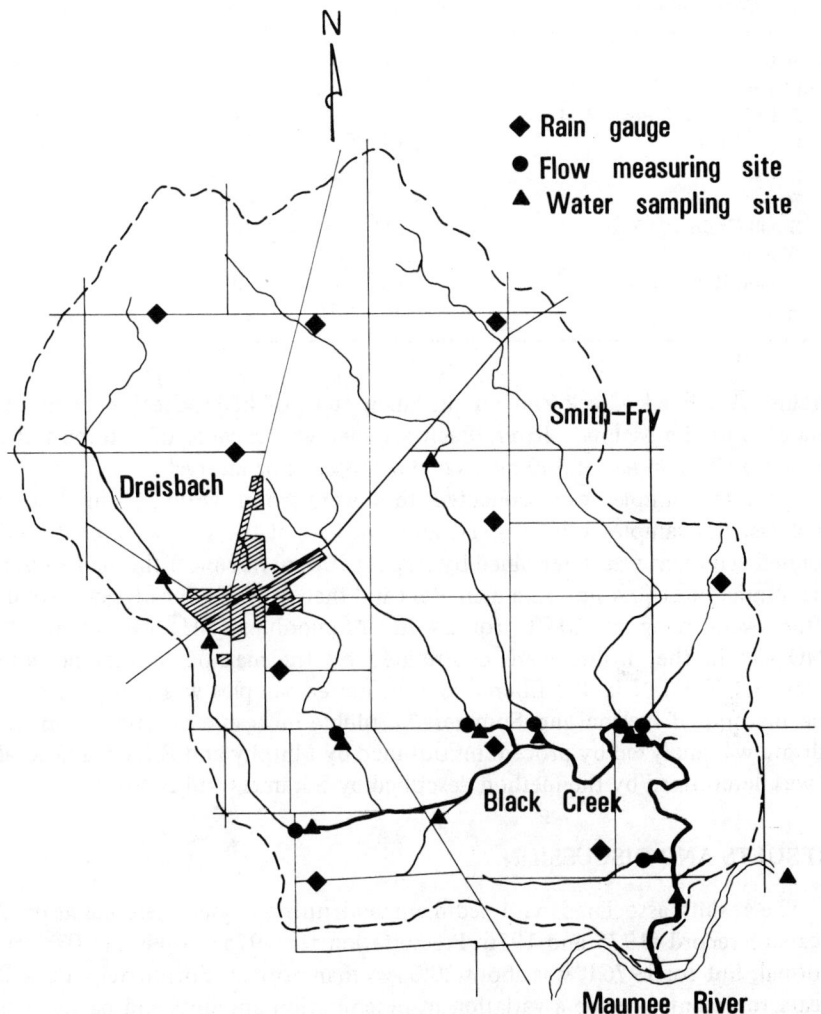

Figure 1. Map of the Black Creek study area.

stage fell below the set minimum stage. The water samples were normally collected and the automated sampler reset before the sample storage was exceeded.

Three automated pumping samplers (PS-69, U.S. Interagency Sedimentation Project) were installed at the junctions of the Dreisbach and Smith-Fry

Table I. Characteristics of the area studied.

Characteristics	Black Creek Watershed	Smith-Fry Drain	Dreisbach Drain
Drainage Area	4950 ha	942 ha	714 ha
Soil Groups:			
Lake Plain and Beach Ridge	64%	71%	26%
Glacial Till	36%	29%	74%
Land Use:			
Row-Crops	58%	63%	35%
Small Grain and Pasture	31%	26%	48%
Woods	6%	8%	5%
Urban, Roads, etc.	5%	3%	12%
Homes	--	28	143

drains with Black Creek and on the main stem of Black Creek near its entrance into the Maumee River. Each sampler was capable of automatically collecting 72 samples of 500 ml each at a chosen time interval.

After the samples were collected, they were frozen within 24 hr. Before analysis, the samples were thawed and one-half of the sample filtered. Suspended sediment was determined by passing 200 ml of runoff through a tared membrane filter (0.4-μm pore diameter) and then weighing the collected solids after oven-drying at 105°C for 24 hr. Ammonium (NH_4^+-N) and nitrate (NO_3^--N) in the filtrate were determined by the method of Bremner and Keeney.[3] Total N in the filtered and unfiltered samples was determined by the method of Nelson and Sommers.[4] Soluble inorganic phosphorus in the filtrate was analyzed by procedures outlined by Murphy and Riley,[5] and total P was determined by the method described by Sommers and Nelson.[6]

RESULTS AND DISCUSSION

The results associated with sediment and nutrient yields are based on 2 years of record—1975 and 1976. Precipitation for 1975 was about 20% over normal, but for 1976 it was about 20% less than normal. Fortunately, these 2 years represent as wide a variation in precipitation amounts and patterns as will likely occur over a lengthy period of record.

Rainfall, Runoff, Sediment, Total Nitrogen, Total Phosphorus Yields

The rainfall, runoff, sediment yields and total N and P yields for the Black Creek watershed and the drainage areas of the Dreisbach and Smith-Fry drains are given in Table II. In the Black Creek watershed, a 40% reduction in rainfall from 1975 to 1976 resulted in 60% reduction in runoff, a greater than

Table II. Rainfall and runoff amounts and yields of sediment and nutrients from the study areas during 1975-1976.

Parameter	Year	Black Creek Watershed	Smith-Fry Drain	Dreisbach Drain
Rainfall	1975	112 cm	112 cm	112 cm
	1976	70 cm	70 cm	70 cm
Runoff	1975	27.5 cm	29.1 cm	26.0 cm
	1976	11.2 cm	12.4 cm	10.1 cm
Sediment	1975	2370 kg/ha	2130 kg/ha	3740 kg/ha
	1976	530 kg/ha	640 kg/ha	380 kg/ha
Total N	1975	48.7 kg/ha	53.2 kg/ha	44.1 kg/ha
	1976	8.6 kg/ha	10.3 kg/ha	6.6 kg/ha
Total P	1975	5.2 kg/ha	5.4 kg/ha	5.0 kg/ha
	1976	1.1 kg/ha	1.1 kg/ha	1.0 kg/ha

fourfold reduction in sediment yield, a greater than fivefold reduction in total N discharged, and about a fivefold reduction in total P discharged from the watershed. These data clearly demonstrate the adverse effects of above-normal rainfall on sediment and nutrient discharges.

The runoff and sediment yield data for the drainage areas of the Smith-Fry drain and the Dreisbach drain present an interesting comparison. The runoff from the drainage area for the Smith-Fry drain was greater than that for the Dreisbach drain in both 1975 and 1976. Although the drainage area for the Smith-Fry drain was more level than the drainage area for the Dreisbach drain, it also had better subsurface drainage and base flow and also seemed to be sustained for longer periods of time from interflow through the ditch banks.

In 1975, the year with above-normal rainfall, the sediment yield at the outlet of the Dreisbach drain was about twice that at the outlet of the Smith-Fry drain. However, in 1976 the reverse was true. Although all 1976 values were much lower, the sediment yield at the outlet of the Smith-Fry drain was about twice that of the outlet of the Dreisbach drain. The greater proportion of small grains and pasture crops in the drainage area of the Dreisbach drain was apparently important, retarding runoff and subsequent erosion as compared to the drainage area of the Smith-Fry drain which contained a high proportion of row-crops. However, when rainfall was excessive, as in 1975, greater runoff per hectare of watershed occurred and subsequent erosion was greater on the drainage area of the Dreisbach drain than on that of the Smith-Fry drain because of steeper slopes in the Dreisbach area.

Nutrient Transport

The total nutrient yields for 1975 and 1976 from the Black Creek watershed and the drainage areas of the Smith-Fry and Dreisbach drains are given in Table II. Losses of various forms of N and P are given in Table III. All of the forms of N and P transported greatly decreased from 1975 to 1976 due primarily to decreased runoff. The amounts transported from one drainage area are roughly similar to the other except for soluble inorganic P and NO_3^--N.

Table III. Amounts of nutrients leaving the watersheds during 1975 and 1976.

Component	Smith-Fry Drain		Dreisbach Drain	
	1975	1976	1975	1976
	(kg/ha)			
Soluble Inorganic P	0.14	0.06	0.34	0.18
Soluble Organic P	0.10	0.03	0.12	0.04
Sediment P	5.2	0.98	4.5	0.73
Ammonium N	1.5	0.60	1.8	0.85
Nitrate N	19.0	5.5	12.0	2.4
Soluble Organic N	1.7	0.31	2.3	0.53
Sediment N	31.0	3.9	28.0	2.8

The larger amounts of soluble inorganic P from the drainage area of the Dreisbach Drain were caused by the larger number of houses discharging domestic sewage into the Dreisbach drain as compared to the Smith-Fry drain. On the other hand, larger amounts of NO_3^--N on an area basis were being discharged from the Smith-Fry drain in comparison to the Dreisbach drain, very likely because the Smith-Fry drain has more extensive subsurface drainage and a higher percentage of land in fertilized row-crops as compared to the Dreisbach drain.

Percentages of N and P forms which were transported past the outlets of the two drains are given in Tables IV and V. Of interest in Table IV is the shifting which takes place between the various forms from 1975 to 1976. Sediment N was high in 1975 because excess rainfall caused more runoff during that year. In 1976, the relatively dry year, NH_4^+ and NO_3^--N constituted a greater percentage of the total N transported than in 1975. With regard to P transport, from 77 to 96% of total P transported was sediment-bound, as shown in Table V. Higher percentages of total P transported as soluble P forms in the Dreisbach drain during 1976 reflect the greater importance of septic contributions to water quality during dry years.

Table IV. Proportions of total N leaving the watershed comprised of various N forms.

Drain	Year	Form of N Transported			
		NH_4^+-N	NO_3-N	Soluble Organic N	Sediment N
Smith-Fry	1975	2.8	35.7	3.2	58.3
	1976	5.8	53.4	3.0	37.8
Dreisbach	1975	5.1	27.2	5.2	63.5
	1976	12.9	36.5	8.1	42.5

Table V. Proportions of total P leaving watersheds comprised of various P forms.

Drain	Year	Form of P Transported		
		Soluble Inorganic P	Soluble Organic P (% of total transport)	Sediment P
Smith-Fry	1975	2.6	1.8	95.6
	1976	5.9	3.2	90.9
Dreisbach	1975	6.9	2.4	90.7
	1976	19.0	4.2	76.8

Partitioning of Runoff and Transported Sediment and Nutrients

The runoff and transported sediment and nutrients were separated into different categories dependent essentially on the size of the runoff event. These categories were: base flow, small events, large events, and snowmelt. Snowmelt was a lumped flow event of base flow and small events which occurred during thaw periods. In many instances, the outlet drains were then clogged by ice and snow and the recorded water stages were high. Under these conditions, the snowmelt flow reported is our best estimate based on the depth of snow over the drainage area and the amount of rainfall which may have occurred simultaneously with the thaw.

Base flow can easily be identified from the runoff hydrograph because it is a long-time, steady (depth does not vary with time) flow event. A large event was arbitrarily established as an event which produced more than 2.5 cm (1 in.) of runoff from an entire drainage area. Small events were those flow events which occurred between base flow and large flow events.

The results from partitioning the runoff and transported sediments and nutrients for Dreisbach drain are given in Table VI. Similar results were obtained for the Smith-Fry drain. The large flow events only occurred a few times during either 1975 or 1976. In 1975 only three storms produced over

Table VI. Partitioning of runoff and transported sediment and nutrients for the Dreisbach drain during 1975-1976.

Component	Base Flow		Small Events		Large Events		Snowmelt	
	1975	1976	1975	1976	1975	1976	1975	1976
	(% transport)							
Runoff	10	22	37	11	38	56	15	11
Sediment	1	6	14	4	78	84	7	6
Soluble Inorganic P	6	22	49	5	26	41	19	32
Soluble Organic P	8	23	42	12	33	52	17	13
Sediment P	2	7	32	4	58	78	8	11
NH_4^+-N	13	20	35	29	28	31	24	20
NO_3^--N	8	10	37	7	34	75	21	8
Soluble Organic N	9	31	27	18	48	36	16	15
Sediment N	1	4	23	3	68	83	8	10

2.5 cm of runoff from an entire drainage area and in 1976 only two such storms occurred. Yet these storms accounted for the major sediment and sediment-bound nutrients transported from the two drainage areas. Less than 6% of sediment was transported by base flow and over 70% by large flow events.

With small watersheds, the runoff period is relatively short. The time from the beginning of a storm to the time peak runoff occurs at the outlets of the two drains is between 2 and 3 hours. With only a grab sample program it would be very likely that all or parts of the major storm events would have been missed. As a consequence, the sediment and nutrient yields from a small watershed may be grossly underestimated if based on grab sample data. In our case, if the large events had been completely missed by a sampling program, only about one-third of the actual sediment yield would have been reported. High concentrations in grab samples, if sustained, will still indicate a serious pollution program. However, an anlysis such as we are presenting here would be impossible because of data gaps.

During certain snowmelt processes, it was noticed that high loadings of soluble inorganic P occurred frequently. This may have been caused by a release from decayed vegetative matter at this time. Whatever the exact causes, however, the levels of soluble inorganic P as well as the other soluble forms of P and N were higher during snowmelt events than the percentage of runoff during these events would indicate.

Comparison of Sediment and Nutrient Concentration in Black Creek with Maumee River

It is difficult to compare sediment and nutrient transport data from the Black Creek with those for the Maumee River. In 1976, the sediment loss from the Black Creek watershed was about the same as the long-term average from the Maumee River as measured at Waterville, Ohio. In 1975, however, the sediment loss from the Black Creek watershed was over four times the long-term average recorded at the station in Waterville. The total P concentration in the Maumee River at New Haven (which is just above the entrance of Black Creek into the Maumee River) as measured weekly by the Water Pollution Control Plant of Fort Wayne averaged about 0.45 mg/l for 1975-76. Their average measured sediment concentration for these 2 years was about 80 mg/l. Measurements of total P concentrations at the State Route 101 bridge across the Maumee River and just below the entrance of Black Creek into the river were 0.48 mg/l for 1975 and 1976, respectively. Average measured sediment concentrations were 240 and 140 mg/l for these 2 years. All of the water samples (those collected by the Water Pollution Control Plant of Ft. Wayne and by the Black Creek project) were grab samples and the concentrations were not flow-weighted. Weekly grab samples and additional samples were collected during storm events. This difference in sampling frequency could account for much of the difference between two measuring stations in terms of sediment concentrations.

The total P concentrations measured at the two stations agree closely. Both sets of data also indicate that total P concentrations in the Maumee River are not closely correlated to sediment concentrations. Our data for the Maumee River show that 21% and 42% of the total P was composed of soluble P forms in 1975 and 1976, respectively. The total P concentrations in the Maumee River seem to be greatly influenced by soluble P levels. On the other hand, the total P concentrations in the Black Creek watershed are greatly influenced by sediment-bound phosphorus levels.

The average suspended sediment concentration at Waterville, Ohio, near the entrance of the Maumee River into Lake Erie is approximately 200 mg/l. This value corresponds to the sediment concentrations we observed in Dreisbach and Smith-Fry drains during base flow and some part of our small flow events. However, our sediment concentrations during high flow events are much higher than the average values reported in the Maumee River.

Sources of Runoff and Transported Sediments and Nutrients

The probable sources of runoff and transported sediment and nutrients in the Dreisbach drain for 1975 and 1976 are given in Table VII. Similar results were obtained for the Smith-Fry drain although the contribution from sewage

Table VII. Sources of runoff and transported sediment and nutrients in the Dreisbach drain during 1975-1976.

Component	Year	Source				
		Tiles	Interflow	Sewage	Runoff	Rainfall[a]
		(% of total transported)				
Discharge	1975	23	8	1	68	
	1976	12	13	3	72	
Sediment	1975	1	--	<1	98	
	1976	2	--	7	91	
Soluble Inorganic P	1975	7	2	37	54	(46)
	1976	3	3	71	23	(54)
Soluble Organic P	1975	15	5	14	66	
	1976	10	11	44	35	
Sediment P	1975	1	--	5	94	
	1976	1	--	35	64	
NH_4^+-N	1975	7	2	13	78	(280)
	1976	3	4	28	65	(380)
NO_3^--N	1975	25	9	6	60	(56)
	1976	24	26	30	20	(170)
Soluble Organic N	1975	13	5	--	82	
	1976	11	12	--	77	
Sediment N	1975	1	--	1	98	
	1976	2	--	16	82	

[a]Percentage of soluble nutrients as measured in the stream which could have been supplied by rainfall. For example, for 1975 the amount of soluble inorganic P leaving Dreisbach drain divided by the amount of soluble inorganic P in precipitation falling on the Dreisbach watershed times 100 equals 46%.

was much lower because of fewer houses in the Smith-Fry area. The waste treatment facility for almost all non-Amish homes on the watershed is a septic tank which is discharged directly into a tile drain or a stream.

The column for sewage reflects the contribution of outfalls for septic tanks. Calculations were made by assuming the discharge from each home to be 100 gpd. The column for rainfall shows those chemical components which were measured in rainfall as a percentage of the total amount which was discharged for that particular year. It does not represent the proportion of the soluble forms which was contributed by rainfall, but it merely indicates that rainfall is also a potential source for these P and N forms. The contribution of soluble inorganic P by rainfall would be included in the column for surface runoff and the contribution of NH_4^+ or NO_3^--N by rainfall might be included under the tiles, interflow or surface runoff.

Runoff from the land surface was the major source of the sediment and nutrients which were transported in the Smith-Fry and Dreisbach drains during 1975. Over 90% of the sediment and sediment-bound nutrients originated from the land surface during that year. This also held true for the Smith-Fry drain in 1976. However, in the Dreisbach drain during 1976, sewage outflow which constituted only 3% of the discharge contained substantial percentages of the soluble nutrients as well as some sediment-bound nutrients. Relatively high amounts of soluble inorganic P originated from septic tank outfalls in all cases, and relatively high amounts of NO_3^--N were associated with tile drainage or interflow into the streams.

The contribution of the discharge from subsurface drainage systems to sediment yield was not large and certainly not at all approaching the amounts reported elsewhere in the Maumee Basin.[7] Based on the sampling of discharge from selected tile outlets it was estimated that subsurface drains accounted for only 1 to 2% of the total sediment yield from the Black Creek watershed.

Estimate of Added Nutrients Lost

An estimation of nutrients which were added as commercial fertilizer or manures lost from the Black Creek Watershed is a tenuous undertaking but certainly one which should be addressed. Estimates of the application rates for commercial fertilizers and manures were based on a questionnaire.[8]

A nitrogen balance for the entire watershed subdivided between Amish and non-Amish farms is shown in Tables VIII and IX. The amount of N applied as commercial fertilizer and manure and fixed by legumes was regarded as the same for both 1975 and 1976. We assumed that mineralization of nitrogen was balanced by that which was immobilized during the year. The input of N associated with that portion of the precipitation which infiltrated into the soil was added to the subtotal for both years. The total applied and fixed N was then estimated as 329,500 kg for 1975 and 316,500 kg in 1976.

The applied or fixed N lost from the watershed was the measured amount of soluble inorganic N discharged from the watershed minus the contribution from septic tank outfalls and that portion of the precipitation which ran off the land surface. The applied plus fixed N lost was estimated as 66,000 kg for 1975 and 10,700 kg in 1976.

If we assume that the same proportion of the applied and fixed N was lost, then 20% of the N applied as commercial fertilizer and manure was lost in 1975, a very wet year, and 3% was lost in 1976, a relatively dry year. The loss of soluble inorganic P originating from commercial fertilizer and manure was estimated to average about 0.3% for both years.

The estimated loss of soluble inorganic N from commercial fertilizers or manures is in line with estimated losses from other areas with similar

Table VIII. Estimate of annual applied and fixed nitrogen in the Black Creek watershed.

Nitrogen Source	Amish	Non-Amish	Total
	(kg of N)		
Commercial Fertilizer or Manure:			
Corn	12,200	153,200	165,400
Soybeans	--	5,700	5,700
Small Grain	5,400	23,400	28,800
Hay and Pasture	13,800	3,000	16,800
Fixation:			
Soybeans	--	57,900	57,900
Forage Legumes	11,000	1,200	12,200
Total	42,400	244,400	286,800

Table IX. Estimate of applied and fixed nitrogen lost from the Black Creek watershed for 1975 and 1976.

Nitrogen Input or Loss	1975	1976
	(kg)	
Applied and Fixed Nitrogen Input	286,800	286,800
Input Due to Infiltrated Precipitation	42,700	29,700
Total Nitrogen Input	329,500	316,500
Total Nitrogen Loss (measured)	87,400	22,700
Loss Due to Precipitation Runoff	14,100	5,800
Loss Due to Septic Tank Discharge	7,300	6,200
Applied and Fixed Nitrogen Loss	66,000	10,700
Percent Applied and Fixed Nitrogen Lost	20%	3%

application rates. The loss of added P in a soluble form is relatively low and suggests that most of it was immediately tied up in the soil complex. However, more P could be expected to be lost through transport on eroded soil particles. We did not attempt to evaluate this because of our imprecise knowledge of where erosion was occurring in relation to the areas which received P fertilization.

The total nitrogen loss per hectare from the Smith-Fry drain was about 20% higher than from the Dreisbach drain (see Table II). However, the percentage of the drainage area of the Smith-Fry drain in row-crops was about double that for the Dreisbach drain. It is probably reasonable to expect that most of this difference was due to the increased usage of commercial fertilizers on the lands contributing runoff to the Smith-Fry drain. Improved fertilizer management techniques may reduce the amounts of soluble inorganic N reaching the streams in the Black Creek watershed, but the level to which it can be reduced is certainly bounded, and the effect of improved techniques may not be all that noticeable in loadings from the watershed into the Maumee River.

Effect of Agricultural Nonpoint Source Pollution on the Maumee River and Lake Erie

The actual impact of sediment P and N discharged from the Black Creek watershed and other agricultural watersheds in the Maumee basin on the eutrophication process in the Maumee River and Lake Erie still is speculative and needs further study. Phosphorus being discharged is largely sediment-bound and would have to be released in order to enter the nutrient cycle of the algal biomass. Other investigations conducted as a part of the Black Creek project suggested that only about 15 to 20% of the P which was bound to sediments could ultimately become available for algae growth. Regarding N, a large percentage of the total N discharge occurs in the late winter and spring months during high flows when algae growth is minimal. How much N remained and became available later in the year for usage by the algal biomass is also largely unknown at this time.

Nonpoint source pollution from the Black Creek watershed occurs primarily from large storm events. These large storm events usually occur during the spring and early summer months. Also the chances are that the Maumee River is at a moderately high stage during the period when the large storm events occur. On the other hand, the input of point source pollution is most critical in the Maumee River during its low-flow periods. While there may be some residual nutrient remaining from high-flow periods, the effect of agricultural nonpoint source pollution by-and-large should not be evaluated using low-flow criteria as is characteristic for point source pollution.

CONCLUSIONS

1. Amounts of runoff, sediment and nutrients discharged from a small watershed are greatly affected by rainfall. Reductions in rainfall give a greater percentage reduction in runoff which in turn gives a still greater percentage reduction in sediment and nutrient transport.

2. During years with above-average rainfall, land slope is clearly the dominant factor affecting sediment yields. However, with below-average rainfall,

the effects of land use on sediment yields becomes relatively more important. This reflects the natural sequence of rainfall-runoff events because rainfall must first meet the storage capabilities of the soil and land surface before runoff begins.

3. With average or above average rainfall amounts, more than 90% of the total P transported is sediment-bound. However, only about 50% of the total N is sediment-bound. Models of P-transport can be based on models of the erosion-sedimentation process. We recommend that models of N-transport should not be based on models of the erosion-sedimentation process only.

4. Transport of sediments and sediment-bound nutrients is strongly associated with large storms which occur only a few times during a year.

5. During snowmelt, the transport of soluble nutrients may be disproportionally high when compared with snowmelt runoff.

6. Average concentrations of sediment and nutrients discharged from the Maumee River are in line with measured concentrations in the Maumee River.

7. Losses of soluble inorganic P, NH_4^+-N and NO_3^--N from the watershed are partially due to the input of these chemical forms by precipitation.

8. Most of the sediment and sediment-bound nutrients originate in runoff from the land surface. The discharge of NH_4^+-N is also largely associated with runoff from the land surface. However, the discharge from septic tank outlets contributes substantial amounts of soluble inorganic P into the streams in the watershed.

9. In order to characterize nutrient and sediment loadings from small watersheds, automated sampling is required.

10. The effects of agricultural nonpoint source pollution and point source pollution on our water resources are sufficiently different that direct comparisons between them cannot be made and separate objectives for their evaluation and control are in order.

ACKNOWLEDGMENTS

This paper is a contribution of the Indiana Agricultural Experiment Station, Purdue University, West Lafayette, Indiana 47907. The study was supported in part by Grant No. G005103 from the Environmental Protection Agency, (Region V). Purdue Univ. Agric. Exp. Sta. Paper No. 7174.

REFERENCES

1. U.S. Geological Survey. "Water Resources Data for Ohio, Part 1, Surface Water Recores," U.S. Department of the Interior, Washington, DC (1971).

2. Monke, E. J., D. B. Beasley and A. B. Bottcher. "Sediment Contributions to the Maumee River," EPA-905/9-75-007, Proc. Non-Point Source Pollution Seminar, November 20, 1975, Chicago, IL (1975), pp. 71-85.
3. Bremner, J. M., and D. R. Keeney. "Steam Distillation Methods for Determination of Ammonium, Nitrate, and Nitrite," *Anal. Chim. Acta* 32:485-495 (1965).
4. Nelson, D. W., and L. E. Sommers. "Determination of Total Nitrogen in Natural Waters," *J. Environ. Qual.* 4:465-468 (1975).
5. Murphy, J., and J. P. Riley. "A Modified Single Solution Method for Determination of Phosphate in Natural Waters," *Anal. Chim. Acta* 27:254-267 (1962).
6. Sommers, L. E., and D. W. Nelson. "Determination of Total Phosphorus in Soils: A Rapid Perchloric Acid Digestion Procedure," *Soil Sci. Soc. Am. Proc.* 36:902-904 (1972).
7. Schwab, G. O., E. O. McLean, A. C. Waldron, R. K. White and S. W. Michner. "Quality of Drainage Water from a Heavy Textured Soil," *Trans. ASAE* 16(6):1104-1107 (1973).
8. Brooks, R. M., and D. L. Taylor. "Leadership Important to Acceptance of Conservation," EPA-905/9-75-006, Environmental Impact of Land Use on Water Quality (1975), pp. 155-196.

35

WATER QUALITY MODELING IN THE DELAWARE COASTAL PLAIN REGION

W. F. Ritter
    Agricultural Engineering Department
    University of Delaware
    Newark, Delaware

P. A. Jensen
    College of Marine Studies
    University of Delaware
    Newark, Delaware

INTRODUCTION

Since the 1972 Federal Water Pollution Control Act Amendments were passed, there has been considerable interest in water quality modeling. Many hydrologic models have been developed over the past 20 years. In recent years, models for the transport of nonpoint source pollutants have been coupled with hydrologic models.[1-3]

Coastal Sussex County, Delaware, was designated an EPA 208 area. The designated Coastal Sussex 208 area includes the Indian River Bay, Rehoboth Bay, Little Assawoman Bay and Broadkill River drainage basins (Figure 1). A land use study by Ritter and Scheffler[4] indicated 44% of the land is in agricultural cropland, 39% in forestland and 6% in urban development. Most of the urban areas are resort-type development with a large number of summer homes. One of the major tasks of the 208 Coastal Sussex Water Quality Management Program was to evaluate the effects future development and different land use plans would have on water quality. The University of Delaware College of Marine Studies and the Agricultural Engineering Department of the College of Agricultural Sciences were given the task of quantifying and

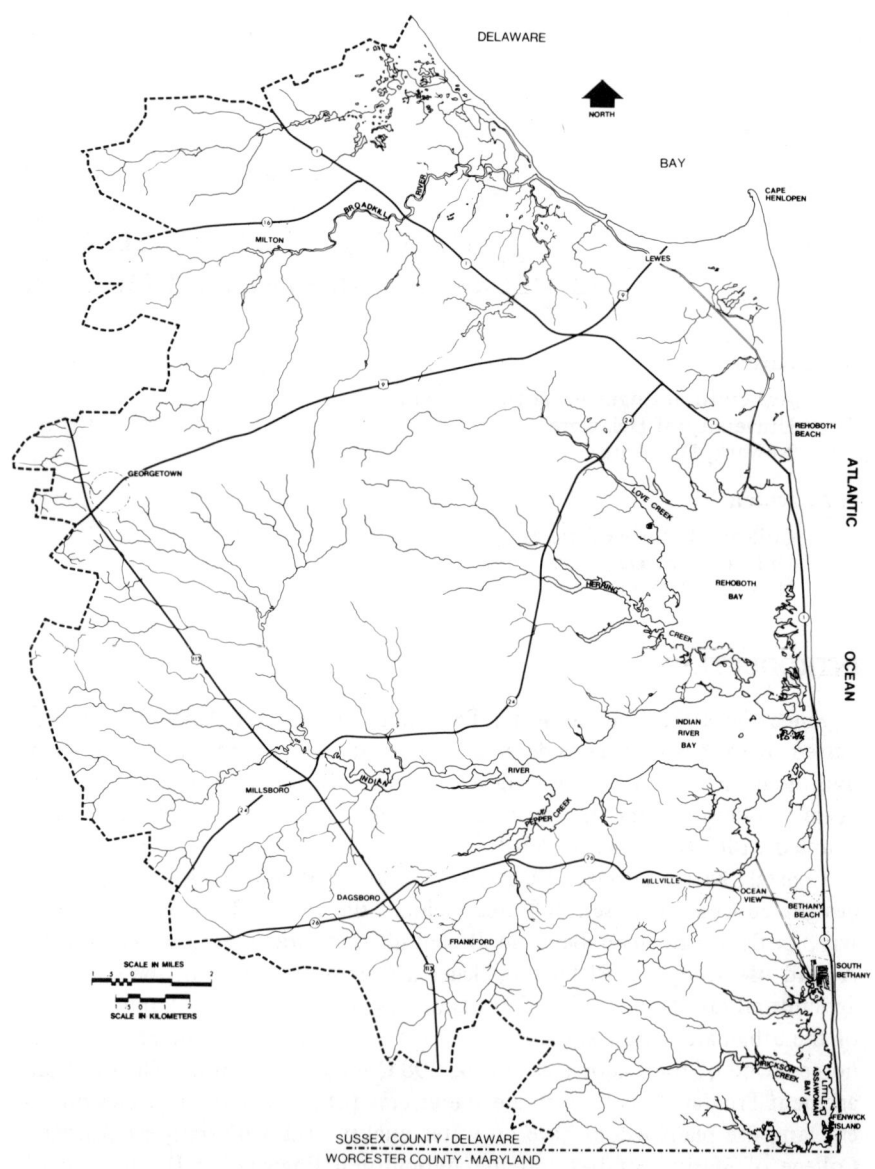

Figure 1. Map of Coastal Sussex 208 area.

describing point source and nonpint source loads in each drainage basin and what effect land use changes would have on water quality. The major technique used to evaluate what effect any changes in future point and nonpoint source loads would have on water quality was by modeling. The EPA Agricultural Runoff Model[2] was used in conjunction with a one-dimensional water quality model.[5] This paper summarizes the calibration of the two models and some of the results obtained.

AGRICULTURAL RUNOFF MODEL (ARM)

Model Description

The Agricultural Runoff Model (ARM) developed by Hydrocomp for EPA simulates runoff, sediment, pesticides and nutrient contributions to stream channels from both surface and subsurface sources. The operation and general structure of the ARM model are shown in Figure 2. The major components of the model individually simulate runoff (LANDS), sediment production (SEDT), pesticide adsorption/desorption (ADSRB), pesticide degradation (DEGRAD) and nutrient transformations (NUTRNT). Only the runoff and nutrient subroutines of the model were used in this study since sediment production was extremely low and pesticide pollution was not considered in the 208 program.

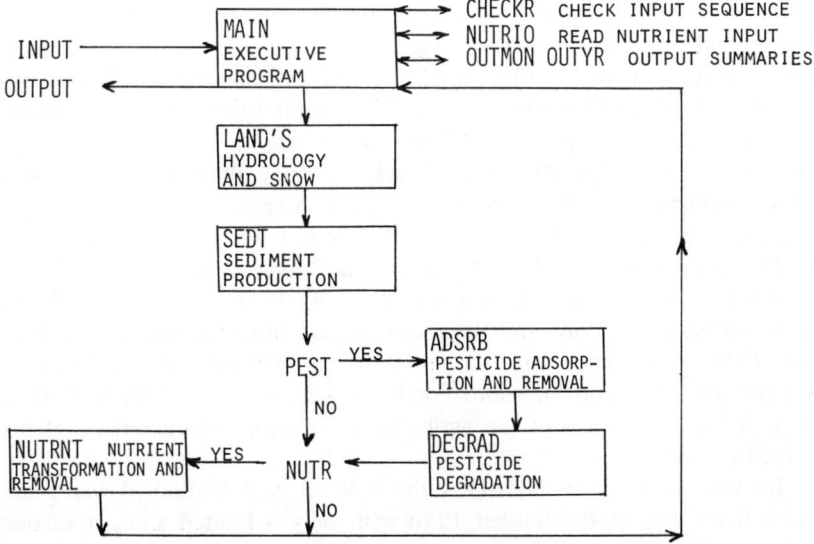

Figure 2. ARM model major program structure and operation.[2]

510 BMPs FOR AGRICULTURE AND SILVICULTURE

The runoff subroutine (LANDS) will simulate runoff continuously for an entire year or can be used to simulate individual storms. In the present study only individual storms were simulated, since the major objective of the study was to determine the effect a storm with a certain duration and return period would have on water quality. The simulation is basically a moisture accounting procedure on the land surface for water. The subroutine operates on a 5- or 15-min time interval with 5- or 15-min time interval precipitation data required for input along with evapotranspiration; 15-min interval rainfall data were used in this study.

The nutrient subroutine (NUTRNT) simulates nitrogen mineralization and microbial immobilization, nitrification, denitrification, plant uptake and ammonia adsorption-desorption. Organic nitrogen, solution ammonia, adsorbed ammonia, nitrate nitrogen, nitrogen gas and plant nitrogen are the different forms of nitrogen considered in the model through a series of coupled differential equations involving ten reaction rates.

Phosphorus is also simulated in NUTRNT. It is assumed to exist as organic phosphorus, solid inorganic phosphate, dissolved inorganic phosphate and phosphorus absorbed by plants in the model. Mineralization, immobilization, adsorption-desorption and plant uptake reactions are modeled in NUTRNT.

Nonpoint Source Monitoring

A nonpoint source monitoring program was conducted for the Coastal Sussex County Water Quality Management Program from April 1976 to September 1977. Four agricultural watersheds and two urban watersheds were monitored for chemical oxygen demand (COD), organic nitrogen, nitrate-nitrite nitrogen, orthophosphorus and total phosphorus.[4] The agricultural watersheds (Stockley Branch, Blackwater Creek, Beaverdam Creek, and the drainage area above Millsboro Pond) were monitored for base flow and storm runoff. On Blackwater Creek, Stockley Branch and Beaverdam Creek, automatic samplers were used for storm events, and base flow samples were collected by hand. Only grab samples were taken from Millsboro Pond. Monitoring was done at the U.S. Geological Survey (USGS) gaging stations on Stockley Branch and Beaverdam Creek, while stage recorders were installed on Millsboro Pond and Blackwater Creek. Stockley Branch and Blackwater Creek were sampled for the entire 18 mo, while Beaverdam Creek was sampled for 9 mo and Millsboro Pond was sampled from June to October 1976. A brief summary of the agricultural watershed characteristics is presented in Table I.

The two urban watersheds (Rehoboth Beach and Millsboro) were monitored from August to October 1976 with only a limited amount of data collected. The Rehoboth Beach watershed (14 ac) was a new development

Table I. Characteristics of agricultural watersheds monitored.

Characteristic	Stockley Branch		Blackwater Creek		Millsboro Pond		Beaverdam Creek	
Drainage Area (mi^2)	5.2		5.6		6.0		6.1	
Land Use (%)	Cropland	45	Cropland	57	Cropland	42	Cropland	52
	Forestland	47	Forestland	37	Forestland	50	Forestland	38
	Urban	4	Urban	2	Urban	4	Urban	2
Major Soil Types	Evesboro loamy sand		Evesboro loamy sand		Evesboro loamy sand		Evesboro loamy sand	
	Fallsington sandy loam		Fallsington sandy loam		Fallsington sandy loam		Rumford loamy sand	
	Rumford loamy sand		Pocomoke loamy sand		Rumford loamy sand		Sassafras sandy loam	
Slope (%)	0.17		0.17		0.15		0.18	

composed of summer houses, while the Millsboro watershed (6 ac) drained part of a main highway and a portion of a trailer park. In both urban watersheds, stage recorders and automatic samplers were placed in storm sewers.

Model Calibration

The procedure in calibrating the ARM model is to initially select values for each parameter and then adjust these values during the calibration process. Besides rainfall and evaporation data, 22 other parameters are required for LANDS to describe the watershed and establish initial conditions. Only summer rainfall and runoff data from Stockley Branch and Blackwater Creek were used to calibrate LANDS. The input parameters length of overland flow, average overland flow slope and drainage area were obtained from topographic maps, while other values were selected on judgment and adjusted during the calibration process.

Reasonable agreement was obtained between simulated and recorded storm hydrographs on Stockley Branch as shown by the storm on August 9, 1976 (Figure 3). The simulated peak always occurred before the recorded peak runoff. Better hydrograph simulation was obtained for Stockley Branch than Blackwater Creek as indicated by a comparison of Figures 3 and 4 for

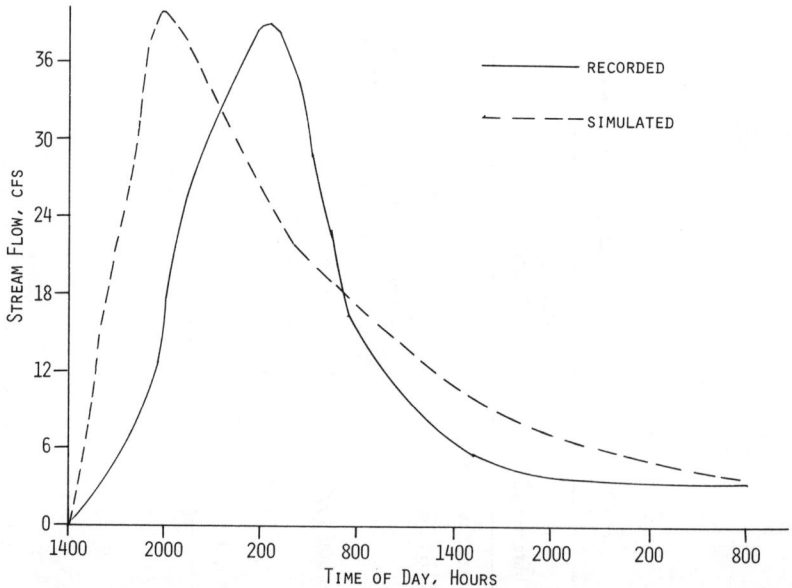

Figure 3. Simulated and recorded runoff for Stockley Branch for storm on 8-9-76.

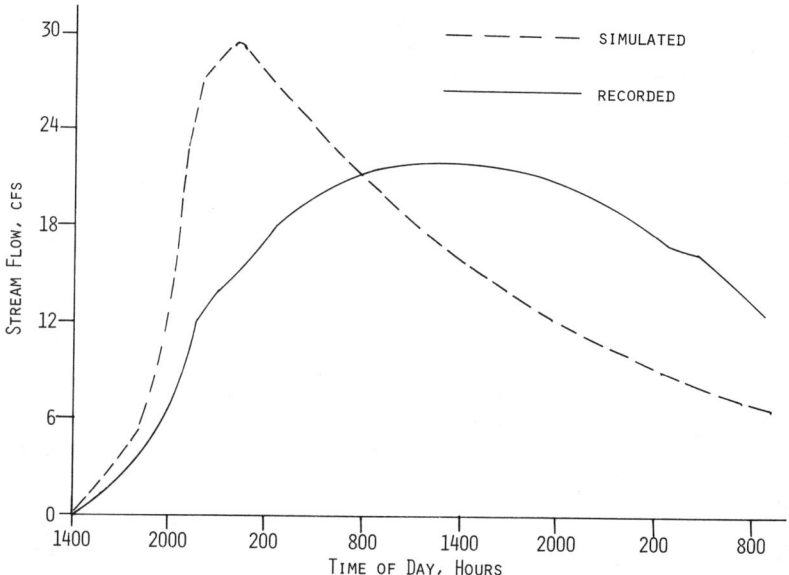

Figure 4. Simulated and recorded runoff for Blackwater Creek for storm on 8-9-76.

the August 9 storm. The various parameters used in the storm simulation for August 9 are presented in Table II. For most storms on Blackwater Creek, the runoff rate increased slowly and maintained a peak for a considerable time interval. There was reasonable agreement between the total simulated runoff volume and recorded peak runoff rates. There is a series of drainage ditches on Blackwater Creek which may provide some channel storage and delay runoff. Since LANDS is an overland flow model and does not account for channel storage, this may explain why there is poor agreement between the recorded and simulated hydrographs. Stockley Branch does not have any drainage ditches.

The nutrient subroutine could not be calibrated to produce reasonable results. In their initial testing of the Model, Davis and Donigian[6] reported they could simulate the nitrogen and phosphorus of the sediment satisfactorily but did not obtain satisfactory results in simulating nutrients in the runoff. Since there was no initial flush of nutrients or definite relationship between nutrient concentration and time since runoff started, average concentrations of nutrients during the storm were used as input to the water quality model. The porous nature of the soil and the fact that most of the runoff occurred as interflow probably contributed to the lack of an initial flush.

Table II. Input parameters for ARM model for simulated storm on 8-9-76.

Parameter	Stockley Branch	Blackwater Creek
A	0.0	0.0
EPXM	0.12	0.12
UZSN	1.0	1.0
UZS	0.5	0.5
LZSN	20.0	23.0
LZS	11.0	8.0
K3	0.4	0.4
K24L	0.6	0.6
K24EL	0.0	0.0
INFIL	1.0	1.0
INTER	1.7	1.7
L	3000	5000
SS	0.0017	0.0017
NN	0.20	0.20
IRC	0.3	0.5
KK24	0.3	0.5
KV	1.0	1.0
GWS	0.5	0.5
SGW	0.0	0.0
ICS	0.0	0.0
IFS	0.0	0.0
AREA	3350	3600
OFS	0.0	0.0

WATER QUALITY MODEL (WQM)

Model Description

The model developed by Jensen and Tyrawski[7] is a one-dimensional, finite difference, solution to a system of equations for 11 parameters. The first eight are components of a simplified estuarine nitrogen cycle -NH_4-N, NO_2-N, NO_3-N, phytoplankton-N, benthic-N, herbivore-N, nonphytoplankton particulate organic nitrogen (PON) and dissolved organic nitrogen (DON). The nitrogen portion of the model is similar in several ways to the one developed by Najarian and Harleman[8] and also to those of O'Connor et al.[9] and Chen and Orlob.[10]

The purpose of the nitrogen portion of the model is to analyze the effects of nutrient enrichment and potential requirements for nutrient removal from point sources. Nitrogen is employed because it is the nutrient most likely to be limiting in an estuarine environment.[11,12] This was confirmed in field data collected for Coastal Sussex County.[13]

The nitrogen cycle model must be regarded only as a crude approximation of actual conditions. Estuarine biochemical cycles are extraordinarily complex and only the simplest relationships are understood with any confidence. Even with the recognition that there are substantial shortcomings in our understanding of estuarine biochemical cycles, it was agreed that analysis of nitrogen limitation would be essential. Despite these necessary "hedgewords," the nitrogen cycle model does appear to correctly represent the gross variations in all the parameters in response to both low- and high-flow conditions during summer months.[7]

The next two components of the model are the traditional biochemical oxygen demand (BOD) and dissolved oxygen (DO). BOD is a very crude parameter developed for river systems without significant "onsite" production of organic material. However, the decision was made early in the modeling process to retain this parameter because the BOD_5 information is a large portion of the point source loadings and stream data. To allow for organic matter produced in the waterway, the phytoplankton-N is converted to oxygen demand by the average C to N ratio (7.8) of phytoplankton[6] and the BOD_u to C ratio and added to BOD_u in the model. To convert this sum to BOD_5 stream data, it is reduced by the percentage of 5-day to ultimate BOD (BOD_u) measured in field experiments.[13] This is only an approximate representation of the true BOD_5 which would be measured with standard test procedures. It is true that phytoplankton cells will be killed by the process of sample preservation (icing) and thus become part of the organic oxygen demand. However, there is probably a significant lag period for freshly killed cells. At the same time, waste material generated by respiration of all components of the ecosystem is not accounted for in the BOD_u term. The two factors appear to balance each other to a first approximation, and the phytoplankton correction to BOD generally gave good results in all areas modeled in Coastal Sussex County.[7]

The DO balance is maintained by a combination of BOD decay, respiration, benthic demand, reaeration and photosynthetic DO production. A series of 24-hr stations monitoring DO, chlorophyll-*a*, nutrients and hydrographic parameters provided data which confirmed that the model represents diurnal DO variations with reasonable accuracy.[7]

The last parameter in the model is total coliform bacteria (TC). This is a parameter of great importance to water quality managers because of its use by public health agencies to regulate areas of shellfish harvesting. The traditional view is that high TC levels are an indication of high bacterial levels usually associated with human or animal pollution. In most coliform modeling studies performed to date, TC have been assumed to die when introduced to surface waters at a rate proportional to temperature.[14,15] However, a great deal of evidence has accumulated suggesting that TC survival is strongly influenced by light intensity[16,17] and organic substrate concentration.[18,19] In addition, there is considerable evidence that both total and fecal coliform populations will increase, or at least reduce very slowly, if organic substrate levels are sufficiently high. Field work with isolation chambers set up in the mixing zones of point source discharges confirmed that one to two orders of magnitude increases in both TC and FC concentrations were common events.[20] This same study also found that TC and FC levels were quite high in wetland areas.

The TC populations were modeled with a first-order die-off rate which increases with light intensity and decreases with organic substrate (BOD_u) concentrations. Regrowth at point sources was handled by increasing the TC loading rate by an appropriate factor in the range of 200-400.

The empirical relation developed for the coliform die-off rate, $K7$ (hr^{-1}) is

$$K7 = B7 + B7(L4) - (B7) BOD_u/BOD_7 + N8) \qquad (1)$$

where $B7 = 0.05\ hr^{-1}$

$L4$ = the vertically integrated relative light level, a function of solar intensity, light extinction coefficient and water depth, and

$N8$ = a Michaelis-type constant with a value of 3.0 mg/l.

Again, this relation can only be regarded as qualitatively correct, and then with limitations. The values produced are in the range of 0.02-0.08 hr^{-1} (0.48 - 1.9 day^{-1}), which are in reasonable agreement with the broad range of literature values. However, as noted by Mitchell and Chamberlain[21] in their presentation of a general model for the survival of enteric organisms in seawater, there are at least five factors—sedimentation, light, predation, nutrients, and physiochemical effects—which are likely to have a strong influence on coliform survival and none are quantified to any reliable degree. In addition, there are a host of possible interactions between these factors. In view of the complexities of bacterial dynamics in natural conditions, it is likely that coliform modeling will remain a largely empirical undertaking. However, this is not to say that significant improvements in coliform modeling are not possible. Equation 1 represents a substantial step in this direction, but it is by no means a final answer.

Inputs of various parameters modeled are from upland runoff, point sources, septic systems and wetlands. Upland runoff data were developed from nonpoint source monitoring.[4,22] Point source loadings were developed from averages of monthly monitoring data collected by the state over the last several years.[23] While the long-term average is probably reasonably correct, variations in treatment plant performance at any given time are likely to be substantial. Shoreline septic system loading rates were developed from literature values and limited field studies. Here again there is likely to be considerable variation, but the magnitude of these loadings is relatively small compared to other sources. Ritter and Scheffler[4] estimated that the nitrogen load from nonpoint sources in Indian River Bay was 1.5×10^6 lb/yr while the load from septic tanks was 59,000 lb/yr.

Calibration and Verification

The WQM was calibrated to low-flow conditions in Indian River Bay. This estuary was employed because of data availability resulting from a separate 2-yr study of thermal effects.[24] As it was not possible, due to computer limitations, to conduct full season simulations, calibration and verification were performed using flow variations during summer months, the time of greatest water quality concerns. The model coefficients were adjusted within the range of reasonable literature values (see Jensen and Tyrowski[7] for full discussion of the WQM) for low summer flow conditions. Low flow was defined as the range of 1.5-3.0 cfs flow at Stockley Branch, or 17 cfs at Millsboro Dam.

Once the model parameters agreed reasonably well with data collected at low flow (Figure 5), the calibration was checked by increasing model freshwater flow to 48 cfs at Millsboro Dam, the approximate long-term average flow at the dam. This flow was representative of the flow when several EAI and State Department of Natural Resources and Environmental Control

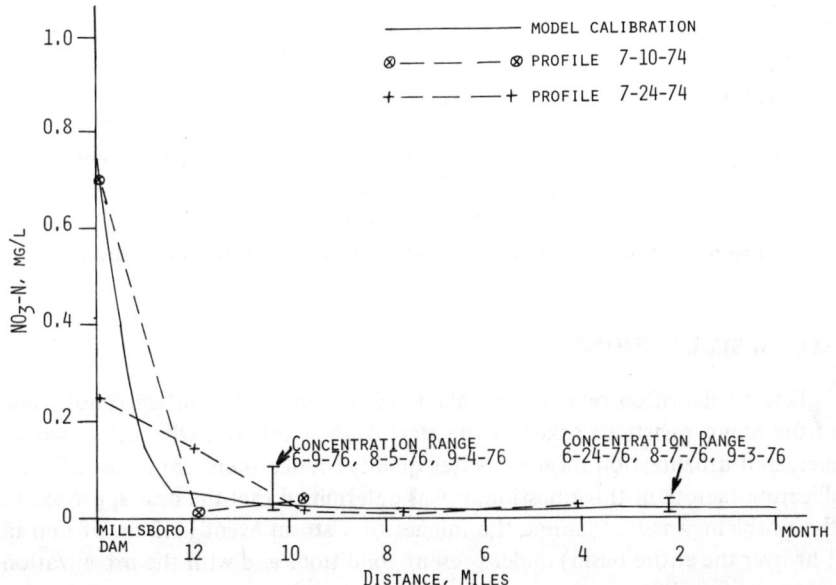

Figure 5. Low flow nitrate data and model calibration for Indian River.

518 BMPs FOR AGRICULTURE AND SILVICULTURE

(DNREC) data sets were collected. Model predictions were compared to these higher flow data sets and were found to be in reasonably close agreement as shown by the nitrate data in Figure 6. This indicats the WQM does represent the gross variations in the major water quality parameters with acceptable accuracy. A further indication of the model's utility was its ability to be applied to the other, substantially different, estuaries in the Coastal Sussex County 208 area with only very slight tuning required.

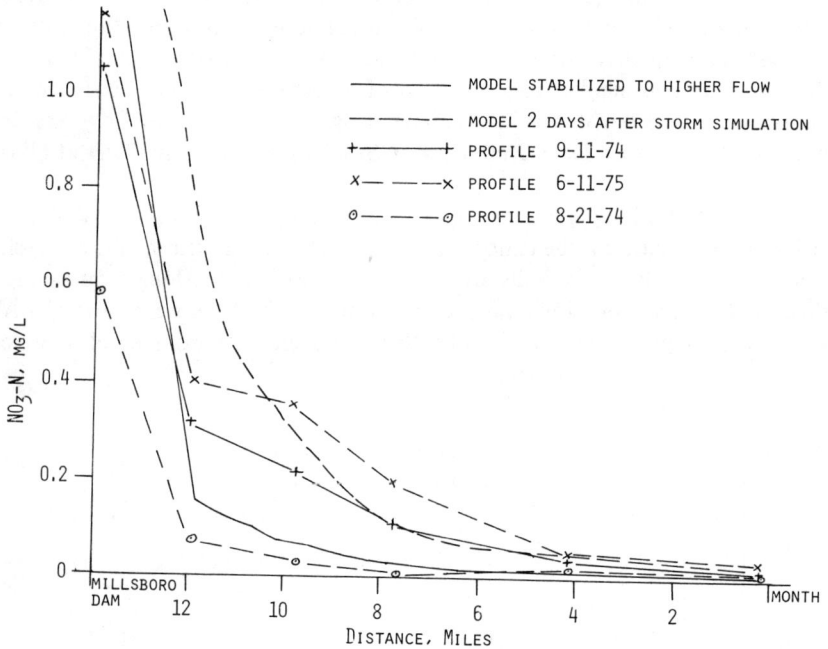

Figure 6. Higher flow nitrate data and model calibration for Indian River.

STORM SIMULATIONS

Due to the often reported problems with poor quality urban runoff, one of the major questions asked at the start of the study was what effect would increased urbanization have on water quality? Since there are a host of complicating factors in this question, it was determined that the best approach to the problem was to examine the impact of a storm event (1.2 in. of rain in 1 hr over the entire basin) under present conditions and with the urbanization projected for 1995.

Based on the parameters used in calibration of the ARM model for Blackwater Creek and Stockley Branch, runoff hydrographs were generated for upland and lowland agricultural watersheds in Indian River Bay, Rehoboth Bay, Little Assawoman Bay and the Broadkill River drainage basins for the 1.2-in. storm (Figure 7). A storm hydrograph was also simulated for a typical low-density and medium-density urban watershed. Since only limited data were collected from urban watersheds, which prevented calibrating the model for urban runoff, most of the parameter selection was based on judgment and calibration experience with the agricultural watersheds. Since urban runoff was not a major source of pollution because of the limited urban area and because of large point source loads and other nonpoint source loads, even a large error in the urban storm simulation would have little effect on the WQM results.

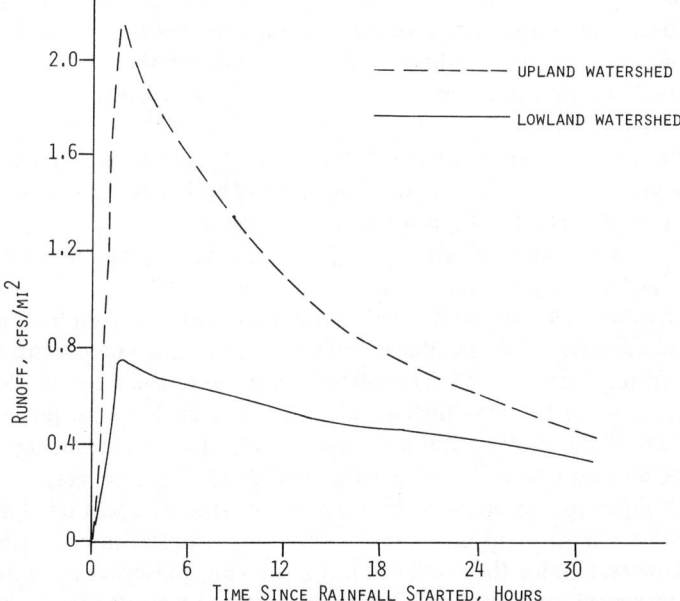

Figure 7. Simulated hydrographs for Indian River Bay for 1.2 in. storm.

The WQM was modified to accept the hydrographs and concentrations developed in the nonpoint source monitoring program for the storm simulation. Before the storm simulation, the system was allowed to come to equilibrium by holding climatic and other variables constant. The system generally stabilized in approximately 1 wk. Water quality parameters were printed for each day of the simulation at 11:00 AM, which was the time when most water quality data were collected. The storm was started at midnight and cloudy conditions (20% of peak sunlight) were maintained throughout the day of the storm. On the next day, conditions were returned to those existing before the storm. These conditions were assumed to be clear weather and a temperature of 77°F.

The storm simulation showed several interesting features. For brevity, only Indian River will be discussed, but the results are indicative of the other basins studied. The upper Indian River peak nitrate-N levels increased dramatically as would be expected since the runoff nitrate-N concentrations were approximately 2.0 mg/l. Ammonia concentrations also increased in the upper river but not to as great an extent as the nitrate-N levels, since ammonia concentrations in the runoff were only 0.15 mg/l. The increase in net flow downstream and the reduced light levels associated with the storm caused phytoplankton (chlorophyll-a) to decrease dramatically on the first day after the storm. By the second day after the storm when streams flowing into Indian River had returned to base flow conditions, the higher nutrient concentrations caused phytoplankton levels to increase quickly. Three days after the storm chlorophyll-a concentrations were 125 μg/l. The dissolved oxygen decreased sharply to 6.7 mg/l (Table III) at 11:00 AM on the day of the storm, but increased to even more supersaturated conditions than before the storm as phytoplankton production increased. The BOD_5 decreased slightly during the storm because of phytoplankton washout.

Small increases in total coliform (TC) occurred during the storm and several days following the storm. Stormwater runoff will affect TC numbers but other factors such as variations in wastewater loads and light level are also important factors. Examination of coliform data and antecedent rainfall revealed little correlation.[20] The highest measured values occurred during dry weather, while the next highest data were collected at high flows after a storm. Most of the increase during the storm was due to reduced light levels which reduces coliform mortality. The actual TC increase may be much greater than the model predicts since reduced light levels and a small increase in nutrient concentrations may cause a very large increase in coliform populations. However, since the model does not simulate coliform regrowth in the stream, only small increases in TC were recorded as a result of lower die-off rates and increased loadings. Coliform dynamics are not understood well enough to model the regrowth process.

There were not any great differences in most parameters for the 1995 and 1976 simulations as can be seen in Table III. The only major change was in TC, which was a result of reduced point source loadings. Reductions in nutrient inputs lowered the phytoplankton and degree of DO supersaturation slightly in the 1995 simulation results. In 1976 point source loads supplied

Table III. Peak values for Indian River water quality parameters during storm simulation.

Day and Year	NH$_4$-N (mg/l)	NO$_3$-N (mg/l)	Chlorophyll-a (µg/l)	BOD$_5$ (mg/l)	DO (mg/l)	TC (col/100 ml)
1 Day Before Storm, 1976	0.02	0.15	102	15.2	11.0	1331
1 Day Before Storm, 1995	0.02	0.15	98.0	13.0	10.4	185
Day of Storm, 1976	0.13	0.40	68.8	14.0	6.7	1490
Day of Storm, 1995	0.11	0.38	65.0	13.0	6.9	235
1 Day After Storm, 1976	0.15	1.10	53.5	12.2	13.7	1550
1 Day After Storm, 1995	0.09	1.10	48.3	11.0	11.0	244
2 Days After Storm, 1976	0.04	0.36	102	13.9	14.3	1540
2 Days After Storm, 1995	0.02	0.41	82.0	11.3	13.4	214

approximately 20% of the nitrogen, while for the 1995 projections, nitrogen loads were reduced 50% from point sources. Although the peak equilibrium value (1 day before storm) for chlorophyll-a and DO changed only slightly for 1995, the location moved upstream near the major nonpoint source nutrient source and away from the major industrial nutrient source.

Since the urbanized area within 1.3 mi of the estuary was only 1.3% of the basin area, the urban nonpoint source pollution load was extremely small. Therefore, a 46% increase in near-water urbanized area was not a significant factor in the water quality simulations. The major factor in the 1995 water quality simulation results was point sources. Since the planning agency anticipated improvements in the waste treatment from the wastewater treatment plants that were to remain and an increase in the percentage of the population that was to be served by a regional treatment plant discharging to an ocean outfall, the point source loadings on the estuary dropped significantly.

CONCLUSIONS

1. The EPA ARM model can be used to simulate runoff from agricultural watersheds on coastal plain soils where overland flow is the major hydrologic process and significant channel storage does not occur.
2. The first version of the EPA ARM model nutrient subroutine cannot be used to simulate nutrient transport from coastal plain soils in Delaware.
3. The one-dimensional water quality model (WQA) could be used by planners to model water quality in the Delaware coastal plain estuaries.
4. Projected increases in urbanization by 1995 in Coastal Sussex County will have little effect on the water quality of the estuaries in the region. Total coliform concentrations will decrease, but other water quality parameters will be similar to present-day conditions.

ACKNOWLEDGMENTS

Published with the approval of the Director of the Delaware Agricultural Experiment Station as Miscellaneous Paper No. 825.

REFERENCES

1. Tubbs, L. J., and D. A. Haith. "Simulation of Nutrient Losses from Cropland," paper presented at American Society of Agricultural Engineers Winter Meeting, Chicago, IL, December 13-16, 1977, ASAE Paper No. 77-2502 (1977).
2. Donigian, S. S., and N. H. Crawford. "Modeling Pesticides and Nutrients on Agricultural Lands," U.S. Environmental Protection Agency, Pub. No. EPA 600/2-76-043 (1976).

3. Frere, M. H., C. A. Onstad and N. H. Holtan. "ACMO, An Agricultural Transport Model," U.S. Department of Agriculture, ARS, Report No. ARS-1-1-3 (1975).
4. Ritter, W. F., and G. Scheffler. "Monitoring of Nonpoint Source Pollution in Coastal Sussex County," Report on Task 2332 208 Program, The Colleges of Marine Studies and Agricultural Science, University of Delaware, Newark (1977).
5. Jensen, P. A., and J. M. Tyrawski. "Water Quality Model for Coastal Sussex County," Report on Task 2359 208 Program, The Colleges of Marine Studies and Agricultural Science, University of Delaware, Newark (1976).
6. Davis, H. H., and A. S. Donigian. "Simulating Nutrient Movement and Transformations with the ARM Model," Agronomy Abstracts of 1977 Annual Meeting, Los Angeles, CA, November 13-18, 1977, American Society of Agronomy, Madison, WI (1977).
7. Jensen, P. A., and J. M. Tyrawski. "Water Quality Modeling and Analysis," Report on Task 2359C 208 Program, The Colleges of Marine Studies and Agricultural Science, University of Delaware, Newark (1977).
8. Najarian, T. O., and D. R. F. Harleman. "A Real Time Model of Nitrogen Cycle Dynamics in an Estuarine System," Department of Civil Engineering Dept. #204, Massachusetts Institute of Technology, Cambridge (1975).
9. O'Connor, D. J., R. V. Thomann and D. M. DiTora. "Dynamic Water Quality Forecasting and Management," Project No. R800369, Program Element 1BA023, Office Research and Development U.S. EPA, Washington, DC (1973).
10. Chen, C. W., and G. T. Orlob. "Ecologic Simulation for Aquatic Environments," Nat. Tech. Info. Serv. Washington, DC (1972).
11. Ryther, J. H., and W. M. Dunstan. "Nitrogen, Phosphorus and Eutrophication in the Marine Environment," *Science* 171:1008-1013 (1971).
12. Hobbie, J. E., B. J. Copeland and W. G. Harrison. *Estuarine Research,* Vol. I, L. Eugene Cronin, Ed. (New York: Academic Press, Inc., 1975), pp. 287-302.
13. Jensen, P. A. "Analysis of Water Quality Data and Land Use Water Quality Relationships," Report on Task 2355 208 Program, The Colleges of Marine Studies and Agricultural Science, University of Delaware, Newark (1977).
14. Orlob, G. T. "Viability of Sewage Bacteria in Seawater," *Sew. Ind. Wastes* 28:1147-1167 (1956).
15. Canale, R. P. "Model of Coliform Bacteria in Grant Traverse Bay," *J. Water Poll. Control Fed.* 45:2358-2371 (1973).
16. Gameson, A. L. H., and I. J. Gould. "Effects of Solar Radiation on the Mortality of Some Terrestrial Bacteria in Sea Water," *Proc. Int. Symp. Discharge of Sewage from Sea Outfalls,* August 27-September 2, 1974 (Elmsford, NY: Pergamon Press, 1974), pp. 209-219.
17. Bellair, J. T., G. A. Parr-Smith and I. G. Wallis. "Significance of Diurnal Variations in Fecal Coliform Die-Off Rates in the Design of Ocean Outfalls," *J. Water Poll. Control Fed.* 49(9):2022-2030 (1977).
18. Nusbaum, T., and R. M. Garver. "Survival of Coliform Organisms in Pacific Coastal Waters," *Sew. Inc. Wastes.* 27:1383-1390 (1955).

19. Won, W. D., and H. Ross. "Persistence of Virus and Bacteria in Sea Water," *J. Environ. Eng. Div. ASCE.* 99:204-211 (1973).
20. Jensen, P. A., W. F. Ritter and J. M. Tyrawski. "Coliform Bacteria Loadings and Dynamics," Report on Task 2359B 208 Program, The Colleges of Marine Studies and Agricultural Science, University of Delaware, Newark (1977).
21. Mitchell, R., and C. Chamberlain. "Factors Influencing the Survival of Enteric Microorganisms in the Sea: An Overview," *Proc. Int. Symp. Discharge of Sewage from Sea Outfalls.* August 27-September 2, 1974 (Elmsford, NY: Pergamon Press, 1974), pp. 237-251.
22. Ritter, W. F. "Modeling of Nonpoint Source Pollution in Coastal Sussex County," Report on Task 2351 208 Program, The Colleges of Marine Studies and Agricultural Science, University of Delaware, Newark (1977).
23. Jensen, P. A., and J. J. Eschleman. "Review of Existing Point Source Waste Loadings," Report on Task 2353 208 Program, The Colleges of Marine Studies and Agricultural Science, University of Delaware, Newark (1976).
24. Ecological Analysts. "Ecological Studies in the Vicinity of Indian River Power Plant for the Period June, 1974 Through August, 1975," Demonstration Vol. 4, prepared for the Delmarkva Power & Light Co. (1976).

36

ESTIMATION OF AGRICULTURAL NONPOINT LOADS TO THE WAKARUSA RIVER BASIN USING THE "NONPOINT CALCULATOR"

M. J. Davis, J. W. Nebgen
    Midwest Research Institute
    Kansas City, Missouri

INTRODUCTION

The estimation of nonpoint loads is becoming an increasingly important aspect of developing cost-effective measures for maintenance of surface water quality. Nonpoint loads arise from diffuse sources and, therefore, are closely associated with land management. Thus, improved management of the land resource will necessarily impact nonpoint pollutant loads with a concomitant improvement in surface water quality. In order to proceed with effective land management to control nonpoint loads, one must first ascertain the magnitude of the nonpoint loads and how such loads affect water quality. One must also have knowledge of sources from which nonpoint loads arise. With this information, it is possible to develop management estimations for nonpoint source control, and to determine the cost-effectiveness of each.

The "nonpoint calculator" developed by Midwest Research Institute (MRI) is a tool for estimating nonpoint loads. It is based upon "loading functions" developed for use in the nonpoint aspects of 208 studies.[1] The nonpoint calculator is a computerized package of loading functions which can use existing information to estimate quantity of nonpoint loads and the effect of these loads on surface water quality. Output from the calculator can be examined to pinpoint areas of nonpoint problems, and to quantify the severity of each of the many nonpoint souces within an area. Thus, it is possible to identify quickly which regions within a study area represent the most serious sources of nonpoint pollution, and to develop strategies for coping with them.

Because every study area will have specific nonpoint problems, the design of the nonpoint calculator is such that it can be used in both small areas (100 mi^2 or less) and large areas (entire states or groups of states). The size of the area is limited to the degree of detailed information available as input; the more detail, the smaller the area that can be considered. The output from the calculator can be formatted to meet the specific needs of the problem being investigated. For example, one can consider annual average nonpoint loads for specific sources, or loads generated during specific storm events. A single important nonpoint source can be studied in depth, *e.g.,* agriculture, or a complex mixture of sources, *e.g.,* agriculture, urban and mining.

This paper will discuss the use of the nonpoint calculator for the Wakarusa River basin in Kansas. This basin was used to develop the nonpoint calculator package, and dealt with one principal land use—agriculture. We will describe what information is needed for input, and discuss inferences that can be drawn from the output.

DESCRIPTION OF THE WAKARUSA BASIN

The Wakarusa River basin comprises an area of about 475 mi^2 in east-central Kansas, and empties into the Kansas River at Eudora. It is primarily an agricultural area with very little urbanization. The major urban center in the basin is Lawrence (population 46,000), drainage from which is split between the Wakarusa and Kansas Rivers (Figure 1).

The first step in using the nonpoint calculator is to designate "regions." Regions are areas of presumably homogeneous land uses which tend to be uniformly distributed in the area. The regions are defined in terms of river miles similar to those used for standard water quality models. The river mile designations needed to describe a stream "reach" for water quality modeling purposes are also defined. Because information available for nonpoint "regions" is not necessarily consistent with the "reach" segments for water quality modeling, the nonpoint calculator treats the two designations independently. The definitions of regions and reaches for the Wakarusa are shown in Table I. Because of the congruence of Shawnee and Osage county lines, the two regions represented by each of the counties reduce to one reach in the stream model.

A second piece of input information is the number of runoff-producing events in an average year. The number of events can be determined by an analysis of precipitation records and corresponding stream flow records. In the case at hand, we evaluated 8 years of daily streamflow data for the gauging station on the Wakarusa and estimated that there are about 13

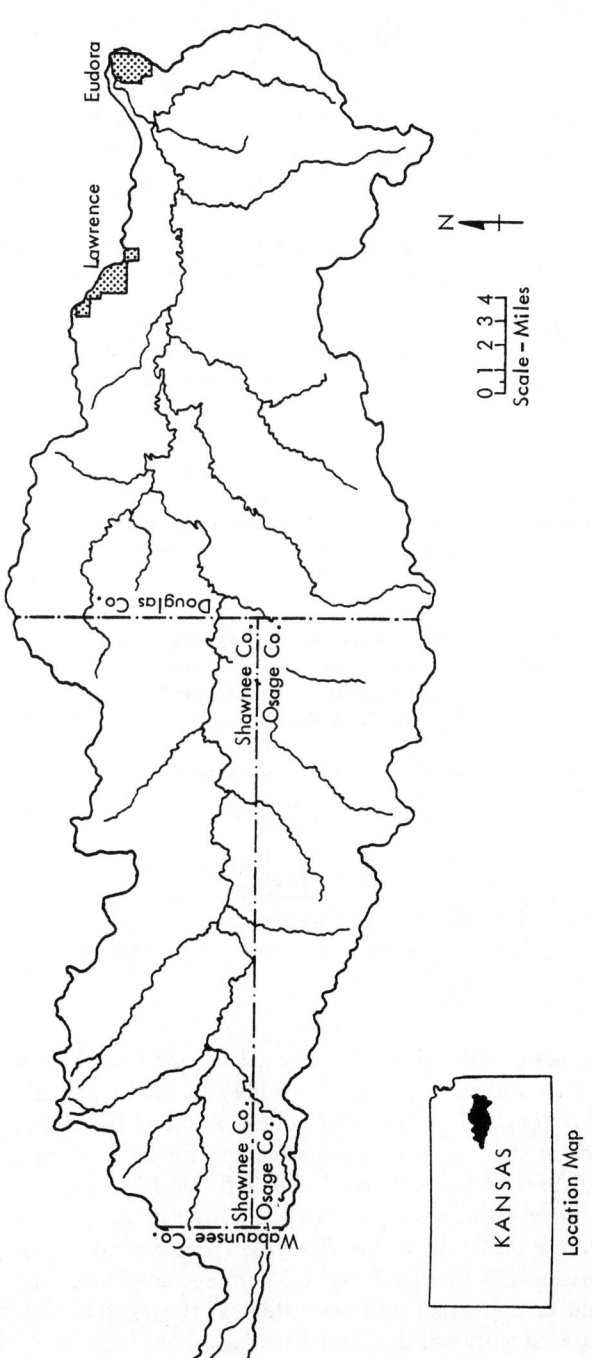

Figure 1. Wakarusa River watershed.

runoff-producing events every year with average duration of 106 hr. The average instantaneous stream flow per event was estimated to be 1,230 cfs for the total basin. In order to estimate concentrations of nonpoint pollutant concentration in runoff after nonpoint levels are calculated, the runoff is distributed among the areas of the contributing regions as shown in Table II.

Table I. Nonpoint calculator regions and reaches for the Wakarusa River basin.

Nonpoint Regions
Wabaunsee County	Mile 67 to mile 60
Shawnee County	Mile 60 to mile 30
Osage County	Mile 60 to mile 30
Douglas County	Mile 30 to mile 0

Nonpoint Reaches
Segment 1	Mile 67 to mile 60
Segment 2	Mile 60 to mile 30
Segment 3	Mile 30 to mile 0

Table II. Runoff-producing events in the Wakarusa River basin.

Region	Number of Annual Events: 13 Runoff by Region per Event: Runoff	Duration
Wabaunsee	30 cfs	106 hr
Shawnee	330 cfs	106 hr
Osage	230 cfs	106 hr
Douglas	640 cfs	106 hr
Total Basin	1,230 cfs	

The third set of descriptive data needed for the basin is land use and land capability. This information is most readily obtained from the *Conservation Needs Inventories* (CNI) for the states in the country. There are provisions for 16 land uses in the nonpoint calculator, and for 29 land capability classes. The 16 land uses in the CNI are listed in Table III along with the 29 land capability classes. One does not need a value for each land use or land capability class. During our study of the Wakarusa River basin, the Kansas Conservation Commission felt that land uses (1) corn and sorghum, and (2) other row crops, should be combined into one category. Having done this, there was an extra land use category which might have been used for a land use other than those defined in the CNI.

Table III. Land use and land capability class.

Land Use (LU)	Land Capability Class (LCC)
1. Corn and Sorghum	1. LCC 1
2. Other Row-Crops	2. LCC 2E
3. Close Grown Crops	3. LCC 2S
4. Summer Fallow	4. LCC 2W
5. Rotated Hay and Pasture	5. LCC 2C
6. Hay Only	6. LCC 3E
7. Conservation Use Only	7. LCC 3S
8. Temporarily Idle	8. LCC 3W
9. Orchards	9. LCC 3C
10. Open Formerly Cropped	10. LCC 4E
11. Pasture	11. LCC 4S
12. Range	12. LCC 4W
13. Other Farmland	13. LCC 4C
14. Other Nonfarmland	14. LCC 5E
15. CNI Commercial Forest	15. LCC 5S
16. CNI Noncommercial Forest	16. LCC 5W
	17. LCC 5C
	18. LCC 6E
	19. LCC 6S
	20. LCC 6W
	21. LCC 6C
	22. LCC 7E
	23. LCC 7S
	24. LCC 7W
	25. LCC 7C
	26. LCC 8E
	27. LCC 8S
	28. LCC 8W
	29. LCC 8C

Finally, the acres of land in the various uses listed in Table III are arranged with respect to land capability classes within the land uses. The arrays for four Kansas counties in which the Wakarusa River basin is located are presented in Tables IV-a through IV-d. The acreages are presented in 100-ac units for the entire county. In order to get acreages within the basin, the 100-ac units are multiplied by the percentage of the county contained within the Wakarusa basin. This procedure assumes that the distribution of land use and land capability class is uniform throughout the entire county, an assumption which is not entirely valid. However, errors introduced by this assumption are likely within the error ranges of the loading functions and hence have little effect on the estimate of nonpoint loads.

Table IV-a. Acres of land by use and capability class: Wabaunsee County (100-ac units).

	LCC 1	LCC 2E	LCC 2W	LCC 3E	LCC 4E	LCC 6E	LCC 6W	LCC 7E
Percentage of Land in Region	1	1	1	1	1	1	1	1
Corn and Sorghum and Other Row-Crops	75	57	57	553	115	0	0	0
Not Used	0	0	0	0	0	0	0	0
Close Grown Crops	72	50	0	237	162	0	0	0
Summer Fallow	0	0	0	0	0	0	0	0
Rotated Hay and Pasture	13	20	0	111	79	0	0	0
Hay Only	0	0	0	99	33	39	0	0
Conservation Use Only	13	6	0	72	138	0	0	0
Temporarily Idle	0	0	0	0	0	0	0	0
Orchards	0	0	0	0	0	0	0	0
Open Formerly Cropped	0	0	0	20	20	0	0	0
Pasture	0	19	16	50	69	3	0	0
Range	6	113	6	1,135	443	2,926	154	248
Other Farmland	3	0	6	26	0	9	11	0
Other Nonfarmland	0	0	0	0	0	0	0	0
CNI Commercial Forest	155	0	38	0	0	0	100	0
CNI Noncommercial Forest	0	0	0	0	0	5	0	16

Table IV-b. Acres of land by use and capability class: Shawnee County (100-ac units).

	LCC 1	LCC 2E	LCC 2W	LCC 3E	LCC 3W	LCC 4E	LCC 4S	LCC 5W	LCC 6E	LCC 6W	LCC 7E	LCC 7W
Percentage of Land in Region	28	28	28	28	28	28	28	28	28	28	28	28
Corn and Sorghum and Other Row-Crops	212	42	74	172	2	32	5	0	0	0	0	0
Not Used	0	0	0	0	0	0	0	0	0	0	0	0
Close Grown Crops	68	36	5	158	0	27	0	0	0	2	0	0
Summer Fallow	0	0	0	7	0	0	0	0	0	0	0	0
Rotated Hay and Pasture	16	14	7	52	0	9	0	0	0	0	0	0
Hay Only	0	14	0	133	0	7	0	0	29	0	0	0
Conservation Use Only	8	67	30	225	0	45	15	0	15	0	0	0
Temporarily Idle	0	0	0	5	0	7	0	0	0	0	0	0
Orchards	0	0	0	9	0	0	0	0	0	0	0	0
Open Formerly Cropped	0	5	0	0	0	0	0	0	0	0	0	0
Pasture	3	25	8	140	0	23	0	0	15	2	0	0
Range	13	53	0	415	0	145	0	0	245	25	36	0
Other Farmland	9	5	3	27	0	3	0	3	5	0	0	1
Other Nonfarmland	5	0	0	2	0	0	0	0	0	0	0	5
CNI Commercial Forest	18	4	18	4	0	18	0	0	32	7	0	103
CNI Noncommercial Forest	0	0	0	0	0	0	0	0	7	0	0	0

532 BMPs FOR AGRICULTURE AND SILVICULTURE

Table IV-c. Acres of land by use and capability class: Osage County (100-ac units).

	LCC 1	LCC 2E	LCC 2S	LCC 2W	LCC 3E	LCC 3W	LCC 4E	LCC 6E	LCC 6W	LCC 7E	LCC 7W
Percentage of Land in Region	13	13	13	13	13	13	13	13	13	13	13
Corn and Sorghum and Other Row-Crops	145	116	37	86	593	21	29	13	0	0	3
Not Used	0	0	0	0	0	0	0	0	0	0	0
Close Grown Crops	28	26	12	31	221	0	12	5	0	0	0
Summer Fallow	0	0	0	0	0	0	0	0	0	0	0
Rotated Hay and Pasture	11	11	2	0	125	0	0	5	0	0	0
Hay Only	5	2	0	2	148	0	9	71	2	2	0
Conservation Use Only	3	12	3	0	215	0	9	12	0	0	0
Temporarily Idle	7	2	0	0	18	0	2	11	0	0	0
Orchards	0	0	0	0	2	0	0	0	0	0	0
Open Formerly Cropped	0	2	0	0	16	0	9	2	0	2	0
Pasture	8	17	2	0	316	0	8	30	7	2	0
Range	17	44	2	20	768	10	51	515	15	27	2
Other Farmland	3	4	1	1	23	0	1	6	2	0	2
Other Nonfarmland	0	0	0	0	26	0	0	13	0	0	0
CNI Commercial Forest	57	0	0	8	0	0	0	0	16	0	171
CNI Noncommercial Forest	0	0	0	0	1	0	0	4	1	3	1

Table IV-d. Acres of land by use and capability class: Douglas County (100-ac units).

	LCC 1	LCC 2E	LCC 2S	LCC 2W	LCC 3E	LCC 3W	LCC 4E	LCC 5W	LCC 6E	LCC 6W	LCC 7E	LCC 8S
Percentage of Land in Region	58	58	58	58	58	58	58	58	58	58	58	58
Corn and Sorghum and Other Row-Crops	130	60	0	131	373	25	69	0	32	9	0	0
Not Used	0	0	0	0	0	0	0	0	0	0	0	0
Close Grown Crops	40	0	0	110	260	0	40	20	30	0	0	0
Summer Fallow	0	0	0	0	0	0	0	0	0	0	0	0
Rotated Hay and Pasture	19	17	2	55	82	0	36	0	2	0	0	0
Hay Only	0	2	0	1	32	0	3	0	14	0	0	0
Conservation Use Only	0	1	0	3	16	4	5	0	2	0	0	0
Temporarily Idle	0	0	0	0	4	0	0	0	0	0	0	0
Orchards	0	0	0	0	1	0	0	0	0	0	0	0
Open Formerly Cropped	0	0	0	0	0	0	0	0	0	0	0	0
Pasture	2	32	0	49	271	6	75	0	172	8	10	0
Range	0	1	0	0	42	0	12	0	25	0	0	0
Other Farmland	2	3	0	1	17	1	2	1	6	1	1	0
Other Nonfarmland	4	0	0	6	2	0	0	0	0	0	0	2
CNI Commercial Forest	11	16	0	166	27	0	11	0	11	59	27	0
CNI Noncommercial Forest	0	0	0	0	2	0	0	0	7	0	2	0

ESTIMATION OF POLLUTANT LOADS

The key pollutant in agricultural nonpoint loads is sediment. In addition to being a pollutant, the loading functions also treat sediment as a carrier of other nonpoint pollutants, *e.g.*, nitrogen, phosphorus, BOD, pesticides and metals. Thus, the first step in estimating pollutant loads is to estimate the annual soil loss on a per unit area basis. This estimation can be done using the Universal Soil Loss Equation (USLE) developed by Wischmeier and Smith[2]:

$$Y = RKLSCP \qquad (1)$$

where Y = annual soil loss, ton/ac
        R = rainfall factor
        K = erodibility factor
        LS = slope length factor
        C = cover factor
        P = practice factor

The rainfall factor, R, is dependent upon the amount and intensity of storm events accumulated over an annual period. For the Wakarusa River basin, an R value of 210 was chosen for all four regions.

The erodibility factor, K, is dependent upon the textural characteristics and composition of soils in the regions. In the nonpoint calculator, K is distributed as a function of land capability class. The K factors for each of the four regions in the Wakarusa basin are tabulated in Table V. A value of zero indicates that the land capability class is not present in the region.

The slope length factor (LS) is determined by relating the slope, S_j (arrayed by land capability class), and the slope length, L_{ij} (arrayed by both land use and land capability class). The relationship is:

$$LS = \left[\frac{L}{72.6}\right]^n \left[\frac{430 \sin^2\theta + 30 \sin\theta + 0.43}{6.574}\right] \qquad (2)$$

where L = slope length (ft)
        θ = slope angle
        n = 0.3 for slopes $\leq$ 3%
        n = 0.4 for slopes between 3% and 4%
        n = 0.5 for slopes > 5%

The slope array by land capability class is of identical format to that of the K factor (Table V), and the length array by land use and land capability class is the same as that used for the arrays as described in Tables IV-a to IV-d. The LS factor computations for each of the four regions are done automatically

Table V. Erodibility factors (K) for the Wakarusa River basin.

Land Capability Class	Wabaunsee County	Shawnee County	Osage County	Douglas County
1	0.32	0.32	0.32	0.32
2E	0.37	0.37	0.37	0.37
2S	0	0	0.37	0.43
2W	0.32	0.32	0.37	0.32
2C	0	0	0	0
3E	0.37	0.37	0.37	0.37
3S	0	0	0	0
3W	0	0.28	0.28	0.28
3C	0	0	0	0
4E	0.37	0.37	0.37	0.37
4S	0	0.43	0	0
4W	0	0	0	0
4C	0	0	0	0
5E	0	0	0	0
5S	0	0	0	0
5W	0	0.37	0	0.37
5C	0	0	0	0
6E	0.37	0.28	0.37	0.37
6S	0	0	0	0
6W	0.32	0.32	0.32	0.32
6C	0	0	0	0
7E	0.28	0.28	0.28	0.28
7S	0	0	0	0
7W	0	0.32	0.32	0
7C	0	0	0	0
8E	0	0	0	0
8S	0	0	0	0.15
8W	0	0	0	0
8C	0	0	0	0

by the nonpoint calculator program using the arrays inputted for slope and length.

The cover factor, C, is dependent upon land use, since different kinds of use will affect the amount and intensity of erosive rainfall falling on soil. As with K and LS, cover factors are chosen for each of the regions considered in the study. The C values used for the Wakarusa basin are shown in Table VI. A value of zero indicates that the particular land use is not present within the region.

Table VI. Cover factors (C) for the Wakarusa River basin.

Land Use	Wabaunsee County	Shawnee County	Osage County	Douglas County
Corn and Sorghum and Other Row-Crops	0.370	0.400	0.360	0.390
Not Used	0	0	0	0
Close Grown Crops	0.220	0.220	0.220	0.220
Summer Fallow	0	0.640	0	0
Rotated Hay and Pasture	0.020	0.020	0.020	0.020
Hay Only	0.008	0.008	0.008	0.008
Conservation Use Only	0.015	0.015	0.013	0.015
Temporarily Idle	0	0.450	0.450	0.450
Orchards	0	0.100	0.100	0.100
Open Formerly Cropped	0.450	0.450	0.450	0
Pasture	0.015	0.015	0.015	0.015
Range	0.008	0.008	0.008	0.008
Other Farmland	0.200	0.200	0.200	0.200
Other Nonfarmland	0	0.250	0.250	0.250
CNI Commercial Forest	0.025	0.025	0.025	0.025
CNI Noncommercial Forest	0.050	0.050	0.050	0.050

The practice factor, P, is the measure of the degree of land conservation practices utilized within the area under study. If no conservation practice is in use, then a P factor of 1.0 was used. If complete conservation practices were in use, a P factor of 0.5 was used. For partial conservation treatment, the P factor varied between 0.5 to 1.0 according to the percentage of acres under conservative treatment.

An array of P factors by land use and land capability class (similar to the acres shown in Tables IV-a through IV-d were developed for each region in the Wakarusa basin. These arrays, along with the others for K, LS and C, were then used along with the R factor of 210 to estimate soil loss in each region of the Wakarusa basin (Tables VII-a through VII-d).

At this point, the nonpoint calculator has calculated soil loss, *i.e.*, the quantity of soil moved offsite during a year of storm events. It has not estimated sediment loads delivered to streams. A delivery ratio is needed to translate soil loss into loads. The choice of delivery ratio depends greatly upon what the goals of the particular study are, and the size of the area under consideration. If one is concerned with a main stem situation, a smaller delivery ratio would be used than would be the case if one were concerned with tributaries to the main stem. Thus, the nonpoint calculator requires that the user choose a delivery ratio which reflects the nature of his program concerns.

Table VII-a. Annual soil loss by land use and land capability class: Wabaunsee County (ton/ac/yr).

	LCC 1	LCC 2E	LCC 2W	LCC 3E	LCC 4E	LCC 6E	LCC 6W	LCC 7E	Average
Corn and Sorghum and Other Row-Crops	4.5	8.5	4.5	17.4	21.9	0	0	0	15.4
Not Used	0	0	0	0	0	0	0	0	–
Close Grown Crops	2.7	5.1	0	10.4	13.0	0	0	0	9.6
Summer Fallow	0	0	0	0	0	0	0	0	–
Rotated Hay and Pasture	0.2	0.5	0	1.2	1.5	0	0	0	1.2
Hay Only	0	0	0	0.7	0.8	1.2	0	0	0.8
Conservation Use Only	0.2	0.4	0	0.9	1.1	0	0	0	1.0
Temporarily Idle	0	0	0	0	0	0	0	0	–
Orchards	0	0	0	0	0	0	0	0	–
Open Formerly Cropped	0	0	0	37.5	47.1	0	0	0	42.3
Pasture	0	0.5	0.2	1.3	1.6	2.3	0	0	1.2
Range	0.1	0.3	0.1	0.7	0.8	1.2	0.1	2.8	1.1
Other Farmland	2.4	0	2.4	16.7	0	30.1	2.1	0	13.6
Other Nonfarmland	0	0	0	0	0	0	0	0	–
CNI Commercial Forest	0.3	0	0.3	0	0	0	0.3	0	0.3
CNI Noncommercial Forest	0	0	0	0	0	0	0	17.3	13.2
Average	1.7	3.0	2.3	6.2	6.0	1.3	0.2	3.7	

Table VII-b. Annual soil loss by land use and land capability class: Shawnee County (ton/ac/yr).

	LCC 1	LCC 2E	LCC 2W	LCC 3E	LCC 3W	LCC 4E	LCC 4S	LCC 5W	LCC 6E	LCC 6W	LCC 7E	LCC 7W	Average
Corn and Sorghum and Other Row-Crops	4.8	9.1	4.8	18.8	4.2	46.4	7.1	0	0	0	0	0	12.1
Not Used	0	0	0	0	0	0	0	0	0	0	0	0	—
Close Grown Crops	2.7	5.0	2.7	10.4	0	25.5	0	0	0	2.4	0	0	9.1
Summer Fallow	0	0	0	30.1	0	0	0	0	0	0	0	0	30.1
Rotated Hay and Pasture	0.2	0.5	0.2	1.2	0	2.6	0	0	0	0	0	0	1.0
Hay Only	0	0.3	0	0.7	0	1.5	0	0	1.2	0	0	0	0.8
Conservation Use Only	0.2	0.4	0.2	0.9	0	1.9	0.3	0	2.3	0	0	0	0.9
Temporarily Idle	0	0	0	26.5	0	58.0	0	0	0	0	0	0	44.9
Orchards	0	0	0	5.9	0	0	0	0	0	0	0	0	5.9
Open Formerly Cropped	0	14.0	0	0	0	0	0	0	0	0	0	0	14.0
Pasture	0.2	0.5	0.2	1.3	0	2.7	0	0	2.3	0.2	0	0	1.3
Range	0.1	0.3	0	0.7	0	1.5	0	0	1.2	0.1	3.7	0	1.0
Other Farmland	2.4	6.2	2.4	16.7	0	36.5	0	2.0	30.0	0	0	79.2	15.3
Other Nonfarmland	3.0	0	0	20.8	0	0	0	0	0	0	0	98.9	46.0
CNI Commercial Forest	0.3	0.8	0.3	2.1	0	4.6	0	0	3.8	0.3	0	9.9	6.1
CNI Noncommercial Forest	0	0	0	0	0	0	0	0	7.5	0	0	0	7.5
Average	3.6	2.8	2.7	4.9	4.2	10.0	2.0	2.0	2.1	0.3	3.7	14.6	

Table VII-c. Annual soil loss by land use and land capability class: Osage County (ton/ac/yr).

	LCC 1	LCC 2E	LCC 2S	LCC 2W	LCC 3E	LCC 3W	LCC 4E	LCC 6E	LCC 6W	LCC 7E	LCC 7W	Average
Corn and Sorghum and Other Row-Crops	4.4	7.8	5.5	5.0	17.0	3.4	21.3	54.2	0	0	100.8	13.3
Not Used	0	0	0	0	0	0	0	0	0	0	0	—
Close Grown Crops	2.7	4.7	3.4	3.1	10.4	0	13.0	33.1	0	0	0	8.8
Summer Fallow	0	0	0	0	0	0	0	0	0	0	0	—
Rotated Hay and Pasture	0.2	0.5	0.3	0	1.2	0	0	3.0	0	0	0	1.1
Hay Only	0.1	0.3	0	0.1	0.7	0	0.8	1.2	0.1	2.8	0	0.8
Conservation Use Only	0.2	0.3	0.2	0	0.8	0	1.0	2.0	0	0	0	0.8
Temporarily Idle	5.4	11.4	0	0	26.5	0	33.3	67.7	0	0	0	33.7
Orchards	0	0	0	0	5.9	0	0	0	0	0	0	5.9
Open Formerly Cropped	0	11.4	0	0	26.5	0	33.3	67.7	0	155.9	0	38.5
Pasture	0.2	0.5	0.2	0	1.3	0	1.6	2.3	0.2	5.2	0	1.3
Range	0.1	0.3	0.1	0.1	0.7	0.1	0.8	1.2	0.1	2.8	3.2	0.9
Other Farmland	2.4	6.2	3.1	2.8	16.7	0	20.9	30.1	2.1	69.3	79.2	18.3
Other Nonfarmland	0	0	0	0	20.8	0	0	37.6	0	0	0	26.4
CNI Commercial Forest	0.3	0	0	0.4	0	0	0	0	0.3	0	9.9	6.8
CNI Noncommercial Forest	0	0	0	0	4.2	0	5.2	7.5	0.5	17.3	19.8	10.7
Average	2.7	4.8	4.2	3.6	6.3	2.3	9.5	4.8	0.3	12.6	12.2	

Table VII-d. Annual soil loss by land use and land capability class: Douglas County (ton/ac/yr).

	LCC 1	LCC 2E	LCC 2S	LCC 2W	LCC 3E	LCC 3W	LCC 4E	LCC 5W	LCC 6E	LCC 6W	LCC 7E	LCC 8S	Average
Corn and Sorghum and Other Row-Crops	4.7	7.7	0	4.7	18.4	3.0	23.1	0	54.7	4.2	0	0	14.5
Not Used	0	0	0	0	0	0	0	0	0	0	0	0	—
Close Grown Crops	2.7	0	0	2.7	10.4	0	13.0	2.2	30.9	0	0	0	9.2
Summer Fallow	0	0	0	0	0	0	0	0	0	0	0	0	—
Rotated Hay and Pasture	0.2	0.5	0.4	0.2	1.2	0	1.5	0	2.8	0	0	0	0.9
Hay Only	0	0.2	0	0.1	0.7	0	0.8	0	1.1	0	0	0	0.8
Conservation Use Only	0.2	0.4	0	0.2	0.9	0.1	1.1	0	2.1	0	0	0	0.8
Temporarily Idle	0	0	0	0	26.5	0	0	0	0	0	0	0	26.5
Orchards	0	0	0	0	5.9	0	0	0	0	0	0	0	5.9
Open Formerly Cropped	0	0	0	0	0	0	0	0	0	0	0	0	—
Pasture	0.2	0.4	0	0.2	1.3	0.1	1.6	0	2.1	0.2	7.0	0	1.5
Range	0	0.2	0	0	0.7	0	0.8	0	1.1	0	0	0	0.8
Other Farmland	2.4	5.8	0	2.4	16.7	1.5	20.9	2.0	28.0	2.1	93.5	0	17.6
Other Nonfarmland	3.0	0	0	3.0	20.8	0	0	0	0	0	0	0	5.1
CNI Commercial Forest	0.3	0.7	0	0.3	2.1	0	2.6	0	3.5	0.3	11.7	0	1.6
CNI Noncommercial Forest	0	0	0	0.6	4.2	0	5.2	0.5	7.0	0.5	23.4	0	9.5
Average	3.6	3.9	0.4	1.9	9.3	2.1	9.4	2.2	11.1	0.7	13.2	0	

The delivery ratio is applied at later steps in the nonpoint calculator computations where one estimates concentrations of nonpoint pollutants in runoff. Suggested delivery ratios for various sized drainage basins developed by the Soil Conservation Service (SCS) are shown in Table VIII.

Table VIII. Delivery ratios for various-sized drainage basins.

Drainage Area (mi^2)	Drainage Area (ac)	Sediment Delivery Ratio
0.01	6	0.65
0.0156	10	0.60
0.05	32	0.50
0.18	115	0.40
0.5	320	0.33
1	640	0.30
5	3,200	0.22
10	6,400	0.18
50	32,000	0.12
100	64,000	0.10
200	128,000	0.08
600	384,000	0.05

Once Soil loss has been estimated, the nonpoint calculator can also compute the percentage of the soil loss for individual cells defined by land use and land capability class, and the percentage of land areas from which the soil loss is being generated. These computations for the four regions in the Wakarusa basin are presented in Tables IX-a through IX-d. Here one can see that 35.4% of the soil loss in Wabaunsee County (Table IX-c) arises from only 7.3% of the land area as defined by the cell corresponding to "row-crops" and "LCC 3E." An evaluation of the other counties indicates row-crops grown on land capability class 3E land are the most pronounced nonpoint sources in the Wakarusa basin, and suggests erosion control on these acres will do much for alleviating nonpoint problems in the Wakarusa basin.

OTHER NONPOINT LOADS

As stated earlier, sediment is a nonpoint pollutant carrier as well as a nonpoint pollutant. Nutrients (nitrogen and phosphorus), organic matter, pesticides and metals are attached to sediment and transported in runoff-carrying sediment. The loading functions used in the nonpoint calculator have "piggy-backed" these pollutants onto the sediment loads in order to make estimates of their quantities delivered to streams. As with sediment, the initial estimate is made using the soil loss without a delivery ratio. Thus, the loads will be factored down in the program.

Table IX-a. Percent of soil loss by land use and land capability class (percents of land area in parentheses): Wabaunsee County.

	LCC 1	LCC 2E	LCC 2W	LCC 3E	LCC 4E	LCC 6E	LCC 6W	LCC 7E	Average
Percentage of Land in Region									
Corn and Sorghum and Other Row-Crops	1.2 (1.0)	1.8 (0.8)	0.9 (0.8)	35.4 (7.3)	9.2 (1.5)	0 (0)	0 (0)	0 (0)	48.6 (11.3)
Not Used	0 (0)	0 (0)	0 (0)	0 (0)	0 (0)	0 (0)	0 (0)	0 (0)	— (—)
Close Grown Crops	0.7 (0.9)	0.9 (0.7)	0 (0)	9.0 (3.1)	7.7 (2.1)	0 (0)	0 (0)	0 (0)	18.4 (6.9)
Summer Fallow	0 (0)	0 (0)	0 (0)	0 (0)	0 (0)	0 (0)	0 (0)	0 (0)	— (—)
Rotated Hay and Pasture	0 (0.2)	0 (0.3)	0 (0)	0.5 (1.5)	0.4 (1.0)	0 (0)	0 (0)	0 (0)	1.0 (2.9)
Hay Only	0 (0)	0 (0)	0 (0)	0.2 (1.3)	0.1 (0.4)	0.2 (0.5)	0 (0)	0 (0)	0.5 (2.3)
Conservation Use Only	0 (0.2)	0 (0.1)	0 (0)	0.2 (0.9)	0.6 (1.8)	0 (0)	0 (0)	0 (0)	0.8 (3.0)
Temporarily Idle	0 (0)	0 (0)	0 (0)	0 (0)	0 (0)	0 (0)	0 (0)	0 (0)	— (—)
Orchards	0 (0)	0 (0)	0 (0)	0 (0)	0 (0)	0 (0)	0 (0)	0 (0)	— (—)
Open Formerly Cropped	0 (0)	0 (0.3)	0 (0)	2.8 (0.3)	3.5 (0.3)	0 (0)	0 (0)	0 (0)	6.2 (0.5)
Pasture	0 (0)	0 (0.3)	0 (0.2)	0.2 (0.7)	0.4 (0.9)	0 (0)	0 (0)	0 (0)	0.7 (2.1)
Range	0 (0.1)	0.1 (1.5)	0 (0.1)	2.8 (14.9)	1.4 (5.8)	12.9 (38.5)	0 (2.0)	2.5 (3.3)	19.8 (66.2)
Other Farmland	0 (0)	0 (0)	0.1 (0.1)	1.6 (0.3)	0 (0)	1.0 (0.1)	0.1 (0.1)	0 (0)	2.8 (0.7)
CNI Commercial Forest	0.2 (2.0)	0 (0)	0 (0.5)	0 (0)	0 (0)	0 (0)	0.1 (1.3)	0 (0)	0.3 (3.9)
CNI Noncommercial Forest	0 (0)	0 (0)	0 (0)	0 (0)	0 (0)	0 (0.1)	0 (0)	1.0 (0.2)	1.0 (0.3)
Average	2.2 (4.4)	2.9 (3.5)	1.0 (1.6)	52.7 (30.3)	23.3 (13.9)	14.1 (39.2)	0.2 (3.5)	3.5 (3.5)	

Table IX-b. Percent of soil loss by land use and land capability class (percent of land area in parentheses): Shawnee County.

	LCC 1	LCC 2E	LCC 2W	LCC 3E	LCC 3W	LCC 4E	LCC 4S	LCC 5W	LCC 6E	LCC 6W	LCC 7E	LCC 7W	Average
Corn and Sorghum and Other Row-Crops	7.6 (7.1)	2.6 (1.4)	2.4 (2.5)	22.1 (5.8)	0.1 (0.1)	10.2 (1.1)	0.2 (0.2)	0 (0)	0 (0)	0 (0)	0 (0)	0 (0)	44.7 (18.1)
Not Used	0 (0)	0 (0)	0 (0)	0 (0)	0 (0)	0 (0)	0 (0)	0 (0)	0 (0)	0 (0)	0 (0)	0 (0)	— (—)
Close Grown Crops	1.2 (2.3)	1.2 (0.12)	0.1 (0.2)	11.2 (5.3)	0 (0)	4.7 (0.9)	0 (0)	0 (0)	0 (0)	0 (0.1)	0 (0)	0 (0)	18.5 (9.9)
Summer Fallow	0 (0)	0 (0)	0 (0)	1.4 (0.2)	0 (0)	0 (0)	0 (0)	0 (0)	0 (0)	0 (0)	0 (0)	0 (0)	1.4 (0.2)
Rotated Hay and Pasture	0 (0.5)	0 (0.5)	0 (0.2)	0.4 (1.7)	0 (0)	0.2 (0.3)	0 (0)	0 (0)	0.2 (1.0)	0 (0)	0 (0)	0 (0)	0.7 (3.3)
Hay Only	0 (0)	0 (0.5)	0 (0)	0.6 (4.5)	0 (0)	0.1 (0.2)	0 (0)	0 (0)	0.2 (0.5)	0 (0.1)	0 (0)	0 (0)	0.9 (6.1)
Conservation Use Only	0 (0.3)	0.2 (2.2)	0 (1.0)	1.4 (7.5)	0 (0)	0.6 (1.5)	0 (0)	0 (0)	0 (0)	0 (0)	0 (0)	0 (0)	2.4 (13.6)
Temporarily Idle	0 (0)	0 (0)	0 (0)	0.9 (0.2)	0 (0)	2.8 (0.2)	0 (0)	0 (0)	0 (0)	0 (0)	0 (0)	0 (0)	3.7 (0.4)
Orchards	0 (0)	0 (0)	0 (0)	0.4 (0.3)	0 (0)	0 (0)	0 (0)	0 (0)	0 (0)	0 (0)	0 (0)	0 (0)	0.4 (0.3)
Open Formerly Cropped	0 (0)	0.5 (0.2)	0 (0)	0 (0)	0 (0)	0 (0)	0 (0)	0 (0)	0 (0)	0 (0)	0 (0)	0 (0)	0.5 (0.2)
Pasture	0 (0.1)	0.1 (0.8)	0 (0.3)	1.2 (4.7)	0 (0)	0.4 (0.8)	0 (0)	0 (0)	0.2 (0.5)	0 (0.1)	0 (0)	0 (0)	2.0 (7.2)
Range	0 (0.4)	0.1 (1.8)	0 (0)	1.9 (13.9)	0 (0)	1.4 (4.9)	0 (0)	0 (0)	2.0 (8.2)	0 (0.8)	0.9 (1.2)	0 (0)	6.4 (31.3)
Other Farmland	0.1 (0.3)	0.2 (0.2)	0 (0.1)	3.1 (0.9)	0 (0)	0.7 (0.1)	0 (0)	0 (0.1)	1.0 (0.2)	0 (0)	0 (0)	0.5 (0)	5.8 (1.9)
Other Nonfarmland	0.1 (0.2)	0 (0)	0 (0)	0.3 (0.1)	0 (0)	0 (0)	0 (0)	0 (0)	0 (0)	0 (0)	0 (0)	3.4 (0.2)	3.8 (0.4)
CNI Commercial Forest	0 (0.6)	0 (0.1)	0 (0.6)	0.1 (0.1)	0 (0)	0.6 (0.6)	0 (0)	0 (0)	0.8 (1.1)	0 (0.2)	0 (0)	7.0 (3.5)	8.5 (6.8)
CNI Noncommercial Forest	0 (0)	0 (0)	0 (0)	0 (0)	0 (0)	0 (0)	0 (0)	0 (0)	0.4 (0.2)	0 (0)	0 (0)	0 (0)	0.4 (0.2)
Average	8.6 (11.8)	5.0 (8.9)	2.7 (4.9)	44.9 (45.3)	0.1 (0.1)	21.7 (10.6)	0.3 (0.7)	0 (0.1)	4.9 (11.7)	0.1 (1.2)	0.9 (1.2)	10.9 (3.7)	

Table IX-c. Percent of soil loss by land use and land capability class (percent of land area in parentheses): Osage County.

	LCC 1	LCC 2E	LCC 2S	LCC 2W	LCC 3E	LCC 3W	LCC 4E	LCC 6E	LCC 6W	LCC 7E	LCC 7W	Average
Corn and Sorghum and Other Row-Crops	2.5 (3.4)	3.5 (2.7)	0.8 (0.9)	1.7 (2.0)	39.6 (13.8)	0.3 (0.5)	2.4 (0.7)	2.8 (0.3)	0 (0)	0 (0)	1.2 (0.1)	54.8 (24.2)
Not Used	0 (0)	0 (0)	0 (0)	0 (0)	0 (0)	0 (0)	0 (0)	0 (0)	0 (0)	0 (0)	0 (0)	— (—)
Close Grown Crops	0.3 (0.7)	0.5 (0.6)	0.2 (0.3)	0.4 (0.7)	9.0 (5.1)	0 (0)	0.6 (0.3)	0.7 (0.1)	0 (0)	0 (0)	0 (0)	11.6 (7.8)
Summer Fallow	0 (0)	0 (0)	0 (0)	0 (0)	0 (0)	0 (0)	0 (0)	0 (0)	0 (0)	0 (0)	0 (0)	— (—)
Rotated Hay and Pasture	0 (0.3)	0 (0.3)	0 (0)	0 (0)	0.6 (2.9)	0 (0)	0 (0)	0.1 (0.1)	0 (0)	0 (0)	0 (0)	0.7 (3.6)
Hay Only	0 (0.1)	0 (0)	0 (0)	0 (0)	0.4 (3.4)	0 (0)	0 (0.2)	0.3 (1.6)	0 (0)	0 (0)	0 (0)	0.8 (5.6)
Conservation Use Only	0 (0.1)	0 (0.3)	0 (0.1)	0 (0)	0.6 (5.0)	0 (0)	0 (0.2)	0.1 (0.3)	0 (0)	0 (0)	0 (0)	0.8 (5.9)
Temporarily Idle	0.2 (0.2)	0.1 (0)	0 (0)	0 (0)	1.9 (0.4)	0 (0)	0.3 (0)	2.9 (0.3)	0 (0)	0 (0)	0 (0)	5.3 (0.9)
Orchards	0 (0)	0 (0)	0 (0)	0 (0)	0 (0)	0 (0)	0 (0)	0 (0)	0 (0)	0 (0)	0 (0)	— (—)
Open Formerly Cropped	0 (0)	0.1 (0)	0 (0)	0 (0)	1.7 (0.4)	0 (0)	1.2 (0.2)	0.5 (0)	0 (0)	1.2 (0)	0 (0)	4.7 (0.7)
Pasture	0 (0.2)	0 (0.4)	0 (0)	0 (0)	1.6 (7.3)	0 (0)	0 (0.2)	0.3 (0.7)	0 (0.2)	0 (0)	0 (0)	2.0 (9.1)
Range	0 (0.4)	0 (1.0)	0 (0)	0 (0.5)	2.0 (17.8)	0 (0.2)	0.2 (1.2)	2.4 (12.0)	0 (0.3)	0.3 (0.6)	0 (0)	5.0 (34.2)
Other Farmland	0 (0.1)	0.1 (0.1)	0 (0)	0 (0)	1.5 (0.5)	0 (0)	0.1 (0)	0.7 (0.1)	0 (0)	0 (0)	0.6 (0)	3.1 (1.0)
Other Nonfarmland	0 (0)	0 (0)	0 (0)	0 (0)	2.1 (0.6)	0 (0)	0 (0)	1.9 (0.3)	0 (0)	0 (0)	0 (0)	4.1 (0.9)
CNI Commercial Forest	0.1 (1.3)	0 (0)	0 (0)	0 (0.2)	0 (0)	0 (0)	0 (0)	0 (0)	0 (0.4)	0 (0)	6.7 (4.0)	6.8 (5.9)
CNI Noncommercial Forest	0 (0)	0 (0)	0 (0)	0 (0)	0 (0)	0 (0)	0 (0)	0.1 (0.1)	0 (1.0)	0.2 (0.1)	0.1 (0)	0.4 (0.2)
Average	3.1 (6.6)	4.4 (5.5)	1.0 (1.4)	2.1 (3.4)	61.0 (57.4)	0.3 (0.7)	4.9 (3.0)	12.8 (16.0)	0 (1.0)	1.8 (0.8)	8.6 (4.2)	

Table IX-d. Percent of soil loss by land use and land capability class (percent of land area in parentheses): Douglas County.

	LCC 1	LCC 2E	LCC 2S	LCC 2W	LCC 3E	LCC 3W	LCC 4E	LCC 5W	LCC 6E	LCC 6W	LCC 7E	LCC 8S	Average
Corn and Sorghum and Other Row-Crops	3.2 (4.8)	2.4 (2.2)	0 (0)	3.2 (4.8)	35.6 (13.7)	0.4 (0.9)	8.3 (2.5)	0 (0)	9.1 (1.2)	0.2 (0.3)	0 (0)	0 (0)	62.4 (30.4)
Not Used	0 (0)	0 (0)	0 (0)	0 (0)	0 (0)	0 (0)	0 (0)	0 (0)	0 (0)	0 (0)	0 (0)	0 (0)	— (—)
Close Grown Crops	0.6 (1.5)	0 (0)	0 (0)	1.5 (4.0)	14.0 (9.5)	0 (0)	2.7 (1.5)	0.2 (0.7)	4.8 (1.1)	0 (0)	0 (0)	0 (0)	23.8 (18.4)
Summer Fallow	0 (0)	0 (0)	0 (0)	0 (0)	0 (0)	0 (0)	0 (0)	0 (0)	0 (0)	0 (0)	0 (0)	0 (0)	— (—)
Rotated Hay and Pasture	0 (0.7)	0 (0.6)	0 (0.1)	0.1 (2.0)	0.5 (3.0)	0 (0)	0.3 (1.3)	0 (0)	0 (0.1)	0 (0)	0 (0)	0 (0)	0.9 (7.8)
Hay Only	0 (0)	0 (0.1)	0 (0)	0 (0)	0.1 (1.2)	0 (0)	0 (0.1)	0 (0)	0.1 (0.5)	0 (0)	0 (0)	0 (0)	0.2 (1.9)
Conservation Use Only	0 (0)	0 (0)	0 (0)	0 (0.1)	0.1 (0.6)	0 (0.1)	0 (0.2)	0 (0)	0 (0.1)	0 (0)	0 (0)	0 (0)	0.1 (1.1)
Temporarily Idle	0 (0)	0 (0)	0 (0)	0 (0)	0.6 (0.1)	0 (0)	0 (0)	0 (0)	0 (0)	0 (0)	0 (0)	0 (0)	0.6 (0.1)
Orchards	0 (0)	0 (0)	0 (0)	0 (0)	0 (0)	0 (0)	0 (0)	0 (0)	0 (0)	0 (0)	0 (0)	0 (0)	— (—)
Open Formerly Cropped	0 (0)	0 (0)	0 (0)	0 (0)	0 (0)	0 (0)	0 (0)	0 (0)	0 (0)	0 (0)	0 (0)	0 (0)	— (—)
Pasture	0 (0.1)	0.1 (1.2)	0 (0)	0 (1.8)	1.8 (10.0)	0 (0.2)	0.6 (2.8)	0 (0)	1.9 (6.3)	0 (0.3)	0.4 (0.4)	0 (0)	4.7 (23.0)
Range	0 (0)	0 (0)	0 (0)	0 (0)	0.1 (1.5)	0 (0)	0.1 (0.4)	0 (0)	0.1 (0.9)	0 (0)	0 (0)	0 (0)	0.3 (2.9)
Other Farmland	0 (0.1)	0.1 (0.1)	0 (0)	0 (0)	1.5 (0.6)	0 (0)	0.2 (0.1)	0 (0)	0.9 (0.2)	0 (0)	0.5 (0)	0 (0)	3.2 (1.3)
Other Nonfarmland	0.1 (0.1)	0 (0)	0 (0)	0.1 (0.2)	0.2 (0.1)	0 (0)	0 (0)	0 (0)	0 (0)	0 (0)	0 (0)	0 (0.1)	0.4 (0.5)
CNI Commercial Forest	0 (0.4)	0.1 (0.6)	0 (0)	0.3 (6.1)	0.3 (1.0)	0 (0)	0.1 (0.4)	0 (0)	0.2 (0.4)	0.1 (2.2)	1.6 (1.0)	0 (0)	2.7 (12.0)
CNI Noncommercial Forest	0 (0)	0 (0)	0 (0)	0 (0)	0 (0.1)	0 (0)	0 (0)	0 (0)	0.3 (0.3)	0 (0)	0.2 (0.1)	0 (0)	0.5 (0.1)
Average	3.9 (7.6)	2.7 (4.8)	0 (0.1)	5.2 (19.2)	54.8 (41.5)	0.4 (1.3)	12.3 (9.3)	0.2 (0.8)	17.4 (11.1)	0.3 (2.8)	2.7 (1.5)	0 (0.1)	

Nitrogen loads are divided into two parts: that associated with soil erosion, and that arising in rainfall. The yield of available nitrogen is given by:

$$Y_N = 20[N] \, Y r_N f_N + \frac{Q_{or}}{Q_{pr}} N_{pr} b \qquad \text{(lb/ac/yr)} \qquad (3)$$

where 20 = dimensional constant
$[N]$ = soil nitrogen concentration (%)
Y = sediment yield (ton/ac/yr), from soil loss computations
r_N = nitrogen enrichment ratio (enrichment due to preferential erosion of small soil particles)
f_N = ratio of available N to total N in sediment (fraction of nitrogen available for biological process)
Q_{or} = overland flow from precipitation (in./yr)
Q_{pr} = total precipitation (in./yr)
N_{pr} = nitrogen load in precipitation (lb/ac/yr)
b = attenuation factor for N in rainfall

For phosphrous:

$$Y_P = 20[P] \, Y r_p f_p \qquad \text{(lb/ac/yr)} \qquad (4)$$

where 20 = dimensional constant
$[P]$ = soil phosphorus concentration (%)
r_p = phosphorus enrichment ratio
f_p = ratio of available P to total P in sediment

For biochemical oxygen demand (BOD) it is assumed that the BOD concentration equals 0.1 times the soil organic matter. Soil organic matter is assumed to be 20 times the soil nitrogen concentration. Therefore:

$$[BOD] = 20[N] \qquad (5)$$

and

$$Y_B = 20Y[BOD] r_N \qquad \text{(lb/ac/yr)} \qquad (6)$$

The enrichment ratio is assumed to be the same for BOD and nitrogen. BOD concentrations and enrichment ratios may also be supplied to the program as data.

Pesticides are treated in the same manner as nutrients. They are assumed to be carried along with sediment and the yield of pesticides is given by:

$$Y_{PT} = 0.002 \, [PEST] \, Y r_{PT} \qquad \text{(lb/ac/yr)} \qquad (7)$$

where [PEST] = pesticide concentration in soil in ppm
r_{PT} = enrichment ratio for pesticide in sediment

Any pesticide can be considered. However, the procedure is most appropriate for pesticides such as paraquat which are strongly absorbed to the sediment. The input information for nutrients and pesticides used for Wakarusa basin are shown in Table X.

Table X. Input information for nutrients and pesticides.

Input	Wabaunsee County	Shawnee County	Osage County	Douglas County
[N] (%)	0.12	0.14	0.12	0.12
r_n	2.0	2.0	2.0	2.0
f_n	0.1	0.1	0.1	0.1
Q_{or} (in./yr)	3.0	3.0	3.0	3.0
Q_{pr} (in./yr)	36.1	29.2	38.5	38.5
N_{pr} (lb/ac/yr)	2.0	2.0	2.0	2.0
b	0.75	0.75	0.75	0.75
[P] (%)	0.05	0.07	0.04	0.04
r_p	1.5	1.5	1.5	1.5
f_p	0.1	0.1	0.1	0.1
[PEST] (ppm)	0.01	0.01	0.01	0.01
r_{PT}	1.0	1.0	1.0	1.0

COMPILATION OF RESULTS

After all loads have been estimated (without delivery ratios), the nonpoint calculator will sum and average the total nonpoint loads for each region. It will also print out the average USLE factors for each region used in the study. The results for the Wakarusa River basin are presented in Table XI.

In the Wakarusa River basin study, it was decided that nonpoint loads should be delivered to the nearest stream offsite as a result of runoff-producing events. Therefore, a sediment delivery ratio of 0.6 corresponding to a 10-ac drainage area was used. This choice implies that the nearest gully is a receiving stream as a consequence of storm events. The annual delivery nonpoint loads for each of the four regions are shown in Table XII.

The nonpoint calculator will also estimate the amount of nonpoint pollutant delivered to the receiving waters on a per event basis by dividing the annual loads by the number of annual runoff events. From the incremental flow rates and the duration of events, the concentration of each pollutant in the runoff can also be estimated. Since concern at this point is with water

Table XI. Nonpoint pollutant loads per acre generated within the Wakarusa River basin (delivery ratio equals 1.0).

Load (lb/ac/yr)	Wabaunsee County	Shawnee County	Osage County	Douglas County	Basin Average
Area (ac)	7,598	83,468	55,965	157,943	304,965
Soil Loss (ton/ac/yr)	7,168	9,814	11,796	14,124	12,344
Nitrogen	1.7	2.7	2.8	3.4	3.1
Phosphorus	0.5	1.0	0.7	0.8	0.9
BOD	3.4	5.5	5.7	6.8	6.1
Pesticides	0.7×10^{-4}	1.0×10^{-4}	1.2×10^{-4}	1.4×10^{-4}	1.2×10^{-4}
USLE Factors					
R	210	210	210	210	
K	0.36	0.35	0.36	0.35	
LS	1.44	1.31	1.21	0.85	
C	0.07	0.11	0.12	0.17	
P	0.97	0.97	0.96	0.94	

Table XII. Annual nonpoint loads delivered to receiving waters in the Wakarusa Basin (lb/yr).

Pollutant	Wabaunsee County	Shawnee County	Osage County	Douglas County	Basin Total
Sediment	0.3×10^8	4.9×10^8	4.0×10^8	13.4×10^8	22.6×10^8
Nitrogen	0.1×10^5	1.5×10^5	1.0×10^5	3.4×10^5	6.0×10^5
Phosphorus	0.0×10^5	0.5×10^5	0.2×10^5	0.8×10^5	1.6×10^5
BOD	0.0×10^6	0.3×10^6	0.2×10^6	0.6×10^6	1.1×10^6
Pesticides	0.3	4.9	4.0	13.4	22.6

quality rather than load, it is more realistic to report concentration in terms of stream reaches rather than regions (Table I). The results for the Wakarusa River basin are presented in Table XIII for the three stream reaches.

Finally, the nonpoint calculator will present the reuslts in an output compatible with standard water quality models. The results for the Wakarusa basin have been made compatible with the QUAL-II model and are shown in Table XIV. Thus, nonpoint loads have been reduced to a format where water quality models can treat them as point sources.

Table XIII. Nonpoint loads and concentrations delivered per storm event in the Wakarusa River basin.

Pollutant Loads (lb/event)	Reach 1 Mile 60-Mile 67	Reach 2 Mile 30-Mile 60	Reach 3 Mile 0-Mile 30
Sediment	0.03×10^8	0.68×10^8	1.03×10^8
Nitrogen	0.07×10^4	1.94×10^4	2.61×10^4
Phosphorus	0.19×10^3	5.80×10^3	6.18×10^3
BOD	0.12×10^4	3.58×10^4	4.94×10^4
Pesticide	0.03	0.68	1.03
Pollutant Concentration (mg/ℓ)			
Sediment	3,500	5,100	6,800
Nitrogen	0.9	1.5	1.7
Phosphorus	0.3	0.4	0.4
BOD	1.7	2.7	3.2
Pesticide (μg/ml)	3.5×10^{-5}	5.1×10^{-5}	6.8×10^{-5}

Table XIV. Nonpoint loads for the Wakarusa basin expressed in compatible format for QUAL-II Water Quality Model.

Parameter	Reach 1 Mile 60-Mile 67	Reach 2 Mile 30-Mile 60	Reach 3 Mile 0-Mile 30
Flow (CFS)	30.0	560.0	640.0
Temperature (°F)	57.2	55.1	57.2
DO (mg/ℓ)	8.3	9.1	8.3
BOD (mg/ℓ)	1.7	2.7	3.2
Nitrogen (mg/ℓ)	0.95	1.45	1.71
Phosphorus (mg/ℓ)	0.26	0.43	0.41

SUMMARY

The nonpoint calculator is a simple tool which can be used to estimate nonpoint loads. Output from the calculator can pinpoint the areas in a region which have significant nonpoint problems, and estimate the severity of the problem in a semiquantitative sense. Output from the calculator can be used further to develop alternatives for managing nonpoint problems to fit the specific needs of particular problem areas. Water quality problems associated with nonpoint pollutants, particularly sediment, should be alleviated by control or management of the source rather than by a posttreatment after nonpoint loads have been generated. The nonpoint calculator stresses the source generation of nonpoint loads so that data become available which can be used to address the control and management of the sources.

The nonpoint calculator can also reduce the different nonpoint loads to a "point" load so that the impact of nonpoint loads on in-stream processes can be estimated. Although this feature has not been evaluated in depth, we are confident that nonpoint calculator output can serve this purpose.

The nonpoint calculator has been designed as a very flexible tool. The user has complete control over the input, and can adjust its output to fit his particular needs. Use of the nonpoint calculator in areas other than the Wakarusa basin described above clearly shows that different applications are associated with different study plans. The nonpoint calculator thus becomes a method of readily accomplishing simple, but tedious calculations and frees the user to address the real issue of nonpoint pollution—what can be done to alleviate the problem?

REFERENCES

1. McElroy, A. D., S. Y. Chiu, J. W. Nebgen, A. Aleti and F. W. Bennett. "Loading Functions for Assessment of Water Pollution from Nonpoint Sources," Report No. EPA-600/2-76-151, U.S. Environmental Protection Agency, Washington, DC (1976).
2. Wischmeier, W. H., and D. D. Smith. "Predicting Rainfall-Erosion Losses from Cropland East of the Rocky Mountains," Agr. Handbook No. 282, USDA, Washington, DC (1965).

37
APPROACH FOR ANALYZING AND MANAGING AGRICULTURAL NONPOINT SOURCES IN THE STATE OF MARYLAND

R. F. Schoenhofer
  State of Maryland Department of Natural Resources
  Water Resources Administration
  Annapolis, Maryland

W. A. Knight
  Soil Conservation Service
  College Park, Maryland

C. V. Hancock
  Soil Conservation Service (on assignment to
  State of Maryland Department of Natural Resources,
  Water Resources Administration, Annapolis, Maryland)

INTRODUCTION

This paper describes the development of a management program for agricultural nonpoint sources of pollution in the State of Maryland and contains the following sections:

1. a description of the State of Maryland, its 208 planning areas and agencies, and a brief history of nonpoint source control efforts before the onset of 208 planning;
2. a description of the present organizational structure of the agencies involved in 208 planning as it relates to agriculture;
3. a description of the philosophy and the bench-mark decisions relating to the development of an agricultural nonpoint source management program; and
4. a description of an information management system developed by the 208 program in cooperation with the Soil Conservation Districts and its application in support of program development.

BACKGROUND INFORMATION ON THE STATE OF MARYLAND AND ITS 208 PROGRAM

Physical

Maryland is quite diversified for its relatively small land area of 6.3 million acres. Three main physiographic provinces which section the state are the Coastal Plain, the Piedmont, and the Appalachian. These are further broken down into the Lower and Upper Coastal Plains, the Eastern and Western Piedmont and the Blue Ridge, Great Valley, Valley and Ridge, and finally, the Allegheny Plateau in the western portion of the state. A variety of topography exists in combination with 750 soil types. Within most areas of the state, soils will be found to have a variety of physical and chemical properties. Approximately 45% of the state is forested with some metropolitan counties still having as much as 40% of their land surface in forest cover. Agricultural land in Maryland accounts for almost an equal share, namely 42% of the state as of the recent 1974 Census of Agriculture. Precipitation in Maryland averages 43 in. annually, ranging from 55 in. at the western end of the state to a low of about 36 in. just 35 mi to the east. Maryland has an estimated population of 4.2 million citizens that impacts this physical environment.

Institutional

Maryland responded to the planning requirements of PL 92-500 in 1973 through the development of a Continuing Planning Process for Water Quality Management Planning.[1] Soon thereafter the state started the development of 303(e) River Basin Plans for 18 river basins. The state took this charge seriously and developed in-depth documents focusing, however, on point source control, incorporating nonpoint source considerations only to a very minor degree. These 18 river basin plans were completed and adopted throughout the period extending from 1975 through 1977.

Section 208 planning was begun by two designated agencies in 1975 and 1976. One of these designated agencies is the Metropolitan Washington Council of Governments which has six counties and the District of Columbia under its jurisdiction.[2] Two of the six counties are located in Maryland. The second designated planning agency is the Baltimore Metropolitan Regional Planning Council.[3] It was designated by the Governor of Maryland in June of 1975. The planning area consists of Baltimore City and its surrounding five suburban counties.

The remaining 16 counties of Maryland comprise the nondesignated planning area. The Water Resources Administration, an agency of the State Department of Natural Resources, is the planning agency for these 16 counties. The Water Resources Administration applied for a 208 grant of $3.3 million in 1976[4] but received only approximately $150,000 from EPA.[5] This funding situation for the nondesignated area has not changed since, with the exception of minor amounts being acquired as supplementary 106 funds.

Since most of the urbanized acreage in the state is located in the designated areas, the Water Resources Administration focused on the refinement of existing programs to manage the agricultural and silvicultural land uses, while the designated agencies studied urban problems in more detail. To facilitate cooperation between the three planning agencies in development of a uniform statewide program for agriculture, the U.S. Department of Agriculture agreed to assign one staff member to each agency under Intergovernmental Personnel Act (IPA) arrangements at the beginning of the planning period.

Legal

It is illegal, by virtue of the general water pollution control statutes of the State of Maryland,[6] to discharge any pollutant into Maryland's waters except as permitted by the Water Resouces Administration. A pollutant is defined as any waste or wastewater discharge from publicly owned treatment works or industrial sources and all other substances which will pollute the waters of the state.

The Legislature enacted Maryland's Sediment Control Law in 1970, the first state program in the nation.[7] This law requires that counties and municipalities adopt grading and sediment control ordinances that state that before land is cleared, graded, transported or otherwise disturbed the proposed changes shall first be submitted to and approved for sediment control by the local Soil Conservation District or the Water Resources Administration in the case of projects on state or federal land. Agriculture is exempted from the sediment control requirements of this act. However, by virtue of the general pollution control statute, agriculture is subject to enforcement action in cases of pollution of the waters of the state.

PRESENT ORGANIZATIONAL STRUCTURE OF THE AGENCIES INVOLVED IN AGRICULTURAL 208 PLANNING

The organizational relationship between the various agencies involved in agricultural 208 planning can best be explained with the assistance of the chart in Figure 1. The chart depicts all involved agencies organized vertically by local, state and federal levels of government. The local level is placed at the top since it is this level of government which most frequently interacts with the citizens, including the farmers. The agencies shown in Figure 1 are grouped into three categories: (1) agencies providing technical assistance and

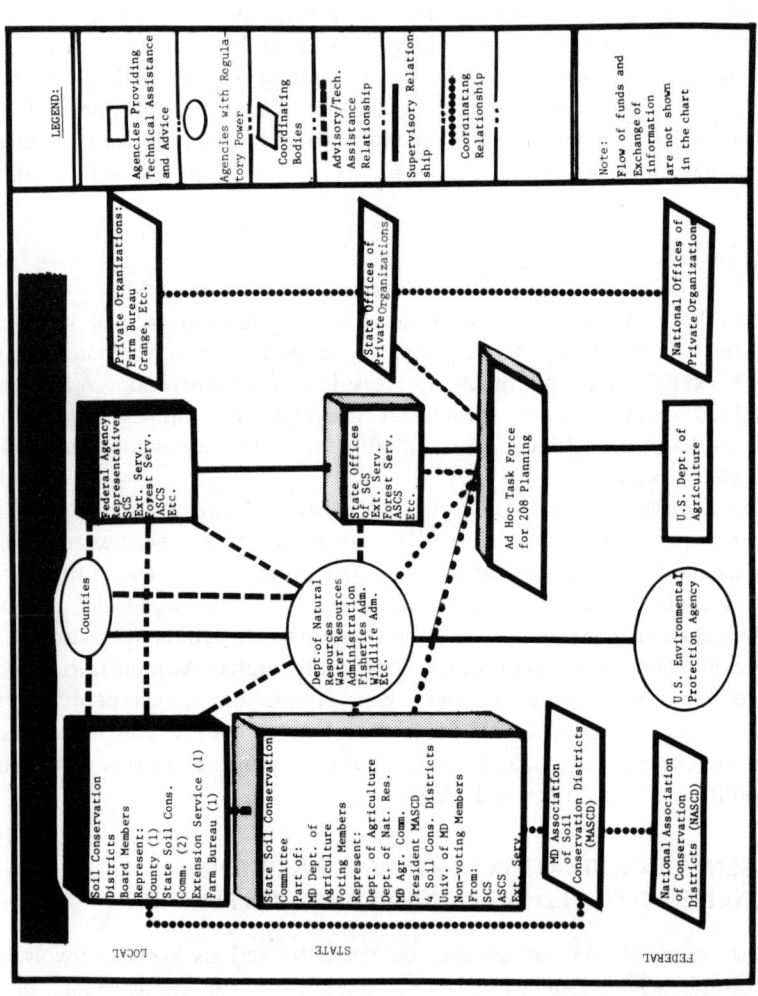

Figure 1. Present organizational structure of agencies involved in agricultural 208 planning.

advice, (2) agencies with regulatory power, and (3) coordinating bodies. The interaction between the varying agencies are similarly grouped into three categories: (1) advisory/assistance relationship, (2) supervisory relationship and (3) coordinating relationship. Keeping these categories in mind, the chart becomes self-explanatory. What remains to be done is an explanation of the interactions particularly important to the development of an agricultural implementation program.

There are two sets of interactions important to the program. One is a triangular relationship between Soil Conservation Districts, the Water Resources Administration and the federal agency representatives at the local level. It is through these channels of communication that much of the day-to-day coordination and decision-making is carried on. A previous District Conservationist from the Soil Conservation Service (SCS) who is on IPA assignment to Water Resources Administration assists in maintaining this triangular relationship. Hand in hand with this relationship goes, of course, close coordination between the Water Resources Administration and the state offices of the involved federal agencies, especially the SCS, the Extension Service, and the Agricultural Stabilization and Conservation Service (ASCS).

The second important set of interactions is somewhat more complicated in nature. Its key component is an "Ad Hoc Task Force for 208 Planning" which was formed under the leadership of the Vice-President for Agricultural Affairs of the University of Maryland and is made up of the leaders of the agricultural agencies and organizations of the state, plus representatives of designated and nondesignated 208 planning agencies. This Ad Hoc Task Force was instrumental in formulating and discussing with the Water Resources Administration the basic concepts of the implementation program. Furthermore, the Ad Hoc Task Force was instrumental in resolving the divergent views on 208 planning held by the various members of the agricultural community. The Ad Hoc Task Force makes recommendations to the Maryland State Soil Conservation Committee which in turn is communicating these recommendations to the districts and coordinates district activities relative to 208 planning as is consistent with its charge under state law.

PHILOSOPHY AND BENCH-MARK DECISIONS RELATIVE TO THE DEVELOPMENT OF THE AGRICULTURAL NONPOINT SOURCE MANAGEMENT PROGRAM

The agencies involved in program development as depicted in the last section decided in the early program stages that the initial effort should, as far as possible, be voluntary. This is consistent with the legislative intent as expressed in the Sediment Control Act which exempts agriculture. The Water Resources Administration, however, will retain ultimate enforcement authority in accordance with the state's general water pollution laws.

A second important decision made in the early program stages was to attempt control of sediment and animal waste initially and to integrate other pollutants such as nutrients and pesticides into the developing framework as more information on these pollutants becomes available. It is well known that sediment acts as a transport agent for nutrients and pesticides. Therefore, the first stage of the management program is expected to also be effective in their control to a degree.

Third, the existing organizational framework and history of the soil conservation movement made the districts appear to be the logical implementation agencies. The districts had in the very early stages, when 208 planning became a statewide issue, voted at one of their annual conventions to participate in the 208 program to the fullest extent.

Based on these three fundamental decisions, the Ad Hoc Task Force proceeded to develop a proposal for an implementation approach, which has since been endorsed by the State Soil Conservation Committee. This proposal incorporates active and passive features. The active feature is a forward-looking implementation effort through the Soil Conservation Districts on a selective basis treating the worst contributing areas first. This involves the selection of critical areas for sediment and animal waste control and channeling resources, manpower and educational efforts into these priority areas. The State Soil Conservation Committee appointed a technical team consisting of various advisory agency representatives and members of the farming community at large to formulate criteria for critical area delineation. The technical team will be developing the criteria for priorities of two types: (1) statewide priorities resulting in the delineation of watersheds by the Water Resources Administration, and (2) priorities within individual farm properties defining areas of greatest potential for erosion and animal waste discharge.

The technical team has further been charged with developing a format for what has become known as a "water quality plan." Water quality plans will be very similar to existing conservation plans, or will be components of these plans. They will draw on Best Management Practice (BMP) Manuals which also will be developed and assembled from existing material by the technical team. The manuals will take account of the diversified nature of the state as explained at the beginning of the paper. Separate manuals will be prepared for each region of the state to accomplish this.

A basic requirement in the technical team's work is that all prepared material be simple enough so it can be understood by individual land owners. The team is trying to develop a simplified system for assessing field-level erosion potential based on the Universal Soil Loss Equation (USLE). Through a combination of factors and simplified tabular display of the remaining inputs to the equation, its complexity has been reduced to the point that farmers can select the appropriate input variables and execute the required simple calculations. A check list for judging the adequacy of animal waste management is also under preparation. Using these assessment tools and the BMP manuals, farmers should be able to formulate their own water quality plans. Districts will assist in this effort and approve plans for technical adequacy.

The second feature of the proposal relates to the development of procedures which will allow fast and satisfactory response to enforcement action which might be taken against land owners by the Water Resources Administration. The intent is to execute the same planning and implementation process as described above. A local pollution task force will investigate the problem and make recommendations for its correction to the district. The district, in turn, will work with the land owner and Water Resources Administration to achieve satisfactory correction.

INFORMATION MANAGEMENT SYSTEM TO SUPPORT THE IMPLEMENTATION EFFORT

In support of the concepts developed in the last section, an information management system is needed which satisfies the following five objectives:

1. assists in identifying sediment and animal waste problem areas for priority area delineation;
2. works through the future implementing agencies, the Soil Conservation Districts, and strengthens cooperation between the Water Resources Administration and the districts;
3. provides districts with an opportunity for improving existing in-house operations independent of the 208 planning process;
4. supplies the basis for continued refinement of the implementation program and progress reporting on accomplishments; and
5. supplies the necessary information to execute technical assessments of agricultural nonpoint source pollution problems.

In developing the information management system, four considerations had to be made: definition of information needs, analysis of existing information flow, identification of information sources, and establishment of the system itself to store, manipulate and retrieve the information. The following is an explanation of the activities carried out under each of these steps.

Definition of Information Needs

From an analysis of the five stated objectives, information needs of two principal types were determined[8]:

1. Information on the conservation treatment status of agricultural land and related managerial information. Examples of this type are acreage adequately treated, number of cooperators and acreage owned by

cooperators, number of conservation plans developed and acreage covered, implementation status of plans, and need for implementation follow–up.
2. Detailed agricultural land use and other physical information required by nonpoint source assessment tools including the USLE and hydrologic models. Examples of this type are field level land use, rotations that can be utilized to define the C factor, information on influence of management on nonpoint source discharge, especially information that can be utilized to define the P factor, and detailed soils and slopes information.

Analysis of Existing Information Flow

There is at present no system available capable of providing these items compatible with program goals. The SCS record keeping and reporting system includes many of the relevant facts. However, accomplishments are mostly reported in their inherent units rather than in relation to the treated land area. The 208 program needed a new system compatible with the SCS system based on readily available sources to the extent possible.

Identification of Information Sources

It was found that most of the desired information was readily available in Soil Conservation District offices and in the files of the local ASCS offices. When this was recognized, the decision was made to implement a manual transfer of the required items from the cooperator folders and ASCS files onto soil survey photography to create a unified information base. This required the negotiation of agreements with the Soil Conservation District and transfer of funds. Initial contacts through the State Soil Conservation Committee were made, a satisfactory framework agreement negotiated and the package then submitted to the 1977 summer meeting of the Maryland Association of Conservation Districts (MASCD) for endorsement which was given. After the general go-ahead from MASCD and the State Soil Conservation Committee, negotiations with district boards were started. Successively inventories were started in most of the State's Conservation Districts.

There were several issues complicating the negotiation process. One was the privacy of the information contained in the cooperator folders and the fear on the part of the district supervisors that the Water Resources Administration might utilize the farm-level information for enforcement activities. This obstacle was overcome by the Water Resources Administration assuring the districts that information would be utilized for management planning and technical work only and furthermore agreeing that the base maps would remain the property of the Soil Conservation Districts. Information would be handed over to the Water Resources Administration only summarized on a small watershed basis. A sample of the resulting base maps is shown in Figure

2. A listing of codes and symbols utilized for mapping is available from the authors upon request.

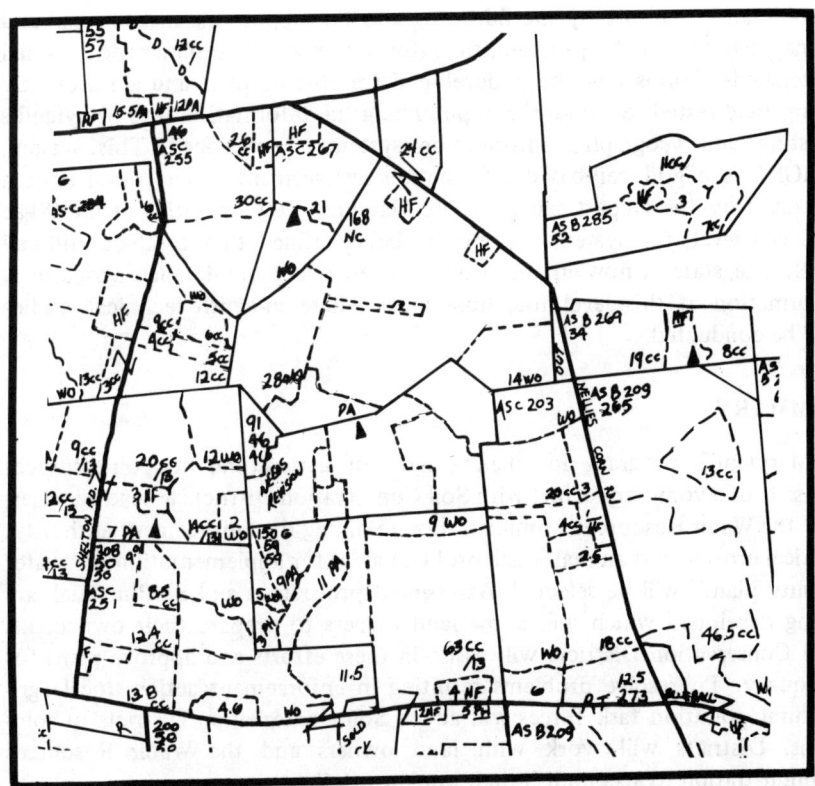

Figure 2. Sample of information base map.

Initial inventory efforts were directed at the establishment of base maps. However, continuing upkeep of the maps is envisioned. This could be accomplished through the assistance of district conservationists who will make entries on the base maps simultaneously with filling in the SCS record keeping and reporting system. The Water Resources Administration is presently attempting to acquire the necessary funds for the continued support of this activity. One of the long-term objectives of the 208 program is to make the map inventory and the SCS record keeping and report systems more compatible.

Establishment of the Information Management System

In a very restricted sense, the base maps themselves constitute an information system since through use of appropriate color codes an instantaneous overview of, for example, the conservation treatment status of any watershed is possible. However, the information will be summarized in two ways for manipulation and use by the 208 program. To support the institutional and managerial part of the program, the information will be summarized by small watersheds. Forms have been developed for this purpose and are presently being field-tested. Secondly, computerizing the information into Maryland's existing and geographic information system is envisioned. This system, MAGI,[9] is a grid cell-based information management and retrieval system operated by the Department of State Planning. Now operating at the 90-ac grid cell level, the system is presently being refined to the 4.5-ac grid cell level. The state is now in the process of encoding the detailed agricultural information at this level for those areas where in-depth technical studies will be conducted.

SUMMARY

Maryland's program for management of agricultural nonpoint sources consists of a voluntary effort with Soil Conservation Districts as lead agencies, and the Water Resources Administration retaining the enforcement authority. Critical erosion and animal waste problem areas for implementation of "water quality plans" will be selected. Assessment procedures and BMP manuals are being developed which will allow land owners to prepare their own plans. Soil Conservation Districts will assist in these efforts and approve plans for adequacy. To resolve problems resulting in enforcement action, local agricultural pollution task forces will advise Soil Conservation Districts of solutions. Districts will work with land owners and the Water Resources Administration to accomplish their implementation.

To identify critical areas and to facilitate refinement of the implementation program and progress reporting on accomplishments, an agricultural information management system has been developed. This system consists of a mapped inventory of managerial and physical information on the farmland in the state. The inventory is executed by Soil Conservation Districts under agreements with the Water Resources Administration. Statistical summaries by small watersheds will satisfy managerial requirements, while computerization of the inventoried information through the state's existing Automatic Geographic Information System will support the analysis of agriculture's contribution to nonpoint source loadings of the state's waters.

ACKNOWLEDGMENTS

The work reported in this paper was partially supported by a planning grant from the U.S. Environmental Protection Agency under Section 208, Federal Water Pollution Control Act Amendments of 1972.

The authors gratefully acknowledge the cooperation of the members of the Maryland State Soil Conservation Committee, Soil Conservation District Supervisors across the state, the staff of the Soil Conservation Service state office in College Park, MD, and the staff of the Planning Section, Water Resources Administration. Special mention is made of Mr. Charles R. Bostater, Jr., who was instrumental in refining and field-testing the information management system.

REFERENCES

1. State of Maryland. "Continuing Planning Process for Water Quality Management Planning," Annapolis, MD (May 1976).
2. Metropolitan Washington Council of Government. "Project Control Program for Metropolitan Washington Water Resources Planning Program," Washington, DC (December 1975).
3. Regional Planning Council. "Project Control Program, Section 208 Areawide Waste Treatment Management Program," Baltimore, MD (April 1976).
4. State of Maryland. "Section 208 Water Quality Management Plan Grant Application," Annapolis, MD (April 1976).
5. State of Maryland. "208 Work Program," Annapolis, MD (September 1976).
6. Section 8-1413(a), Natural Resources Article, Annotated Code of Maryland (1974 Volume, as amended).
7. Section 8-1101 *et sequ,* Natural Resources Article, Annotated Code of Maryland (1974 Volume, as amended).
8. Water Resources Administration. "Methodology for Task 2(e)—Inventories to Support Nonpoint Source Load Assessments," Annapolis, MD (1977).
9. Maryland Department of State Planning. "Maryland Automatic Geographic Information System," Baltimore, MD (December 1973).

38

DEVELOPMENT OF A "208 PLAN" FOR AGRICULTURAL NONPOINT POLLUTION SOURCES IN ILLINOIS

D. H. Vanderholm
 Agricultural Engineering Department
 University of Illinois at Urbana-Champaign

J. F. Frank, A. G. Taylor
 Illinois Environmental Protection Agency
 Springfield, Illinois

INTRODUCTION

The Illinois Environmental Protection Agency (IEPA) was designated as the lead agency in Illinois to produce a comprehensive water quality plan mandated by Section 208 of PL 92-500. The IEPA organized several statewide advisory bodies for consultation on socioeconomic and technical matters related to planning. An Agriculture Task Force of 76 members representing various agricultural and environmental groups was formed to assist IEPA staff in writing the agriculture, forestry and fruit production sections. In a series of 22 monthly meetings from July 1976 through April 1978, the task force developed recommendations for controlling water pollution generated from Illinois farming operations. The task force was divided into six subcommittees, namely, soil erosion, livestock wastes, pesticides, fertilizers, forestry and fruit production. Each subcommittee has studied water quality issues related to its respective area.

A Problem Assessment Report, published by the task force in May 1977, identified agriculture-related water quality problems in Illinois. Recommendations were outlined in a Best Management Practice (BMP) report completed in October 1977. Legal, financial and institutional arrangements for implementing BMPs were proposed in a report completed in February 1978. Five public information meetings were held to obtain public comment before completion of the report.

The task force will forward its final recommendations to the IEPA, which will then write a state water quality plan. Views of the task force may not necessarily parallel those of IEPA. However, the agency will not alter the contents of the task force report. IEPA staff will prepare its draft of an agriculture plan for distribution and comment in early summer 1978.

The sections that follow contain summaries of the findings and recommendations of the six task force subcommittees.

SOIL EROSION

Soil erosion is an agriculture-related water quality problem in Illinois. Sediment produces undesirable conditions in aquatic environments and also transports pesticides and plant nutrients such as phosphorus and organic nitrogen. Erosion estimates for Illinois rural land, based on acreages in the 1967 Illinois Conservation Needs Inventory (CNI) and results of other studies, indicate the following:

1. Most sediment in all counties of Illinois comes from cropland. About 76% of the gross sheet and rill erosion takes place on cropland, which occupies approximately 68% of Illinois rural land. The average annual erosion rate on cropland managed with current practices is estimated to be 5.7 ton/ac/yr.

2. Most cropland erosion occurs in gentle slopes (2 to 5%), class IIe. About 27% of Illinois land is in this category, and most of the land is farmed to nearly continuous corn and soybeans.

3. The highest erosion rates are found on the rolling to steep cropland, class IIIe and above. Where current treatment is inadquate, the rates on this land vary from 16 to 60 ton/ac-yr. Exposed subsoil on much of this land bears testimony to a long history of excessive erosion.

4. On approximately 12 million acres of Illinois cropland, the average annual soil erosion losses do not exceed the soil loss tolerance for sustained crop production. Soil erosion does, however, occur on these lands and may contribute to water quality problems.

5. An estimated 5.7% of the total gross sheet erosion in Illinois comes from pastureland. About 10% of Illinois rural land is used for pasture. Very little erosion occurs on properly managed pastures, but pastures on steeper slopes with only 80% cover averaged as high as 6.5 tons annually. The average for all slopes is 3.1 ton/ac.

6. Statewide, 3.4% of the total gross sheet erosion comes from woodland. A little over 10% of our rural land is in woodland. Erosion rates vary from nearly 0.5 to 9 ton/ac-yr on sloping, grazed woodland, in contrast to rates well below 1 ton for well-managed and well-stocked woodland.

7. Rural land not used for cropland, pastureland or woodland ("other land") contributes an estimated 2.4% of the total gross sheet erosion in Illinois. Strip mines and other highly erosive areas, as well as protected areas such as farmsteads, are included in this category. Erosion rates vary widely, from near zero to over 50 ton/ac-yr. The average annual erosion in this category is 2.6 ton/ac-yr.

8. An estimated 158 million tons of eroded soil is produced annually by sheet and rill water erosion on Illinois' 32.8 million acres of rural land. This is an average of 4.8 ton/ac. Comparing this rate with permissible soil loss standards of 2 to 5 tons for the major soils of Illinois, one might incorrectly conclude that there is no serious erosion problem in Illinois. If one looks beyond the state average, which includes all soils and slopes and excludes the nearly level land where soil erosion is slight, then the erosion rate nearly doubles, to about 9 tons. Erosion from sloping land is about double the commonly accepted permissible soil loss standards and is a serious problem in maintaining soil productivity.

9. An additional 23 million tons of soil from other sources, such as gullies, federal and urban land, and stream banks, brings the total annual erosion for Illinois to 181 million tons.

On the basis of these estimates, the subcommittee feels that most sediment found in Illinois waters comes from sheet erosion on sloping cropland. The greatest potential for lessening sediment therefore lies in establishing BMPs on cropland. Sheet and rill erosion from agricultural land would be reduced to 48 million tons annually—a 70% reduction—if erosion losses on farmlands were reduced to soil loss tolerances set for maintaining soil productivity.

The soil erosion subcommittee is somewhat doubtful that meeting soil loss tolerances on agricultural land will result in significant water quality improvement. Research data prove beyond question that a soil erosion control program will benefit society through sustained soil productivity. With the recommended program, we will be moving in the right direction to achieve water quality improvement and sustained soil productivity. But the subcommittee is concerned that not enough data exist to justify a regulatory program. For this reason, they urge that the programs outlined in the soil erosion goals section be adopted on a voluntary basis.

The concept of Best Management Systems recommended by the subcommittee may involve single practices or combinations selected for specific land or fields. Alternative Best Management Systems for a field take into consideration the soil, slope and topographic characteristics, as well as the farm management needs of the land user. Farm management includes factors such as the type of farming (for example, cash grain or livestock), land ownership, and markets. The land user selects the Best Management System from alternatives appropriate to land conditions, the user's needs and goals, and water quality requirements. Specific practices include conservation tillage, conservation cropping systems, terracing, contour farming, crop residue management, waterways, grade stabilization structures, as well as other practices that have been researched and accepted as effective erosion control measures.

For implementation of BMPs, the subcommittee would use existing federal, state and local agencies wherever possible. No new agencies need to be created, but existing agencies will have additional responsibilities and fiscal and manpower needs. The subcommittee supports a voluntary 208 program. The added responsibility for enforcement would prove to be more harmful than beneficial for most agencies.

The soil erosion and sedimentation subcommittee recommended that soil erosion be reduced to the long-term average annual soil loss tolerance of 2 to 5 ton/ac. The goal can be achieved by implementing Best Management Systems. The program will be costly and will require additional contractors and technical assistance, and a greatly expanded educational program. Therefore, the subcommittee recommended several interim goals, with complete implementation by the year 2010.

Cost estimates of erosion control practices and the consequent changes in land use were obtained by surveying district soil conservationists in Illinois. At 1977 prices, the total bill amounts to $1.3 billion, with 79% allocated to control sheet and rill erosion on cropland. It is estimated that 700,000 ac should be converted to hay and pasture and 131,000 ac to woodland. Per-acre costs would be highest for class IIIe lands, running $299/ac in northern Illinois and $165/ac in southern Illinois.

Practices for controlling sheet and rill erosion on cropland, for which costs are estimated, include contouring and terracing. Tree planting costs are included when planting is required for converting cropland to woodland. Other practices included in an average per-acre cost are grade stabilization structures, grassed waterways, diversions and field borders.

LIVESTOCK WASTES

Current systems for beef, dairy and swine production include pastures, open uncovered feedlots, solid-floor confinement, slotted-floor confinement, and flush-system confinement with a lagoon. Methods of handling wastes include spreading by animals, conventional solid spreaders, tankwagon spreaders, and irrigation. The major water pollutants from manures are oxygen-demanding material (principally organic matter), plant nutrients, and infectious agents. Color and odor are potential pollutants of secondary importance. Review of livestock production and home sewage systems in Illinois produced the following assessment:

1. *Livestock Feedlots.* Animals produced in feedlots, pens and other uncovered enclosures in densities that prevent vegetative cover can present pollution hazards due to runoff of particulate and soluble contaminants. Roofed confinement offers the possibility of a high degree of control over manure disposal. Rainfall runoff is prevented, and discharges from confinement buildings or adjacent manure facilities such as lagoons are highly unlikely except in the case of extreme mismanagement or unusual circumstances. When properly built and managed, roofed confinement facilities present essentially no pollution hazard. The main hazard is pollution from improper manure disposal.

2. *Land Application.* In Illinois, land application of livestock wastes has not been sufficiently monitored to make a defensible assessment of the effects on water quality. Research on winter-spreading has shown that this practice may be a pollution hazard or may actually reduce pollution, depending upon subsequent hydrologic events. Winter-spreading offers several advantages to the producer—less odor, more available labor, more available acres, and less soil compaction. Incorporation has been shown to reduce the pollution potential from surface runoff.

3. *Livestock on Pastures.* Livestock raised on pastureland do not present a major threat to water quality. Isolated cases of water pollution may occur as a result of excessive stocking densities or direct stream access by large numbers of animals.

4. *Milkhouse and Milking Parlor Wastes.* If allowed to directly enter streams, the discharges from milking centers would have a high pollution potential. Some direct discharges are made in Illinois, but this is not a common practice. As dairy facilities are rebuilt or remodeled, these discharges are usually eliminated.

5. *Silage Leachate.* Silage leachate is highly polluting and has the potential to create water quality problems if allowed to reach ground or surface waters. Since 1965 four fish kills have been attributed to silage leachate involving high moisture silage from cannery wastes of peavines and sweet-corn husks. Silage leachate is rarely produced in large quantities in those locations where it can readily enter ground or surface water. In the few instances in which silage leachate finds its way into state waters, the impact on water quality is significant.

6. *Poultry Watering.* Most poultry operations use some type of cup watering system which eliminates the problem of contaminated overflow water. Effluent from continuous-flow systems does not appear suitable for discharge into state waters. The extent of water pollution problems resulting from egg processing facilities is not known.

7. *Private Home Sewage Systems.* Unsatisfactory home sewage disposal systems constitute a health and nuisance hazard in some areas of Illinois.

The problems are caused by poor design, improper location and poor maintenance, and by effluent entering farm drainage tiles, road ditches, drainage ditches, and other surface water courses. Disposal of sludge from septic tanks can also present problems if improperly handled.

Recommended BMPs to prevent water pollution from livestock operations include:

1. *Land Application.* Applications should not exceed the agronomic nitrogen rate. It may be advisable to apply wastes at the agronomic phosphorus rate in some instances to prevent excessive phosphorus buildups and to use fertilizer nutrients more efficiently. Avoid applications where adequate soil erosion control practices do not exist or where soil loss tolerances are exceeded. Use injection or rapid incorporation on slopes greater than 5% unless runoff and erosion are adequately controlled.

If wastes are applied to frozen or snow-covered land, limit applications to land with slopes of 5% or less or to areas with adequate erosion control practices. Do not apply within 150 ft of any water well, on land subject to flooding once in 10 years without incorporation, within 200 ft of surface water unless the water is upgrade or diking is adequate, or in waterways. Do not apply livestock wastes under conditions that would cause surface runoff.

2. *Manure Storage and Handling.* Solid manure storage areas should be big enough so that land application can be done when conditions are favorable. Storage areas should be protected from runoff and should not be constructed on porous soils without effective sealing. Temporary stacks should be protected from runoff and leaching and should not be constructed within 100 ft of a water well. Provide adequate liquid manure storage capacity and management so that overflow does not occur and hauling during unsatisfactory conditions is unnecessary. Pump lagoon effluent onto cropland as necessary to prevent lagoon overflow.

3. *Open, Uncovered Feedlots.* Diversions, channels and roof gutters should be used to minimize the amount of water traversing feedlot surfaces. Prevent discharge of feedlot runoff by collecting and directing it through runoff control devices such as settling basins, holding ponds, and vegetative filter systems. The vegetative filter concept is being researched by University of Illinois and Southern Illinois University personnel to develop effective design criteria.

4. *Pasture Production.* Maintain a stocking density that does not overtax vegetation under normal conditions. Maintain a good vegetative cover through pasture rotation and by locating feeding stations, salt licks, and so forth at reasonable distances from watercourses; follow good pasture fertilization practices. Where practical, use fences along streams. Where the situation

warrants, fence to provide limited access for watering. Downstream water use should be considered when deciding the feasibility of stream fencing. When planning new pastures where severe hazards due to soil conditions or topography exist, consider locating pastures away from streams and, if necessary, install fences to prevent direct access or to provide limited access. More research is needed to quantify the effects of direct animal access to streams.

5. *Milkhouse and Milking Parlor Wastes.* Where possible, adapt facilities so that milking center wastes are stored and handled with other wastewaters, such as liquid manure or lot runoff, for eventual land application.

6. *Silage Leachate.* Ensile at the proper moisture content, preferably 65% wet basis or less. Locate silos away from surface waters, underground springs, or highly permeable soils. Seal silo floors to minimize the possibility of groundwater pollution. Divert surface runoff from trench and bunker silos. Direct the silage leachate that is produced to a collection device, such as a pit or lagoon, for eventual application to cropland.

7. *Poultry Watering and Egg Processing.* Use cup waterers when possible. If a continuous-flow system is used, adjust watering troughs and flow rates to minimize overflow. Dispose of wastewaters from poultry watering or egg processing systems by lagooning and land application, through vegetative filter systems, by septic tank seepage field systems, or by some other acceptable method.

8. *Private Home Sewage Disposal.* Use only home sewage systems approved by the Illinois Department of Public Health. Evaluate site characteristics thoroughly, size the system properly, and adapt the system if necessary for satisfactory performance. Where site conditions permit, the conventional septic tank system with a seepage field is recommended. Ensure that recommended maintenance procedures are carried out.

The existing organizational structure and regulatory authority are adequate for implementing recommended BMPs. Major changes in traditional roles and organizational responsibilities are unnecessary, but changes in program emphasis and level of activity may be essential. Most agencies and organizations performing a major role in 208 planning implementation will require higher levels of funding to increase staff and expand activities.

Regulatory jurisdiction for livestock waste control exists within the IEPA and the Illinois Pollution Control Board under the Illinois Environmental Protection Act as amended, Title 3: Water Pollution. Illinois also has a National Pollution Discharge Elimination System (NPDES) regulatory authority. The Private Sewage Disposal Licensing Act provides authority to the Illinois Department of Public Health "to license private sewage disposal contractors and to estabish and enforce a minimum code of standards for design, construction, materials, operation and maintenance of private sewage disposal systems, for the transportation and disposal of wastes therefrom, and for private sewage disposal systems serving equipment."

Cost estimates for structural measures necessary to control runoff from Illinois feedlots indicate that implementation is expensive. Total costs for holding-pond systems with irrigation could be as high as $86 million, or for vegetative filter systems as high as $73 million. These estimates reflect installation costs without maintenance or operation costs. Bearing the full expense seems an unreasonable burden to place on livestock producers, so it is recommended that cost-sharing programs be greatly expanded in this area. For major pollution control devices, cost-sharing with 75% minimum is recommended with no annual limitation on the amount received by any one cooperator. Other economic incentives may be used as well.

PESTICIDES

Pesticide use can affect water quality in several major ways: (1) by the presence of pesticides in water and in aquatic organisms, (2) by the disposal or recycling of used pesticide containers, and (3) by the disposal of dilute rinsate solutions from sprayers. The 1976 Pesticide Use Survey for Illinois, a Southern Illinois University research project on Disposal and Recycling of Used Pesticide Containers, and a University of Illinois research project on Disposal of Dilute Rinsate Solutions from Commercial Pesticide Spray Equipment were generated by the pesticide subcommittee to provide information for assessment. From data currently available, the subcommittee concluded the following:

1. Pesticides are essential to agriculture as we know it today.

2. All citizens should be concerned about clean water as a valuable natural resource, and every reasonable effort must be made to create and maintain high standards of water quality.

3. Some of the persistent organochlorine insecticides, especially DDT and dieldrin, and the fungicide mercury continue to contaminate water and aquatic organisms. Insecticide contamination is due to residues persisting in the soil from previous use. Mercury is also a contaminant, but the source of contamination is unknown.

4. Residues of agricultural pesticides in water or in aquatic organisms are not being reported for one or more of the following reasons: (1) the pesticides are very biodegradable, (2) they tend not to magnify in the aquatic environment, (3) monitoring for currently used pesticides in water or in aquatic organisms is insufficient.

5. The pesticides that have historically caused problems in water and aquatic organisms have low water solubility, and most are no longer used.

6. The most significant nonpoint sources of water contamination from pesticides are wind or water erosion. Since many pesticides are tightly bound to organic matter and clay in soils, most movement from the site of application in the field to water probably occurs through soil erosion.

7. Cleanup procedures for used containers, as practiced by farmers and commercial applicators, generally follow recommended EPA guidelines. Containers are triple-rinsed (the rinse water being added to the spray material), punctured and temporarily stored in or near the mixing site.

8. Used pesticide containers and drums are sometimes disposed of in woodlots, along stream banks, and in privately owned dump sites.

9. It is not apparent that present disposal of dilute rinsate solutions from pesticide spray equipment is affecting state water quality. However, commercial applicators are extremely interested in a more satisfactory method of handling these waste materials.

10. Research is needed on the fate and effect of pesticides in water and aquatic organisms, on the disposal or recycling of pesticide containers, and on the disposal of rinsate solutions from spray tanks.

11. Greater emphasis should be placed on pest-management research that focuses on integrating pest management and crop production systems and user implementation of these systems.

To assure a productive agriculture and a viable terrestrial and aquatic environment the pesticide subcommittee recommends that:

1. A nonsalaried state pesticide monitoring board be appointed by the Illinois Interagency Committee on Use of Pesticides to coordinate establishment and supervision of a state pesticide monitoring program. Limited to seven members including a person actively engaged in farming, a representative of the agricultural chemical industry, an IEPA agent, and a person from the Illinois Department of Agriculture serving as chairman, the board will report findings and recommendations annually to the Interagency Committee on Use of Pesticides.

A board appointee will serve as project leader of the program. He shall not be an Illinois Department of Agriculture or IEPA employee. Activities under his supervision will include: (1) systematic monitoring for pesticides in water, sediment, fish and other appropriate indicator organisms to detect seasonal or annual levels of pesticide residues in Illinois' aquatic environment, at minimum startup costs of about $200,000; (2) evaluation of all pesticides used in Illinois in the Metcalf laboratory model aquatic ecosystem to identify compounds with potential for persistence and biomagnification in the aquatic environment, at an annual cost of $40,000; (3) an annual survey of Illinois pesticide use in agriculture, costing up to $40,000 per survey; (4) economic analysis, only when necessary, of the impact on Illinois agriculture, the environment, and health and welfare of Illinois citizens whenever regulation is proposed to prohibit or restrict major use of a specific pesticide in Illinois, at estimated costs per study of about $100,000.

Best use should be made of existing state facilities, personnel and agencies to reduce annual costs, and activities should be cost-shared whenever possible with the federal government. The Interagency Committee on Use of Pesticides will review the state pesticide monitoring program every 3 years. The tenure of board members and the need to continue, discontinue or initiate new activities will be at the committee's discretion.

2. Used metal pesticide containers should be recycled or disposed of in a manner to prevent human hazard and contamination of Illinois' water resources. Preventive measures should include: (1) intensified instructional programs to educate pesticide users on disposal of pesticide containers; (2) distribution of informational circulars at point of sale, describing proper preparation of pesticide containers for disposal, alternate disposal methods, and identification of nearby sites for recycle or landfill disposal; (3) encouraging private contractors to pick up and recycle used metal pesticide containers for profit.

Designated as the implementing agency, the Illinois Department of Agriculture should seek educational assistance from Cooperative Extension Service staff and guidelines on container pick-up and sites for disposal or recycling from the IEPA. Implementation will require about $25,000 for producing educational materials.

3. The subcommittee did not make recommendations on disposal of dilute rinsate water from sprayers, pending reports on a special research project in progress at the University of Illinois.

4. The state should identify agronomic and soil conservation practices and provide economic incentives to reduce soil erosion to acceptable limits. Subcommittee on erosion and sedimentation recommendations designed to reduce cropland erosion should be followed. Practices that reduce movement of soil to water will reduce pesticide movement into aquatic systems. The state of Illinois should exclude from local real property taxation all unimproved pasture not used for grazing, agricultural croplands previously devoted to annual crops and converted to grass waterways, grass barrier strips, tree or shrub windbreaks, and all existing grass waterways, grass barrier strips, and windbreaks taxed as cropland but intentionally established by man (vs natural means) for reducing soil erosion. Implementation will require appropriate legislation.

5. Illinois should fund pest management research that focuses on integrating pest control and crop production systems and on educational programs that aid user implementation of these techniques. Implementation, financed

by appropriated state funds, should be directed by the University of Illinois, College of Agriculture. Research would be a coordinated, interdisciplinary effort of University of Illinois, College of Agriculture and Illinois Natural History Survey personnel. Related educational programs essential to grower implementation and use should be conducted by extension specialists and county advisers in the Illinois Cooperative Extension Service. About $600,000 annually, equivalent to $.025/ac of agricultural cropland in Illinois, is needed for basic and developmental research to implement the pest management program.

FERTILIZERS

Eighteen public water supplies in Illinois periodically exceed the nitrate level recommended to prevent methemoglobinemia in infants. A majority of Illinois surface waters violates phosphrous and heavy metal standards at intervals throughout the year. To determine the effects of fertilizer use on these water quality parameters, the fertilizer subcommittee conducted an extensive literature review; compared available data for sales, water quality monitoring, crop production and so forth; and consulted with specialists in hydrology, geology and agronomy. The results were inconclusive. There are conditions that may degrade, may have no effect on, or in some instances may benefit water quality. Observations of specific issues were as follows:

1. *Nitrogen Fertilizer Use.* Watersheds and counties having higher than average fertilizer use do not account for the majority of incidents related to high nitrate levels in municipal water supplies. Further, in these areas there are more communities without high nitrate levels than there are with problems. Data available on nutrients in river impoundments in Illinois show no correlation with fertilizer use in those same watersheds.

Additional use of fertilizer should lead to an increased loss of nitrates from farmland, but the amount of increase cannot be easily determined. A direct connection between fertilizer use and water quality cannot be supported by any available data. Neither do the data eliminate fertilizer use as a contributing factor. Nitrate-contaminated water is more likely to result from excessive application rates or improper timing of application than from recommended use of fertilizers to achieve target yields.

2. *Phosphorus Fertilizer Use.* While phosphorus levels are a legitimate concern, there is no evidence that agricultural use of fertilizer is the major culprit. Certain practices, such as surface application on no-till or minimum-till land and application on frozen soils, do appear to increase nutrient content of runoff water. Where soil test results indicate that only maintenance phosphate fertilization is needed, excessive rates may contribute to this

problem. Because of low solubility, the movement of phosphorus into waterways is most closely associated with soil erosion.

3. *Sewage Sludge Utilization.* Sewage sludge, a valuable nutrient and soil conditioning resource, can be used on Illinois farmlands without creating undue problems.

4. *Heavy Metal Violations.* Micronutrients are used to a very limited extent in Illinois agriculture and cannot account for heavy metal violations in the state's waters.

5. *Other Considerations.* Besides the truly nonpoint sources of pollution there are thousands of small unregulated point sources, which by their very nature cannot be practically regulated as point sources; hence their effects must be considered in total as a nonpoint source.

The fertilizer subcommittee concluded that substantial water quality problems resulting from fertilizer use do not prevail and are not likely to develop in the foreseeable future on a statewide basis. A potential for water quality degradation does exist when recommended practices are not followed.

The subcommittee recommended the following BMPs to minimize possible effects of fertilizers on water quality:

1. Soil testing, used on a regular basis, is the most important available guide to proper use of fertilizer. Every farm operator should have soil tests made at least once every 4 years and should use the results of this diagnosis in planning a lime and fertilizer program.

2. Soil productivity information, based on both soil type and previous cropping history, should be combined with soil test results when choosing the amount and kind of fertilizer to use.

3. The current issue of the *Illinois Agronomy Handbook,* available through the University of Illinois Cooperative Extension Service, should be used as a guide for determing lime and fertilizer needs. Do not exceed the levels of fertilizer application recommended in this handbook.

4. Proper timing of fertilizer applications to agronomic crops is very important, especially for nitrogen applications. Nitrate forms are subject to leaching and, along with other forms, are subject to denitrification. The following points should be considered when planning any nitrogen fertilizer application that does not meet an immediate need for growing crop plants: (1) The most efficient means of applying nitrogen is side-dressing or at planting time. (2) The safe soil temperature for application of non-nitrate forms is 50°F at a 4-in. depth. Significant losses can occur at higher temperatures. (3) Nitrification inhibitors may reduce the potential for nitrogen loss when conditions are favorable for leaching and denitrification. The economic and environmental advantages of inhibitor use have not been fully established,

although research has shown they may be beneficial. (4) Fall application of nitrate forms of nitrogen are not recommended.

5. Do not apply fertilizer on the surface of barren or frozen soil having more than a 5% slope nor leave fertilizer on the surface without proper incorporation.

6. Proper measures to control soil erosion should be taken to prevent loss of applied nutrients in surface runoff.

7. To avoid excessive nutrient losses, special attention should be given to proper fertilizer amounts and to timing and methods of application on sandy soils, highly organic soils, or poorly drained areas.

8. "Buildup" applications of fertilizer should be used only where and when soil test results clearly show a need for such applications.

9. Farm manures and sewage sludges should be applied only in accordance with current pending guidelines. Current recommendations of the IEPA and instructions in the *Illinois Agronomy Handbook* should be followed. Amounts of plant nutrients in manure or sludge should be considered when planning a fertilizer program, with rates of chemical fertilizers reduced accordingly.

10. In Illinois nitrogen fertilizer is used primarily for corn production. A simple corn-soybean rotation appreciably reduces the added nitrogen required per acre of corn.

The subcommittee also listed the following recommendations that, while not BMPs in the strict sense, were considered important in addressing overall problems in this area:

11. Educational programs need to be intensified (1) to inform plant operators, dealers and distributors about acceptable procedures for avoiding dangers associated with accidental leaks and spills or with deliberate dumping of fertilizers and other agricultural chemicals, including rinsewater from cleaning operations, and (2) to inform users about proper disposal methods for used agricultural chemical containers.

12. Additional research is needed on the movement of applied nutrients from treated soil into waters of the state.

13. Alternate solutions to the problem of high nitrate levels in public water supplies should be considered and costs estimated for providing alternate water sources or treatment facilities to reduce excessive nitrate levels to acceptable levels.

Recommendations 1 through 10 can best be implemented by expanding the program of the Illinois Agricultural Extension Service and by employing specialists in farm planning, fertilizer needs evaluation, and soil conservation. The specialists could both educate farmers in BMPs and measure the results of BMPs used in the area. Cost of this program will be a minimum of $500,000.

The fertilizer subcommittee feels that implementation can be satisfactorily approached through an educational program designed to get more farmers to adopt recommended BMPs. Educational programs in recommendation 11 can be handled through existing networks. The 4-year agricultural colleges are already incorporating pertinent parts of these recommendations into existing courses. Through the articulation process, 2-year community colleges will follow. The Illinois Cooperative Extension Service will emphasize these needs in some of their regular programs. A central coordinating effort is needed to make sure that the state's educational institutions, the Illinois Department of Agriculture, and the IEPA remain actively interested, are alert to needed changes, and are working toward the same goals. One additional state specialist with the Illinois Cooperative Extension Service could handle these program needs. Costs would be about $50,000 per year.

Research mentioned in recommendation 12 will be expensive, but the results are sorely needed. Costs could be as much as $2 million to $3 million in one-time expenditures for land and research facilities, and $1 million annually to conduct research projects. Such a program should be operated by the Illinois Institute for Environmental Quality and contracted to agricultural colleges having research capabilities in areas where information is needed.

The subcommittee did not recommend any regulatory programs, because regulations restricting certain fertilizer use practices would require a large staff and would be very difficult to enforce. Further, the high cost of restricting fertilizer use, increased food prices, and taxes to support regulatory programs would tend to obscure the benefits from improved water quality. Moreover, available research does not provide enough information to determine the effects of various levels of fertilizer restriction on water quality.

FORESTRY

Basing its conclusions on limited qualitative observations and on the results of a survey of all Illinois district foresters, the forestry subcommittee feels that overall water quality degradation in Illinois due to forestry activities is relatively small. Localized problems from timber harvesting are few and are attributed to poor management.

Sediment is assumed to be the major water quality degrading factor from Illinois' forested areas. Fertilizers and pesticides are not extensively used on Illinois timber. Activities that generate sediment are: (1) methods of forest product removal, (2) conversion of forested areas to other uses, (3) livestock grazing, and (4) inappropriate use of recreational vehicles. Only the first of these is considered a silvicultural activity. The subcommittee concluded that: (1) conversion of forested land to any other use produces a negative impact on water quality, and many previously converted areas should be reestablished in timber; (2) livestock should be excluded from land specifically managed for forest and wood production, especially where their presence may increase erosion or the potential for sediment or other pollutants reaching streams; (3) recreational vehicles should be allowed on and within forested lands wherever terrain and soils are stable enough to withstand such use. However, such vehicles should not be allowed in sensitive areas where they may cause environmental degradation.

Methods for removing forest products are considered in the subcommittee's comprehensive Best Management Practice report. Recommendations range from techniques in planning timber harvest to constructing stream crossings and drainage systems. Vegetative filter strips are also suggested. The forestry subcommittee recommended that the Illinois 208 plan certify that while localized and sometimes severe instances of water quality degradation due to silvicultural activities may exist, overall adverse effect on water quality within the state is minimal and not significant enough to warrant a regulatory program.

A voluntary, nonregulatory program will be effective and will be the most practicable method of assuring that nonpoint pollution associated with silvicultural activities is minimized. However, it is recognized that the voluntary, nonregulatory program may not fully achieve desired water quality goals, and additional courses of action may have to be implemented in the future. Accordingly, the subcommittee developed a three-phase program designed to meet such contingencies.

Phase I: Voluntary, Nonregulatory Program. This program consists of two parts. The first part emphasizes information, education and training, with assistance to land owners and timber operators. The second part emphasizes legislative and program changes needed to provide incentives for forestland owners and timber operators.

Phase I should begin by October 1, 1978, or as soon as feasible. Every effort should be provided to allow the voluntary, nonregulatory program to be effective. It is recommended that the voluntary, nonregulatory program be allowed to function unhindered until December 31, 1980. Effectiveness of this program should then be evaluated in 1981 by the implementing agency and the IEPA and a decision made about proceeding with Phase II. If judged essential, Phase II should be implemented by January 1, 1982.

Phase II: Voluntary, Semiregulatory Program. Similar to the voluntary, nonregulatory program, this phase has specific guidelines which the land owner or timber operator will be expected to follow. These guidelines include mandatory training and certification of timber operators and monitoring and reporting of performance. Evaluation of the effectiveness of Phase II should take place in 1984 and Phase III implemented—only if absolutely necessary—by July 1, 1985.

Phase III: Mandatory, Regulatory Program. This program consists of well-defined regulations which must be followed. Failure to follow the prescribed regulations may result in legal sanctions such as stiff fines or other penalties.

The forestry subcommittee wishes to go on record as being in strong opposition to a mandatory, regulatory program for silvicultural activities. The committee feels that such a program would have an adverse impact on forestry in the state and would quite probably cause many presently forested areas to be converted to nonforest usage, with an accompanying greater potential for water quality degradation.

The forestry subcommittee also recommends that funding be provided: (1) to develop background and baseline water quality data associated with natural, undisturbed forested areas; (2) to evaluate the effectiveness of existing and proposed BMPs used alone or in combination; (3) to evaluate the economic implications of BMPs used alone or in combination; (4) to develop more effective and acceptable BMPs; (5) to develop effective and equitable means of implementing recommended BMPs; and (6) to adequately support the implementing agency's efforts to administer the 208 program.

It is recommended that the Illinois Department of Conservation, Division of Forestry be the lead agency in developing and guiding all nonpoint-source pollution programs designed to maintain and improve the quality of Illinois streams flowing through forested areas.

Costs of implementation estimated by the subcommittee emphasize the fact that funding a regulatory program is much more expensive than funding a nonregulatory program. Minimum annual costs for implementing Phase I the first 3 years approach $260,000 and for Phase II, $320,000. Costs for Phase III increase from $1.35 million the first year to $2.35 million for the third year.

Members of the forestry subcommittee are of the opinion that, since the development and carrying out of 208 plans has been mandated by Congress through the EPA, funds should be provided through EPA to facilitate carrying out the required program.

FRUIT PRODUCTION

The fruit production subcommittee concluded that fruit growing in Illinois is a very insignificant source of nonpoint pollution of the state's waters. Apple and peach orchards occupy only 15,499 ac or 0.1%, of Illinois cropland. Orchards are very sparsely distributed on sloping lands bordering watershed drainage basins in Illinois. The County Assessor's Farm Census Reports show that orchard acreage had decreased 20% in the 7 years ending in 1975. The 1976 report states that orchard acreage further declined 39% from 1975 to 1976.

Nearly 90% of the orchard acreage is managed so as to prevent water pollution. Preliminary summaries of the subcommittee's fruit grower survey show that apple orchards, occupying 75% of the total acreage, are maintained in continuous sod cover. About 50% of the peach orchards, occupying 25% of the total orchard acreage, are grown in continuous sod or cover crops. Continuous sod cover is considered the most effective management practice in controlling soil erosion and potential water pollution from fruit growing activities. Sod plant surfaces also dilute pesticide residues and increase exposure of residues to degradation by weathering and biological organisms.

As a result of its studies, the subcommittee has concluded that a separate water quality control program for fruit growing is unnecessary for Illinois' 208 plan. Where localized water quality problems exist due to soil erosion or agricultural chemical use, recommendations of the subcommittee addressing those issues should be adequate.

SUMMARY

These guidelines are a very brief summary of extensive, well-documented reports by the various subcommittees of the Agriculture Task Force. Except in a few instances where regulatory programs already exist, the subcommittees unanimously recommended voluntary programs. The subcommittees also agreed that existing agencies and organizations should have the major responsibilities for implementing these programs. The development and maintenance of a successful program will require the cooperation of many organizations, such as the Soil Conservation Service, the Illinois Cooperative Extension Service, the Agriculture Stabilization and Conservation Service, and the Soil and Water Conservation Districts, among others.

Each organization will play a vital role in the success or failure of the program. Expertise of the Soil Conservation Service and Soil and Water Conservation Districts should be fully utilized. The Illinois Cooperative Extension Service should be given major responsibility for conducting special training sessions and continuing education programs, and for developing educational workshops, radio programs, news releases, demonstrations, and so forth to get the message of responsible forestry to land owners, farmers, timber operators, and the general public. The organization of the Agricultural Stabilization and Conservation Service should be fully utilized to administer cost-sharing and incentive programs. The University of Illinois and Southern Illinois University forest research units should become deeply involved in research. Many other organizations will also play direct or indirect roles in the implementation phase.

The cost of implementing the recommended programs will be high. Because these costs are too high to be placed on the agricultural industry alone, they must be shared by the public. If the objectives of clean water are to be met, Congress and the public must realize their responsibilities in supporting these activities. If they do not, there is no hope of accomplishing these goals, and the extensive planning currently in progress will be an exercise in futility.

39

DEVELOPMENT OF BMPs FOR AGRICULTURE—
NEW YORK STATE STRATEGY

P. D. Robillard, M. F. Walter
R. Gilmour
    Department of Agricultural Engineering
    Cornell University
    Ithaca, New York

INTRODUCTION

Any water quality management program dealing with agriculture will require joint planning and assistance from a number of governmental units. The New York State Department of Environmental Conservation (DEC), Soil Conservation Service (SCS), New York State Soil and Water Conservation Committee and the Department of Agriculture and Markets are currently assisting in planning efforts in the designated and nondesignated areas of the state. Cornell University and the Applied Forestry Research Institute at Syracuse are assisting DEC in the development of BMPs to control nonpoint source pollution from agriculture and silviculture. This paper outlines some important considerations and criteria which might be used in developing an agricultural nonpoint source water quality management program.

THE EFFICIENCY OF NONPOINT SOURCE CONTROLS

A practice (or combination of practices) that reduces the load of a specific water pollutant from concentrations which are unacceptable to concentrations which are acceptable, however defined, is an effective practice for the control of that pollutant. Adoption of that practice, however, may augment problems with another pollutant(s). If full information were available, decisions could be based on the decreased damage and associated treatment costs of alternative water quality improvement programs. Figure 1 gives an example of this trade-off. Treatment costs are assumed to increase at an increasing rate for higher levels of treatment while the social cost of pollution is assumed to decrease at lesser rates for higher levels of treatment. Social costs include those subjective and often intangible phenomena associated with degradation of the physical and social environment resulting from pollutant discharges. Examples of these social costs include the siltation of road ditches and harbors, increased water supply treatment costs and lost recreational opportunities. A base concentration of pollutant is a level of concentration that occurs naturally and cannot be decreased by any level of treatment.

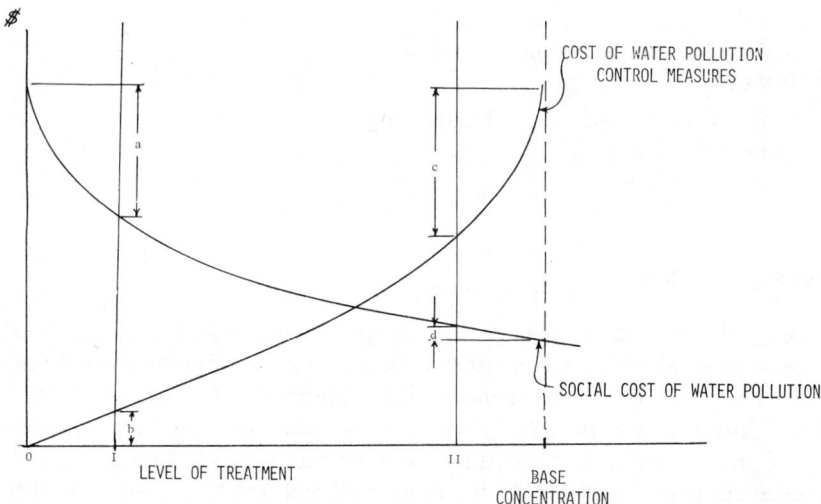

Figure 1. The social cost of water pollution and the associated cost of treatment measures.

Based on the relationships depicted in Figure 1, the reduction in the social cost of water pollution per dollar spent on treatment increases from 0 to I, for example, much less than the corresponding relationships when the level of treatment is increased from II to Base Concentration; that is a/b > d/c.

In addition, efficient expenditures for pollution control practices would require knowing the amounts of a particular pollutant attributable to nonpoint sources or point sources as well as a distribution within nonpoint sources. Figure 2 provides a hypothetical example of load distribution for a pollutant

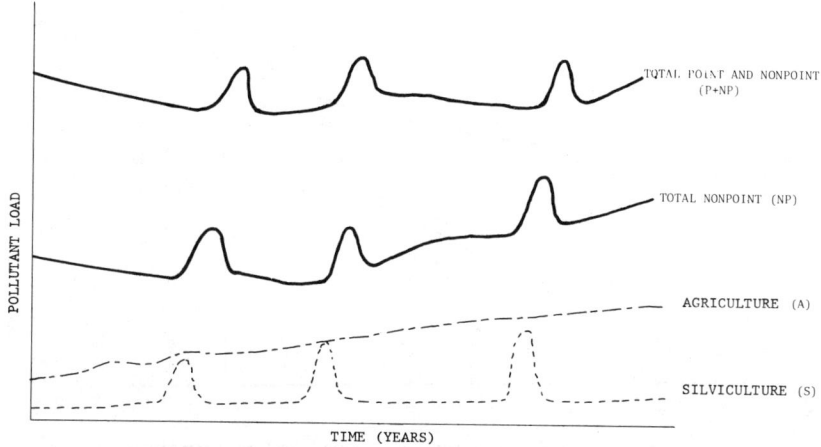

Figure 2. Hypothetical example of a pollutant load distribution for a watershed.

within a watershed. This distribution will likely vary for different watersheds, depending on the intensity of each activity and the overlying physical processes which affect pollutant availability and transport to watercourses. It might be reasoned that expenditures for pollutant control should be proportional to the magnitude of respective pollutant categories as contributors to the total pollutant load in the watershed. However, considerable information about types, sources, magnitudes and movements of pollutants would be necessary. The thrust of this approach would be that point source controls for a given watershed might be allocated $P/(P + NP)$ of program funding. Likewise, silviculture practice would be allocated S/NP of the nonpoint funding and $S/(NP + P)$ of total funding for the watershed. Implicit in this distribution scheme is the assumption that a dollar expended on a practice associated with a nonpoint source would be as cost-effective as a dollar spent on a point source control.

More appropriately, the marginal cost of reducing pollutant load for various practices should be examined. For example, the marginal costs for two practices to reduce the pollutant load from nonpoint agricultural sources might be represented by the marginal cost curves in Figure 3. If this were the case, the cost of reducing the load from $0 \rightarrow I$ using practice A would be much greater than the cost of using practice B. Beyond load reduction II, the marginal costs of practice B to reduce a unit of the pollutant load are higher than for practice A.

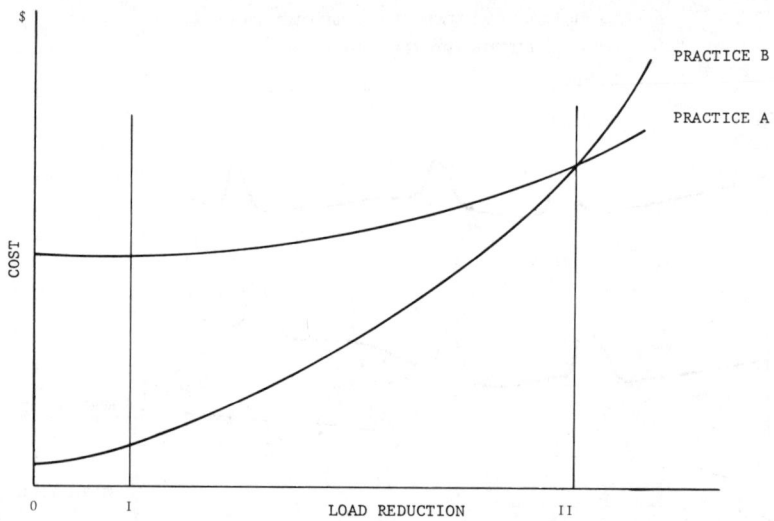

Figure 3. Marginal cost of reducing agricultural pollutant load.

CONSTRAINTS ON DEVELOPING STRATEGIES FOR REDUCING POLLUTION FROM NONPOINT SOURCES

In many cases the above relationships are not known. That is, specific pollutant loads, origins and movements of these pollutants, cost of action and inaction to the public are not known. Also, the distribution of load between point and nonpoint sources is a dynamic process and knowledge of these relationships is generally not available. Finally, the marginal cost of reducing pollutant levels for a given activity is difficult to estimate because the effectiveness of certain practices in reducing pollutant loads is not clearly understood.

Given this uncertainty, some guidelines are provided by EPA.[1] Emphasis should be on practices which manage pollutants at their source and are not capital-intensive structural measures. In addition, where the cost of errors in program development is great, stream monitoring may be needed. In addition to EPA guidelines, two aspects of program development which could be emphasized are:

1. The intensity of different agricultural activities on a watershed basis might be used to develop priorities for where and how to begin a nonpoint source (NPS) control program. For example, a watershed cropped intensively in vegetables might have greater potential nutrient and pesticide losses than a watershed where dairy farms are more common, given similar topography and soils.

2. Emphasis should be placed on practices which are cost-efficient and of relatively low cost to farmers. This approach would encourage the use of practices which reduce pollutant load in the 0 → I phase of Figure 1.

STAGES OF A BMP PROGRAM

Given the limitations of current NPS studies (problem identification, load distribution and cause-effect relationships) a reasonable strategy might include the following stages of NPS controls to achieve 1983 water quality objectives.

Stage I. Implementation of practices which are management intensive and relatively low cost and, where possible, which are cost-efficient to the farmer should be emphasized. During this period, watersheds which have the highest potential problems would be given highest priority for implementation of practices and stream monitoring. The monitoring system should be designed such that the distribution between point and nonpoint load can be calculated.

Stage II. Selected watersheds where concentrations of specific pollutants are relatively high should be treated more intensively with management and, if necessary, structural practices. The effectiveness of certain measures in reducing specific NPS pollutant concentrations should be estimated.

PROBLEM IDENTIFICATION—
A FIRST STEP IN THE DEVELOPMENT OF BMPs

In concentrating on stage I practices, the effectiveness, cost and practicability of a measure should be considered. Figure 4 outlines this selection process.

Few practices are effective in controlling all potential agricultural pollutants. As a first step in the BMP selection process, the pollutant of interest is classified as soluble or adsorbed to soil. Identifying a pollutant's soil adsorption characteristics (adsorption partition coefficient) will determine, in large part, which candidate measures would be effective control measures.[2]

Control measures can affect pollutant yield to streams through three basic mechanisms; availability, detachment and transport. Some examples of candidate measures for each phase of pollutant control are given in Figure 5.

To determine whether a practice is effective in controlling a specific pollutant, the pathway(s) by which pollutants are transported from cropland to streams must be determined. Figure 6 outlines pollutant pathways and the overlying physical variables which provide the energy to detach and transport a pollutant. Table I summarizes the means by which some pollutants become available for transport, the transport medium and their major pathways.

For example, terraces and grassed waterways will have a different effect on pollutant pathways. Generally, terraces will decrease soil loss and runoff

586 BMPs FOR AGRICULTURE AND SILVICULTURE

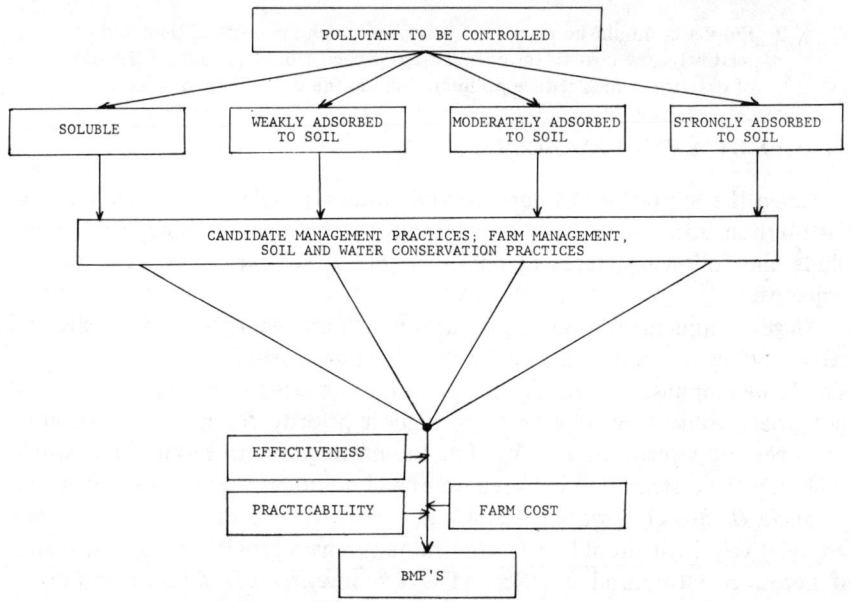

Figure 4. Outline of BMP selection process.

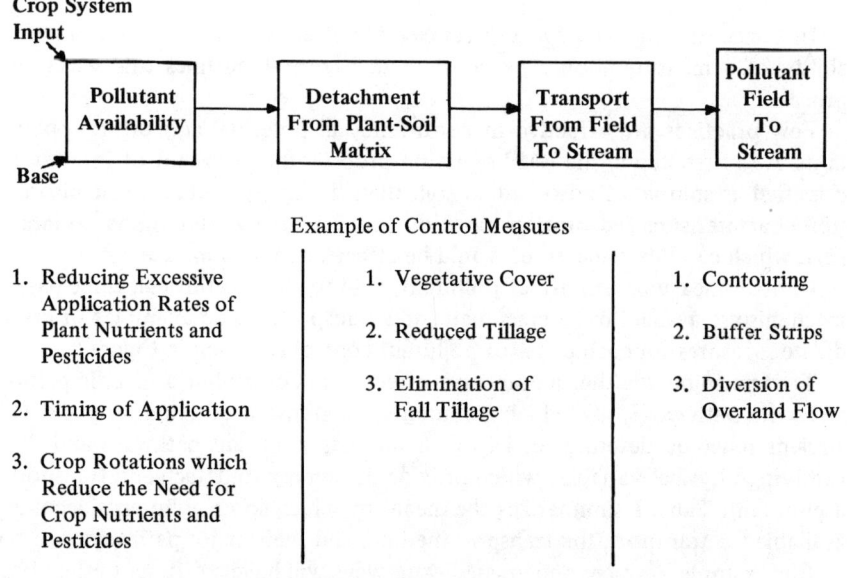

Figure 5. Phases of pollutant availability and movement to streams.

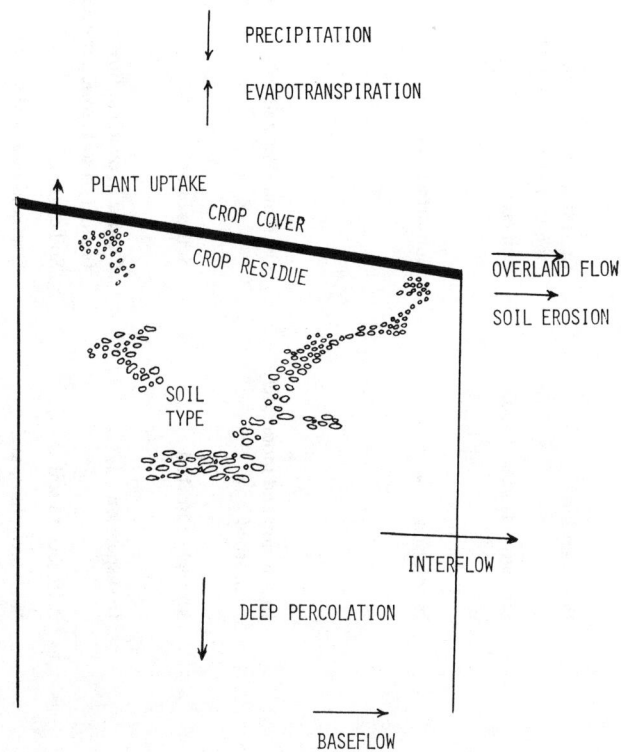

Figure 6. Pollutant pathways and the physical variables which influence pollutant movement.

volume while grassed waterways will have little or no effect on infiltration and overland flow. Table II summarizes the effect of terraces and grassed waterways on potential pollutant losses.

Thus, the appropriateness of a given measure in physically controlling pollutant movement and yield should reflect:

1. the extent to which the measure reduces the amount of pollutant made available for transport; and
2. the extent to which the measure partitions water flow between overland flow and subsurface flow components.

AN EXAMPLE OF THE EVALUATION OF CANDIDATE MEASURES FOR TWO CASE STUDY FARMS

To demonstrate this selection process, BMPs were evaluated for two case study dairy farms. Farm A located in central New York on Kendaia/Lansing/

Table I. Pollutant availability, adsorption/solubility characteristics, and like pathways from cropland.

Pollutant	Availability	Solubility/Adsorption[a]	Likely Pathways
Sediment			Soil erosion
Nitrogen:			
Particulate Organic	Soil mineralization Manure disposal	In suspension	Carried in overland flow (similar to soil erosion process)
Ammonium	Soil mineralization Manure disposal Fertilizer applications	Strongly adsorbed to soil	Soil erosion
Nitrate	Soil mineralization Manure disposal Fertilizer applications Rainfall	In solution	Subsurface flow
Phosphorus:			
Dissolved Inorganic	Soil Manure disposal Fertilizer applications	In solution and moderately adsorbed	Overland flow and soil erosion
Particulate Inorganic	Soil	Strongly adsorbed	Soil erosion
Manure:			
Organic Matter	Manure disposal	In suspension	Carried in overland flow (similar to soil erosion process)
Pathogens	Manure disposal	Adsorbed to soil	Soil erosion
Pesticides	Crop applications	Most are soluble and moderately adsorbed	Overland flow and soil erosion

[a]Loehr et al.[2]

Table II. Effect of terraces and grassed waterways on potential pollutant losses.[3]

				Effect On:			
Practice	Surface Runoff	Base Flow and Interflow	Soil Erosion	Strongly Adsorbed Substances	Moderately Adsorbed Substances	Phosphate Losses	Nitrate Losses
Terraces	Reduced volume and velocity	Increased	Reduced transport capacity Prevents gully erosion	Reduced to the extent fine particles and organic matter settle out	Reduced to the extent that runoff volume is decreased	Generally reduced	Increased proportional to decrease in runoff volume
Grassed Waterways	Very little	Very little	Prevents gully erosion	No effect	No effect	No effect	No effect

Ovid/Conesus soils (3-5% slopes) and farm B in northern New York on Rhinebeck/Benson/Swanton/Elmwood soils (0-3% slopes). In each case three potential pollutants were considered; sediment, nitrate-nitrogen and manurial organic matter. The candidate measures considered include source management and structural measures:

    Sediment Management   Contouring, crop rotations
                                    Chisel plowing, no-till planting
                                    Diversion ditches
    Crop Nutrient Management   Accounting for plowed sod and manurial sources of nitrogen and phosphorus
    Management of Manurial Organic Matter   Storage of manure with periodic spreading
                                                              Field priority and spreading schedule

A summary of the findings include:

Sediment Management

A linear programming model[4] was used to compare changes in net farm income (Y) for different soil erosion constraints. To isolate changes in income due to soil erosion restrictions from changes in income due to improved farm management, the base solution is an optimal combination of resources for unlimited soil erosion.

Reducing soil erosion from the unlimited case to an average of 10 MT/ha resulted in a change in income of less than $300 for farm A and almost no change for farm B (Table III). In terms of soil erosion reductions, the cost on farm A was $0.17/MT and $0.04/MT on farm B. To reduce soil erosion further, to 5 MT/ha, a change in income of $898 or $2.53/MT of soil was estimated for farm A. To reduce soil erosion from 10 MT/ha to 5 MT/ha on farm B would further reduce income only slightly, $0.10/MT.

To accomplish these soil erosion reductions, farm A would grow more hay in rotation on the contour and buy more corn grain than in the base solution.

Table III. Changes in farm income (ΔY) to meet soil loss (SL) restrictions.

	From Unlimited SL to 10 MT/ha		From 10 MT/ha to 5 MT/ha	
	Δ SL	ΔY/Unit	Δ SL	ΔY/Unit
Farm A	-1746 MT	-$0.17/MT	-355 MT	-$2.53/MT
Farm B	-84 MT	-$0.04/MT	-250 MT	-$0.10/MT

For farm B to meet these soil erosion constraints it would have to increase the amount of hay in rotation for some fields while increasing corn production on less erosive fields.

Finally, to establish which fields are contributing the most sediment to streams, one of the many analytical forms of sediment delivery rates (SDR) could be used. For example, if a cost-sharing constraint limited the number of fields that could be treated, those fields with the highest delivery of sediment could be treated first. Delivery rates can be related to the distance between a field and stream.[5] Figure 7 shows this relationship while Table IV gives an example of how this relationship could be used to establish field priorities for farm B.

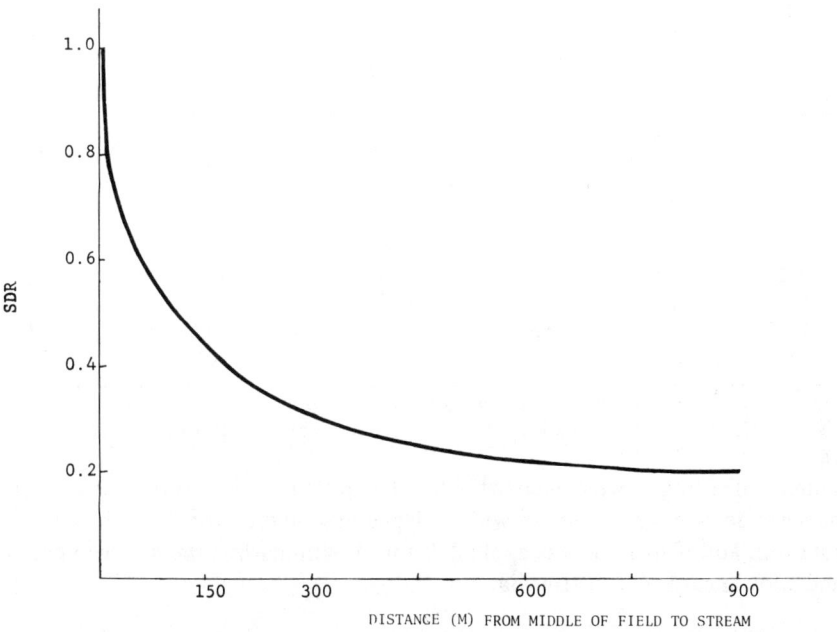

Figure 7. Sediment delivery ratios (SDR). Source: Modified from Renfro.[5]

Nitrate-Nitrogen

This form of nitrogen is soluble and its movement in soil is essentially unaffected by soil and water conservation practices.[3] Source control of nitrate-nitrogen requires estimation of the amount of nitrogen needed by plants and the amount that can be supplied through natural soil mineralization, plowed under sod and applied animal manure, if any. For both case study farms, A and B, increases in farm income could be expected if sod and manurial

Table IV. Sediment delivery from cropland.

		Farm B			
Field No.	Area (ha)	Base Soil Loss[a] (MT/ha)	Sediment Delivery Ratio (SDR)	Sediment Delivery Rate	Sediment Delivered (MT/yr)
1	1.2	0	0.6	0	0
2[b]	3.8	46.4	0.4	18.6	70.5
2a	0.8	51.0	0.6	30.6	24.5
3	2.0	27.2	0.5	13.6	27.2
6	3.0	0	0.4	0	0
7[b]	3.2	41.7	0.4	16.7	53.4
8	1.2	15.8	0.5	9.5	11.4
9	1.6	22.5	0.4	9.0	14.4
10	1.2	11.0	0.6	6.6	7.9
11	2.4	16.8	0.4	6.7	16.1
12	1.8	33.4	0.4	13.4	24.0
13	1.2	15.2	0.5	7.6	9.1
14	3.2	8.3	0.3	2.5	8.0
18	0.8	10.5	0.3	3.2	2.5
21	6.9	0	0.4	0	0
22	5.9	0	0.3	0	0
22b[b]	1.4	66	0.6	39.6	55.4
23a	4.0	0	0.3	0	0
23b[b]	1.2	56	0.6	33.6	40.3
24	2.0	0	0.5	0	0

[a]Base soil loss; chisel plowing, continuous corn, up-down slope, no conservation.
[b]Priority fields, 2, 7, 22b and 23b, deliver 60% of sediment from 19% of crop acreage.

sources of nitrogen were accounted for. The potential savings from decreased nitrogen fertilizer rates varied widely depending on crop rotation, the year in rotation, and if manure was applied. Table V summarizes the nitrogen needs and allowances for corn after hay sod.[6]

Table V. Estimated nitrogen (N) requirements for corn following hay.[6]

Year After Sod	Total N Required (kg/ha)	N Supplied by Sod (kg/ha)	N Supplied by 33.7 MT/ha Manure Applied Annually (kg/ha)	Net N Required (kg/ha)
1	89.8	100	67.2	0
2	89.8	50	100.8	0
3	89.8	25	117.6	0
4	89.8	12	125.4	0
5	89.8	0	125.4	0
6	89.8	0	125.4	0

Actual Savings = Δ Application Rate + Decrease in Field Spreading Cost.

Manurial Organic Matter

Reductions in the load of manurial organic matter to streams involve possible restrictions on the time, rate and location of spreading operations. The two principal candidate measures investigated were:

1. 180-day storage of manure with concentrated spreading operations in the fall and early spring. The spring-spreading would usually coincide with corn planting and take full advantage of the plant nutrient content of manure.
2. Development of a field priority and spreading schedule keyed to expected manure delivery from fields. The system incorporates current cropping rotations and short-term 7- to 14-day temporary storage.

Because farm A is presently using a tractor scraper to collect barn manure, the least-cost storage system is stacking. Farm B uses an alley scraper, and the least-cost system would be an earthern storage structure. The approximate cost to farmer A would be $1400/yr or $18/milking cow. The least-cost storage system for farmer B would be $2840/yr or $42/cow.

A manure-spreading schedule for each farm would establish which fields should be spread at different times of the year (Table VI). This approach minimizes losses of organic matter to streams for daily spreading systems. The cost to each farmer is reflected in increases in average hauling distance.

Table VI. Field Priority and Spreading Schedule (farm B).

Field	Area (ha)	Manure Delivery (Index 1977)	Priority	Manure Capacity (MT)	Season to Apply
21	6.9	0	1	223	Winter
22a	5.9	0	1	191	Winter
23a	4.9	0	1	158	Winter
24	2.0	0	1	65	Winter
17	1.2	1.0	5	39	Summer
9	1.6	3.6	6	52	Spring or fall
8	1.2	3.8	7	39	Spring or fall
14	3.2	4.0	8	103	Spring or fall
7	3.2	6.3	9	103	Spring or fall
10	1.2	8.8	10	39	Spring plow down
13	1,2	12.2	11	39	Spring plow down
3	2.0	12.2	11	65	Spring plow down
22b	1.4	61	13	45	None
23b	1.2	70	14	39	None

Increases in average hauling distance were difficult to estimate but are probably less than $540/yr for farm A ($7/cow), while the total cost to farmer B would be less than $840/yr or $13/cow. A summary of the cost for storage and spreading schedule is given in Table VII.

Table VII. Farm cost of two manure management systems to reduce nutrient losses.

	180 Days of Storage	Field Priority and Spreading Schedule
Farm A	$1400 ($18/cow)	$540 ($7/cow)
Farm B	$2840 ($42/cow)	$840 ($13/cow)

It is unclear how much of a load reduction would be achieved if either system were implemented. But using the technique cited in section B above, the storage system would have to reduce a load by 260 100 x ($1500/$540) units for every 100 units of load decreased by the field priority and spreading schedule to be equally cost-effective. In addition, the storage measure is a structural solution and should have a high degree of effectiveness as a water quality improvement measure before implementation is considered.[1]

THE INTEGRATION OF BMPs INTO FARM MANAGEMENT SYSTEMS

The use of farm conservation plans as an implementation vehicle is appealing. The following points encourage modification and use of farm conservation plans as a water quality management tool:

1. Farm plans allow for a wide range of physical variables and farm management systems upon which the effectiveness of BMPs rely.
2. They provide a sound basis for stage I practices which are essentially good farm management practices.
3. As specific problem pollutants are identified, many of the control practices would exist in farm plans and be familiar to the farm operator.
4. Farm plans are developed and administered by county conservation personnel familiar with farming and farm management systems. In this respect many of the practicability factors mentioned on page 585 would be assured.
5. If maintained as a farm management tool they provide an opportunity to satisfy the continuous planning requirements on nonpoint source water quality management programs.

Examples of how farm plans might incorporate water quality management practices include:

1. a field priority and manure spreading schedule which specifies which fields should be spread during different seasons of the year;
2. annual crop nutrient recommendations based on soil tests and available sources of nutrients in soil organic matter, plowed under sod and manure applications; and
3. management-intensive soil erosion control practices which are keyed to sediment delivery from cropland.

REFERENCES

1. U.S. Environmental Protection Agency. "Guidelines for State and Areawide Water Quality Management Program Development," Washington, DC (1976).
2. Loehr, R. C., et al. "Effectiveness of Soil and Water Conservation Practices for Pollution Control," unpublished progress report, Department of Agricultural Engineering, Cornell University, Ithaca, NY (November 1977).
3. Walter, M. F., T. S. Steenhuis and D. A. Haith. "Soil and Water Conservation Practices," ASAE Paper No. 77-2506, St. Joseph, MI (December 1977).
4. Lang, E., and E. Smith. "Effectiveness of Soil and Water Conservation Practices for Pollution Control," unpublished linear programming model currently being developed in conjunction with a research project sponsored by the U.S. Environmental Protection Agency, Athens, GA, at Cornell University, Ithaca, NY (1978).
5. Renfro, G. W. "Use of Erosion Equations and Sediment-Delivery Ratios for Predicting Sediment Yield," in *Present and Prospective Technology for Predicting Sediment Yields and Sources,* ARS-S-40/USDA, USDA Sedimentation Laboratory, Oxford, MS (1975).
6. Klausner, S. D. Department of Agronomy, Cornell University, Ithaca, NY, personal communication (1977).

SECTION VII

MODELING STUDIES

40

EVALUATION OF CONTROLS FOR AGRICULTURAL NONPOINT SOURCE POLLUTION

**J. J. Wineman, W. Walker, J. Kühner,
D. V. Smith, P. Ginberg, S. J. Robinson**
   Meta Systems Inc.
   Cambridge, Massachusetts

INTRODUCTION

It is now widely recognized that the goals of the Water Pollution Control Act Amendments of 1972 (FWPCA, PL 92-500) will be achieved only if nonpoint source, as well as point source, pollution is controlled. Authority exists under the FWPCA and the Clean Water Act of 1977 (PL 95-217) for the Environmental Protection Agency (EPA), in conjunction with individual states, to devise policies and initiate control programs to manage nonpoint source pollution. Progress, however, has been slow. Many reasons for slow progress can be cited, including: (1) strong economic forces that are in conflict with attempts at environmental control, and (2) the lack of detailed knowledge of physical, chemical and biological processes associated with the environmental impacts of pollutants from nonpoint sources.

The subject of this paper is a policy analysis methodology that addresses the problem of developing effective strategies for the management of agricultural nonpoint source pollution. The paper is based upon a feasibility study undertaken by Meta Systems Inc. for the EPA.[1] Specifically, the proposed methodology is designed to assess both the water quality and socioeconomic impacts of agricultural practices and specific government policies aimed at encouraging agricultural practices which decrease nonpoint source pollution. The methodology allows the simultaneous examination of (1) the water quality impacts of selected agricultural practices, and (2) the economic effects which alternative practices and nonpoint source pollution control policies have upon the farmer. The nonpoint source pollution control problems which the methodology addresses are limited to those which are amenable to solution by incremental on-farm adjustments for damage reduction. Such problems would include, for example, the generation and transport of sediment and nutrients. In addition, socioeconomic impacts on downstream users of water from an agricultural watershed are examined qualitatively.

Figure 1 is a flowchart of the proposed methodology showing: (1) the farm model, which accepts as exogenous inputs alternative agricultural practices available to the farmer and determines the net revenues resulting from each alternative; (2) the water quality model, which analyzes the water quality impacts of the selected agricultural practices and which is composed of (a) a watershed model which describes the pollutants generated by the farming practices and their impact on river water quality and which evaluates soil loss, and (b) an impoundment model which evaluates the impoundment water quality effects of the watershed pollutants; and (3) a qualitative approach for the assessment of the socioeconomic impacts of water quality changes on downstream water users. Each of these components will be described in more detail below. As Figure 1 indicates, the methodology is designed to facilitate the comparison of alternative agricultural practices for the purpose of identifying best management practices (BMPs).

Figure 2 shows how the methodology may be applied to evaluate government nonpoint source pollution control policies and the effects of alternative agricultural futures. The control policies and alternative futures are inputs to the methodology. Examples illustrating the use of the methodology for these purposes will be discussed below.

To be effective, potential methods for the assessment of environmental and socioeconomic impacts of agricultural practices should exhibit the following characteristics:

1. compatibility between data availability and requirements,
2. robustness under a wide range of alternative agricultural futures,
3. capability to evaluate major policy options,
4. ease of understanding,
5. usefulness at the state level, and
6. applicability to the full range of on-farm options.

In the present study, it appeared that the best way to evaluate the proposed methodology on the basis of the above characteristics was to test it in a case study. The Black Creek watershed in northeastern Indiana, a USEPA-USDA demonstration project[2,3] was selected as the subject of the case study and combined with a synthesized downstream impoundment with characteristics typical of those found in the corn belt.

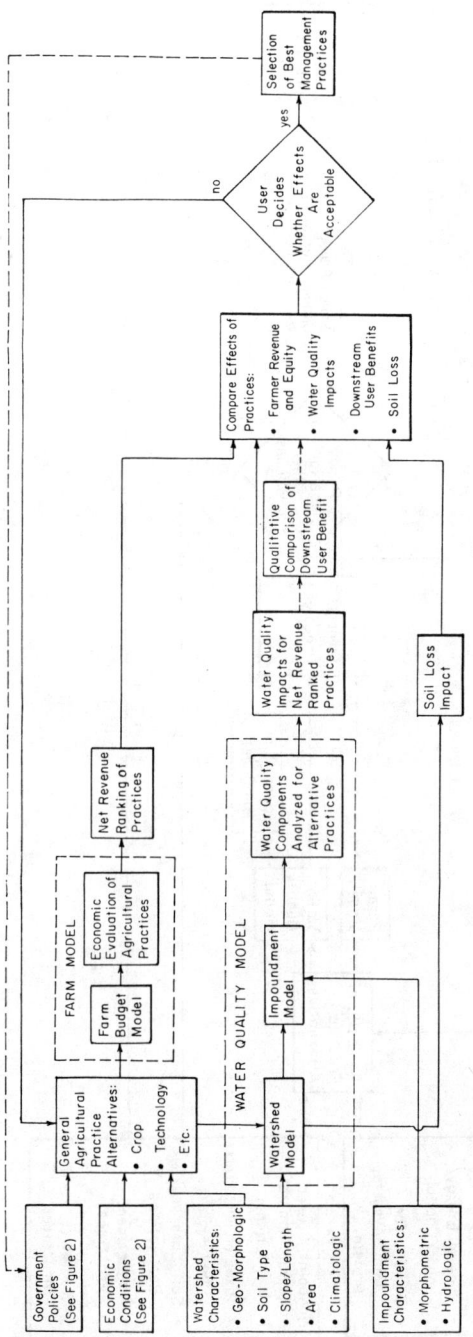

Figure 1. Methodology for assessment of water quality and socioeconomic impacts of agricultural practices.

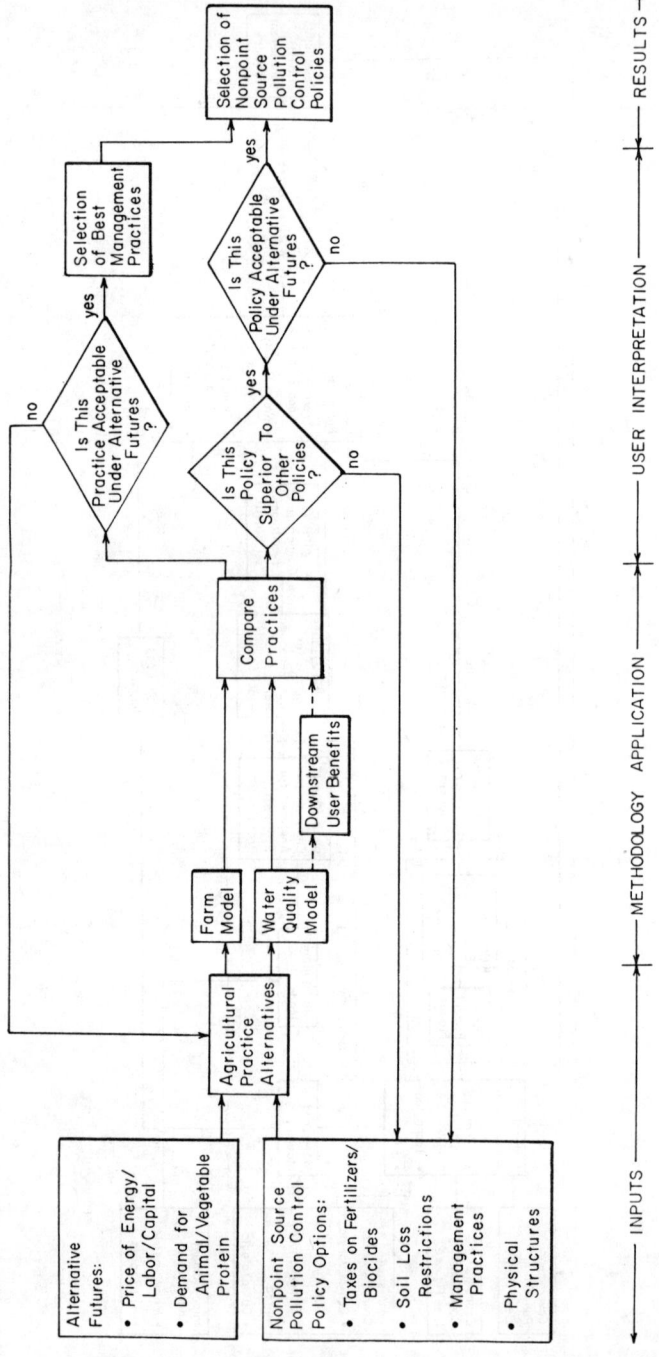

Figure 2. Use of methodology for assessment of nonpoint source pollution control options under alternative futures.

METHODOLOGY DEVELOPMENT

Agricultural Practices and the Farm Model

While major market and regulatory pressures such as prices, taxes, subsidies and government regulations are exerted at a regional or national level, it is the farmer who responds by choosing both his crops and methods of farming. For this reason, the methodology starts with a farm budget.

It is assumed that the farmer desires to maximize net revenues. Accordingly, he chooses a set of agricultural practices that include: (1) crop rotation, (2) tillage practices, (3) structural erosion and drainage control practices, and (4) levels of chemical application. These choices are represented as inputs to the farm model for the calculation of a variety of costs associated with operating the farm in the specified manner. This required developing a data base for the model. The procedure described by Dr. Klaus Alt in Appendix C of *Control of Water Pollution from Cropland*[4] was used. Each element of cost was updated for 1977 prices and modified where necessary to adapt the model for the Black Creek area. Additional inputs to the model specify expected yields and market prices for each crop.

Table I identifies the major categories of cost and revenue data incorporated in the model. In the case study, eleven practices (plus two modifications to include custom hiring) are selected, and farm budgets are developed for a uniform farm of 250 ac on each of three soil types (upland, ridge and lowland) characteristic of the Black Creek watershed. Table II identifies and describes the eleven farm practices considered in the analysis. (The definitions of the farm practices and variations associated with soil type were developed by Meta Systems in consultation with agricultural experts associated with the Black Creek project at Purdue University.) Table III shows the revenues, costs and net returns for three farms, each on one soil type, assuming uniform adoption of each of the eleven farm practices and existing government policies.

As noted above, the main purpose of constructing a farm model is to evaluate agricultural practices for their impacts on farm income, with the objective of identifying BMPs. However, an important additional feature of the model is its potential usefulness for public policy analysis. Investigation of government policies designed to encourage BMPs may be carried out by simply modifying, within the farm model, the appropriate cost or revenue factors affected by such policies, and recomputing the net revenues. These results are then evaluated together with the soil loss and water quality impacts as estimated by the water quality models. This overall evaluation procedure will be illustrated below after the water quality models are described in the following section.

While the farm budget model, as presented here, captures the major elements important for assessing the economic impacts of alternative nonpoint

Table I. Farm model—elements of cost and revenue.

Costs	Revenues
Terracing	*Corn*
Construction	Yield
Maintenance	Price
Machinery	*Soybeans*
Fixed Cost	Yield
Maintenance and Repair	Price
Tractor	*Wheat*
Fixed Cost	Yield
Maintenance and Repair	Price
Fuel	*Hay*
Tractor	Yield
Combine	Price
Seed	
Corn	
Soybeans	
Wheat	
Meadow	
Fertilizer	
Nitrogen	
Phosphorus	
Potassium	
Equipment Rental	
Biocides	
Herbicides	
Insecticides	
Labor	
Direct Labor	
Overhead	
Other Costs	
Grain Drying	
Interest on Operating Capital	

source pollution control policies on the farmer, further modifications would be necessary before it could be used effectively in a planning context. Most importantly, the model should be automated, perhaps employing a revenue-maximizing linear programming model for policy analysis. In addition, a much broader range of agricultural practices must be considered, including variations in fertilizer applications, organic farming and livestock integration.

Water Quality Model

The methodology includes a quantitative framework to estimate the water quality impacts of agricultural practice/soil type combinations. Examples of

Table II. Major features of a selected set of farm practices in the Black Creek area.

Crops	Tillage Practice	Soil Conservation Practice	Abbreviated Designation of Farm Practice
Continuous Corn (CC)	Conventional tillage, fall turn plow (CV)	Without terracing	CC-CV
Continuous Corn (CC)	Conventional tillage, fall turn plow (CV)	With terracing (T)	CC-CVT
Continuous Corn (CC)	Fall shred stalks, chisel plow, spring disk (CH)	Without terracing	CC-CH
Continuous Corn (CC)	Fall shred stalks, chisel plow, spring disk (CH)	With terracing (T)	CC-CHT
Continuous Corn (CC)	Fall shred, no-till planting (NT)	Without terracing	CC-NT
Corn-Soybean Rotation (CB)	Conventional tillage, fall turn plow (CV)	Without terracing	CB-CV
Corn-Soybean Rotation (CB)	Fall shred, chisel plow, spring disc (CH)	Without terracing	CB-CH
Corn-Soybean Rotation (CB)	Fall shred, no-till planting (NT)	Without terracing	CB-NT
Corn-Soybean Rotation (CB)	Fall shred, no-till planting (NT)	With terracing (T)	CB-NTT
Corn-Soybean-Wheat-Hay Rotation (CBWH)	Conventional tillage fall turn plow for corn; no-till planting for soybeans, wheat hay	Without terracing	CBWH* CBWH
Corn-Soybean-Wheat-Hay Rotation (CBWH)	Fall shred stalks, no-till planting for all crops, increased use of herbicides (NT)	Without terracing	CBWH*-NT CBWH-NT

Note: Entry in parenthesis used where needed to distinguish specific component of farm practices. Asterisk (*) indicates farmer-owned equipment for wheat and meadow planting and for hay mowing, baling, rather than custom hiring for these operations.

Table III. Summary of farm model output—1977 dollars, in thousands (under existing government policies).

Farm Practice and Cost	Tillage Practices						Rotations				Terraces		
	Corn, Conventional Tillage (CC-CV)	Corn, Chisel Plow (CC-CH)	Corn, No-Till (CC-NT)	Corn, Soybean, Conventional Tillage (CB-CV)	Corn, Soybean, Chisel Plow (CB-CH)	Corn, Soybean, No-Till (CB-NT)	Corn, Soybean, Wheat, Hay, Partial Use of Herbicides (CBWH)*	Corn, Soybean, Wheat, Hay, No-Till (CBWH*-NT)		Corn, Soybean, Wheat, Hay, No-Till (CBWH-NT)	Corn Conventional Tillage (CC-CVT)	Corn Chisel Plow (CC-CHT)	Corn Soybean No-Till (CB-NTT)
Gross Revenue													
A. Upland Soil	52.5	52.5	49.9	46.3	46.3	44.4	43.0	43.0		43.0	56.0	56.0	47.4
B. Ridge Soil	65.0	65.0	65.0	59.1	59.1	57.9	51.8	51.8		51.8	68.5	68.5	60.9
C. Lowland Soil	65.0	65.0	52.0	59.1	59.1	50.7	49.9	49.0		49.0	68.5	68.5	53.7
Costs													
A. Upland Soil	39.7	39.1	43.0	32.9	32.6	32.3	34.4	34.2		30.3	46.4	45.8	38.9
B. Ridge Soil	41.4	40.9	44.9	33.3	33.1	32.7	34.7	34.4		30.7	48.2	47.6	39.3
C. Lowland Soil	42.7	42.1	45.5	34.8	34.5	34.1	35.4	35.1		31.4	49.4	48.9	40.7
Net Return													
A. Upland Soil	12.8	13.4	6.9	13.5	13.7	12.2	8.5	8.8		12.8	9.6	10.2	8.6
B. Ridge Soil	23.6	24.1	20.1	25.8	25.1	25.1	17.4	17.4		21.1	20.3	20.9	21.5
C. Lowland Soil	22.3	22.9	6.5	24.4	24.6	16.6	14.5	13.9		17.6	19.1	19.6	13.0

Note: Columns may not add due to rounding. Asterisk (*) indicates farmer-owned equipment for wheat and meadow planting and for hay mowing, baling, rather than custom hiring for these operations.

watershed and water quality analyses are based on the assumption of a homogeneous watershed, as assumed in the farm model. At this preliminary stage of methodology development, this approach was considered to be more appropriate than one dealing with aggregate economic and environmental impacts in a heterogeneous watershed. A more realistic evaluation of these practices, using a heterogeneous watershed (incorporating variations in soil, slopes, farm sizes, and other characteristics) is recommended as the next step in the development of an operational methodology.

Figure 3 illustrates the separation of the water quality analysis into two major sections.

1. the *watershed,* which is characterized as generating different loadings of pollutants, depending upon agricultural activities and watershed characteristics, and
2. the *impoundment,* where water quality is dependent upon the type and quality of loadings from the watershed and upon impoundment characteristics.

In this scheme, the river is represented as a medium for transporting pollutant loadings from the watershed to the impoundment. Water quality conditions in the river reflect these loadings, which enter the river in surface runoff and groundwater base flow and are transported in dissolved and sediment-bound phases.

The methods developed for the watershed analysis are of an empirical nature and are concerned with long-term average emissions, consistent with the Universal Soil Loss Equation (USLE).[5] The use of long-term average time scales precludes direct assessment of responses under extreme meteorologic conditions, effects of the timing of various agricultural operations, seasonal variations in water quality, and analysis of relatively short-lived compounds. As a compromise, modification of the methodology to permit assessments of average seasonal responses would be feasible without losing many of the advantages of a long-term-average approach. In general, given currently available data and limited knowledge of the relevant physical processes, we feel that a framework built from complex models would not be feasible or useful at the regional planning level.

Pathways involved in the watershed model are summarized in Figure 4. Average annual export rates of the following substances are evaluated in the watershed analysis: (1) sediment (sand, silt and clay fractions) (2) phosphorus (biologically available), (3) dissolved nitrogen, and (4) dissolved color. The computed concentrations of these components are assumed to be representative of average water quality conditions in rivers draining the agricultural watershed. This part of the methodology is appropriate for linking with downstream models for the purpose of evaluating quality impacts in impounded waters, as discussed below. Dissolved oxygen, biocide residues and

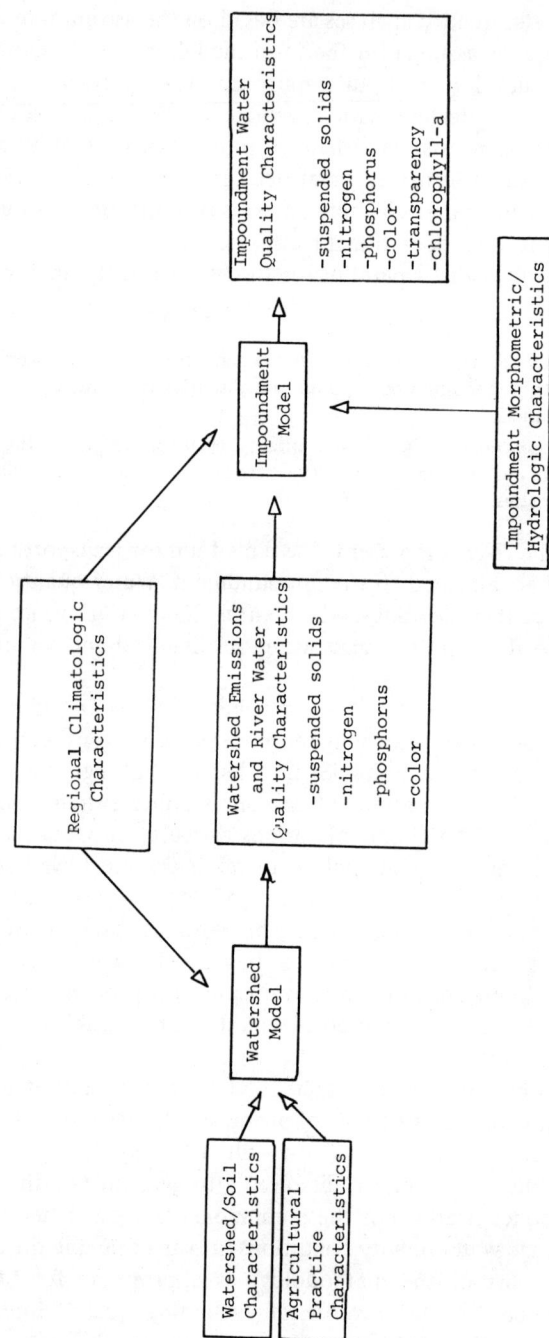

Figure 3. Schematic view of the watershed/impoundment water quality analysis.

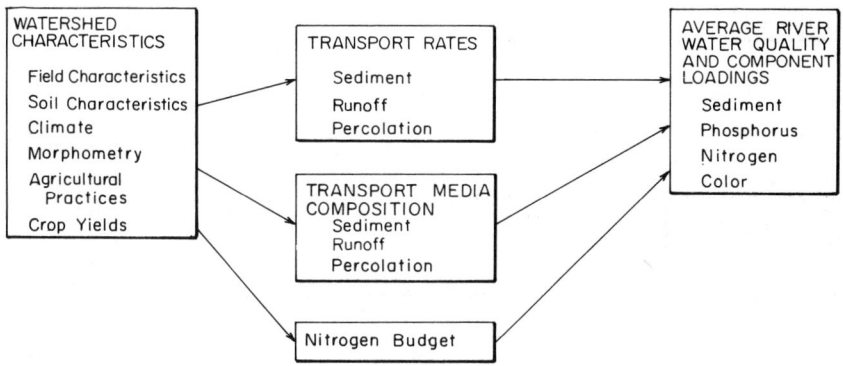

Figure 4. Pathways in the watershed analysis.

biocides are additional relevant water quality components that have not been included in the framework at this time. Modification to account for certain aspects of these factors is left for future work.

The framework developed for assessing impoundment water quality impacts consists of a group of empirical models which are designed to predict steady-state, seasonal, or long-term-average conditions. Pathways in the impoundment water quality analysis are summarized in Figure 5. Water quality components considered are suspended solids, phosphorus, nitrogen, transparency, and epilimnetic chlorophyll-a concentration. Chlorophyll-a concentration is used as an index of eutrophication and represents the extent of the algal growth in the surface waters of impoundment during the summer. The model developed for predicting chlorophyll-a levels considers the possible limitation of algal growth by light, phosphorus and/or nitrogen. Transparency is computed as a function of suspended solids, dissolved color and chlorophyll-a concentrations. Models are formulated for each of the above water quality components, based upon theoretical considerations and the results of previous modeling efforts. They are calibrated and tested empirically using a data base characterizing the behavior of these components in corn belt impoundments and compiled from various sources.[6-9]

While the model described above includes what we consider to be those functional aspects of the watershed/impoundment system which are required for evaluation of the impacts of farm management practices on water quality, it is currently at a preliminary stage of development and requires additional analysis, calibration and testing before it can be applied in a planning context. A preliminary sensitivity analysis has been conducted in order to identify aspects of the model framework which are most critical to the predictions and which therefore may need additional refinement.

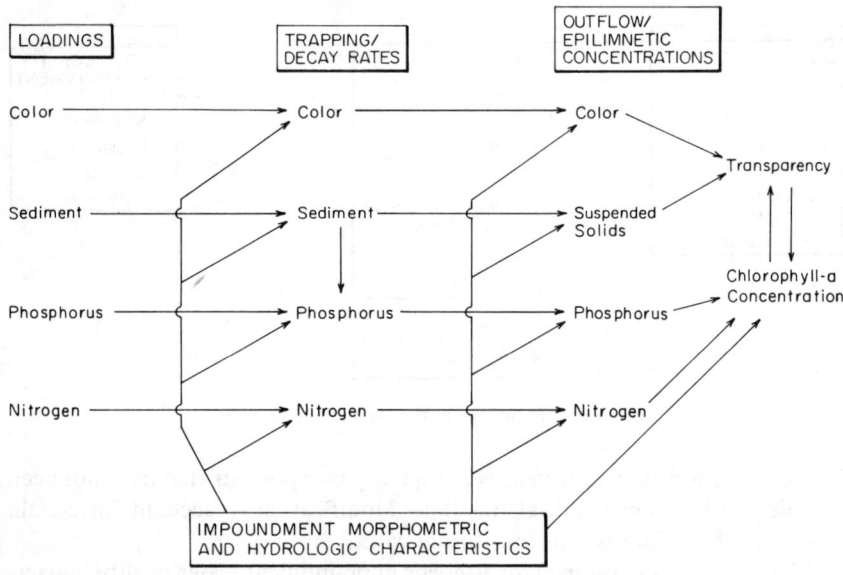

Figure 5. Pathways in the impoundment water quality analysis.

Important additional independent variables include total watershed area, impoundment surface area, and impoundment mean depth. Assumed values for the examples discussed below are 200 km^2, 5 km^2 and 4 m, respectively. It should be noted that the evaluations of the relative impacts of the agricultural practices on impoundment water quality may be somewhat sensitive to the choice of watershed/impoundment configuration.

USE OF THE FARM AND WATER QUALITY MODELS

Comparison of Practices

The methodology described above can be used to illustrate: (1) how agricultural practices can be evaluated in terms of water quality impact to facilitate selection of BMPs (under existing government policies); (2) how government policies which encourage the implementation of practices that are conducive to water quality improvements (*i.e.*, BMPs) may be examined; and (3) how changes in economic conditions can affect agricultural practices and water quality impacts.

The evaluation of agricultural practices under current policies uses the eleven selected farm practices listed in Table II (as if they constituted a comprehensive set of alternatives currently available to farmers). In addition to soil loss, six variables related to water quality (discussed earlier in the context of the water quality model) were analyzed for the three soil types (upland, ridge, lowland) and the eleven farm practices. For illustrative purposes, the results (together with net revenues) for one soil type, the lowland, are displayed as a set of bar graphs (Figure 6). (Refer back to Table II for definitions of the eleven agricultural practices and the symbols used to represent them.) The six water quality components displayed and the dimensions used to quantify them are:

1. impoundment sedimentation (kg/m^2-yr)
2. river nitrogen (g/m^3)
3. river phosphorus (g/m^3)
4. river light extinction coefficient (a measure of resistance to light penetration or decrease in transparency) (m^{-1})
5. impoundment light extinction coefficient (m^{-1})
6. impoundment biomass (g chlorophyll-a/m^3)

The bar graphs are constructed so that increasing pollutant loads or concentrations are shown by higher vertical height of the bar. For net revenue, vertical height increases with higher returns.

In terms of net revenue ranking of the eleven farm practices, Figure 6 shows that the corn-soybean rotation is most profitable based on the prices chosen for these commodities in this example (*i.e.*, corn, $2.00 per bushel, soybeans, $5.00 per bushel, wheat, $2.50 per bushel, hay, $60 per ton). This practice is also most profitable on the other two soil types.

Figure 6 can also be used to examine the differences in soil loss (gross erosion) or any of the six water quality components, for each farm practice on each of the three soil types. For example, on the lowland, the practice which maximizes net revenue ($24,600) is calculated to produce an annual soil loss of 2.2 ton/ac (losses range from 3.5 to 0.4 ton/ac. This can be contrasted with results obtained for the other two soil types. On the ridge, the annual net revenue maximizing ($26,100) practice results in 5.2 ton/ac annual soil loss (range: 9.4 to 0.9 ton/ac). On the upland, the net revenue maximizing ($13,700) practice shows an annual soil loss of 15.2 ton/ac (range: 27.2 to 2.7 ton/ac). On all soil types the no-tillage corn, soybean, wheat, hay rotation (CBWH-NT) minimizes soil loss.

As shown in Figure 6, and in similar analyses for the other two soil types, the water quality impacts of agricultural practices vary with field/soil type, water body (river versus impoundment) and specific pollutant. Use of soil loss alone as the criterion for farm practice evaluations can lead to erroneous conclusions because of the importance of various dissolved components and the interactive effects of different processes (*e.g.*, decay, adsorption/desorption, sedimentation).

612 BMPs FOR AGRICULTURE AND SILVICULTURE

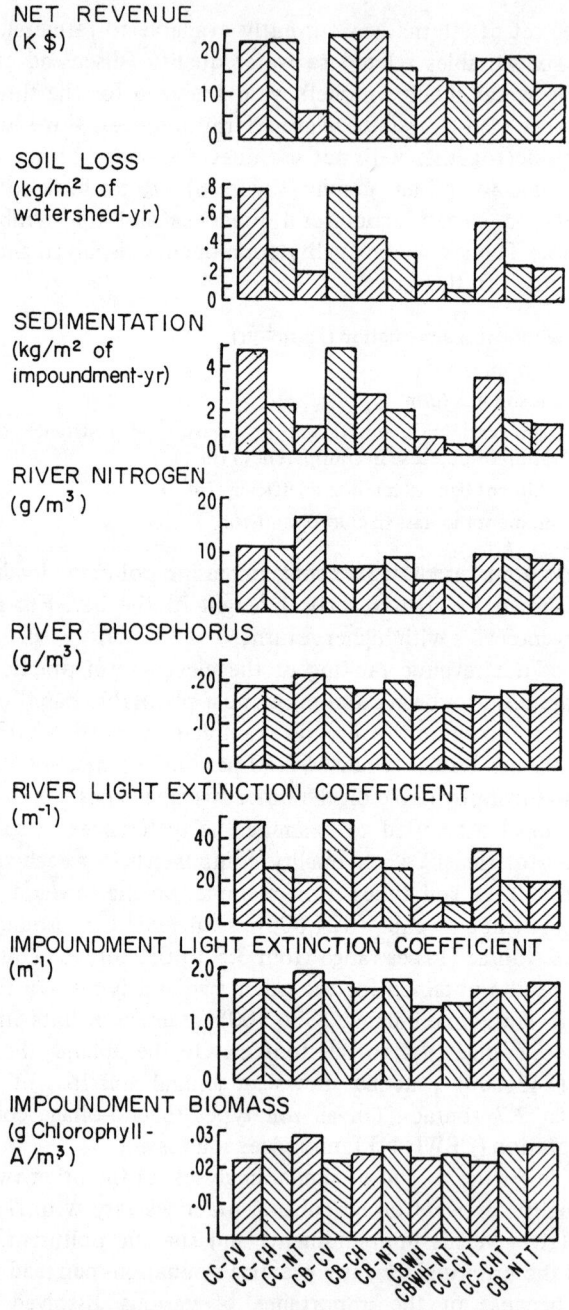

Figure 6. Comparison of practices—lowland.

The types of soil loss controls evaluated here do not appear to have proportionate impacts on phosphorus losses and impoundment eutrophication problems. This results from the influence of the following factors which are considered in the predictive framework: (1) a small fraction of soil phosphorus (5 to 10%) is biologically available; (2) minimum-tillage methods tend to cause phosphorus enrichment of the surface soil layer and create a potential for leaching of phosphorus from crop residues during snowmelt periods; (3) impoundment phosphorus trapping efficiency is correlated with sedimentation rate; (4) reduced erosion rates result in increased availability of light to promote algal growth in downstream impoundments. Because of the combined effects of these factors, the agricultural practices evaluated do not appear to be effective in controlling eutrophication and, in some cases, cause moderate increases in algal growth (Figure 6). In such cases, corresponding increases in the potential for fish production could be interpreted as water quality benefits, although additional analysis would be needed in order to quantify such benefits. Impacts of soil loss controls on suspended solids concentrations and transparency levels are generally in the same direction as impacts on gross erosion rates, but are usually attenuated.

The importance of one pollutant compared to another may also shift from watershed to watershed and, hence, influence the selection of those water quantity components of primary importance to the evaluation of the BMPs. In assessing BMPs, it seems reasonable to rank the different pollutants on the basis of severity of local water quality issues. Comparisons can first be made of farm practices and their net revenues with respect to the primary pollutants and then additional pollutants can be incorporated into the analysis.

Evaluation of Policies

The results of the comparison of agricultural practices under existing government policies, discussed above, provide the reference conditions from which alternative nonpoint source pollution control policies may be evaluated. The three illustrative policies discussed below—cultivation practice prohibition, soil loss restriction, and fertilizer tax—concern reduction of soil loss and river nitrogen. The number and types of control options would, of course, have to be expanded for a more representative evaluation.

Prohibition of Certain Cultivation Practices

Prohibition of certain tillage practices, such as conventional plowing, would have no apparent economic impact on the farms anayzed in this case example, but would reduce soil loss and would have varying effects on water quality. Success of a tillage practice prohibition policy such as this assumes, of course, that moldboard and chisel plows are equally accessible to farmers.

The cost impact on the farmer of prohibiting moldboard plowing in favor of chisel plowing for continuous corn (CC-CH) or corn-soybean rotations (CB-CH) is indicated by comparing the prohibited, but maximum revenue-producing alternative to the permitted maximum revenue alternative.

	Continuous Corn		Corn-Bean Rotation	
	Prohibited	Permitted	Prohibited	Permitted
Upland	$12,800	$13,400	$13,500	$13,700
Ridge	$23,600	$24,100	$25,800	$26,100
Lowland	$22,300	$22,900	$24,400	$24,600

The required shift in tillage practice reduces the soil loss (rounded to the nearest ton) for the two copping systems as follows.

	Continuous Corn	Corn-Bean Rotation
Upland	15 ton/ac	12 ton/ac
Ridge	5 ton/ac	4 ton/ac
Lowland	2 ton/ac	1 ton/ac

The water quality effects of this policy vary with soil type and pollutant. Generally, percentage reduction in impoundment sedimentation rates and river light extinction coefficients approach those of soil loss. Effects are reduced when measured in terms of percentage reduction in river phosphorus concentrations and impoundment light extinction coefficients. Nitrogen concentrations are not influenced. Small increases in chlorophyll-*a* concentrations may result in water quality degradation from increased eutrophication, but may also provide increased potential for fish production.

Gross Soil Loss Restrictions

Restrictions limiting gross soil loss are sometimes suggested as watershed planning goals. Although there are numerous ways to implement such restrictions, for purposes of this discussion we consider them to apply over each acre of a watershed. Such an interpretation maximizes their impact on costs and on erosion.

Consider, for example, a restriction on gross soil loss of 4 ton/ac maximum. This implies the following mandated shifts in cropping activities to comply with the maximum 4-ton/ac soil loss.

For the upland soils, the practice with highest net revenue that meets the soil loss criterion is the corn, soybean, wheat, hay rotation with no-tillage (CBWH-NT). Net revenue decline is:

```
       CB-CH   = $13,700
       CBWH-NT = $12,800
       Decline = $    900 for 250 ac.
```

Reduction in soil loss is about 12 ton/ac (15 ton/ac for the CB-CH practice-3 ton/ac for the CBWH-NT practice). Water quality effects follow the same pattern as those of the previously discussed policy.

For ridge soils, costs to the farmer are somewhat greater and soil loss reductions considerably smaller. Shift is from corn, soybeans with chisel-plowing (CB-CH), to corn, soybeans with no-tillage (CB-NT).

```
       CB-CH   = $24,100
       CB-NT   = $25,100
       Decline = $  1,000 for 250 ac
```

Reduction in soil loss is only 1 ton/ac and water quality effects are not significant.

For the lowlands, no change from the net revenue maximizing farm practice (CB-CH) would be necessary to meet gross soil loss restrictions of 4 ton/ac. As these results indicate, certain policies may have widely differing impacts on individual farmers, depending on their location or other factors. Potential conflicts among farmers and equity questions are thus illuminated.

Fertilizer Control Policies

One possible fertilizer control policy would impose a fertilizer tax to reduce overapplication of fertilizer—especially nitrogen. The rationale behind such a tax is made clear in Figure 7. It can be seen that the corn-nitrogen response curve is relatively flat in the region where farmers now operate (fertilization $\simeq$ approximately 160 lb/ac; yield $\simeq$ 130 bu/ac). Large reductions in nitrogen application result in small reductions in yield. Therefore, a tax which would discourage fertilizer use would be accompanied by only modest declines in crop yield and even smaller proportional reductions in net revenue. However, as Figure 7 shows, reductions in crop nitrogen applications are highly beneficial to water quality because of the nonlinear nature of the estimated response of river nitrogen concentrations to changes in nitrogen fertilization rates. In Figure 7, the river nitrogen concentrations represent increases over baseline levels which are approximately one g/m^3.

If a fertilizer tax could be imposed so that net revenues for the corn, soybean, wheat, hay no-till farm practice (CNWH-NT) (one of the least nitrogen-dependent practices) were greater than those for the corn, soybean chisel plow practice (CB-CH) (the maximum net revenue practice), then river nitrogen would be reduced 28% on the uplands, 24% on the ridge and 26% on the lowlands. On the upland, the fertilizer tax required is $0.13/lb

616 BMPs FOR AGRICULTURE AND SILVICULTURE

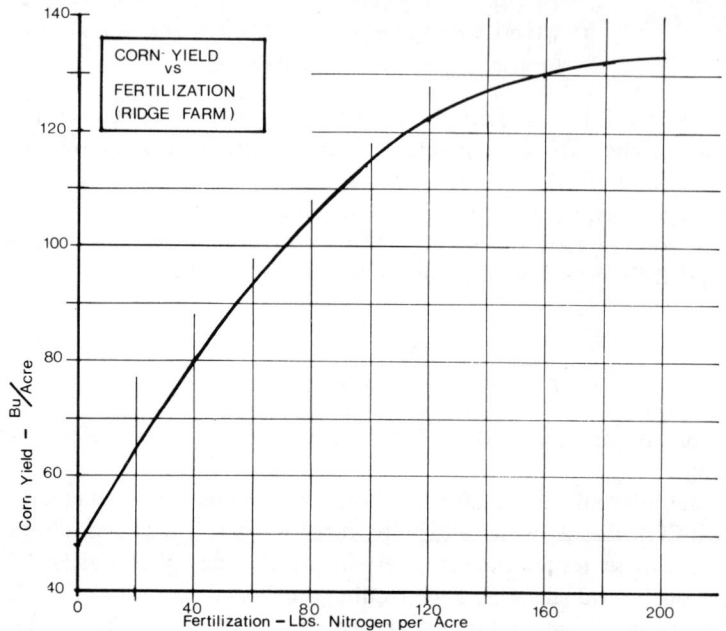

Figure 7a. Corn yield vs fertilization (ridge farm).

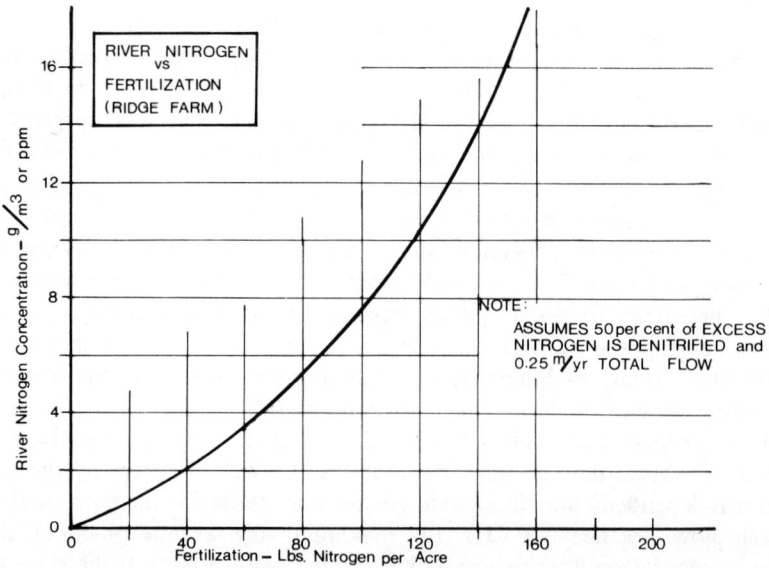

Figure 7b. River nitrogen vs fertilization (ridge farm).

(or 100% increase in the price of nitrogen used in the reference case). On the ridge, the tax required is $0.54/lb and on the lowland, $0.74/lb, representing price increases to the farmer of 415% and 570%, respectively. This comparison further illustrates the differential impacts on farmers in different locations.

Alternative Futures

The rapidly changing nature of U.S. agriculture and the uncertainty regarding future economic conditions affecting the farmer necessitate the introduction of alternative futures into an evaluation of agricultural practices and water quality control policies. Such conditions can be incorporated into the framework by an adjustment to the farm model. As an illustration of the use of the model for this purpose, a future characterized by sharp energy cost increases is postulated.

An economic future for 1985 is assumed in which energy prices will be approximately double the 1977 prices,[10] while prices for other inputs such as labor will remain constant. (A more complete analysis than was possible in our preliminary study, using an automated farm model, would be required to explore more realistic, and necessarily complex, alternative futures.) Prices for tractor and combine fuel, grain drying operations and the various chemicals bought by the farmer will therefore be substantially higher.

As illustrated in Table IV, maintaining the same eleven farm practices previously described results in increased cost of farm operations ranging from $10,000 to $30,000 annually, depending on the practice. This range corresponds to a 30 to 65% increase over 1977 costs. Net returns are, of course, drastically affected. We have not attempted to project prices received by the farmer for corn, soybeans, wheat and hay. Even had this been done, it is possible that some of the farm practices would no longer appear financially viable. Since we are interested in the potential effects of farm practices on water quality as induced by profitability considerations, it is sufficient to evaluate changes in farm costs without attempting to adjust gross revenues.

On all three soil types, the corn, soybean, wheat, hay (CBWH) rotations indicate their lower energy intensity by an upward shift in their net revenue rankings compared to the reference case with 1977 energy prices. The impacts are most dramatic on the uplands farm where the CBWH-NT alternative becomes the highest net revenue producer. This can be compared to the 1977 reference case in which this practice yielded $900 less in net revenue than the corn, soybean rotation using chisel-plowing (CB-CH). As discussed above in the evaluation of alternative policies, the annual soil loss from this farming practice (CBWH-NT) on the uplands soil is about three ton/ac compared to the highest net revenue practice in 1977 (CB-CH) which produces an annual

Table IV. Effect of future energy prices (constant 1977 dollars).

	Uplands					Ridge					Lowlands			
	1985 Net Revenue $	Rank	1977 Net Revenue Rank	1985 Net Revenue $	Rank	1977 Net Revenue Rank	Soil Loss Rank	1985 Net Revenue $	Rank	1977 Net Revenue Rank	Soil Loss Rank
Conventional Corn, Conventional Tillage, without Terracing (CC-CV)	-8,100	8	4	10	+800	7	5	10	-1,500	7	4	10
Continuous Corn, Conventional Tillage with Terracing (CC-CVT)	-11,700	10	9	9	-2,800	9	10	9	-5,000	10	6	9
Continuous Corn, Chisel Plowing without Terracing (CC-CH)	-7,400	7	3	7	+1,500	6	4	7	-800	6	3	7
Continuous Corn, Chisel Plowing with Terracing (CC-CHT)	-11,000	9	8	5	-2,100	8	8	5	-4,400	9	5	5
Continuous Corn No Till Planting without Terracing (CC-NT)	-19,900	11	11	3	-6,000	11	11	3	-22,500	11	11	3
Corn-Soybeans Conventional Tillage without Terracing (CB-CV)	-300	4	2	11	+11,600	2	2	11	+8,800	2	2	11

System												
Corn-Soybeans Chisel Plowing without Terracing (CB-CH)	-80	3	1	8	+11,800	1	1	8	+9,000	1	1	8
Corn-Soybeans No Till Planting without Terracing (CB-NT)	-3,200	5	7	6	+9,300	4	3	6	-400	5	9	6
Corn-Soybeans No Till Planting with Terracing (CB-NTT)	-7,000	6	10	4	-5,300	10	6	4	-4,200	8	10	4
Corn-Soybeans-Wheat-Hay Conventional Tillage for Corn only without Terracing (CBWH)	+50	2	6	2	+8,200	5	9	2	+4,900	4	7	2
Corn-Soybeans-Wheat-Hay No Till Planting without Terracing (CBWH-NT)	+2,600	1	4	1	+10,700	3	7	1	+7,600	3	8	1

Notes: Highest soil loss rank, 1 = minimum soil loss.
Highest revenue rank, 1 = maximum net revenue.

soil loss of 15 ton/ac. The difference in water quality effects of these two practices varies depending on the water quality component of interest, as discussed earlier, and is not always as significant as the soil loss difference.

IMPACTS ON DOWNSTREAM USERS

As discussed above, the results of linking the farm, watershed and impoundment models show that alternative farming practice/soil type combinations have different water quality impacts on the receiving water. To estimate the impacts of these practices on downstream users, changes in water quality must be related to measurements of value to these users. Depending on the use of the water and the land uses surrounding the watershed and the impoundment, certain water quality components are of more or less interest to different groups of people concerned with water quality (users).

Some benefit categories of interest in this case are human health, municipal water supply, dredging (flood control), ecology, recreation, aesthetics, and the local economy. Rigorous quantitative methods of benefit estimation would vary depending upon the particular benefit category of interest. Examples of alternative techniques appropriate for measuring different water quality benefits (impacts) are population surveys to collect time-budget or travel-cost data or to conduct bidding games, development of marginal cost data, an input/output models. It is clear that a comprehensive benefit estimation methodology covering several such measurements techniques would be a major undertaking, requiring significant time and resources to implement and presenting numerous empirical difficulties. As an alternative, a simplified version is presented here which qualitatively assesses the direction of benefits resulting from water quality changes induced by the alternative farming practices.

In order to compare the various agricultural practices from the downstream users' point of view, we need to select a base case. Having assumed that the farmer is a maximizer of net revenue, we choose the practice producing the highest net revenue [the corn-soybean rotation using chisel plowing (CB-CH)]. To illustrate, Figure 8 depicts the relative water quality, soil loss and net revenue impacts (measured as percentage increases or decreases relative to the base case) of the other ten practices on the lowland soil type.

The downstream benefits of alternative farming practices can be qualitatively compared by mapping the quantitative practice/water quality relationships depicted in Figure 8 onto each user or benefit category. A minus sign has been chosen to indicate that an increase in a specific water quality component has a detrimental effect on the specified benefit group. For example, an increase in nitrogen concentration in drinking water is potentially harmful to human health. A zero indicates that an increase in the parameter is of no

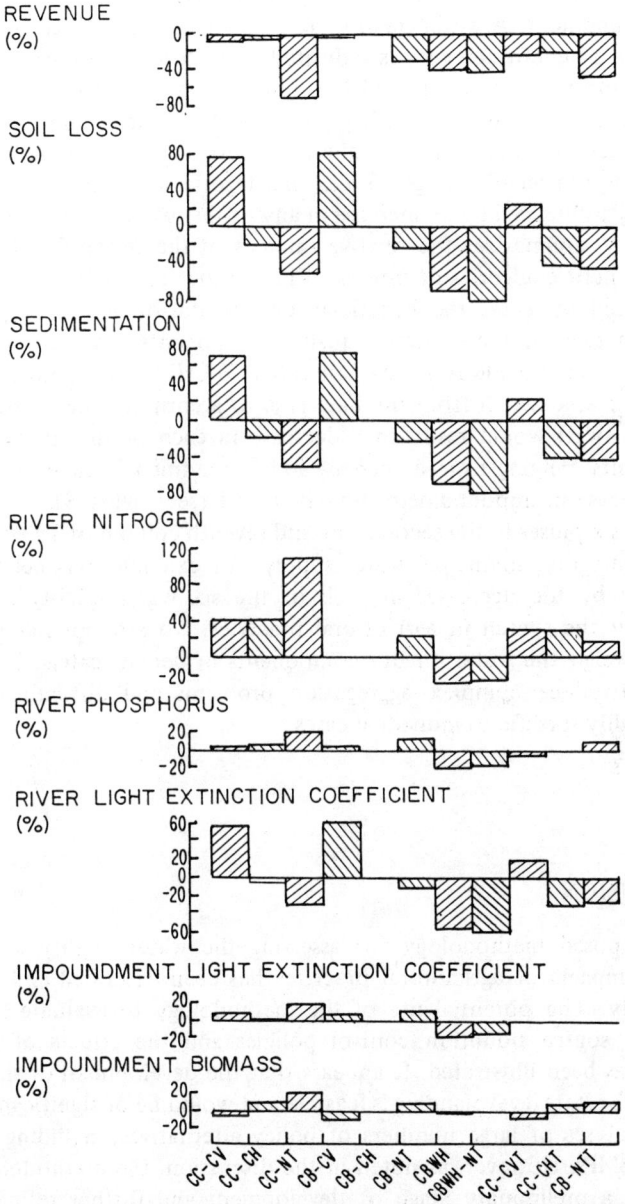

Figure 8. Percent change from highest revenue practice (#5)–lowland.

importance to the benefit category. For instance, the same increase in nitrogen concentration just mentioned would not impact dredging operations in the impoundment. A water quality component increase which has a positive impact on a benefit category is indicated by a plus sign. Increasing impoundment biomass, for example, might improve sport fishing since more food might increase the available fish population. With the possible exception of the beneficial impact of higher biomass levels on sport fishing, no conflicts exist among the benefit categories, *i.e.*, all categories are either not influenced or negatively influenced by an increase in any of the water quality components.

Table V summarizes the relative impacts of the eleven farming practices on the benefit categories of interest. The number of pluses, minuses or zeros in each cell indicates the beneficial, detrimental or neutral impact of the change in each of the six water quality components, displayed in Figure 8, when a switch is made from the base case (CB-CH) to the compared practice. For example, a switch from the base case to a corn, soybean, wheat, hay rotation (CBWH) would result in a decrease in each of the six water quality components, ranging from a decrease in impoundment sedimentation of 70% to a decrease in impoundment biomass of 3% (see Figure 8). It can be seen from the six pluses in the second row and seventh column of Table V that the benefit category, municipal water supply, for example, has been impacted positively by the decreases in each of the six water quality components caused by the switch in agricultural practices. No attempt has been made here to weigh the water quality components or benefit categories since this would introduce complex aggregation problems and difficult value judgments highly specific to individual cases.

CONCLUSION

A proposed methodology for assessing the water quality and socioeconomic impacts of agricultural practices has been described and tested in a case study. The potential use of the methodology to evaluate agricultural nonpoint source pollution control policies and the effects of alternative futures has been illustrated. It appears that the development of such a methodology for state level planning is feasible and would be of significant value for broad analyses of large numbers of policy alternatives, including identification of BMPs. However, as noted in the discussion, the methodology is currently at a preliminary stage of development and further refinements are necessary to make it fully operational.

Table V. Summary of relative impacts of farming practices on benefit categories.

Soil Type: L

Benefit Categories[b]	Farming Practices[a]										
	CC-CV	CC-CH	CC-NT	CB-CV	CB-CH	CB-NT	CBWH	CBWH-NT	CC-CVT	CC-CHT	CB-NTT
Human Health (drinking water)	1(+) 1(0) 4(-)	2(+) 1(0) 3(-)	2(+) 1(0) 3(-)	1(+) 2(0) 3(-)	6(0)	2(+) 1(0) 3(-)	5(+) 1(0)	4(+) 2(0)	1(+) 2(0) 3(-)	2(+) 2(0) 3(-)	2(+) 1(0) 2(-)
Municipal Water Supply	1(+) 5(-)	3(+) 3(-)	3(+) 3(-)	1(+) 1(0) 4(-)	6(0)	3(+) 3(-)	6(+)	5(+) 1(0)	1(+) 1(0) 4(-)	3(+) 1(0) 1(-)	3(+) 3(-)
Dredging (flood control)	5(0) 1(-)	1(+) 5(0)	1(+) 5(0)	5(0) 1(-)	6(0)	1(+) 5(0)	1(+) 5(0)	1(+) 5(0)	5(0) 1(-)	1(+) 5(0)	1(+) 5(0)
Ecology	1(+) 5(-)	3(+) 3(-)	3(+) 3(-)	1(+) 1(0) 4(-)	6(0)	3(+) 3(-)	6(+)	5(+) 1(0)	1(+) 1(0) 4(-)	3(+) 1(0) 2(-)	3(+) 3(-)
Recreation Sport Fishing	1(0) 5(-)	4(+) 1(0) 1(-)	4(+) 1(0) 1(-)	1(0) 5(-)	6(0)	4(+) 1(0) 1(-)	4(+) 1(0) 1(-)	4(+) 2(0)	2(0) 4(-)	4(+) 2(0)	4(+) 1(0) 1(-)
Contact	1(-) 1(0) 4(-)	2(+) 1(0) 3(-)	2(+) 1(0) 3(-)	1(+) 2(0) 3(-)	6(0)	2(+) 1(0) 3(-)	5(+) 1(0)	4(+) 2(0)	1(+) 2(0) 3(-)	2(+) 2(0) 2(-)	2(+) 1(0) 3(-)
Noncontact	1(+) 2(0) 3(-)	2(+) 2(0) 2(-)	2(+) 1(0) 3(-)	1(+) 2(0) 3(-)	6(0)	2(+) 2(0) 2(-)	4(+) 2(0)	3(+) 3(0)	1(+) 3(0) 2(-)	2(+) 3(0) 1(-)	2(+) 2(0) 2(-)
Aesthetics	1(+) 2(0) 3(-)	2(+) 2(0) 2(-)	2(+) 2(0) 2(-)	1(+) 2(0) 3(-)	6(0)	2(+) 2(0) 2(-)	4(+) 2(0)	3(+) 3(0)	1(+) 3(0) 2(-)	2(+) 3(0) 1(-)	2(+) 2(0) 2(-)
Local Economy	1(+) 5(-)	3(+) 3(-)	3(+) 3(-)	1(+) 1(0) 4(-)	6(0)	3(+) 3(-)	6(+)	5(+) 1(0)	1(+) 1(0) 4(-)	3(+) 1(0) 2(-)	3(+) 3(-)

[a] See farm model discussion for definition of farming practices.
[b] See text for explanation of benefit categories.

REFERENCES

1. Meta Systems Inc. "Water Quality Impact and Socio-Economic Aspects of Reducing Nonpoint Source Pollution from Agriculture, Draft Report and Appendices," U.S. Environmental Protection Agency, Athens, GA (February 1978): Appendix A, Farm Model; Appendix B, Methods for Predicting Watershed Loadings; Appendix C, Methods for Predicting Impoundment Water Quality; Appendix D, Water Quality Impact Results: Additional Interpretations and Sensitivity Analysis; Appendix E, A Discussion of Benefit Estimation; Appendix F, Crop Response to Fertilizer.
2. Christensen, R. G., and C. D. Wilson, Eds. "Best Management Practices for Nonpoint Source Pollution Control," EPA-905/9-76-005, U.S. EPA, Region 5, Chicago (November 1976).
3. Lake, J., and J. Morrison, Eds. "Environmental Impact of Land Use on Water Quality," Progress Report, Black Creek Project, Allen County, Indiana, Allen County Soil and Water Conservation District, EPA-905/9-75-006 (November 1975).
4. U.S. Department of Agriculture, U.S. Environmental Protection Agency. Office of Research and Development. "Control of Water Pollution from Cropland, Vol. 1: A Manual for Guideline Development," EPA 600/2-75-0266 (June 1976).
5. Wischmeier, W. H., and D. D. Smith. "Predicting Rainfall—Erosion Losses from Cropland East of the Rocky Mountains," ARS, U.S. Department of Agriculture, Agriculture Handbook No. 282 (1972).
6. U.S. Environmental Protection Agency. "National Eutrophication Survey Series of Working Papers," Corvallis Environmental Research Laboratory. Las Vegas, NV (1975-76).
7. U.S. Department of Agriculture. "Summary of Reservoir Sediment Deposition Surveys made in the United States through 1965," Miscellaneous Publication Number 1143 (May 1969).
8. Indiana State Board of Health. "Reports on Limnologic Investigation of Lakes Martin, Palestine, Sylvan, Waubee, Webster, Crooked, Long and Hamilton," Water Pollution Central Division, Biological Studies and Standards Section, IN (1976).
9. U.S. Army Crops of Engineers. "Miscellaneous Water Quality Data from Indiana and Ohio Reservoirs, 1971-1977," Louisville, KY (1977).
10. Data Resources Inc. "Data Resources Outlook for the United States Energy Sector: Control Case," *Energy Review*, Lexington, MA (summer 1977).

41

MATHEMATICAL MODELING OF WATER QUALITY EFFECTS OF AGRICULTURAL BEST MANAGEMENT PRACTICES

C. Tang
  URS Company
  Seattle, Washington

SCOPE AND OBJECTIVES

This paper is based in part on the report "Water Quality Modeling," a product of the SNOMET/King County 208 Study. The primary objective was to determine the effect of agricultural best management practices (BMPs) on the water quality of the receiving waters. This was investigated by the findings of an earlier study, the "3(c) Snohomish Basin Plan." In that report, it was concluded that: "In general, the water quality of the basin is good to excellent, with some notable exceptions in . . . . These exceptions are generally caused by urban and industrial sources of pollution in consort with nonpoint sources . . . ." It was further stated that: " . . . in the [water quality] problem area . . ., 50 to 95% of the nonpoint contribution to the water quality conditions would have to be removed during worst case [7-day, 10-yr low-flow period] conditions to meet state standards." The most noticeable violation to the standards was the fecal coliform concentration.

The study area (Figure 1), a large part of the Snohomish River basin, is located in the northwest part of the State of Washington. It encompasses 1200 mi^2 of drainage area, and the system is formed by two rivers, the Skykomish and Snoqualmie, and an estuary through which the river travels before discharging into Puget Sound.

In past studies of the Snohomish River basin, agricultural nonpoint waste loads were estimated indirectly by calculating the difference between the

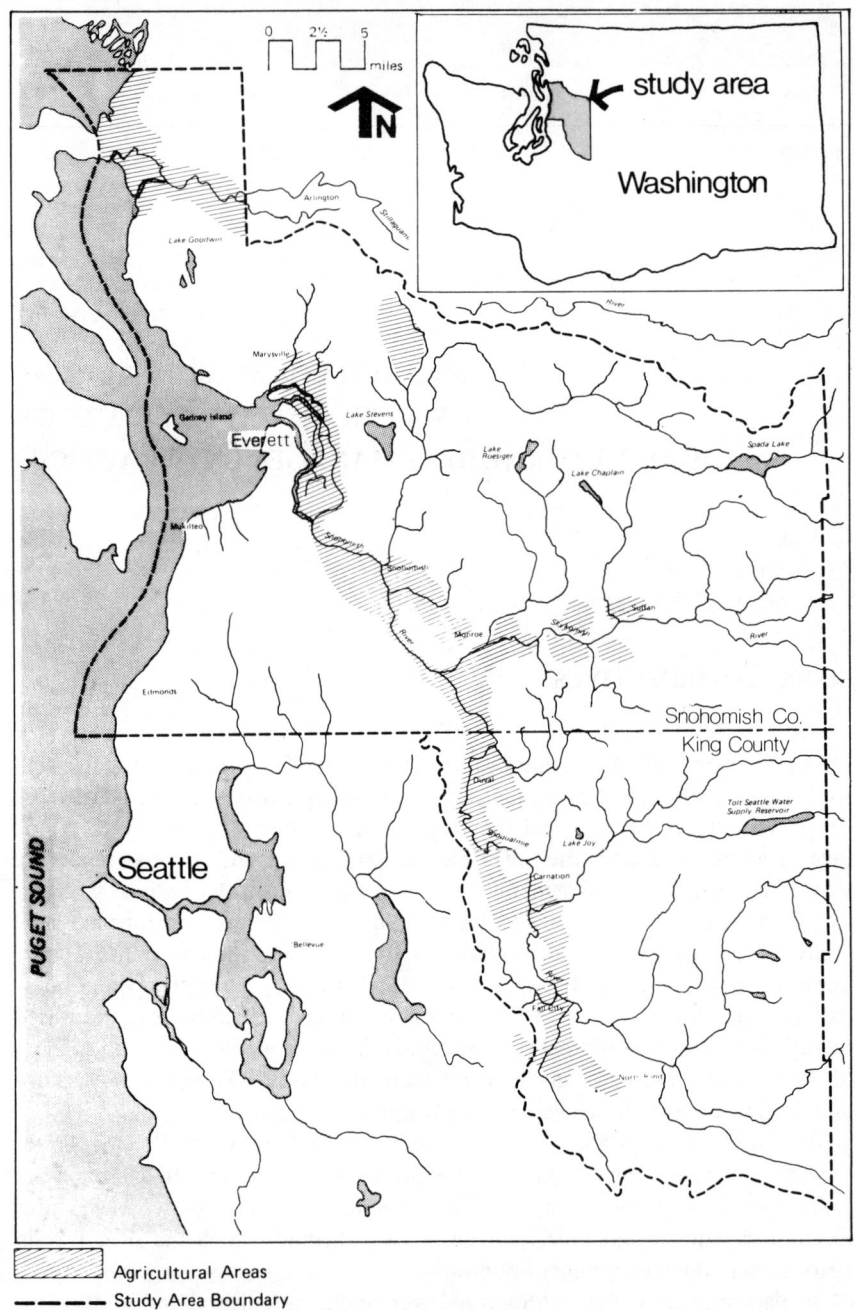

Figure 1. SNOMET/King County 208 planning area.

total load observed in the river and the simulated known point loads. However, for this 208 study, a more elaborate computation scheme was designed to calculate the nonpoint waste loads directly from the field data collected during a series of storm events (SNOMET/King County 208 Technical Appendix III, 1977). With the aid of several computer programs (data base programs and simulation models), this scheme was used to perform more extensive analysis on the impact of the existing nonpoint waste loads. The results of this analysis have not only confirmed the previous conclusion, but helped to elaborate on the significance of the impact of the agricultural nonpoint sources in light of the total basin waste loads and in reference to the water quality standards of the receiving waters. Therefore, because the objective of this project was to determine the would-be waste load reductions resulting from the implementation of various agricultural BMPs and to predict whether the resultant concentrations in the receiving water would meet imposed water quality standards, this analytical process was utilized once again. The process consisted of four major steps:

1. inventory of the agricultural practices in the study area,
2. construction of the analytical scheme,
3. determination of the nonpoint waste loads, and
4. simulation of the resultant receiving water quality conditions.

AGRICULTURAL PRACTICES

Using maps of the land uses in the area in conjunction with unit waste loading values obtained from literature, previous tasks in the 208 study had identified subbasins to which agriculture potentially might contribute significant pollutant loads. In general, these subbasins were found to lie in the floodplain of the Snohomish Basin (Figure 1), with some upland farming occurring.

Seven basic categories of agricultural practices were established, each of which was subdivided into "lowland" and "upland" subcategories to account for the effect of slope and soil conditions on runoff quantity and quality: (1) row-crop, (2) row-crop with manure application, (3) pasture, (4) pasture with manure application, (5) pasture with manure application and animal access to adjacent stream/ditch, (6) barns and other confinement areas, and (7) forest.

Field sampling established the pollutant loads attributed to these categories. These were subsequently used in the modeling effort. As determined by field sampling, the most significant of the above-noted categories is barns, with yearly pollutant loads up to three orders of magnitude greater than any other category.

Various types of BMPs were recommended for use on farms in the area.[1] All of the BMPs were essentially geared to zero discharge of pollutants to surface waters. The BMPs may be divided into several broad categories relating to the practice being managed: (1) animal confinement areas (barns), (2) manure disposal (on fields), (3) animal access to streams, and (4) other.

For the modeling effort, the use of only one set of BMPs was investigated, those for animal confinement areas. There were two major reasons for limiting simulations in this manner. First, the model is not sensitive enough to differentiate among the various BMPs for confinement areas. Second, when testing multiple BMPs, the confinement area loads are of such an order of magnitude larger than those for other agricultural practices that they would mask the effects of nonconfinement area BMPs.

MODELING METHODOLOGY

The two primary objectives of the modeling effort were to develop nonpoint waste load forecasts and to simulate the impact of solution programs on the water quality conditions in the Snohomish River Basin. The following steps outline the general approach utilized in this modeling effort:

1. set up and code areawide land use data base in computer;
2. select and set up appropriate models;
3. reduce the coded data base to form the land/use practice categories identified in the previous section;
4. determine constituents to be modeled based upon problems identified in prevous studies;
5. reduce the data collected in the field investigation program into two matrices: curve number (describing the subbasin runoff characteristics) and waste loading factors (describing the subbasin loading characteristics);
6. determine appropriate design storm and flow conditions;
7. calculate nonpoint waste loads after the implementation of BMPs;
8. calculate the impact of the above loads on the upper basin*;
9. calculate the impact of the above loads on the estuary.

Steps 1 through 6 will be briefly discussed in the following section. A more detailed discussion can be found in the 208 document.[2] Steps 7 through 9 are simply the model simulation steps and will be discussed separately in later sections.

---

*The mouth to 1/2 mile below the confluence of the Skykomish and Snoqualmie Rivers is called the estuary and everything upstream on these two rivers is denoted the upper basin.

Data Base Program

The Snohomish County Primary Data Base Program was used, with the information coded for each 10-ac parcel. Information retrieved from the data base for this study included: Water Exposure Code, Nonpoint Source Subbasin, Existing Land Use, Soil Phases, Slopes, Land Cover, Agricultural Practice, Irrigation Practice, Drainage Practice, Animal Units, Dominant Livestock and Flood Plain.

Mathematical Models

Storm Runoff Model–SNODOB

The SNODOB model of the Snohomish County Planning Department was calibrated especially for the Snohomish River basin. It computes runoff by the Soil Conservation Service (SCS) curve number method and produces hydrographs from synthetic unit hydrographs. Pollutant concentrations in runoff are computed by an exponential decay equation.

Receiving Water Model–SRMSCI

The SRMSCI model of the Snohomish County Planning Department is a nonsteady-state river/estuary model modified from the RECEIVE block of EPA's SWMM (Stormwater Management Model) package.

Data Reduction Program

The Data Reduction Program, REDUC, was developed by the Snohomish County staff. Its function is to provide pre-storm loadings of each pollutant and the average curve number for each subbasin.

Modeling Constituents

The nonpoint loads of biochemical oxygen demand (BOD), ammonia and fecal coliform were modeled as they are closely related to the identified dissolved oxygen (DO) and fecal coliform violations in the study area.

Curve Number and Load Factor Matrices

Curve Number (CN)

A curve number (CN) is a representation of the capacity to allow runoff of storm water from a parcel of land and is a key factor in determining runoff for the SNODOB model. Initially, a matrix was constructed based on the data published by the SCS.[3] This matrix was later calibrated for the 208 area to reflect the measured runoff data.

Load Factor (LF)

The loading factor (LF), expressed as load-unit/acre/number of dry days, is commonly used to describe the prestorm (accumulated) waste loads of a parcel of land. Data collected from the field investigation program, in which six farm plots and several streams were chosen, were used to derive the LF matrix. The validity of the LF matrix was tested with additional observed data from three sites.

Design Storm and Flow Conditions

Since the models being used (SNOBOD and SRMSCI) are centered on a time frame of a single storm event, it is reasonable to choose some approximation of a "worst case condition." Such a condition will generally occur with an initial low flow in the receiving water to which a sizable storm event is added. The 7-day, 10-yr low flow (1080 cfs at the mouth of the river) and the annual summer storm (1.25 in. in 35.5 hr) with an antecedent dry period of 4 days were chosen as the design condition. The probability of mutual occurrence of these "worst case" events ranges from 0.09 to 10.0%. Results presented in this report should be placed in the context of this limited probability of occurrence.

NONPOINT WASTE LOADS

The development of nonpoint waste loads due to a storm involved running two computer programs, REDUC and SNODOB. The outputs of REDUC were subbasin area, average CN, and total prestorm waste loads for each constituent, which served as inputs to the runoff model SNODOB. Other inputs to SNODOB included items such as time of concentration for the hydrograph, washoff coefficients for the pollutograph for each water quality constituent, etc.

The above parameters are subbasin-dependent and normally are obtained through calibration by matching the computer results to a given set of observed data for each subbasin. However, in this 208 program detailed data gathering was limited to a few representative subbasins and the results of the model calibration were extended to the other subbasins. Such an approach was considered valid for several reasons:

1. Within the SNOMET/King County 208 designated area there are only a limited number of agricultural practices which yield distinctly different runoff characteristics. Hence, by calibrating and verifying the model for a subbasin of each distinct type, it was possible to extend the model to the remaining nonpoint subbasins.

2. The shape of the pollutograph of an uncalibrated agricultural subbasin may differ from the shape of the calibrated one, but the total pollutant load (the area under the pollutograph) will still be similar because of the flat response of agricultural hydrographs.
3. Most nonpoint subbasins in the estuary area discharge through a pump station or tide gate which generally operates under steady-state conditions. Hence the sensitivity of discharge to the above hydrograph parameters diminishes significantly.

Some nonpoint sources are discharged in the form of a tributary system which, therefore, presents no problem in identifying their discharge sites. For the other sources, which enter the river in a diffuse manner, a middle point of the respective subbasin boundary with the river was selected as the modeled discharged site. Generally, the length of that boundary was less than a mile.

Among the contributors to nonpoint waste loads, animal confinement areas were found to be dominant. The actual number of animal confinement areas contributing to the wasteload from a subbasin was determined in two ways: (1) through a field survey conducted for Woods Creek, French Creek and the Marshland drainage districts, and (2) by counting the number of soil types which have 10-ac parcels in either the "manured pasture" or "manured row-crop" designation as derived from REDUC outputs. This counting method was used because one barn might be identified by REDUC as occupying more than one 10-ac parcel, but seldom on two soil types. Table I gives the results obtained from these two methods. The results show a very high correlation. Because of this success, the latter method was used to identify the number of barns for other nonpoint subbasins in the upper basin and in the estuary.

Table I. Comparison of barn identification methods.

Basin Name	Subbasin I.D.	Number of Barns Identified	
		By Field Survey	By REDUC
Marshland	LH	1	1
	LX	1	1
	LM	0	2
	LP	6	5
French Creek	MA	7	8
	MG	1	6
	MH	0	1
	MI	2	2
Woods Creek	QA	2	2
	QD	1	1

Note: Correlation coefficient, $r = 0.77$.

All of the nonpoint sources were identified in terms of nonpoint subbasins and were situated among three major drainage basins: Skykomish, Snoqualmie and estuary. The results of the waste load situation in each drainage basin are shown in the following.

Skykomish Basin

Table II gives the existing conditions of the modeled nonpoint subbasins in the Skykomish basin and their respective locations, areas, number of barns, average basin curve number (CN), and loads discharged to the Skykomish River during the design storm. Also given are the reduced waste loads if BMPs are applied to the barns. As noted earlier, it was assumed that the waste loads from animal confinement areas would be reduced to zero upon BMP implementation. The reduction is about 75, 83 and 85% for BOD, ammonia and fecal coliform, respectively. High waste loads were concentrated within 3 mi (RM 28 to RM 22) upstream and downstream of the Woods Creek confluence.

Snoqualmie Basin

Table III gives the simulated loads from the modeled nonpoint subbasins in the Snoqualmie basin under existing land use conditions. Also shown are the loads if BMPs are implemented. The barns were identified in three areas: near Cherry Creek, Carnation and Fall City. The number of barns identified in this basin was greater than that in the Skykomish basin (30 versus 22). However, the major difference between the two basins was the runoff curve numbers. The average CN in the Snoqualmie basin was lower than in the Skykomish due to a difference in soil types. Because of these low CNs which result in lower total runoff volume, the simulation produced lower storm loads. Therefore, despite the higher prestorm loads, the simulated storm loads in the Snoqualmie basin for existing conditions were considerably smaller than in the Skykomish basin. Data from the field survey are insufficient to verify this conclusion.

Estuary

The nonpoint sources in the Snohomish estuary consisted of runoff from the following major drainage basins: Everett, French Creek, Marshland district, Cavalero, Quilceda, Estuary (Slough) and Upper Snohomish (Figure 2).

Combined Sewer Overflow

Runoff from the Everett basin to the estuary is routed through the combined sewer system and then through treatment facilities. It becomes a source

Table II. Modeled storm loads in the nonpoint subbasins within the Skykomish basin.

RM June	Nonpoint Subbasin	Acre	Barns	CN	BOD (lb)	Ammonia-N (lb)	Fecal Coliform Organisms x 10^6
47.5 35	SK 09 Gorge	510	--	79	13	8	368
46.4 34	SK 11 Austin CR	840	--	78	17	11	978
42.9 30	SK 16 Gold Bar E	460	--	81	7	3	4673
40.5 27	SK 17 Gold Bar	640	--	82	17	6	12072
39.0 25	SK 18 Forks	4920	--	78	120	76	7725
37.3 23	SK 19 Gold Bar W	740	--	81	18	12	1645
36.1 21	SK 20 Chappel	460	--	78	11	7	1310
35.2 20	SK 23 Sultan E	3670	1	81	323 (122)[a]	204 (67)[a]	301556 (45561)[a]
25.2 20	SK 24 Voss	2210	--	79	57	38	20717
32.6 16	PO Hill	1640	4	77	585 (49)	378 (15)	435382 (70824)
32.6 16	PN Elwell E	610	--	77	13	8	1627
30.5 14	PL Wilner	1950	--	78	37	17	14790
29.3 13	Fern Bluff	2260	1	78	101 (23)	60 (7)	127496 (15338)
27.3 11	PH Sky RV 7	2260	1	78	195 (13)	128 (3)	224196 (24600)
26.3 10	PF Sky RV 5	1180	1	73	91 (13)	56 (3)	136757 (24599)
25.1 9	QA WX 1	1610	2	74	216 (34)	130 (7)	305704 (47856)
25.1 9	QB WC 2	6100	--	71	61	26	30543
25.1 9	QC WC 3	3750	--	76	74	29	33414
25.1 9	QD WC 4	2090	--	71	84 (28)	45 (8)	105242 (24248)
24.8 8	PJ Hasket Slough	930	2	78	327 (27)	206 (3)	434546 (33366)
24.8 8	PE Sky RV 4	1070	6	76	763 (50)	485 (2)	1049168 (64385)
22.8 5	PI Riley Slough	4360	3	78	566 (116)	353 (48)	691071 (89299)
22.8 5	PD Sky RV 3	190	1	71	57 (2)	38 (1)	82529 (1534)
	Total				3753 (922)	2324 (405)	4023509 (592472)
	Reduction After BMP Application				75%	83%	85%

[a]Loads after the application of BMPs.

Table III. Modeled storm loads in the nonpoint subbasins within the Snoqualmie basin.

RM June	Selected Nonpoint Basin	Acre	No. of Barns	CN	BOD (lb)	Ammonia-N (lb)	Fecal Coliform Organisms x 10^6
26.2 45	SQ 2B High Bridge Road	1400	7	71	491 (105)[a]	265 (4)[a]	791,000 (224,000)
27.2 47	SBø4 Cherry Creek	10540	4	68	178 (67)	103 (27)	227,000 (58,000)
28.7 49	SQ 5a Duvall N	510	2	67	33 (12)	15 (0)	65,000 (21,000)
31.9 53	SQ6C Gravel Pit	720	--	63	1	0	943
32.4 54	SQ 5C Cemetery	1340	--	68	24	2	32,000
37.4 60	SQø7 Ames Lake	4210	6	61	0 (0)	0 (0)	500 (100)
39.5 63	SQ 8C Gilead No	780	3	77	534 (131)	275 (3)	734,000 (187,000)
44.8 69	SQ 9A Carnation	19000	1	67	35 (14)	17 (3)	53,000 (21,000)
47.7 73	SQ10 Griffin Cr	14480	3	63	14 (6)	7 (3)	16,000 (3,000)
48.6 74	SQ8B Stickney Slough	2170	--	68	15	4	22,000
54.9 81	SQ 13 Fall City	920	4	73	344 (32)	214 (3)	532,000 (84,000)
61.6 90	Q17B Snoq Falls W	1716			13	3	11,000
	Total				1682 (420)	905 (52)	2484443 (663943)
	Reduction After BMP Application				69%	86%	73%

[a] Loads after the application of BMPs.

in itself only when the capacity of the system is exceeded and overflow occurs. The Lower Snohomish Basin 201 Facility Plan study conducted an analysis of Everett's combined sewer overflows. The estimated waste loads used in this analysis were obtained directly from the 201 project. The quantity of the overflow was simulated with an annual storm from a sewer model. The quality characteristics of the overflows were taken from sampled overflow data. Figure 2 shows the location of the overflow sites and other nonpoint sources in the river model SRMSCI.

Pump Stations

Runoff from French Creek and the Marshland drainage districts is detained behind pump stations. The accumulated volume of water is pumped continuously after a threshold volume is reached. Therefore, these two sources may

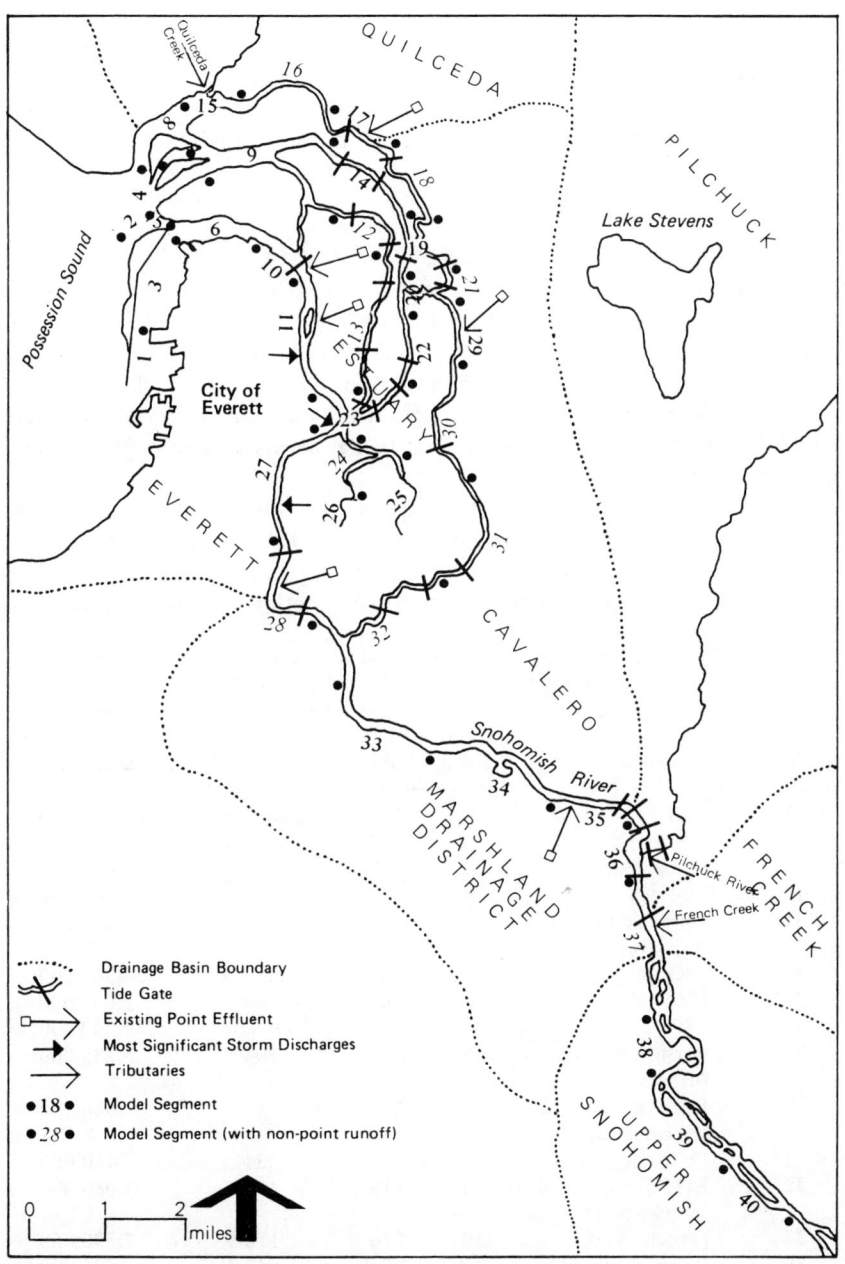

Figure 2. Modeled lower basin.

continue to contribute a waste load after a storm event has actually ended. The flows discharged from the pump station were calculated from the power consumption records and design pump rates. The pollutant concentrations were determined from samples taken from the pools behind the pump stations. A field survey identified ten barns in the French Creek district and seven barns in the Marshland district as contributors of pollutants.

Tide Gates

Runoff from Cavalero, Quilceda and Slough basins is discharged through tide gates. The quantity of the flow through the tide gates is dependent upon the degree to which the gate is open, which in turn is dependent upon the water level on both sides of the gate. The detained water is drained out during a low tide period. Figure 2 shows the locations of the tide gates and associated drainage subbasins. Through the data base reduction program, REDUC, a total of 16 barns were identified as discharging through tide gates in the drainage basins.

Figure 2 shows how the estuary system has been divided into various junctions and segments for modeling purposes. Table 4 gives the nonpoint waste loads (BOD, ammonia and fecal coliform) in the estuary with the respective location expressed by the modeled junction numbers. These loads

Table IV. Modeled storm loads discharged from the nonpoint subbasins to the estuary in 4.5-hr simulation period.

Junction No.	Nonpoint Subbasin	No. of Barns	BOD (lb)	Ammonia-N (lb)	Fecal Coliform Organisms x 10^6
3	CSO	--	285	no data	13,000,000
11	CSO	--	1653	no data	75,500,000
13	JA 12 (slough)	1	86	31	800,000
16	Quilceda	4	70	41	600,000
21	JA04 (slough and JC (Cavalero)	1	56	33	500,000
23	CSO	--	1187	no data	54,300,000
24	JA06 (slough)	1	189	42	1,650,000
27	CSO	--	5153	no data	232,900,000
28	Marshland Drainage District	8	554	104	3,280,000
30	JA03 (slough) and JF (cavalero)	1	100	55	880,000
31	JA02 (slough)	1	49	34	460,000
32	JA15 (slough) and JK (cavalero)	3	118	64	1,080,000
37	French Creek	10	846	116	2,890,000
39	Upper Snohomish	2	64	41	620,000

were calculated over a time period of 4.5 hr. This time scale represents the duration of the combined sewer overflow (CSO) event mentioned earlier. During this time span, the waste loads from the French Creek and Marshland districts were found to be relatively constant for both observed and simulated data. The period of low tide is about 4-6 hr in the first tidal cycle. It is, therefore, reasonable to assume that during the 4.5-hr time span, flows will be continuously discharged through the tide gates, provided that a large volume of water has built up behind the tide gates during the high tide period. For a "worst case" condition, as was assumed in this study, these assumptions would not be unreasonable. Since flow measurements could not be made for all of the tide gates, an assumption was made that during the 4.5 hr of the low tide period, the discharge from the tide gate would be equal to the runoff reaching it. Based on this assumption, the simulated flow ranges from 2 to 37 cfs through the tide gates during the peak of the flow. The results were found to be very close to what was observed during the 3(c) Basin Plan.

RECEIVING WATER QUALITY

As was done in the 3(c) Basin Plan,[4] the effect of the nonpoint waste loads on receiving water quality was evaluated in terms of whether the loads met the proposed revised Washington State Standards.[5]

Both the Skykomish and Snoqualmie Rivers are classified as Class A waters. The proposed standards for DO and fecal coliform concentration are 8 mg/l and median value of 50 MPN/100 ml, respectively. The estuary is classified as Class A, with a special condition given to the slough that raises the fecal coliform concentration to a median value of 100 MPN/100 ml.

The purpose of using the median value is to dampen the large inherent errors found in present coliform sampling and analysis methods. Simulations, however, do not suffer from this problem. Thus, single simulated values at any time are sufficient for analysis. Of course, errors which might be found in the original data from the field program are carried through in the simulations. The standards presume a violation even if only one value of a parameter exceeds the standard or is outside the allowable range for that parameter. For the analysis performed, cases where a parameter is outside the allowable range for a single value are denoted as exceeding the standards. The effects of nonpoint sources on the receiving water quality are seen most noticeably on the fecal coliform concentration from both sampling data and simulation results. Although visible effects also showed on the DO concentrations, they were not shown as seriously violating the standards, as do the fecal coliform concentrations. For the purpose of this paper only the results on fecal coliforms will be discussed and the reader referred to the SNOMET 208 Modeling

Appendix document for further results. Because the SRMSCI model is limited to simulating one system type (river or estuary) at a time, receiving water quality simulation for the study area proceeded in two steps: First the Upper Basin was simulated and then the estuary was simulated, with its upstream boundary condition defined in the preceding run.

Upper Basin Receiving Water Quality

For this analysis, the nonsteady-state river/estuary model (SRMSCI) was set up for the nontidally influenced upper basin. SRMSCI was verified against the field data[5] collected on January 15-17, 1977 from the lower reach of the Skykomish River. The upper basin receiving water quality simulation involved two basic steps:

1. Using REDUC and SNODOB, calculate the waste loads of all point and nonpoint sources.
2. Using SRMSCI, simulate the resultant river water quality with the above inputs waste loads.

The important consideration here was whether the use of BMPs in the non-point basins would result in improved water quality.

Field Sampling and Model Verification

On January 15-17, 1977, river samples were collected from the lower Skykomish and Snoqualmie Rivers following a storm event. The results are given in the SNOMET 208 Technical Appendix II.[6] Figure 3 shows the location of the sample sites, with four stations in the lower Skykomish basin and two stations in the Snoqualmie basin.

The river flows measured at the mouths (*i.e.*, confluence of Skykomish and Snoqualmie) rose from 2-3,000 cfs to 4-5,000 cfs for each of the rivers. Although this was observed in January, this type of flow regime was often recorded in the middle of October from the U.S. Geological Survey gauge stations at both rivers.

The simulated quality results (Figure 4) confirmed the findings, both in the 3(c) Basin Plan and nonpoint source modeling, that nonpoint sources located between the sample sites were contributing significant waste loads. A sharp rise in the fecal coliform concentrations at sites 2, 3 and 6 may be attributed predominantly to the nonpoint waste loads located in the Skykomish River between river miles (RM) 22.0 to 25.5, as seen in Table II and located in the Snoqualmie River between RM 22.0 and 56.0, as seen in Table III.

The fecal coliform concentration was found to be greater than 1,000 MPN/100 ml at site 3. Based on these field data, it can be concluded that the

MODELING STUDIES 639

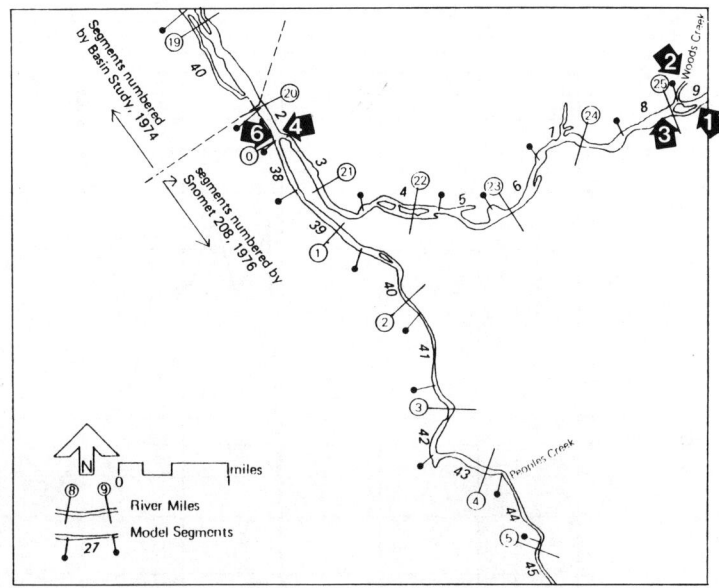

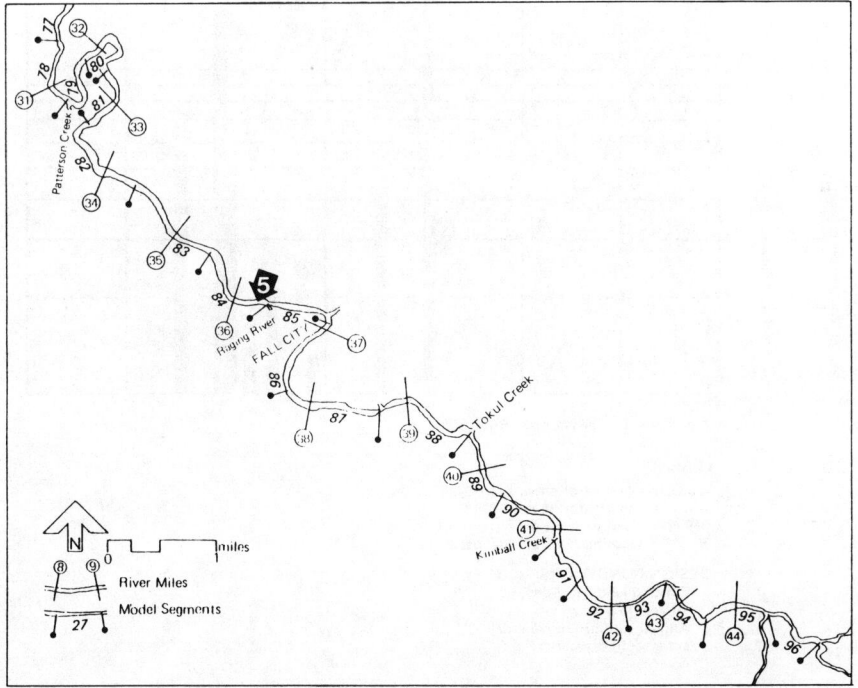

Figure 3. Sampling stations (shown by arrows).

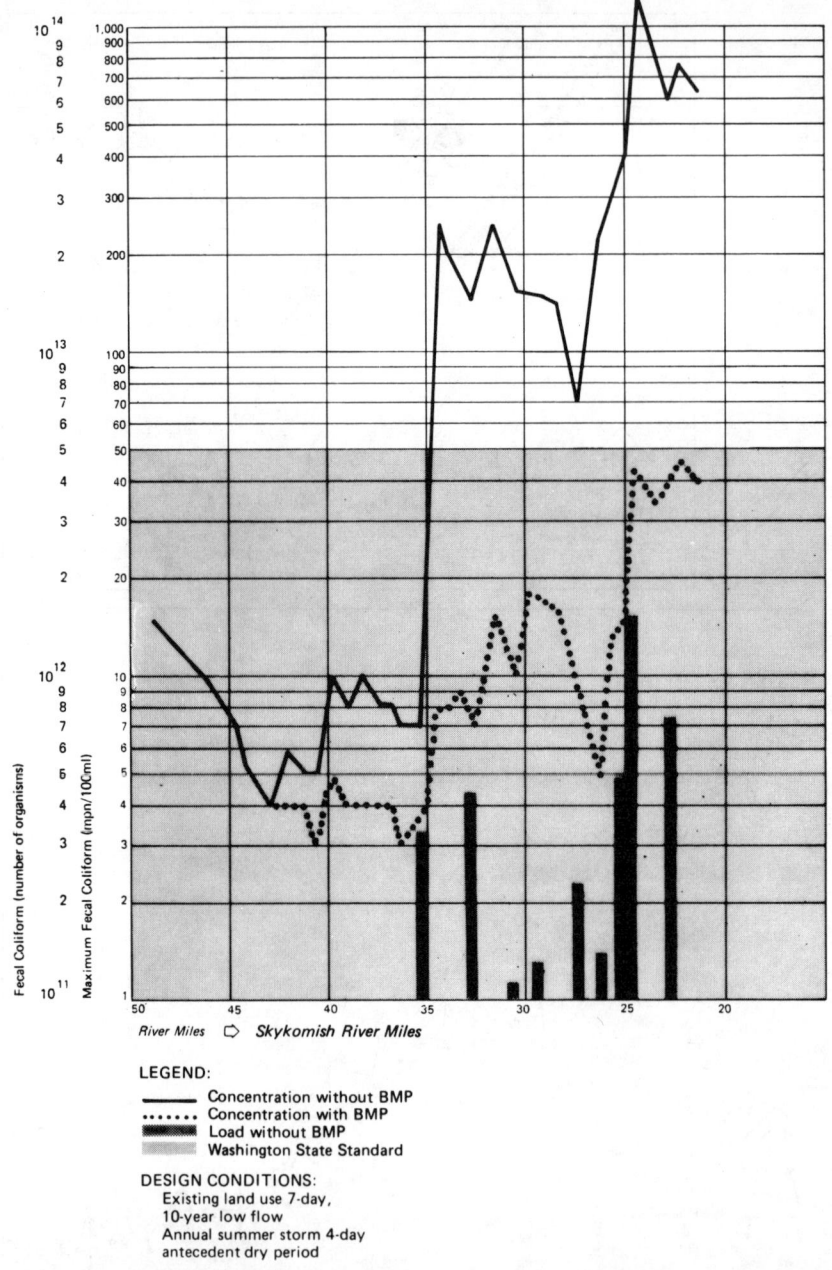

Figure 4. Simulated nonpoint fecal coliform loads and concentration profile in the Skykomish River with and without agricultural BMP implementation.

nonpoint sources in the upper basin indeed have an impact on the local receiving water as suggested in the 3(c) Basin Plan. The impact on the estuary is also evident from the data collected at sites 4 and 6, which are located at the confluence of the two rivers and at the headwater of the estuary. The high fecal coliform concentrations observed at these two sites were about 100-200 MPN/100 ml, exceeding the proposed standards for the estuary (50-100 MPN/100 ml, depending on location in estuary).

Simulation Results

The 7-day, 10-yr low flow for the Skykomish River is 514 cfs, and for the Snoqualmie River it is 466 cfs. After receiving the existing point and nonpoint loads, simulated fecal coliform results of the Skykomish River are plotted in Figure 4. The plot gives the maximum concentrations. The maximum fecal coliform concentrations rose sharply at RM 35 and reached a peak of over 1,000 MPN/100 ml immediately downstream of the Woods Creek confluence, which is located at RM 24.5, in the vicinity of site 3. The same figure shows that the fecal coliform concentrations have improved significantly with the reduced loads of BMP implementation; the maximum concentration at any reach was reduced to less than 100 MPN/100 ml. The average fecal coliform concentrations were less than 50 MPN/100 ml in all areas. Based on the simulated results, the fecal coliform concentrations were on the borderline of violating the standards. It is very difficult to give the exact level of improvement from BMP application with respect to the standards. However, with the implementation of BMPs, standards violations would be much less frequent.

Simulation results indicated similar responses for the Snoqualmie River. Figure 5 shows that the fecal coliform concentrations rose sharply at three distinct locations (RM 53, 39 and 25.5), which related directly to the nonpoint waste loads. It produced three peaks which had concentrations of 230, 260 and 200 MPN/100 ml, respectively. These concentrations were also reduced significantly to 50 MPN/100 ml and less in all areas when the BMP loads were applied in the simulation.

Estuary Receiving Water Quality

The simulation of receiving water quality in the estuary was performed in essentially the same manner as described in the previous section, with one additional step—upstream boundary conditions were introduced (at the confluence of the Skykomish and Snoqualmie Rivers).

The model, SRMSCI, was calibrated and verified for the estuary region in the 3(c) Basin Plan. No additional verification process like the one used in the river region was attempted.

642 BMPs FOR AGRICULTURE AND SILVICULTURE

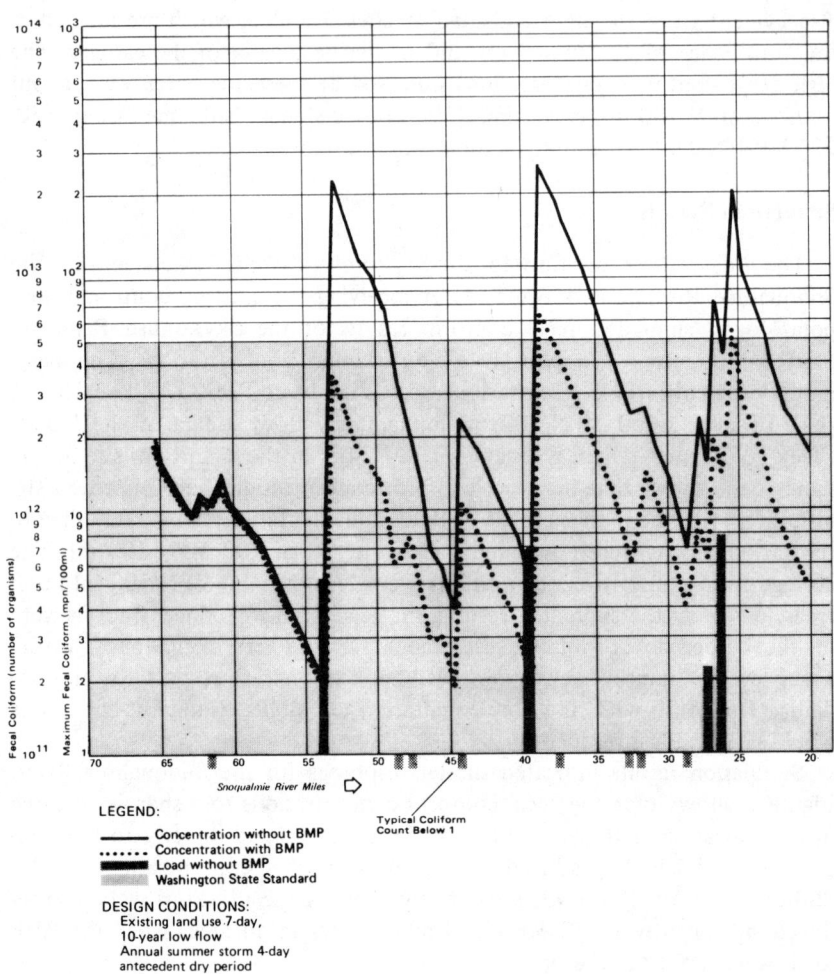

Figure 5. Simulated nonpoint fecal coliform loads and concentration profile in the Snoqualmie River with and without agricultural BMP implementation.

Simulated estuary quality results were displayed at various junctions on the modeled estuary system illustrated in Figure 2. The estuary system was divided into five sections for the convenience of displaying the plots: Snohomish proper, Union Slough, Steamboat Slough, Ebey Slough and Deadwater Slough.

Water Quality Without BMP Use

Figures 6a and 6b show the simulated maximum fecal coliform concentration in the estuary after receiving the existing point and nonpoint loads. Standards are exceeded in almost the entire Snohomish proper. The greatest number was shown at CSO discharge sites, where the resultant concentrations were repeatedly over 1,000 MPN/100 ml and as high as 35,000 MPN/100 ml. The effects of the fecal coliform loads from the other nonpoint sources (from the river and in the sloughs) were not as great as the CSO sources. With cessation of the CSO loads, the maximum fecal coliform concentration in the estuary dropped significantly, from 1,000-10,000 MPN/100 ml to 100 MPN/100 ml. The maximum concentrations in Steamboat and Ebey Sloughs fell lower than the standard of 100 MPN/100 ml. The maximum concentrations in the upper Snohomish proper (junction 28-40) were still exceeding the regular Class A standard, 50 MPN/100 ml, and it was most noticeable in the reaches immediately downstream of the river estuary confluence. This was due predominantly to the effect of the incoming river loads. The maximum concentrations in Union Slough (at junction 13) were over 100 MPN/100 ml. But judging from the average concentration of 27 MPN/100 ml, the standards over this reach could very well be met. Another area with a simulated high level of fecal coliform concentration was in Deadwater Slough (junction 24), over 1,000 MPN/100 ml. This high concentration resulted from a combination of large nonpoint loads in that area and the small volume of water exchanged with other parts of the estuary. Because Deadwater Slough is separated from the rest of the estuary by a tide gate and functions as a lagoon, the simulated impact of the water in Deadwater Slough upon the downstream reaches appears insignificant.

Water Quality With BMP Use

A series of quality simulations were conducted for the estuary to determine the improvement resulting from BMP application. Table V summarizes the results of these simulations with respect to satisfying the standards.

The CSO sources have the predominant effect of the level of fecal coliform concentrations in Snohomish proper, regardless of BMP implementation. This is shown in Table V under simulated condition A (with BMPs applied to the entire study area). The effects of the river loads and the French Creek and Marshland district loads are demonstrated between simulated condition C (BMP applied to the upper basin only) and D (BMP applied to the estuary only). By comparing the results under simulated conditions C and D, one can deduce that the river (upper basin) load dictates the level of DO and fecal coliform concentrations in the Snohomish proper. As was suggested by the results under condition D, the standards for DO and fecal coliform are

644 BMPs FOR AGRICULTURE AND SILVICULTURE

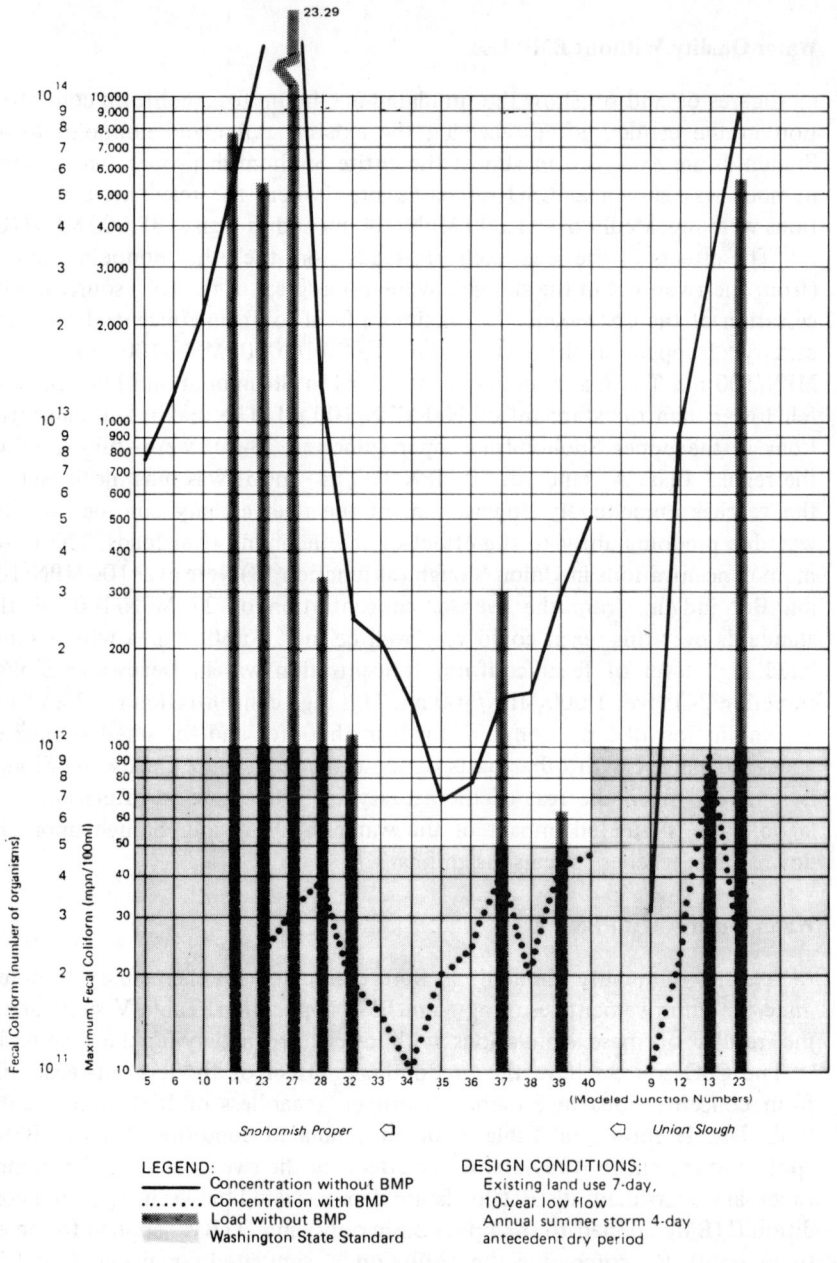

Figure 6a. Simulated nonpoint fecal coliform profile in Snohomish proper and Union Slough with and without agricultural BMP implementation.

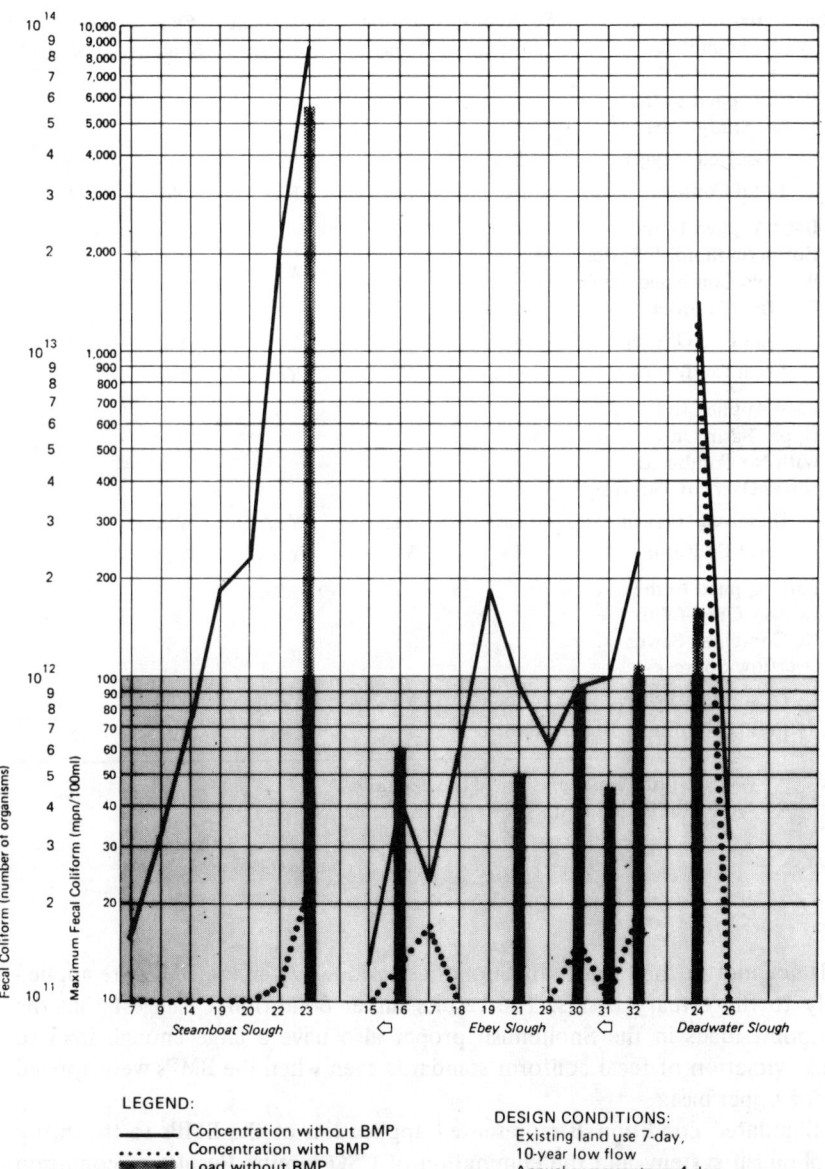

Figure 6b. Simulated nonpoint fecal coliform loads and concentration profile in Steamboat Slough, Ebey Slough and Deadwater Slough with and without agricultural BMP implementation.

Table V. A summary of water quality conditions after the application of BMPs.[a]

Simulated Condition with Parameters	Snohomish Proper	Union Slough	Steamboat Slough	Ebey Slough	Deadwater Slough
A. BMP Applied to the Entire Study Area					
Dissolved Oxygen	Yes	Yes	Yes	Yes	Yes
Fecal Coliform	No	No	No	No	No
B. BMP Applied to the Entire Snohomish System With No Combined Sewer Overflow Sources					
Dissolved Oxygen	Yes	Yes	Yes	Yes	Yes
Fecal Coliform	Yes	Yes	Yes	Yes	Yes
C. BMP Applied to the Upper Basin Only—With No Combined Sewer Overflow Sources					
Dissolved Oxygen	Yes	Yes	Yes	Yes	Yes
Fecal Coliform	No	Yes	Yes	Yes	No
D. BMP Applied to the Estuary Only—With No Combined Sewer Overflow Sources					
Dissolved Oxygen	No	Yes	Yes	Yes	Yes
Fecal Coliform	No	Yes	Yes	Yes	No

[a] Yes = values within limits prescribed in standards.
No = values not within limits prescribed in standards.

still not met in the reaches of Snohomish proper when the BMPs are applied only to the estuary basins. The results under condition C suggest that the nonpoint loads in the Snohomish proper also have a large enough load to cause violation of fecal coliform standards even when the BMPs were applied to the upper basin.

Simulated condition B represented application of the BMPs to the entire Snohomish system, and the elimination of CSO inputs. Results of condition B are also illustrated in Figures 6a and 6b by the resultant fecal coliform profile in the estuary. Standards of fecal coliform were met in all areas of the estuary except Deadwater Slough. Problems in Deadwater Slough were due to a combination of high loads and low flow volumes.

CONCLUSION

The results of this modeling analysis indicate that a combination of abatement of Everett's combined sewer overflow and application of BMPs to the agricultural waste sources should be sufficient to improve river quality and meet the specified standards during the 7-day, 10-yr low-flow event.

ACKNOWLEDGMENTS

I would like to acknowledge the contributions of the Snohomish County Planning Department in conducting this study. Also, the U.S. Environmental Protection Agency should be acknowledged for providing funds for this project, P-D00091.

REFERENCES

1. URS Company. "Farm Water Quality Management Manual," prepared as part of the SNOMET/King County 208 Plan (December 1977).
2. URS Company. "Technical Appendix III—Water Quality Modeling," prepared as part of the SNOMET/King County 208 Plan (December 1977).
3. U.S. Department of Agriculture, Soil Conservation Service. *National Engineering Handbook,* Section 4, "Hydrology," Chapters 7-11 (1972).
4. Department of Ecology. *Washington Administrative Code*, "Water Quality Standards for Waters of the State of Washington," Chapter 173-201, (November 1977).
5. URS Company. "Technical Appendix II—Water Quality Sampling," prepared as part of the SNOMET/King County 208 Plan (December 1977).

42

METHODOLOGY FOR DETERMINING THE OPTIMAL MIX OF BMPs AND AGRICULTURAL PRODUCTION MODIFICATIONS

J. P. Heaney, D. C. Ammon
Department of Environmental Engineering Sciences
University of Florida
Gainesville, Florida

INTRODUCTION

The evaluation of wastewater management alternatives in urban areas has become much more complicated in the past few years. Consideration is being given not only to traditional waste treatment options but also to structural and nonstructural controls to abate wet-weather runoff problems. Last year, we published a procedure which permits the analyst to aggregate a variety of upstream and/or downstream control options into a single overall performance curve.[1] This curve stipulates the total cost of controlling any level of pollution if the most efficient mix of control options is used.

This procedure is extended to permit evaluation of agricultural areas. The main difference between the urban and agricultural problem is that the farmer may consider production modifications in response to an imposed restriction on pollutant discharges. Thus, it is necessary to estimate the response that he might make if confronted with such a situation. The method used to measure this response is an expansion of a farm production model to include constraints on pollutant discharges. With this formulation the allowable pollutant discharge is varied from a large amount down to zero. The farmer is assumed to respond to the restrictions as best he can in that he tries to minimize his losses. A linear programming model of a dairy farm, developed by Haith and Atkinson,[2] is used to illustrate this procedure.

The other control option to be considered as an alternative to production modification is a detention basin which can be installed at the downstream end of the farm. The performance of this basin is estimated based on information presented by Chen et al.[3]

Knowing the farmer's production modification possibilities and the comparative effectiveness of the detention basin, the graphical evaluation procedure is applied to determine the optimal blend of these two options. The result is a single cost function which represents the overall optimal response to any level of pollutant discharge restrictions.

METHODOLOGY

A seven-step graphical procedure was developed to estimate wet-weather pollution control costs in an urban area. The technique is based on production theory and marginal analysis from economics. Production theory states that a limiting technological relationship, known as a production function, exists between the inputs and outputs of any production process. Marginal analysis simply asks the question—will an action result in a sufficient additional benefit to justify the additional cost?

A network illustrating the methodology is shown in Figure 1. Three pollution control options form a configuration in which two options (p = 1 and 2) operate in parallel followed by a downstream option (p = 3) operating in

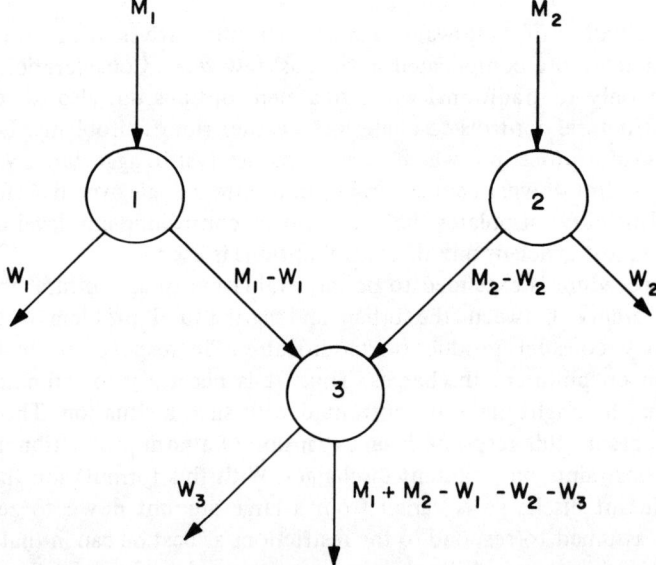

Figure 1. Generalized stormwater pollution control network.

STEP 1: FIND TOTAL COST CURVE FOR EACH OPTION

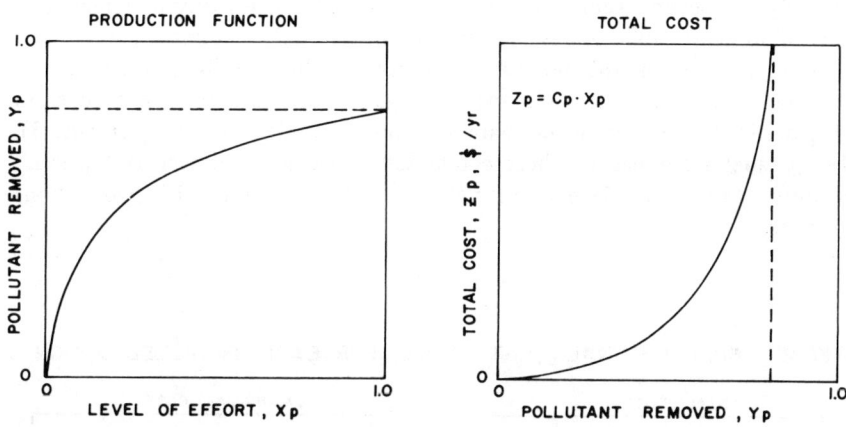

Figure 2. Step 1 of methodology.[1]

series with the parallel group. The methodology is not confined to this example, but may be applied to any number of processes in parallel and/or series. Each control option is a production process described by a production function similar to the one shown in Figure 2, which is part of step 1 of the procedure. The input to each option is the level of effort, X_p $(0 \leq X_p \leq 1)$, and the output is the fraction of pollutant removed, Y_p $(0 \leq Y_p \leq 1)$. Each production function (and the entire methodology) is derived on an annual or seasonal basis. The definition of level of effort depends upon the particular option. For example, the level of effort for street sweeping is the fraction of days per year the streets are swept. The shape of the production function is governed by the "law" of diminishing returns, *i.e.*, the production function exhibits decreasing marginal output with increases in input.

Once the production functions are established, the first step is to construct the total cost curve for each option. If it is assumed that the cost, Z_p, is a linear function of X_p, then the total cost curve is produced by reversing the axes of the production function and applying a unit cost, C_p, to the X_p axis (Figure 2). The second step is to generate the marginal cost curve for each *parallel* option. This curve gives the relationship between the marginal cost per pound of pollutant removed, MC_p, and the pounds of pollutant removed, W_p. The curve is constructed by first converting the abscissa of the total cost curve from the fraction, Y_p, to pounds of pollutant removed, W_p $(= M_p \cdot Y_p$, where M_p is the total pollutant load available to process p) and graphically estimating the instantaneous slope, $\Delta Z_p / \Delta W_p$, at several values of W_p (Figure

3). Once the marginal cost curves are developed for the parallel options, a composite marginal cost curve representing the actions of the entire parallel group may be constructed. This is accomplished by summing the W_p's to obtain W_{12} at several values of MC_p. In other words, for any marginal cost, what amount of pollutant is each parallel option able to control? According to marginal analysis, this represents the optimal "mix" (Figure 4). From the composite marginal cost curve, the composite total cost curve may be derived by graphically integrating the marginal curve with respect to W_{12} (Figure 5). The parallel group has now been condensed into a single equivalent "option." Therefore, the example network has been reduced to one with two options in series.

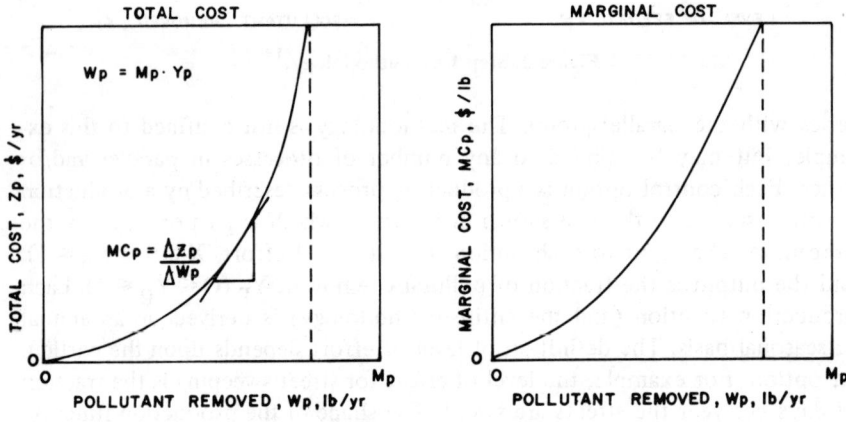

Figure 3. Step 2 of methodology.[1]

A two-option serial operation may be viewed as a production "process" with two inputs and one output. The corresponding three-dimensional production function can be represented with isoquants, *i.e.,* lines of input combinations capable of producing a given output. The inputs are the total costs of the parallel group, Z_{12} and process 3, Z_3. The output is the fraction of pollutant removed by the entire network, Y. Using the total cost curves for the parallel group and option 3, the isoquants of Y may be constructed. These curves must be in terms of the fraction removed due to the nature of the serial operation. The action of the parallel group affects the input to option 3. Therefore, it becomes necessary to obtain the optimal solution with respect to the fraction of incoming pollutant removed rather than the actual

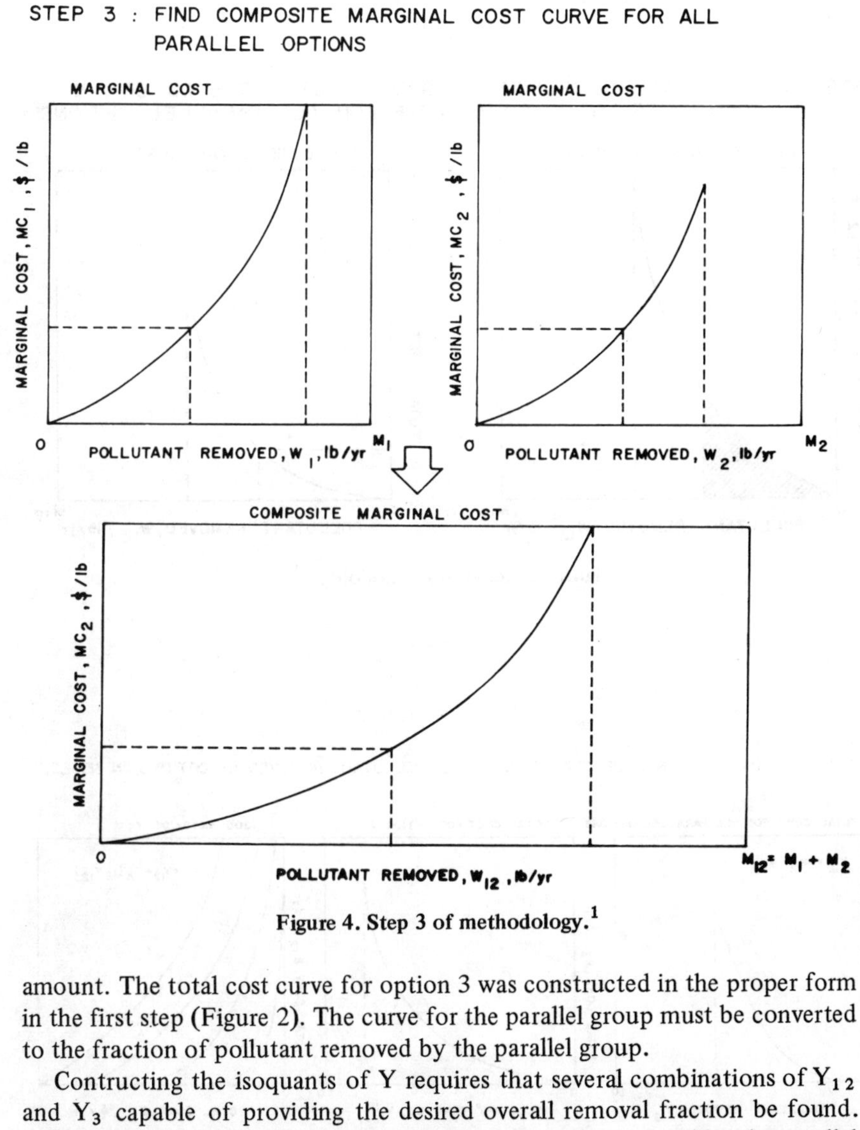

STEP 3 : FIND COMPOSITE MARGINAL COST CURVE FOR ALL PARALLEL OPTIONS

Figure 4. Step 3 of methodology.[1]

amount. The total cost curve for option 3 was constructed in the proper form in the first step (Figure 2). The curve for the parallel group must be converted to the fraction of pollutant removed by the parallel group.

Contructing the isoquants of Y requires that several combinations of Y_{12} and Y_3 capable of providing the desired overall removal fraction be found. This is done by noting that Y is simply the sum of the removal by the parallel group, Y_{12}, and the incremental removal by option 3, $Y_3(1-Y_{12})$. Given an arbitrary Y_3, Y_{12} may be solved for a desired value of Y. By noting the values of Z_{12} and Z_3 corresponding to the combinations of Y_{12} and Y_3 from the total cost curves, the isoquants of Y may be drawn (Figure 6). The next step is to develop the optimal expansion path from these isoquants by

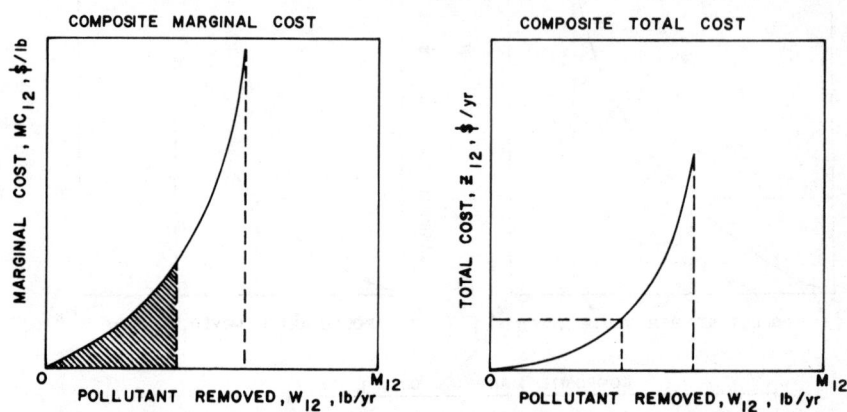

Figure 5. Step 4 of methodology.[1]

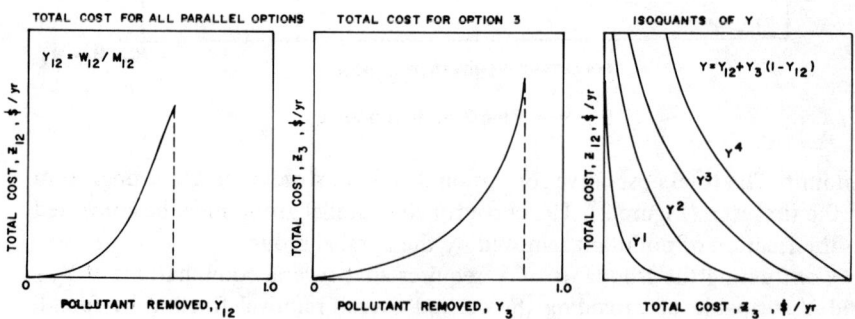

Figure 6. Step 5 of methodology.[1]

locating the points of tangency between the isoquants and isocost lines (given by $Z = Z_{12} + Z_3$), where Z is the total cost for the entire network. The points indicate the least cost combination of Z_{12} and Z_3 in attaining Y—if the isoquants are convex to the origin. If they are not, the optimal solution lies at the point where the isoquant intersects the axis giving the lower cost (Figure 7). The final step is to construct the total cost curve for the entire network. This is done by plotting values of Z against the corresponding values of Y found on the expansion path (Figure 8).

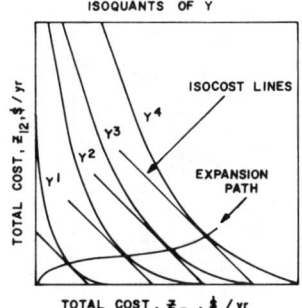

Figure 7. Step 6 of methodology.[1]

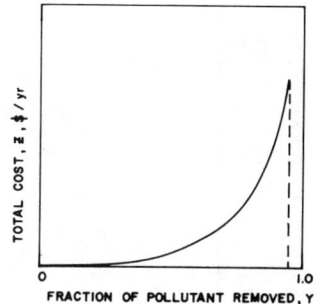

Figure 8. Step 7 of methodology.[1]

APPLICATION TO URBAN AREAS

Before applying this methodology, the production functions and unit costs for the control options must be developed. Information from several sources provides the necessary data to establish production functions for street sweeping, combined sewer flushing, and storage-treatment.[4-9] The first two technologies are commonly known as management practices. There are several others, but these are well known and therefore selected for more detailed analysis. The production functions are in terms of biochemical oxygen demand (BOD) due to the fact that this parameter remains the primary indicator of water pollution. However, any pollutant could be used.

The sweeping of roadways, primarily for debris control, is a long-established practice. However, street sweeping does have potential pollution control capability.[4,9] "Pick-up" efficiencies of 0.50 for brush-type sweepers and

0.95 for vacuum-type sweepers have been reported.[4] The street sweeping production function has as its input the fraction of days per year the streets are swept, X_{SW}, and the fraction of BOD removed, Y_{SW}, as its output. A simple computer model was developed to simulate the conditions of a typical urban area and the actions of sweeping. Running the model (with precipitation data for 1971 and demographic characteristics of Minneapolis) for several "pick-up" efficiencies and sweeping intervals generated the family of curves shown in Figure 9. These curves illustrate diminishing marginal returns. A unit cost of $7.00 per curb mile, based on data provided by APWA, was used to derive the total cost curve.[10]

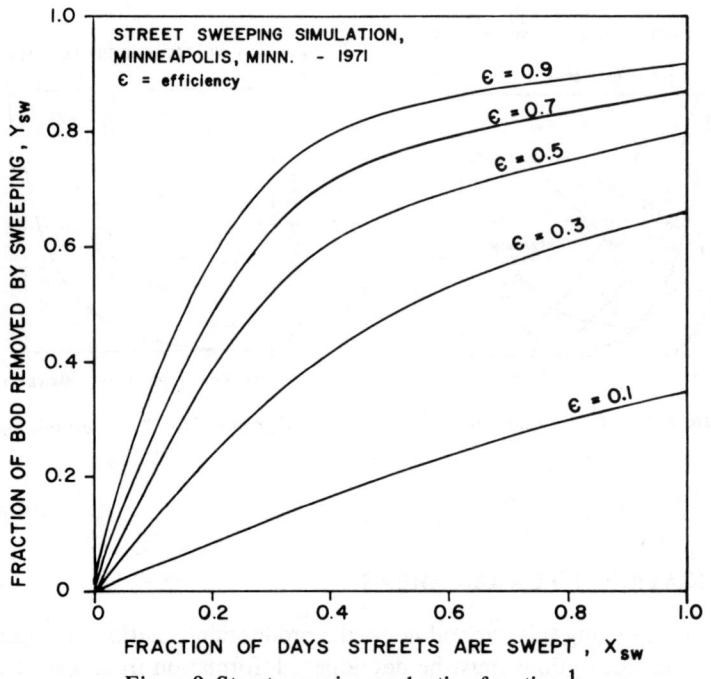

Figure 9. Street sweeping production function.[1]

A second alternative is combined sewer flushing. Combined sewers often experience dry-weather sewage deposition. These solids accumulate until removed by a storm surge or artifical flushing. The material removed by stormwater discharges directly to the receiving water, whereas controlled flushing routes the solids to the dry-weather treatment plant. A study investigating deposition problems in two Boston systems provided the data necessary to develop a production function for sewer flushing.[8] The input to this process is the fraction of combined sewers flushed daily, X_{SF}, and the output is the fraction of BOD removed, Y_{SF}. The resulting function is shown in Figure 10. Again, the "law" of diminishing returns is quite evident. A unit cost of \$11.78 per foot of sewer was used to derive the total cost curves.[8]

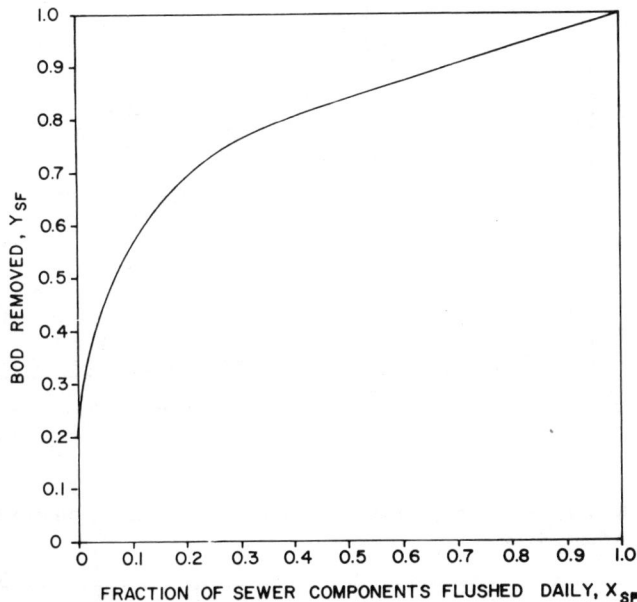

Figure 10. Combined sewer flushing production function.[1]

The methodology to derive production functions and total cost curves for storage-treatment was discussed in an EPA publication by Heaney, Huber and Nix.[6] As an example, a production function for Atlanta, Georgia, is shown in Figure 11. Additionally, the total cost curve for the storm-sewered areas of that city is shown in Figure 12. The interested reader should refer to this report for details.

APPLICATION TO AGRICULTURAL AREAS

The seven-step procedure can be applied to any mix of control options in series, parallel or both in agricultural areas. The difficult part of the analysis is to derive realistic production and cost functions. Due to the availability of a dairy farm production model, the methodology is applied to a 306-ac (tillable area) dairy farm described by Haith and Atkinson.[2] They present a linear programming model which allows determination of the production adjustments that could be made on a dairy farm facing pollution constraints.

658 BMPs FOR AGRICULTURE AND SILVICULTURE

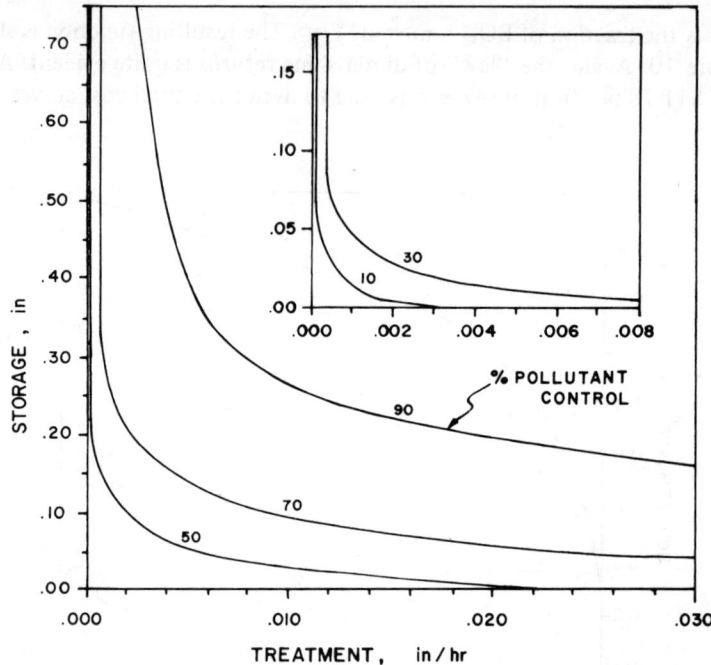

Figure 11. Storage-treatment production function (isoquants); storm-sewered area, Atlanta, Georgia.[6]

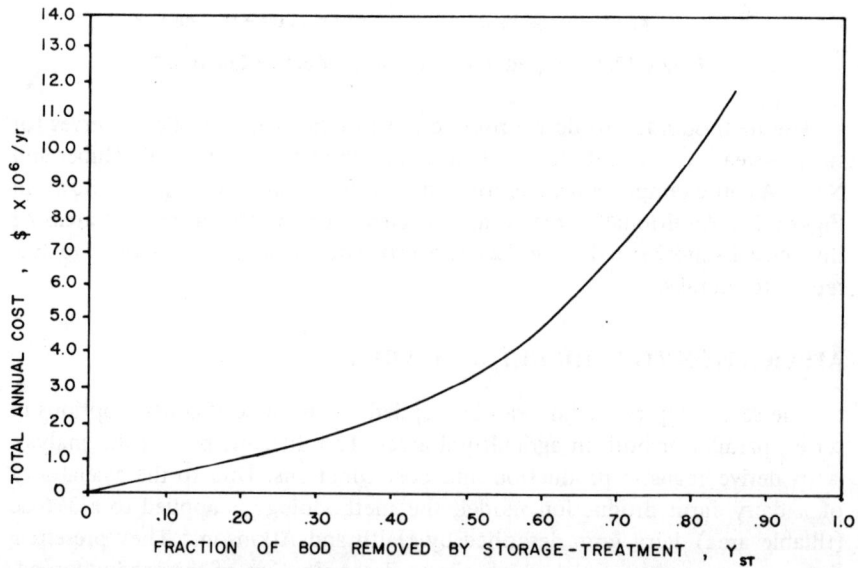

Figure 12. Total cost curve for storage-treatment; storm-sewered area, Atlanta, Georgia.[6]

For this application, production adjustments and a detention basin, excavated downstream on nontillable land, are the options available to the farmer. Sediment washoff is the pollutant discharge to be controlled since sediment loss is often the indicator of nutrient loadings in surface runoff. For example, the NPS model simulates all pollutants as a function of sediment washoff.[11]

The Haith and Atkinson linear programming model is a relatively simple dairy farm model which focuses on the nutrient budgets of a dairy farm. Farm income is maximized subject to constraints on land, manure handling, pollutant loss, and feasible soil-crop combinations. The soil loss constraint is as follows:

$$\sum_{i,j} S_{ij} X_{ij} \leq b \qquad (1)$$

where S_{ij} = annual soil loss from soil type i in land use j as determined by the universal soil loss equation
X_{ij} = acres of soil type i in land use j
b = maximum allowable soil loss

The other pollution constraints, *e.g.*, particulate phosphorus and dissolved nitrogen, are similar.

In order to determine the farmer's response to the pollutant constraint, it is necessary to know the marginal cost of pollutant control at any level (*i.e.*, the $\partial Z/\partial b$ where b is the right-hand side of equation 1 and Z is the net farm income). This information can be found by modifying the objective function and doing postoptimal analysis. The objective function is as follows:

$$\text{Maximize } Z = rH - \text{cost of farm operation} \qquad (2)$$

where r is the average rate of return per cow and H is the herd size. For our analysis, the rate of return per cow is expressed as the marginal rate of return rather than the average rate of return for a given herd size. Table I summarizes the difference between the average and the marginal rate of return. Substituting $\sum_k MR_k H_k$ for rH in Equation 2 gives a piece-wise linear approximation for the total return for a given herd size where MR_k is the marginal rate of return in interval k and $\sum_k H_k$ is equal to the herd size.

The postoptimal analysis examines the effect on farm income of varying the maximum allowable soil loss per unit area from zero to the level necessary to maximize farm income. Profits with the herd size of 110 are $11,776 and the soil losses are 0.57 ton/ac-yr, hence discharge of 0.57 ton/ac-yr represents 0% control. The sensitivity analysis varies the maximum allowable soil loss per unit area from 0.57 ton/ac-yr to zero ton/ac-yr (*i.e.*, 100% control) and gives the resulting profits at desired increments. The total cost of controlling

Table I. Average and marginal rate of return per cow.

Herd Size[a]	Average Return per Cow (r)	Interval (k)	Herd Size in Interval k (H_k)	Approximate Marginal Return per cow (MR_k)
0				
		0-50	50	298
50	298			
		50-70	20	221
70	276			
		70-90	20	146
90	247			
		90-110	20	143
110	228			

[a]From Haith and Atkinson.[2]

the pollutant is the difference between the maximum profits ($11,776) and the profit of the given level of control. The cost to the farmer is the profits foregone due to satisfying the soil loss constraint. The relationship between total costs and various levels of control using production adjustments is shown in Figure 13.

For the downstream option, the production function and cost function are needed for the detention basin. Chen et al.[3] present detention basin capacity vs sediment removal functions for construction site drainage areas. These construction sites generally resemble fallow ground conditions in agricultural practices. For this simplified application, the performance of the detention basin is assumed to be independent of sediment concentration, since the performance must be in terms of the fraction removed due to the nature of the serial operation. For instance, if the basin were to be sized such that it would remove 50% of the sediment washoff resulting from fallow ground condition, the basin is also assumed to remove 50% of the sediment washoff resulting from lesser loadings. The relationship between basin capacity and the fraction of sediment washoff removed for the 306-ac drainage area (i.e., the dairy farm's tillable area) is shown in Figure 14. Applying an annual unit cost of $25/ac-in. of basin capacity gives the total cost curve, Figure 15, for sediment removal using this detention basin.

Since the total cost curve is known for each option acting alone, the isoquants of the fraction of sediment washoff controlled by the serial operation of the two options is found by minimizing the combined costs subject to the following equation:

$$Y = Y_M + Y_{DB}(1 - Y_M) \tag{3}$$

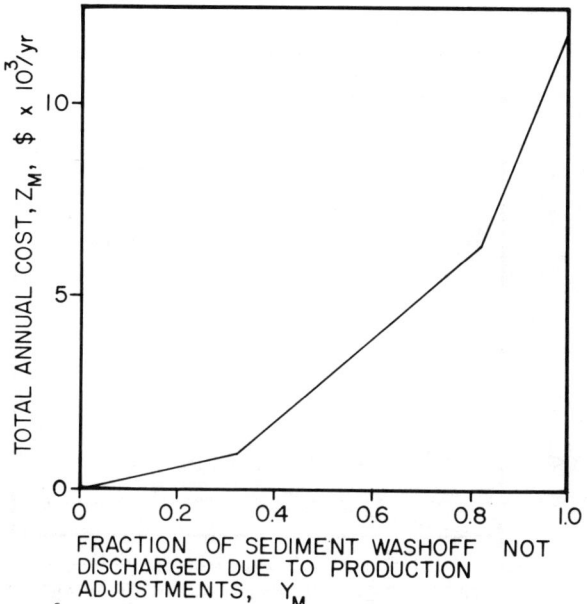

Figure 13. Total cost curve for farm production modification.

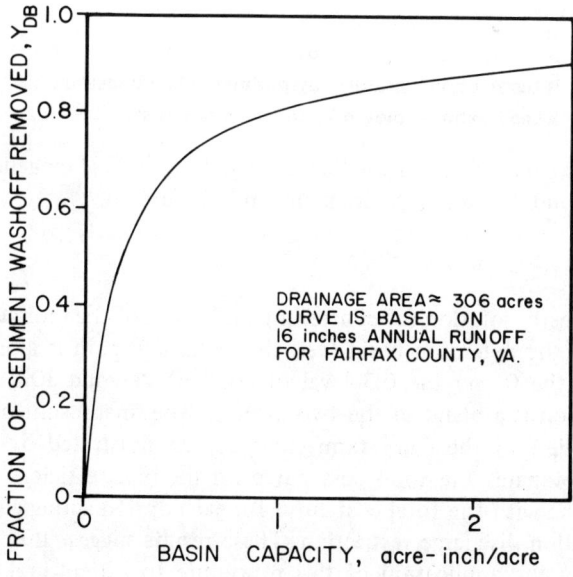

Figure 14. Detention basin production function.[3]

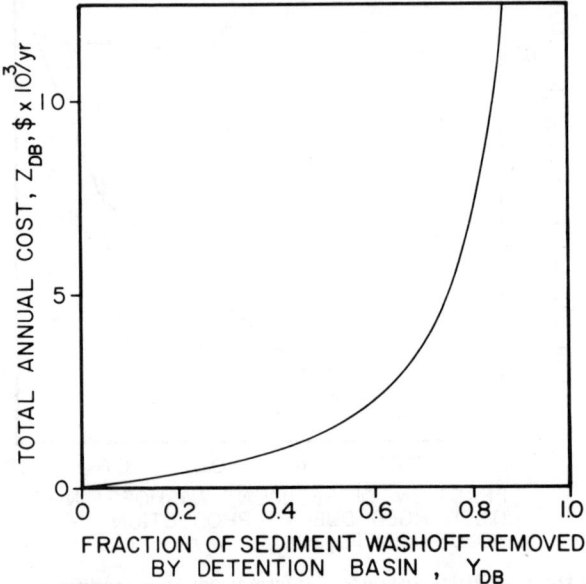

Figure 15. Total cost curve for detention basin.

where Y is the overall fraction removed
Y_M is the fraction controlled by production modification
Y_{DB} is the fraction removed by the detention basin

The optimal expansion path is the line through the points of tangency between the isoquants and the isocost lines, as shown in Figure 16.

Results

The least costly solution for controlling 0.3 or less of the sediment washoff is to only use the detention basin (*i.e.*, the expansion path is along the Z_{DB} axis between the 0 and the 0.3 level of control). Beyond 30% control the optimal solution is a blend of the two options. The final results of the methodology, applied to the dairy farm example, are illustrated by Figure 17. This figure compares the total cost curve for the most efficient mix of the control options with the total cost curve for each option acting alone for any level of pollution discharge restriction. These results suggest that savings can be realized by the application of this procedure to agricultural areas faced with a variety of pollution control options.

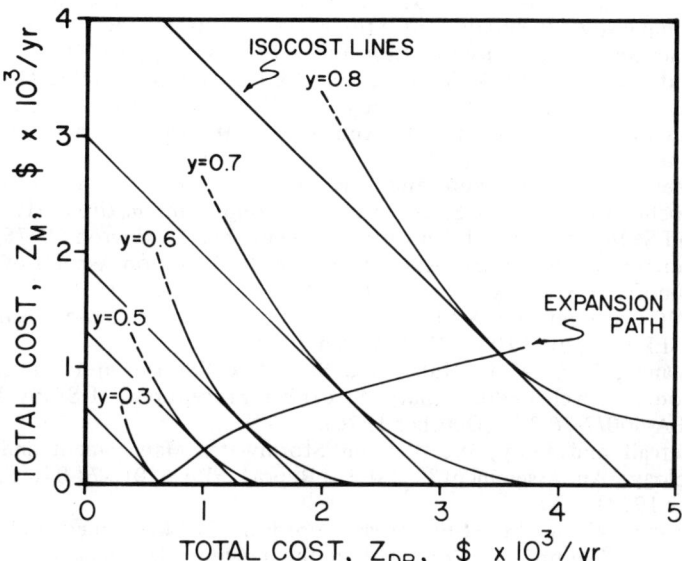

Figure 16. Optimal expansion path for overall sediment control.

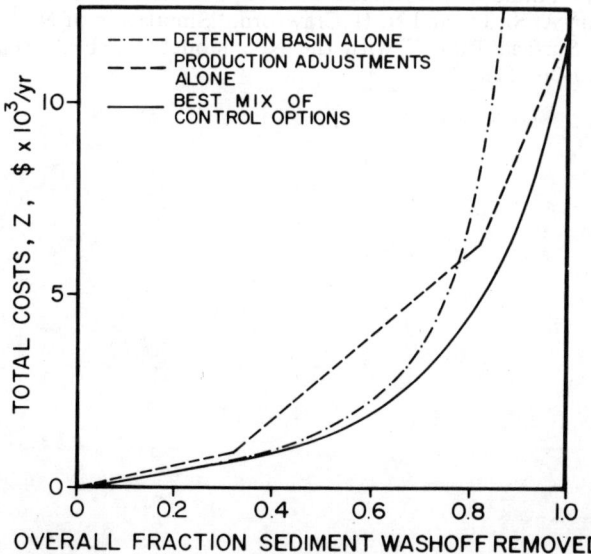

Figure 17. Comparison of total cost curves for most efficient mix of control options, and each control option acting alone.

REFERENCES

1. Heaney, J. P., and S. J. Nix. "Stormwater Management Model: Level I—Comparative Evaluation of Storage-Treatment and Best Management Practices," USEPA Report EPA-600/2-77-083 (April 1977).
2. Haith, D. A., and D. W. Atkinson. "A Lienar Programming Model for Dairy Farm Nutrient Management," in *Food, Fertilizer and Agricultural Residues,* R. C. Loehr, Ed. (Ann Arbor, MI: Ann Arbor Science Publishers, 1977), pp. 319-337.
3. Chen C., F. Santomauro and J. B. Fisher. "Erosion Control System for Pipeline Construction Sites," National Symposium on Urban Hydrology and Sediment Control, University of Kentucky, Lexington (1975).
4. American Public Works Association. "Water Pollution Aspects of Urban Runoff," USEPA Report 11030 DNS 01/69 (January 1969).
5. FMC Corporation. "A Flushing System for Combined Sewer Cleansing," USEPA Report 11020 DNO (March 1972).
6. Heaney, J. P., W. C. Huber and S. J. Nix. "Stormwater Management Model: Level I—Preliminary Screening Procedures," USEPA Report EPA-600/2-76-275 (October 1976).
7. Metcalf and Eddy, Inc. "Urban Stormwater Management and Technology: An Assessment," USEPA Report EPA-670/2-74-040 (December 1974).
8. Pisano, W. A. "Cost-Effective Approach for Combined and Storm Sewer Cleanup," in Proc. Urban Stormwater Management Seminars, USEPA Report WPD-03-76-04 (January 1976).
9. Sartor, J. D., and G. B. Boyd. "Water Pollution Aspects of Street Surface Contaminants," USEPA Report EPA-22-72-081 (November 1972).
10. American Public Works Associatoin. Unpublished data (1976).
11. Donigian, A. S., Jr., and N. H. Crawford. "Simulation of Nutrient Loadings in Surface Runoff with the NPS Model," USEPA Report EPA-600/3-77-065 (1977).

43

MODELING NUTRIENT EXPORT IN RAINFALL AND SNOWMELT RUNOFF

D. A. Haith
Departments of Agricultural Engineering and
 Environmental Enginering
Cornell University
Ithaca, New York

L. J. Tubbs
Department of Agricultural Engineering
Cornell University
Ithaca, New York

INTRODUCTION

Mathematical models can be valuable tools for the analysis of nonpoint source water pollution. Several models are currently available for estimating nutrient inputs to surface waterways from agricultural runoff, and can, in general, be categorized as simulation models[1-3] or loading functions.[4,5] The former are typically rather complex and require large quantities of input data and major investments of time and manpower for their implementation. Loading functions are based on simple predictive relationships for soil loss[5] or runoff[4] and hence, are significantly easier to use. However, since these simple models have not been tested on large watersheds, their value in water quality planning studies is uncertain.

The writers are currently engaged in a research program designed to refine and test a loading function based on the U.S. Soil Conservation Service's (SCS) Curve Number Runoff Equation. An earlier study developed the procedure for rainfall runoff.[4] The current paper extends the methodology to snowmelt runoff and compares predicted and observed nutrient export from a large (335-km^2) watershed. Since agricultural land use is only 42%

of the total watershed area, the testing of the loading function's predictive capability is incomplete (the observed nutrient flux includes other nonpoint sources in addition to agricultural runoff). Nevertheless, the results are promising, and subsequent tests of the model on other watersheds are ongoing and will be reported in the future. The remaining portions of this paper include a general development of the model and its application to Fall Creek watershed in upstate New York.

DESCRIPTION OF THE LOADING FUNCTION MODEL

The general form of the loading function is

$$L = 0.1 \sum_i \sum_j \sum_t C_{jt} Q_{ijt} A_{ij} \qquad (1)$$

where L = pollutant loading to surface waters (kg)
Q_{ijt} = runoff from crop management j on soil category i due to precipitation or snowmelt on day t (cm)
A_{ij} = area of land with soil/cover complex i, j, *i.e.*, soil category i, crop management j (ha)
C_{jt} = pollutant concentration in runoff from crop management j on day t (mg/l)

As applied to a watershed, Equation 1 is essentially a distributed-parameter model which assumes that the pollutant in question is conservative; *i.e.* during a storm event all pollutant losses in runoff from contributing land areas A_{ij} are transported without loss in the stream to the watershed outlet. Since the model estimates pollutant losses for each storm event, it can be applied to arbitrary time periods, providing loss estimates for all the events within the time period of interest.

Concentrations, C_{jt}

In principle, the loading function can be used to estimate losses of a variety of pollutants which might be transported in agricultural runoff. Applications have thus far been limited to the nutrients nitrogen and phosphorus. An earlier version of the loading function[4] was restricted to rainfall runoff. The nitrogen and phosphorus concentrations used previously were obtained from several field studies of nutrient losses in rainfall runoff from cropland. Since nutrient concentrations in snowmelt runoff differ significantly from those in rainfall runoff, new concentration data were necessary. For consistency, it was desirable to use rainfall and snowmelt concentrations obtained from the same fields. The South Dakota field studies described by Harms *et al.*[6] represent one of the few attempts to measure nutrient losses from a variety of crops on large field plots over several years of natural precipitation data. In addition, the published results of these studies[7] include basic data for all runoff events observed, thus permitting calculation of flow-weighted mean concentrations. These concentrations, which are presented in Table I, were used in the current study. Snowmelt concentrations were determined from the winter runoff events while rainfall concentrations were based on nonwinter events.

Table I. Flow-weighted mean concentrations of nitrogen and phosphorus in runoff from rainfall and snowmelt.[7]

Crop	Soluble Nitrogen		Total Nitrogen		Soluble Phosphorus		Total Phosphorus	
	Rain	Snow	Rain	Snow	Rain	Snow	Rain	Snow
	(mg/l)							
Fallow	2.0	4.0	4.0	4.0	0.1	0.2	1.0	0.5
Pasture	2.0	3.0	2.0	4.0	0.3	0.3	0.5	0.6
Hay	1.0	3.0	1.0	4.0	0.2	0.2	0.3	0.4
Small Grains	3.0	1.0	7.0	2.0	0.3	0.3	4.0	0.4
Corn	2.0	3.0	4.0	3.0	0.2	0.3	3.0	0.4

The nutrient losses observed in the South Dakota studies were from crops to which commercial fertilizer had been applied. In dairy farming areas, a portion of a corn crop's nutrient needs is commonly supplied by manure applications, and nutrient losses from such fields may differ from unmanured fields. Nitrogen and phosphorus losses in runoff from manured cornland were obtained from field studies at Aurora, New York, described by Klausner et al.[8,9] Data for individual rainfall and snowmelt events for 0.3-ha field plots to which manure had been applied at the rate of 35 ton/ha-yr in the spring, winter or summer were used to calculate the flow-weighted mean concentrations given in Table II. It can be seen that in most cases nutrient concentrations in runoff from the manured fields exceed the concentrations from unmanured corn given in Table I.

Runoff, Q_{ijt}

Runoff is computed using a modified version of SCS runoff equation:

$$Q_{ijt} = \frac{(R_t + M_t - 0.2 \, S_{ijt})^2}{R_t + M_t + 0.8 \, S_{ijt}} \qquad (2)$$

Table II. Flow-weighted mean nutrient concentrations in rainfall and snowmelt runoff from corn with manure applied at 35 ton/ha.[8]

	Rain	Snow
	(mg/l)	
Soluble Nitrogen	6.0	5.0
Total Nitrogen	9.0	14.0
Soluble Phosphorus	0.3	1.2
Total Phosphorus	1.8	5.2

for $R_t + M_t \geqslant 0.2\ S_{ijt}$; otherwise $Q_{ijt} = 0$. In Equation 2, R_t = rainfall during day t (cm), M_t = snowmelt during day t (cm), and S_{ijt} is a detention parameter which is a function of soil category, crop management and antecedent 5-day rainfall (cm). Rainfall and/or snowmelt events are considered to be of 1-day duration. Thus, a prolonged rainstorm lasting for 3 days would be treated as three separate events. For events consisting exclusively of rainfall on unfrozen soil, Equation 2 reduces to the standard SCS runoff equation. The detention parameter is readily calculated from tabulated curve numbers, CN_{ijt},[10,11] as follows:

$$S_{ijt} = \frac{2540}{CN_{ijt}} - 25.4 \tag{3}$$

The modified runoff equation (Equation 2) has been shown to be a satisfactory means of handling runoff involving snowmelt and/or frozen soil[3] provided the detention parameter S_{ijt} is adjusted. During the winter, the infiltration capacity of the soil is reduced due to freezing and/or high water table. Let $CN_{ij}(III)$ be the tabulated curve number for the wettest antecedent moisture condition. The corresponding detention parameter, $S_{ij}(III)$ is computed by Equation 3. Winter detention parameters are estimated as

$$S_{ijt} = \begin{cases} 0.5\ S_{ij}(III), \text{ frozen soil} \\ S_{ij}(III), \text{ for thawed soil and } M_t > 0 \end{cases} \tag{4}$$

Soil freezing is determined by air temperatures. Soil is assumed frozen if the mean maximum air temperature for the previous 5 days is less than 0°C, unless the soil was unfrozen before the establishment of snow cover and the sum of the average daily air temperatures for the previous 5 days is greater than -25°C. This procedure for prediction of soil freezing was tested by comparison with soil temperatures at Aurora, New York.[3]

Estimation of Snowmelt

Although the loading and runoff calculations (Equations 1 and 2) are substantially the same for rainfall and snowmelt, the extension of the loading function to winter requires a record of snowmelt quantities M_t that may not be readily available. When this is the case, values of M_t must be estimated from other meteorological data. The starting point is a daily inventory equation for snow accumulation:

$$SN_{t+1} = SN_t + 0.9 NS_t - ES_t - M_t \tag{5}$$

where SN_t = water equivalent of snow on the ground at the beginning of day t (cm)
NS_t = water equivalent of new snow falling on day t
ES_t = evaporation from the snowpack on day t

The coefficient "0.9" reflects the fact that due to wind and sublimation not all of falling snow is retained on a field. The fraction of snow retained is highly variable, but Kuzmin's average value of 0.9 for agricultural land is used.[12]

The water equivalent of new snow can be determined directly from precipitation data, using a temperature-based method to partition precipitation between rain and new snow. The variable C_t is defined as the fraction of day t with air temperatures greater than 0°C, such that:

$$C_t = \begin{cases} 0 & , Tmax_t \leq 0°C \\ 1 & , Tmin_t \geq 0°C \\ \dfrac{0.5\, Tmax_t}{(Tmax_t - T_t)} & , \text{otherwise} \end{cases} \tag{6}$$

where $Tmax_t$, T_t, and $Tmin_t$ are maximum, average and minimum air temperatures on day t. This relationship assumes that temperature rises and falls linearly over a day as shown in Figure 1. If P_t is the amount of precipitation falling on day t (cm), snowfall is estimated as

$$NS_t = (1 - C_t) P_t \tag{7}$$

and rain is

$$R_t = C_t P_t \tag{8}$$

Evaporation from the snowpack may be significant during warmer periods. Eggleston et al.[13] have determined that evaporation is predominantly dependent on temperature:

$$ES_t = (0.0356\, T_t^2 + 0.907\, T_t + 4.88) D_t \tag{9}$$

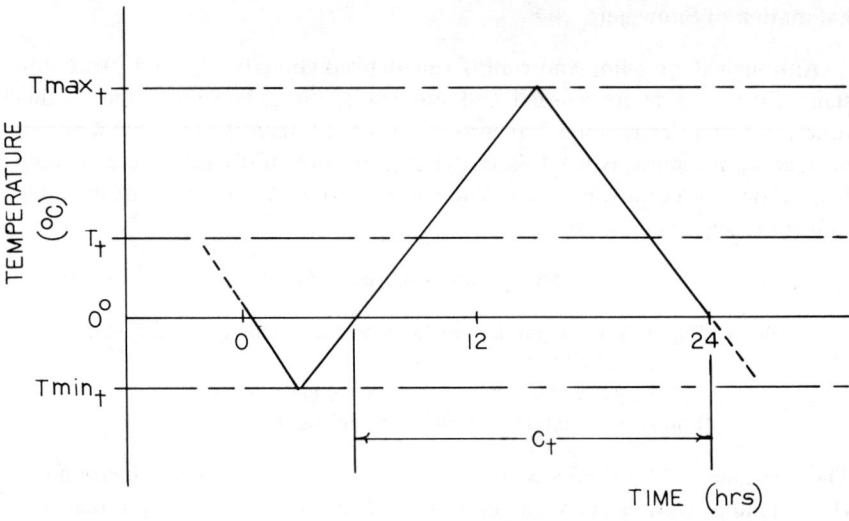

Figure 1. Procedure for apportioning precipitation between rain and snow.

where D_t = possible hours of sunshine on day t, as a fraction of total yearly sunshine.

Two melt rate equations have been developed by the U.S. Army Corps of Engineers[14] which include a number of meteorological factors. For days without rain,

$$M_t' = (0.0129\ I_t)(1 - a_t) + (1 - N_t)(0.097\ T_t - 2.1) \\ + 0.13\ N_t T_t + 0.0022\ v_t (T_t + 3.6\ Td_t) \tag{10}$$

where M_t' = snowmelt rate on day t (cm/day)
I_t = solar short-wave radiation on day t (langleys)
a_t = albedo (reflectivity) of snow surface on day t (dmless)
N_t = fractional cloud cover on day t
v_t = average wind speed on day t (km/hr)
Td_t = dewpoint temperature on day t (°C, approximated by $Tmin_t$)

On rainy days,

$$M_t' = (0.133 + 0.024\ v_t + 0.013\ R_r)\ T_t + 0.23 \tag{11}$$

The albedo or reflectivity of the snow surface has been related to the age of the current snowpack by Riley et al.[15] as:

$$a_t = 0.4\ (1 - e^{-0.2m}) \tag{12}$$

where m = number of days since last snowfall. Melt days are included in the computation of m, whether or not precipitation occurs on that day. In the case where fractional cloud cover data are not available, N_t may be estimated using total daily radiation I_t. Following the reasoning of Hendrick et al.[16] cloud cover and net radiation are related by:

$$I_t = I_{ct} (1 - 0.7 N_t) \tag{13}$$

where I_{ct} = maximum possible clear sky radiation on day t (langleys). It is then possible to solve for N_t in terms of daily radiation I_t:

$$N_t = \text{Min} [1.0, 1.43 (1 - I_t/I_{ct})] \tag{14}$$

Since the melt rate, M_t' determined from either Equation 10 or 11, may exceed the available snow, the actual melt for day t is given by

$$M_t = \text{Min} (SN_t + NS_t, M_t') \tag{15}$$

The extension of the loading function to winter conditions can substantially increase computational requirements when Equations 5-15 must be used to generate daily snowmelt. More significant is the increase in meteorological data requirements. Rainfall runoff estimates require only daily precipitation records, while winter runoff estimates require daily maximum and minimum temperatures, wind speed and solar radiation (Tmax_t, Tmin_t, N_t, I_t). Maximum sunlight (D_t) and radiation (I_{ct}) parameters are also needed, but these are tabulated monthly constants which can be obtained from standard references.[14,17]

Soil/Cover Complexes, Aij

Soil/cover complexes include all agricultural land in the watershed not separated from stream channel by a "buffer" of forest or brushland at least 46 m (150-ft) wide. It is assumed that such buffering would either capture the agricultural runoff or remove nutrients from the runoff before it reaches a stream channel.[4] Soils are divided into four categories according to their hydrologic group, and crop management includes crop type (row-crops, small grains, etc.), supporting practice (contouring, terracing, etc.) and hydrologic condition.

APPLICATION TO A LARGE WATERSHED

Fall Creek is a large (335 km²) rural watershed near Ithaca, New York (Figure 2). Less than 2% of the watershed has nonrural land uses (Table III). Agriculture is the principal land use, consisting mainly of corn, alfalfa, hay

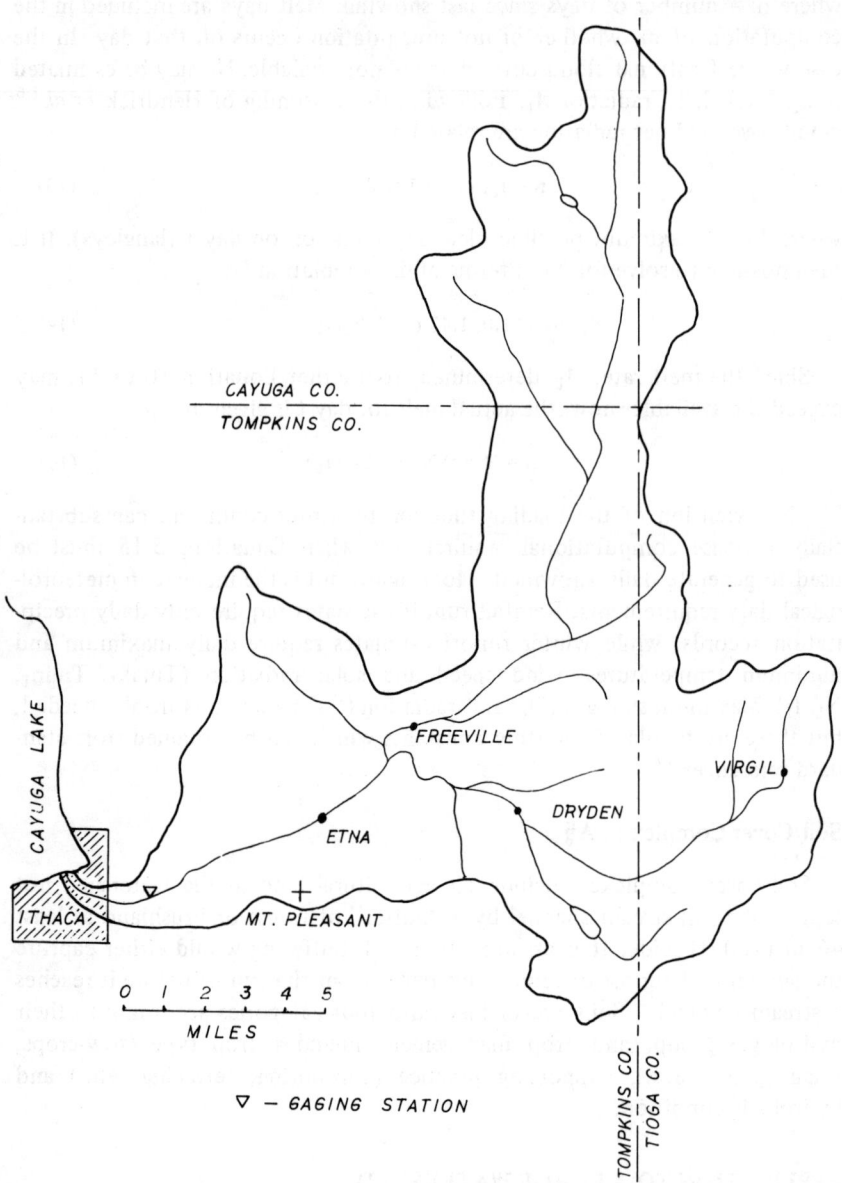

Figure 2. Fall Creek watershed.

Table III. Land uses in Fall Creek watershed.

	Area[a] (ha)	% of Total
Agriculture	14,560	43.4
Idle Land	5,500	16.4
Brushland	8,220	24.5
Forest	3,480	10.4
Waterways	110	0.3
Wetlands	1,040	3.1
Low-Density Residential	210	0.6
Urban	310	0.9
Other (Transportation, Outdoor Recreation, etc.)	130	0.4
Total	33,560	100.0

[a]Above water quality sampling station, based on 1973 aerial photography.

and pasture (Table IV). A water quality sampling program was conducted from September 1972 through April 1974 in order to estimate phosphorus and nitrogen export from the watershed. The sampling procedures are described in Johnson et al.,[18,19] Bouldin et al.,[20] and Johnson.[21] The availability of this water quality data and recent (1973) land use information made Fall Creek a likely first candidate for a test of the loading function.

Table IV. Cropland distribution in Fall Creek.

	Area (ha)	% of Agricultural Land
Corn	2,990	20.5
Alfalfa	2,400	16.5
Hay	3,920	26.9
Small Grains	820	5.6
Sudax	100	0.7
Buckwheat	20	0.1
Pasture	4,200	28.9
Plowed	110	0.8
Total	14,560	100.0

Although previous analysis of the Fall Creek water quality sampling results has indicated that cropland runoff is most likely the major nonpoint source of nitrogen and phosphorus in the watershed,[18,19] other nonpoint sources are present including barnyard, urban and residential drainage, as well as runoff from forest, idle and brushlands. The only major identifiable wastewater point source is the discharge from a secondary sewage treatment plant serving 1,400 people in Dryden.[18] Additional minor point sources include discharges from farm milking centers, mobile home parks and septic tanks.

Testing of the loading function on Fall Creek required the computation of nonpoint source nutrient fluxes in streamflow using water quality and discharge data, estimation of nutrient losses in agricultural runoff using the loading function (Equation 1), and the comparison of loading function estimates with the observed nutrient fluxes in the stream.

Determination of Nutrient Fluxes in Fall Creek Runoff

The measurement of nonpoint source pollution associated with runoff in a large watershed is imprecise. At least in the northeast, a sizable portion of annual streamflow volumes in streams is due to base flow. Hence, even during a storm event, the nutrient flux in stream flow will not be solely due to runoff, and empirical procedures must be used to separate fluxes into base flow and direct runoff components. Such separation can also serve to eliminate the effects of point sources such as sewage and milking center wastes since these sources should not be markedly influenced by storms and hence would be associated with stream base flow. Unfortunately, methods for partitioning nutrient fluxes into runoff and base flow components rely on hydrograph separation techniques which are essentially arbitrary.[22]

A second source of uncertainty in the nutrient flux estimates in runoff involves the water quality sampling. Approximately 700 samples were taken by Johnson et al.[18,19] at the sampling station shown in Figure 2 during the 20-month period. However, even this large number of samples was not sufficient to directly determine the nutrient fluxes during all runoff events. Discharge rates and nutrient concentrations often change rapidly during a storm, and although discharge data were available at 2-hr intervals, water quality sampling frequency was generally much less. Regression equations or flow-weighted averages were used to estimate the nutrient concentrations at 2-hr frequencies during a storm. For example, 7 or 8 samples may have been taken during a 48-hr storm, but the above procedure can be used to estimate 24 distinct nutrient concentrations (at 2-hr intervals) for the storm.

Nonpoint source water pollution in stream runoff is not subject to direct measurement. Even given substantial quantities of discharge and water quality data, indirect methods are needed to infer the magnitudes of such pollutants. The methods used for analysis of Fall Creek data are outlined below.

1. *Runoff:* Two-hour gage height records and rating table were obtained for the permanent Forest Home gauging station on Fall Creek from the U.S. Geological Survey (USGS), Ithaca, New York. Using these data,

storm events were delineated as follows (where Q_n = streamflow in m^3/sec during hour n of a storm event):

a. An event begins when flows rise and eventually double, *i.e.*, if Q_o is the initial rising flow, the hydrograph reaches some $Q_n > 2Q_o$. This condition eliminates very small stream flow increases that may be associated more with random data fluctuations or gradual rises in base flow than with runoff events.

b. An event ends when the hydrograph reaches a point of inflection and the flow is less than $2Q_o$, *i.e.*, the final flow, Q_f is the first Q_n such that $Q_{n-1} \geqslant Q_n \leqslant Q_{n+1}$ and $Q_n \leqslant 2Q_o$. The second condition allows for hydrographs with more than one peak, as shown in Figure 3.

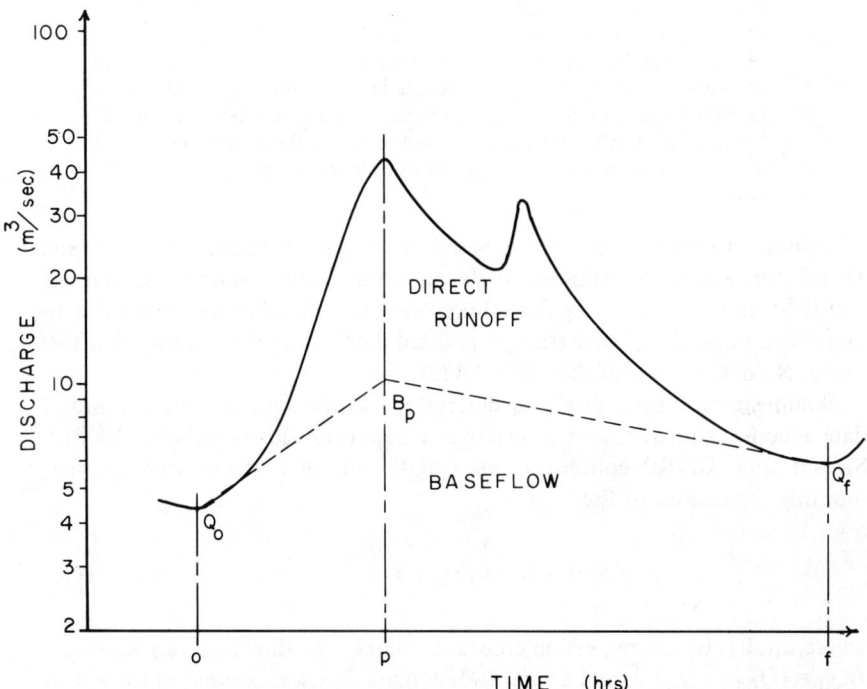

Figure 3. Illustration of hydrograph separation procedure.

The separation of the hydrograph into base flow and direct runoff is based on a procedure outlined in Chow[22] and illustrated graphically in Figure 3. On a semilogarithmic plot, Q_o and Q_f are connected to a peak base flow, B_p and the area under these lines is base flow with the remaining hydrograph area constituting direct runoff. Peak base flow is assumed to occur at the same time as the hydrograph peak, and B_p is estimated from a recession equation:

$$B_n = B_p e^{-a(n-p)} \tag{16}$$

where B_n is the base flow at hour n of the storm (n-p hours after peak). Base flow typically recedes to one-half its peak value in 17 days in Fall Creek[20] and hence a = 0.0017. Assuming that all stream flow is base flow at the end of the storm ($B_f = Q_f$), B_p is computed as

$$B_p = Q_f e^{0.0017(f-p)} \tag{17}$$

2. *Nutrient Fluxes:* Nutrient concentrations were calculated using monthly regressions given in Bouldin et al.[20] and flow-weighted concentrations in those months for which regressions were not available.[20,21] These data are presented in considerable detail elsewhere[18-21] and will not be repeated here. Rather, the following discussion will summarize the ways in which the data were used to estimate nutrient fluxes in direct runoff. As with runoff volume, runoff nutrient calculations were often by difference, *i.e.,* both total and base flow nutrient fluxes were computed and the runoff portion determined by subtracting the latter from the former.

Soluble nitrogen fluxes were based on nitrate-nitrogen concentrations. These concentrations showed little variation with discharge within any month[19] and total and base flow fluxes were calculated as the product of the same mean monthly concentration and total and base flow volumes, respectively. No data were available for total nitrogen.

Soluble phosphorus flux was determined as the sum of dissolved molybdate reactive and dissolved unreactive phosphorus fluxes (DMRP, DUP).[18] Stream flow DMRP concentrations (mg/l), during a storm were given by monthly regressions of the form

$$\text{DMRP} = b_0 + b_1 Q_n + b_2 \frac{\Delta Q_n}{\Delta t} \tag{18}$$

where b_0, b_1, b_2 are regression constants. Since 2-hr flow intervals were used, $\Delta Q_n = (Q_{n+2} - Q_n)/2$, and a total DMRP mass flux was computed for a storm based on flow-weightings of the concentrations given by Equation 18. Concentrations of DMRP in base flow were computed from Equation 18 letting $Q_n = B_n$ and neglecting the final term ($\Delta B_n/\Delta t$ was small). Dissolved unreactive phosphorus concentrations were relatively stable in Fall Creek, and the average value of 12 $\mu g/l$[20] was used for runoff and base flow for all events.

The total phosphorus flux in the stream during a storm was the sum of soluble phosphorus and solid-phase phosphorus (SPP) associated with sediment particles. The latter was given by 0.11% of the suspended solids flux,[20] and the suspended solids concentrations, SS (mg/l) were determined by monthly regressions of the form

$$SS = a_0 + a_1 Q_n + a_2 \frac{\Delta Q_n}{\Delta t} \qquad (19)$$

where a_0, a_1 and a_2 are regression constants. It was assumed that all SSP in stream flow was associated with runoff.

During summer and early fall months, there were insufficient storm events to construct reliable regression equations for either DMRP or SS, and during these months flow-weighted average concentrations were used for both total and base flow nutrient flux estimates. However, the total storm runoff during these months was much less than runoff during the remainder of the sampling period, and the use of these average concentrations had little effect on the total runoff nutrient flux for the 20 months.

The runoff volumes and nutrient fluxes estimated from stream flow and water quality data are summarized in Table V. There were 114 identifiable storm events during the 20-mon period. The 17 largest accounted for 80-85% of the observed nutrient fluxes.

Table V. Runoff and nutrient fluxes in Fall Creek, September 1972-April 1974, estimated from water quality and discharge data.

	Discharge ($10^6 m^3$)	Soluble Nitrogen (10^3 kg)	Phosphorus (10^3 kg)	Total Phosphorus (10^3 kg)
17 Largest Events				
Total Stream Flow	141.6	160.8	4.9	31.8
Base Flow	64.0	74.5	1.6	1.6
Runoff	77.6	86.3	3.3	30.2
All 114 Events				
Total Stream Flow	197.1	223.9	6.4	37.9
Base Flow	102.9	118.8	2.6	2.6
Runoff	94.2	105.1	3.8	35.3

Estimation of Nutrient Fluxes Using the Loading Function

Calculation of nutrient fluxes using the loading function required the collection of cropping, soils and precipitation data, the computation of snowmelt quantities from temperature, wind speed and solar radiation data using Equations 5 through 15, and the selection of appropriate curve numbers for determination of runoff detention parameters as in Equation 3.

Cropping and soils information was determined using overlays of cropping maps based on 1973 aerial photography and soils maps on USGS 7.5-min quadrangle maps. Crops were grouped according to soil hydrologic group (ranging from group A, lowest runoff potential, to group D, highest runoff potential). This distribution for the entire watershed is shown in Table VI. Group A and unclassified soils were neglected, since they were only 3% of the total cropland. Only unbuffered cropland was included in the loading function calculations. As noted earlier, such cropland is adjacent to identifiable stream channels. The distribution of unbuffered cropland is shown in Table VII.

Table VI. Crop distribution by soil hydrologic group in Fall Creek watershed.

Crop Cover	Area (ha) by Soil Hydrologic Group			
	B	C	D	Total
Pasture	720	2,898	429	4,047
Hay[a]	1,202	4,698	251	6,151
Corn[b]	677	2,231	115	3,023
Small Grains[c]	179	690	31	900
Total	2,778	10,517	826	14,121

[a] Includes alfalfa and buckwheat.
[b] Includes plowed land.
[c] Wheat, oats, sudax.

Table VII. Distribution of cropland unbuffered from stream channels by forest or brush.

Crop Cover	Area (ha) by Soil Hydrologic Group			
	B	C	D	Total
Pasture	335	1,536	285	2,156
Hay[a]	455	1,620	57	2,132
Corn[b]	220	681	42	943
Small Grains[c]	61	205	6	272
Total	1,071	4,042	390	5,503

[a] Includes alfalfa and buckwheat.
[b] Includes plowed land.
[c] Wheat, oats, sudax.

Meteorological data were collected for weather stations at Ithaca, Mt. Pleasant, Freeville and Cortland (Figure 2). Land areas associated with each station were determined using Thiessen polygons. Over 94% of the agricultural land was associated with the Freeville (67%) and Cortland (27%) stations. Monthly precipitation measurements for these stations are given in Table VIII.

Table VIII. Monthly precipitation at principal weather stations.

Month	Cortland	Freeville
	(cm)	
9/72	8.0	5.9
10/72	7.0	6.4
11/72	15.6	15.4
12/72	15.2	11.2
1/73	4.4	3.5
2/73	7.9	4.3
3/73	6.5	6.5
4/73	12.4	11.2
5/73	9.7	8.8
6/73	9.6	8.7
7/73	7.4	7.3
8/73	5.9	5.0
9/73	8.6	9.1
10/73	5.6	4.5
11/73	5.5	4.3
12/73	14.4	9.1
1/74	5.4	4.4
2/74	5.4	3.8
3/74	12.7	9.3
4/74	5.8	4.9
Total	173.0	143.6

Selection of curve numbers and nutrient concentrations required assumptions regarding cropping practices. A growing season of June through September was used for the adjustment of antecedent moisture conditions.[10] All field crops were assumed to be planted in straight rows and in poor hydrologic condition. Pasture was assumed to be uncontoured and in good hydrologic condition. The selection of appropriate nutrient concentrations was somewhat arbitrary. Since two alternative sets of concentrations were available corresponding to manured (Table II) and unmanured (Table I) corn, it was necessary to estimate manure applications on cornland. There are approximately 6,800 cows in the watershed,[19] resulting in production of 102,000 MT of manure per year (@ 15 MT per cow). If all manure is spread on corn at 35 MT/ha, this would require 2914 hr. This approximates the total corn land in the watershed (Table VI) and hence it is assumed that all such land is manured with resulting runoff nutrient concentrations given in Table II. It is obvious that this assumption may not be accurate. Portions of the manure may be spread on other crops and idle land, and manure may not be spread at comparable rates on buffered and unbuffered land. However, these two exceptions may cancel each other, since buffered land is typically in upland areas, which are often farther from barns than the lowland unbuffered areas which are considered in the loading function computations. The nutrient concentrations in Table I were used for pasture, hay and small grains.

Runoff and nutrient losses calculated by the loading function model are summarized in Table IX. The major sources of nutrients are seen to be snowmelt runoff from pasture, hay and corn. Snowmelt runoff from corn completely dominates the phosphorus loss estimates. Unforunately, the assumption

Table IX. Loading function estimates of nutrient export from agricultural runoff in Fall Creek, September 1972-April 1974.

Crop	Runoff (10^6 m^3)	Soluble Nitrogen (10^3 kg)	Soluble Phosphorus (10^3 kg)	Total Phosphorus (10^3 kg)
Pasture				
Rainfall Runoff	0.5	1.1	0.2	0.3
Snowmelt Runoff	5.1	15.3	1.5	3.1
Total	5.6	16.4	1.7	3.4
Hay				
Rainfall Runoff	1.4	1.4	0.3	0.4
Snowmelt Runoff	8.5	25.6	1.7	3.4
Total	9.9	27.0	2.0	3.8
Small Grains				
Rainfall Runoff	0.2	0.7	0.1	0.9
Snowmelt Runoff	1.0	1.0	0.3	0.4
Total	1.2	1.7	0.4	1.3
Corn				
Rainfall Runoff	0.8	4.8	0.2	1.4
Snowmelt Runoff	4.4	22.1	5.3	23.0
Total	5.2	26.9	5.5	24.4
Total Rain	2.9	8.0	0.8	3.0
Total Snow	19.0	64.0	8.8	29.9
Grand Total	21.9	72.0	9.6	32.9

of manured cornland is critical to this dominance. Table X shows a comparison of nutrient loadings from cornland snowmelt runoff computed by the loading function for manured and unmanured cases. If all unbuffered cornland in the watershed was unmanured, the total estimated soluble phosphorus and total phosphorus losses would be 4,600 kg and 11,700 kg, respectively, rather than 9,600 kg and 32,900 kg as shown in Table IX. The assumption that most unbuffered cornland is manured is probably not unreasonable for the Fall Creek watershed. Still, it is disturbing that such an assumption would so strongly influence the loading function estimates.

Table X. Comparison of nutrient loadings in cornland snowmelt runoff for manured and unmanured case.

	All Cornland Assumed Unmanured[a]	All Cornland Assumed Manured[b]
	(10^3 kg)	
Soluble Nitrogen	13.3	22.1
Soluble Phosphorus	1.3	5.3
Total Phosphorus	1.8	23.0

[a] Concentrations from Table I.
[b] Concentrations from Table II.

Comparison of Predicted and Observed Nutrient Fluxes

Nutrient fluxes predicted by the agricultural runoff loading function are compared in Table XI to the total runoff fluxes estimated from discharge and water quality data. The extent to which these comparisons constitute a test of the loading function model is limited by three factors:

Table XI. Comparison of loading function predictions of nutrient fluxes in agricultural runoff with observed total runoff nutrient fluxes determined from water quality and discharge data.

	Runoff Volume (10^6 m^3)	Soluble Nitrogen (10^3 kg)	Soluble Phosphorus (10^3 kg)	Total Phosphorus (10^3 kg)
17 Largest Events;				
Predicted Agricultural Runoff (Equations 1 and 2)	15.4	50.2	6.2	21.4
Total Observed Runoff	77.6	86.3	3.3	30.2
All Storm Events				
Predicted Agricultural Runoff (Equations 1 and 2)	21.9	72.0	9.6	32.9
Total Observed Runoff	94.2	105.1	3.8	35.3

1. As noted earlier, although agriculture is the principal source of runoff-related nutrients in the watershed, it is not the only source. These other nonagricultural sources are not accounted for in the modeling procedure.
2. The loading function assumes that all nutrients are conserved in the stream during storm events; *i.e.*, they are transported without loss to the watershed outlet.
3. The estimation of "observed" runoff nutrient flux is an imprecise procedure which relies on arbitrary hydrograph separation techniques and concentration estimates based on regression equations and mean concentrations.

These factors are considered in the following comparisons of predicted and observed runoff, nitrogen and phosphorus losses. Discussion will focus on the comparisons produced from totals of all storm events.

Predicted runoff is 23% of observed. Since unbuffered cropland is 16% of the watershed area, this indicates that this cropland contributes a disproportionate share of total runoff. This is reasonable, since such land is near stream channels and can be expected to produce greater runoff than certain other major land uses such as forest and brushland. Conversely, there remain other major runoff sources including urban and residential drainage, as well as runoff from wetlands, idle land and transportation uses (paved highways run along much of the Fall Creek stream channel).

Predicted soluble nitrogen flux is 69% of observed. This result is probably the most consistent of all the model predictions. The remaining nitrogen flux of 33,100 kg is easily accounted for by background levels. If, for example, it is assumed that the runoff from land uses other than cropland (94.2 - 21.9 = 72.3 10^6 m^3) contains soluble nitrogen at a background concentration of 0.5 mg/l, this would account for 36,200 kg.

Soluble phosphorus is the only nutrient for which the predicted flux is greater than observed. Although the accuracy of the observed fluxes may be questioned, they are so much smaller than predictions that other explanations are necessary. The most obvious is that soluble phosphorus is not conserved in the stream. Thus, while 9,600 kg of soluble phosphorus may have been lost from croplands in runoff during the 20-month period, a large fraction of this phosphorus may have been scrubbed from the streamflow by adsorption on sediment. A second problem may be the assumed phosphorus concentrations in snowmelt runoff from manured cornland. These concentrations are so much greater than those associated with other crops that they could be considered suspect.

Although predicted total phosphorus fluxes are 93% of observed, this is not as encouraging as it may appear. It is difficult to believe that all other sources of total phosphorus in runoff account for less than 7% of the observed flux. In addition, if the soluble phosphorus scrubbing hypothesis is accepted, this should produce greater differences in observed and predicted total phosphorus even if all phosphorus was associated with agricultural runoff.

In spite of the limitations discussed above, the comparisons are rather encouraging. It should be noted that the loading function approach requires no calibration, and it might be considered surprising that a methodology based on an empirical runoff equation and average nutrient concentrations predicts nutrient fluxes that are remotely comparable to those observed in stream flow. Yet this is clearly the case, and although the comparisons do not completely validate the loading functions, neither do they provide any circumstantial evidence that would lead to rejection of the approach. The primary uncertainty that was brought out in the comparisons is the phosphorus concentrations in snowmelt runoff from cornland.

CONCLUSIONS

A simple modeling procedure previously developed for estimating nutrients in rainfall runoff from agricultural land has been extended to snowmelt runoff, permitting year-round application of the methodology. The model is in the form of a loading function which multiplies daily runoff by nutrient concentrations and contributing cropped areas. Runoff is computed by a modified version of the SCS runoff equation and nutrient concentrations are flow-weighted means determined from field studies. Contributing areas are croplands which are unbuffered from stream channels by forest or brushland. Such areas are identified from land use, soils and topographic maps. The loading function relies on readily available data, is computationally straightforward and does not require calibration by water quality sampling.

The loading function was tested by comparison of its predicted loadings with observed nutrient fluxes in stream flow runoff estimated from 20 months of water quality and discharge observations in the Fall Creek watershed. Although the testing was incomplete due to deficiencies in stream water quality data, limitations in the methods used to separate runoff and base flow and the presence of nonagricultural nonpoint sources, comparisons of predictions and observations indicated that the loading function produced reasonable results. A significant uncertainty revealed by the testing was the quantities of phosphorus lost in snowmelt runoff from manured cropland.

As a minimum, the model testing suggested no evidence that would lead to the rejection of the loading function. It is possible to select watersheds and water quality sampling procedures which would minimize the limitations in model testing indicated above. The results of the application of the loading function to Fall Creek have been sufficiently promising to establish the value of such further testing. This work is ongoing and will be reported on in the future.

ACKNOWLEDGMENTS

This research was supported in part by funds provided by the U.S. Department of the Interior, Office of Water Research and Technology. Aerial photography and soils data were assembled by W. T. Tseng as part of an earlier research project sponsored by the Rockefeller Foundation. K. Curry assisted in the identification of unbuffered cropland using this data.

REFERENCES

1. Donigan, A. S., Jr., and N. H. Crawford. "Modeling Pesticides and Nutrients on Agricultural Lands," Report No. EPA-600/2-76-043, U.S. Environmental Protection Agency, Athens, GA (1976).
2. Frere, M. H., C. A. Onstad and H. N. Holtan. "ACTMO, An Agricultural Chemical Transport Model," Report No. ARS-H-3, U.S. Department of Agriculture, Agricultural Research Service, Hyattsville, MD (1975).
3. Tubbs, L. J., and D. A. Haith. "Simulation of Nutrient Losses from Cropland," ASAE Paper 77-2502, presented at 1977 Winter Meeting of the American Society of Agricultural Engineers, Chicago (1977).
4. Haith, D. A., and J. V. Dougherty. "Nonpoint Source Pollution from Agricultural Runoff," *J. Environ. Eng. Div. Am. Soc. Civil Eng.* 102 (EE5):1055-1069 (1976).
5. McElroy, A. D., S. Y. Chiu, J. W. Nebgen, A. Aleti and F. W. Bennett. "Loading Functions for Assessment of Water Pollution from Nonpoint Sources," Report No. EPA-600/2-76-151, U.S. Environmental Protection Agency, Washington, DC (1976).
6. Harms, L. L., J. N. Dornbush and J. R. Andersen. "Physical and Chemical Quality of Agricultural Land Runoff," *J. Water Poll. Control Fed.* 46(11):2460-2470 (1974).
7. Dornbush, J. N., J. R. Anderson and L. L. Harms. "Quantification of Pollutants in Agricultural Runoff," Report No. EPA 660/2-74-005, U.S. Environmental Protection Agency, Washington, DC (1974).
8. Klausner, S. D., P. J. Zwerman and D. R. Coote. "Design Parameters for the Land Application of Dairy Manure," Report No. EPA 600/2-76-187, U.S. Environmental Protection Agency, Washington, DC (1976).
9. Klausner, S. D., P. J. Zwerman and D. F. Ellis. "Nitrogen and Phosphorus Losses from Winter Disposal of Dairy Manure," *J. Environ. Quality* 5(1):47-49 (1976).
10. Ogrosky, H. O., and V. Mackus. "Hydrology of Agricultural Lands," in *Handbook of Applied Hydrology,* V. T. Chow, Ed. (New York: McGraw-Hill Book Company, 1964), Chapter 21.
11. Stewart, B. A., D. A. Woolhiser, W. H. Wischmeier, H. J. Caro and M. H. Frere. "Control of Water Pollution from Cropland—Vol II," Report No. EPA-660/2-75-0266, U.S. Environmental Protection Agency, Washington, DC (1976).

12. Kuzmin, P. P. "Snow Cover and Snow Reserves," in *The Role of Snow and Ice in Hydrology*, Banff Symposia, (Paris: Beauregard Press, 1972), p. 6.
13. Eggleston, K. O., E. K. Israelsen and J. P. Riley. "Hybrid Computer Simulation of the Accumulation and Melt Processes in a Snowpack," Utah State University, Logan (1971).
14. U.S. Army Crops of Engineers. "Runoff From Snowmelt," Manual 1110-2-1400, Washington, DC (1960).
15. Riley, J. P., E. K. Israelsen and K. O. Eggleston. "Some Approaches to Snowmelt Prediction," in *The Role of Snow and Ice in Hydrology*, Banff Symposia (Paris: Beauregard Press, 1972), pp. 956-971.
16. Hendrick, R. L., and B. D. Filgate. "Application of Environmental Analysis to Watershed Snowmelt," *J. Appl. Meteorol.* 10:418-419 (1971).
17. Eagleson, P. A. *Dynamic Hydrology*, (New York: McGraw-Hill Book Company, 1970).
18. Johnson, A. H., D. R. Bouldin, E. A. Goyette and A. H. Hedges. "Phosphorus Loss by Stream Transport from a Rural Watershed: Quantitites, Processes and Sources," *J. Environ. Quality* 5(2):148-157 (1976).
19. Johnson, A. H., D. R. Bouldin, E. A. Goyette and A. M. Hedges. "Nitrate Dynamics in Fall Creek, New York," *J. Environ. Quality* 5(4): 386-391 (1976).
20. Bouldin, D. R., A. H. Johnson and D. A. Lauer. "The Influence of Human Activity on the Export of Phosphorus and Nitrate from Fall Creek," in *Nitrogen and Phosphorus—Food Production, Waste and the Environment*, K. S. Porter, Ed. (Ann Arbor, MI: Ann Arbor Science Publishers, Inc., 1975), pp. 61-120.
21. Johnson, A. H. "Phosphorus Export from the Fall Creek Watershed," Ph.D. Dissertation, Cornell University, University Microfilms, Ann Arbor, MI (August 1975).
22. Chow, V. T. "Runoff," in *Handbook of Applied Hydrology*, V. T. Chow, Ed. (New York: McGraw-Hill, 1964), Chapter 14.

44

MODELING SOIL AND WATER CONSERVATION PRACTICES

D. C. Beyerlein, A. S. Donigian, Jr.
  Hydrocomp, Inc.
  Palo Alto, California

INTRODUCTION

Since 1972 the Environmental Research Laboratory in Athens, Georgia, has sponsored the development and testing of mathematical models to simulate the quantity and quality of agricultural runoff. The overall goal of the program is to provide state-of-the-art models as tools for analyzing agricultural nonpoint pollution and evaluating the impact and effectiveness of alternative land management procedures. This paper describes the use of one of these tools, the Agricultural Runoff Management (ARM) model [1,2] to evaluate the effectiveness of soil and water conservation practices in controlling runoff, sediment loss, and the movement of pesticides and nutrients from agricultural watersheds.

Soil and water conservation practices (SWCPs) impact hydrologic and physical characteristics of a watershed, in addition to the timing and type of agronomic practices. Specific ARM model parameters are related to certain hydrologic and physical characteristics of an agricultural watershed, while other parameters describe agronomic practices. As these characteristics are changed for a specific SWCP, the model parameter values can be adjusted to reflect the use of the SWCP. Output from the model for different SWCPs is compared to the corresponding output for the conventional practice or base condition. The total volume (or mass) and frequency of runoff, sediment loss, and pesticide and nutrient washoff are compared for the different practices. The change in frequency or volume for a particular constituent by a

688 BMPs FOR AGRICULTURE AND SILVICULTURE

SWCP is used to evaluate the effectiveness of that SWCP to control that constituent. This analysis was done for three SWCPs (minimum tillage, contouring and terracing) on a small agricultural watershed in Watkinsville, Georgia, using 10 years of rainfall and evaporation data as input to the model. This work was performed for Cornell University under an EPA research grant to determine the effectiveness of SWCPs for pollution control.

AGRICULTURAL RUNOFF MANAGEMENT (ARM) MODEL

The ARM model is a continuous model that simulates runoff (including snow accumulation and melt), sediment, pesticides, and nutrient contributions to stream channels from both surface and subsurface sources. Figure 1 demonstrates the general structure and operation of the ARM model. The major components of the model individually simulate the hydrologic response (LANDS) of the watershed, sediment production (SEDT), pesticide adsorption/desorption (ADSRB), pesticide degradation (DEGRAD), and nutrient transformations (NUTRNT). The executive routine, MAIN, controls the overall execution of the program: calling subroutines at proper intervals, transferring information between routines, and performing the necessary input and output functions.

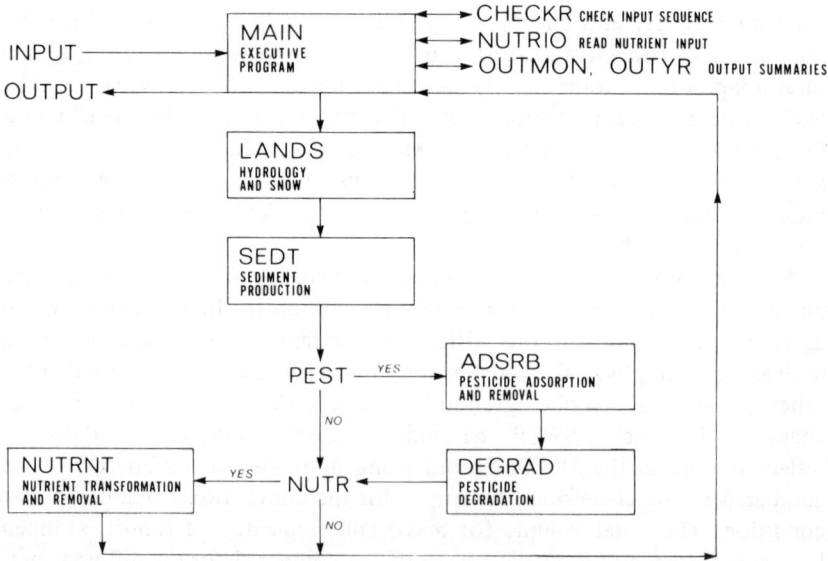

Figure 1. ARM model structure and operation.

In order to simulate vertical movement and transformations of pesticides and nutrients in the soil profile, specific soil zones (and depths) are established so that the total soil mass in each zone can be computed. Total soil mass is a necessary ingredient in the pesticide adsorption/desorption reactions and nutrient transformations. The vertical soil zones simulated in the ARM model include the surface, upper, lower and groundwater zones. The depths of the surface and upper soil zones are specified by the model input parameters, and are generally 0.1-0.3 in. and 3-6 in. respectively. The upper zone depth corresponds to the depth of incorporation of soil-incorporated chemicals. It also indicates the depth used to calculate the mass of soil in the upper zone whether agricultural chemicals are soil-incorporated or surface-applied.

The transport and vertical movement of pesticides and nutrients, as conceived in the ARM model, is indicated in Figure 2. Pollutant contributions to the stream can occur from the surface zone, the upper zone, and the groundwater zone. Surface runoff is the major transport mechanism carrying dissolved chemicals, pesticide particles, or sediment and adsorbed chemicals. The interflow component of runoff can transport dissolved pesticides or nutrients occurring in the upper zone. Vertical chemical movement between the soil zones is the result of infiltrating and percolating water. From the surface, upper and lower zones, uptake and transformation of nutrients and degradation of pesticides is allowed. The groundwater zone is presently considered a sink for deep percolating chemicals.

Model Components

The algorithms, or equations, used to describe the processes simulated by the ARM model are fully discussed in the original model reports.[1-3] A brief presentation of the general methodology is included here.

Hydrologic simulation by the LANDS subprogram is derived from modifications of the Stanford Watershed Model[4] and the Hydrocomp Simulation Program.[5] Through a set of mathematical functions, LANDS simulates continuously the major components of the hydrologic cycle, including interception, surface runoff, interflow, infiltration, and percolation to groundwater. In addition, energy balance calculations are performed to simulate the processes of snow accumulation and melt.

The algorithms for simulating soil loss, or erosion, were initially derived from research by Negev[6] at Stanford University and have been subsequently influenced by the work of Meyer and Wischmeier[7] and Onstad and Foster.[8]

Although Negev simulated the entire spectrum of the erosion process, only sheet and rill erosion are included in the ARM model. The two component processes of sheet and rill erosion pertain to (1) detachment of soil fines (generally the silt and clay fraction) by raindrop impact, and (2) pick-up and

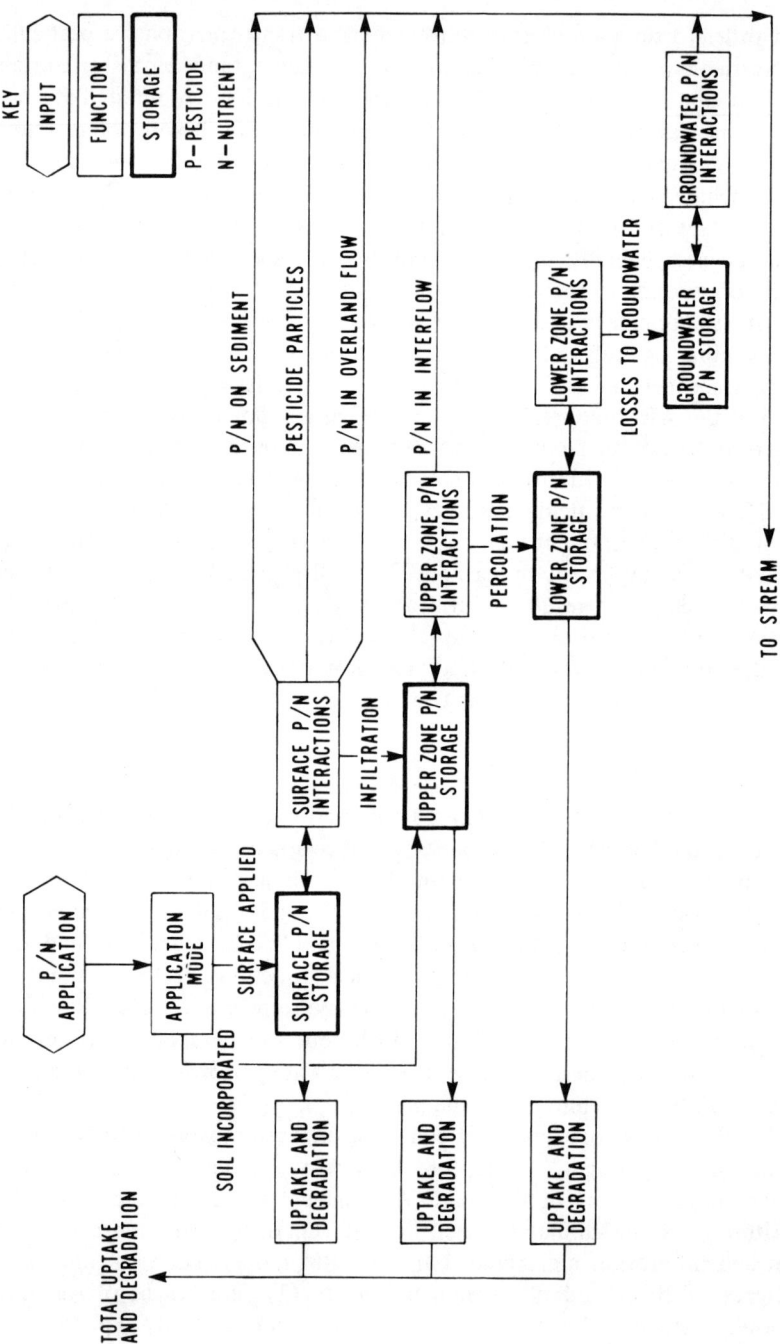

Figure 2. Pesticide and nutrient movement in the ARM model.

transport of soil fines by overland flow. These processes are represented as follows. Soil fines detachment:

$$RER(t) = (1-COVER(T)) * SMPF * KRER(PR(t))^{JRER} \quad (1)$$

Soil fines transport:

$$SER(t) = \begin{cases} KSER*OVQ(t)^{JSER}, \text{ for } SER(t) \leq SRER(t) & (2) \\ SRER(t), \text{ for } SER(t) > SRER(t) & (3) \end{cases}$$

$$ERSN(t) = SER(t)*F \quad (4)$$

where RER(t) = soil fines detached during time interval t, ton/ac
COVER(T) = fraction of vegetal cover as a function of time, T, within the growing season
KRER = detachment coefficient for soil properties
SMPF = supporting management practice factor (P-factor in Universal Soil Loss Equation, USLE, Wischmeier and Smith 1965)
PR(t) = precipitation during the time interval, in.
JRER = exponent for soil detachment
SER(t) = transport of fines by overland flow, ton/ac
JSER = exponent for fines transport by overland flow
KSER = coefficient of transport
SRER = reservoir of soil fines at the beginning of time interval, t, ton/ac
OVQ(t) = overland flow occurring during the time interval, t, in.
F = fraction of overland flow reaching the stream during the time interval, t
ERSN(t) = sediment loss to the stream during the time interval, t, ton/ac

In the operation of the algorithms, the soil fines detachment (RER) during each time (5 or 15 min) interval is calculated by Equation 1 and added to the total fines storage or reservoir (SRER). Next, the total transport capacity of the overland flow (SER) is determined by Equation 2. Sediment is assumed to be transported at capacity if sufficient fines are available, otherwise the amount of fines in transport is limited by the fines storage, SRER (Equation 3). The sediment loss to the waterway in the time interval is calculated in Equation 4 by the fraction of total overland flow that reaches the stream. An overland flow routing technique determines the flow contribution to the stream in each time interval. After the fines storage (SRER) is reduced by the the actual sediment loss to the stream (ERSN), the algorithms are ready for simulation of the next time interval. Thus, the sediment that does not reach the stream is returned to the fines storage and is available for transport in the next time interval. The methodology attempts to represent the major processes of importance in soil erosion so that the impact of land management practices (for example, tillage, terracing, mulching, etc.) can be specified by their effects on the sediment parameters.

Since land cover by growing crops and crop residues has a major impact on sediment loss, the variability in the land surface cover is explicitly represent in the ARM model. The land cover variable in Equation 1, COVER(T), represents the fraction of the land surface effectively protected from the kinetic energy and detachment capability of rainfall. Monthly cover values as of the first day of the month are specified by the user. The model interpolates linearly between the monthly values to evaluate land cover on each day. Figure 3 demonstrates the land cover function in the model. The kinetic energy of rainfall is effectively dissipated by the land cover with values of 90 to 95% of the area. Thus, judicious use of land cover function allows simulation of various land surface conditions for different practices.

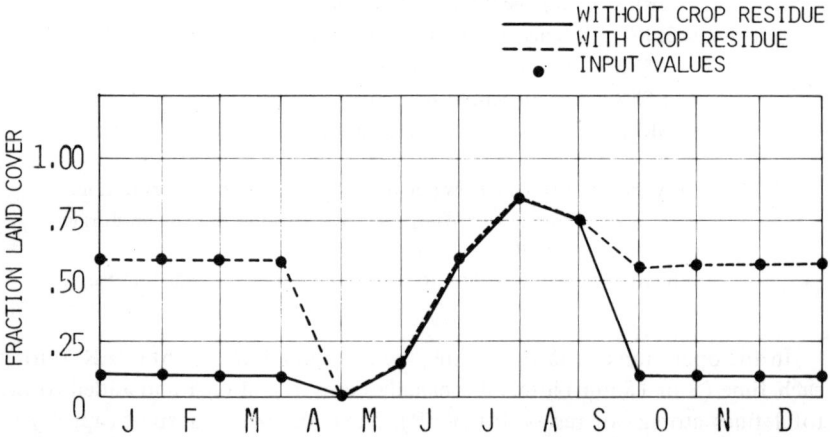

Figure 3. Land cover with and without crop residue remaining on the P2 watershed.

The timing and severity of tillage operations have a controlling effect on the sediment loss from an agricultural watershed. With regard to sediment production, the effect of tillage operations is to increase the mass of soil fines available for transport and produce a reasonably uniform distribution of fines across the watershed. Consequently, the ARM model allows the user to specify the dates of tillage, planting, or other land-surface disturbing operations. For each of these dates the user must specify a new detached soil fines storage resulting from the operation. At the beginning of each tillage day the ARM model resets the fines storage to the new value, resulting in a uniform fines distribution across the watershed. The amount of fines storage produced by different tillage operations is related to the depth and extent of the operation, and edaphic characteristics.

The process of pesticide adsoprtion/desorption onto sediment particles is a major determinant of the amount of pesticide loss that will occur. This process establishes the division of available pesticide between the water and sediment phases, and thus specifies the amounts of pesticide transported in solution and on sediment. To simulate this process in the ARM model, the following equation is used:

$$X/M = KC^{1/N} + F/M \tag{5}$$

where X/M = pesticide adsorbed per unit soil, g/gm
        F/M = pesticide adsorbed in permanent fixed state per unit soil. F/M is less than or equal to FP/M, where FP/M is the permanent fixed capacity of the soil in g/gm for each pesticide.
        C = equilibrium pesticide concentration in solution, mg/l
        N = exponent
        K = coefficient

This algorithm is comprised of an empirical term, F/M, plus the standard Freundlich single-valued (SV) adsorption/desorption isotherm. The empirical term, F/M, accounts for pesticides (for example, paraquat) that are permanently adsorbed to soil particles and will not desorb under repeated washing.

The ARM model includes an option to use a non-single-valued (NSV) adsorption/desorption function because research has indicated that the assumption of single-valued adsorption/desorption is not valid for many pesticides.[9] In these cases, the adsorption and desorption processes result in different pesticide concentrations. The form of the desorption equation is identical to Equation 5 except that K and N values are replaced by K' and N', respectively, with the prime denoting the desorption process. The user specifies the N' value as an input parameter and the ARM model calculates K' as a function of the adsorption/desorption parameters (K, N, N') and the pesticde solution concentration.[9] The NSV function simulates higher pesticde concentrations on sediment than the SV function in order to represent the irreversibility of the adsorption process.

Attenuation of applied pesticides, through volatilization and degradation processes, is critical to the accurate simulation of pesticide runoff because these mechanisms control the amount of chemical available for transport. These processes are not well understood and are topics of continuing research. To approximate the pesticide attenuation following application, the ARM model includes a step-wise first-order attenuation function that allows the use of different degradation rates for separate time periods after application. The function calculates the combined degradation of pesticides by volatilization, microbial degradation, and other attenuation mechanisms. This approach was chosen after evaluating both simpler and more sophisticated degradation models.[2]

Nutrient simulation in the ARM model attempts to represent the reactions and transformations of nitrogen and phosphorus compounds in the soil profile as a basis for predicting the nutrient content of agricultural runoff. The nutrient model assumes first-order reaction rates and is derived from work by Mehran and Tanji,[10] and Hagin and Amberger.[11] The processes simulated include immobilization, mineralization, nitrification/denitrification, plant uptake, and adsorption/desorption. In the model, fertilizer, plant residue, or animal waste is applied in its chemical form (NH_4, NO_3, PO_4, Cl, and organic N and P) to the land surface or into the soil.

The ARM model simulates nutrient movement in the watershed by water or sediment. Transformations of nutrients determine the nutrient forms in each soil zone and their resulting susceptibility to movement. The ARM model imitates nutrient transport processes only to the extent that is needed to predict runoff quality and quantity. The model does not consider the generally secondary movement of soil chemicals by concentration and thermal gradients. However, it does model the lateral and downward transport of chemicals by water from the soil zones. Lateral transport of nutrients toward the stream can occur from the surface and upper zone storages. Groundwater transport is not currently modeled.

Figure 4 diagrams the transformation pathways and storages and gives the names of the reaction rate input parameters. Reaction rates are input on a per day basis for each soil zone. Nitrite (NO_2) transforms so quickly in most agricultural areas that it is not considered separately. The adsorbed phase represents the nutrients in a complex form along with those adsorbed on the soil. The plant uptake rates (KPL) are modified monthly by an input parameter which depends on the stage of crop growth. These monthly input parameters are adjusted to represent the crop uptake of N and P from the soil storages and to distribute it throughout the growing season.

Soil temperature is presently the only environmental factor that is modeled as affecting the reaction rates with the exception that the transformations are stopped entirely at very low moisture levels. Soil temperatures for the surface and upper zone are determined by regression equations based on air temperature, while the daily soil temperature for the lower zone and

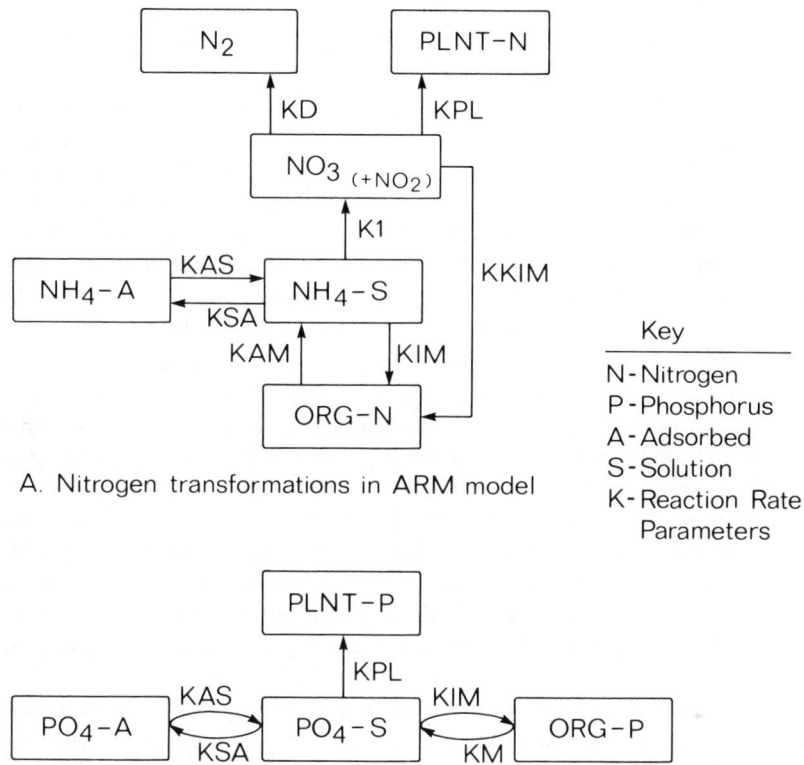

Figure 4. Nutrient transformations.

groundwater is interpolated from average monthly input values. Other factors deemed as having a constant influence during the simulation period, such as soil pH, are represented by adjustments to the input reaction rates.

DESCRIPTION OF SOIL AND WATER CONSERVATION PRACTICES

In a joint study by the EPA and ARS,[1,2] 18 soil and water conservation practices (SWCPs) were determined to be important in controlling runoff and sediment loss from agricultural areas. Of these 18 practices we selected three for detailed study. These three practices are the use of contours, terraces and minimum tillage. These practices were selected based on their relative importance as a SWCP and the ability of the model to represent the particular changes involved in each practice.

To represent a SWCP by changing model parameter values it is important to define how the SWCP differs from the base conditions or standard practice. In addition, definition of the base conditions is necessary, as the base conditions change for each region of the country. The ARM model has been tested on the ARS-EPA P2 watershed near Watkinsville, Georgia,[2] and was found capable of representing the hydrology, sediment movement, and pesticide and nutrient washoff measured on the watershed. This previous work provided initial parameter values for the model. It also provided an initial practice on which to base these parameter values. A comparison of the management practice used on the watershed with the three SWCPs to be studied helped to define the necessary parameter value changes. But before the parameter changes can be discussed it is important to clearly define each soil and water conservation practice relative to the base conditions.

For this study the base conditions represent conventional practices in common use in the region of the test watershed. For our southeast region watershed (P2) this means planting row crops in straight rows parallel to (that is, up and down) the land slope. Tillage of the soil is in the form of discing and is done in the spring to prepare the soil for planting. After harvest the crop residue is removed from the field and the soil is fallow through the winter until a new crop is planted in the spring. In many regions of the country, these practices would correspond to minimum or conservation tillage, but they represent the base conditions for this study of the southeast region.

The minimum-tillage SWCP differs from the base conditions in two ways. There is no tillage prior to spring planting and the crop residue is left on the field after harvest. Planting is done with the crop residue remaining from the previous fall. This protects the soil from erosive forces until the coming spring when a new crop is planted. This would usually correspond to no-tillage in other regions of the country.

The contour SWCP differs from the base conditions only in that the crops are planted in rows perpendicular to the slope of the land, which by definition are contoured rows. For the purposes of this study, the contour SWCP is assumed to use discing in the spring prior to planting and removal of crop residue after harvest.

The terrace SWCP requires physical alteration of the land to establish and maintain terraces. The watershed area is 3.2 ac and is assumed to be divided into two terraces if this SWCP is used. Contoured rows are used with the terraces. However, it is assumed that discing and crop residue removal practices are retained from the base condition practice.

These three SWCPs plus the base condition are modeled by a selection of parameter values which represent each watershed condition. How the parameters and their values were selected for each SWCP is discussed below.

ARM MODEL PARAMETER VALUES AND SWCPs

The selection of the ARM model parameters and their values for different SWCPs were based on the expected physical and agronomic changes on the watershed associated with a particular SWCP. A total of eight input parameters for which adjustments could be reasonably estimated were identified as being related to SWCPs. These parameters and values for each watershed condition are listed in Table I.

Table I. ARM model parameter value changes for SWCPs.

Parameters[a]	Base Condition	Soil and Water Conservation Practices		
		Minimum Tillage	Contours	Terraces with Contours
UZSN	0.42	0.42	0.50	0.65
NN	0.15	0.20	0.20	0.20
L	100	100	100	250
SS	0.025	0.025	0.025	0.015
COVPMO	(see Figure 3)			
SMPF	1.0	1.0	0.5	0.35
KSER	0.6	0.55	0.5	0.5
SRERTL[b]	1.5	0.15	1.5	1.5

[a] Parameters are defined as follows:
- UZSN = nominal upper zone moisture storage, in.
- NN = Manning's n (roughness) for overland flow
- L = length of the overland flow path, ft
- SS = slope of the overland flow path
- COVPMO = fraction of land covered by vegetation, monthly value
- SMPF = supporting management practice factor (equal to the P factor in the USLE)
- KSER = sediment washoff coefficient
- SRERTL[b] = sediment fines produced by tillage operations, ton/ac

[b] Tillage occurs once each year on day 115 (April 25) for the 10-yr simulation.

Determining how to change parameter value to represent a particular SWCP on a watershed is often difficult. For some characteristics we have related parameter value changes to established techniques (for example, USLE, SCS Curve Number Runoff Method) derived from field data. However, for certain parameters, our own experience, judgment and knowledge of the physical processes and their representation in the model was our only guide. The general procedure was to estimate initial parameter changes from the literature (which was nonexistent for some parameters) and then adjust the values to conform with our experience on other watersheds. These estimating procedures allowed us to establish parameter values for each of the three SWCPs included in this study.

UZSN (upper zone nominal moisture storage) can be related to the initial abstraction parameter, Ia, as defined by the SCS procedures for estimating storm runoff volume.[13] The relationship is assumed to be proportional: UZSN should be changed by the same percent as Ia when comparing a SWCP with a base condition. This was done for each of three antecdeent moisture conditions included with SCS method. The average percent change for a particular SWCP was used to determine the UZSN value for that SWCP. The calibrated UZSN value for the P2 watershed with contours was the baseline from which the other values were calculated.

NN (Manning's roughness coefficient for overland flow) is proportional to the amount and type of ground cover and soil type. An inexact, but relative value of NN can be found from the literature[13] for comparison of SWCPs with the base condition value. However, the NN values in Table I are based largely on our own experience and judgment.

L (length of the overland flow path) changes only for the terrace SWCP when compared to the base condition. The modification of the watershed with terraces changes the overland flow path such that instead of draining to a central location to form a stream channel the flow must instead cross the length of the terrace to its outflow point. This increase in length will vary with the terrace configuration. For our use on the P2 watershed the addition of two terraces increased L from 100 ft to 250 ft.

SS (slope of the overland flow path) is affected only by the terrace SWCP. Installation of terraces both increases lengthens the flow path and decreases the slope. For this specific situation the slope is decreased from 0.025 to 0.015.

COVPMO (cover per month) is the fraction of the land surface covered by both live and dead vegetation and effectively protects the soil surface from the erosive force of raindrops. When crop residue is left on the watershed after harvesting, as is done with the minimum-tillage SWCP, the fraction of land covered each month is equal to values measured by a joint ERA-ARS study.[14] Monthly cover values for the period between harvest and planting are close to zero for conditions where the crop residue is removed and the land left without cover. A minimum cover value of 0.1 was used as a reasonable value for this situation (Figure 3).

SMPF (supporting management practice factor) has been added to the ARM model to provide a means of incorporating the effect of management practices on the generation of sediment fines (Equation 1). SMPF is equal to the P-factor of the USLE. As such, data have been collected from which SMPF can be calculated for the contour and terrace SWCPs.[12]

KSER (sediment washoff coefficient) is affected by changes to soil surface conditions brought about by the introduction of SWCPs. Fleming and Leytham[15] show a relationship between KSER, Manning's roughness coefficient, overland flow length and slope, and mean surface particle size, although the relationship's reliability is unknown. The first three of these factors are related to the SWCPs we are investigating. Thus, although we cannot establish any direct correspondence between KSER and a particular SWCP, relative changes can be estimated from past experience with the model. KSER is expected to decrease with practices that retard overland flow and include crop residues. Adjustments to KSER similar to the changes in Table I were included in a previous study[16] and were found to produce reasonable results.

The last parameter, SRERTL (soil fines produced by tillage), is affected by the minimum-tillage SWCP in much the same way as is COVPMO. Model testing on watershed with similar practices provided SRERTL values for spring discing of the P watershed. Discing is a conventional practice for the southeastern section of the United States. Thus, for our study it represents the base condition. Minimum tillage is then effectively defined as a no-tillage condition for this situation and the sediment fines are only produced by planting in the crop residue. SRERTL for the minimum-tillage SWCP is reduced to one-tenth of the base condition value. A comparison of sediment loss data from similar watersheds of which one was tilled and one was not supports this reduction of SRERTL.[14]

WATERSHED SIMULATION AND METHODOLOGY

The P2 watershed is a small (3.2-ac), nonterraced agricultural watershed near Watkinsville, Georgia (Figure 5). From 1973 through 1975 EPA and ARS intensively monitored this watershed.[14] The data collected by these agencies made possible a thorough calibration of the ARM model's hydrology and sediment algorithms and testing of the pesticide and nutrient routines in the model. This has been done[2] and the calibrated model was found to reasonably represent the hydrology, sediment loss, and pesticide and nutrient washoff on P2.

The management of the P2 watershed during the EPA-ARS study was a mixture of SWCPs. Corn was grown on the watershed in contoured rows. The watershed was disced each spring prior to planting, but in the autumn after harvest the residue was left on the field until the coming spring. The ARM model was calibrated to represent this combination of the contour and some minimum-tillage practices; the calibrated parameter values reflect these conditions. Thus, the calibrated parameter values are not identical to the base condition values (which assume no SWCP) or any particular SWCP. However,

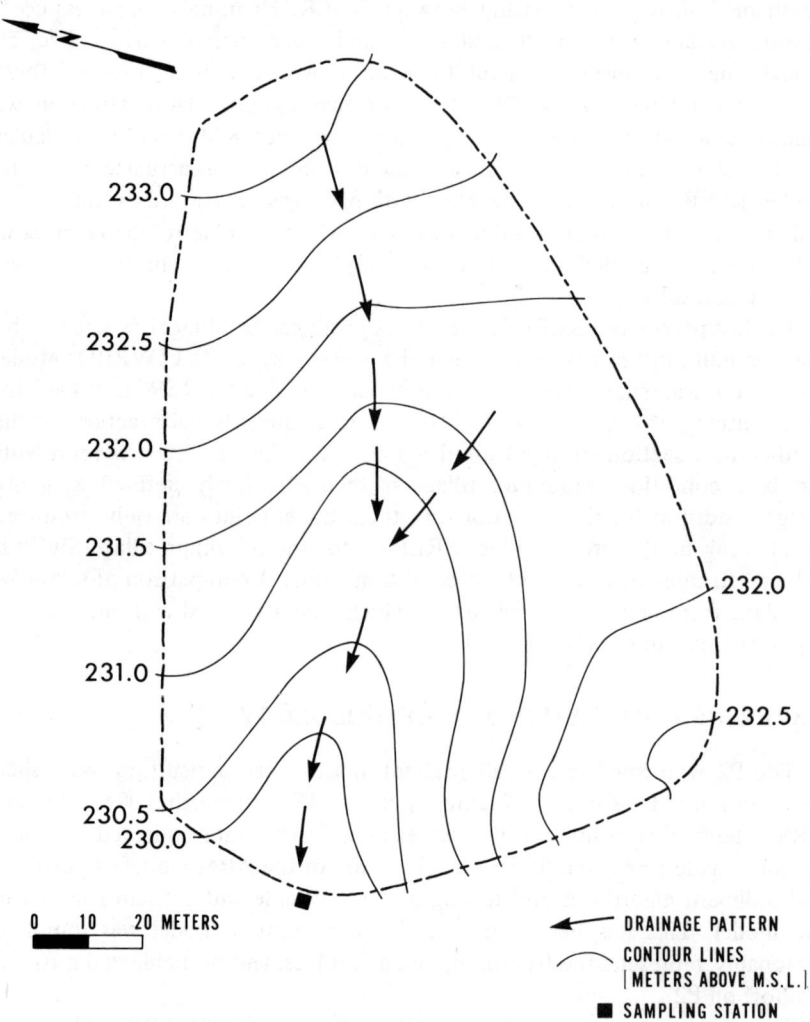

Figure 5. P2 watershed, Watkinsville, Georgia (3.2 ac).

with knowledge of how the actual watershed management practices differ from the base conditions and the SWCPs the appropriate parameter adjustments for each condition were evaluated.

To study the effects of SWCPs on runoff, sediment loss, and pesticide and nutrient washoff the P2 watershed was simulated for the 10-yr period of 1966 through 1975. Simulation of hydrology, sediment and pesticides by the ARM model requires 5- or 15-min rainfall and daily evaporation. Nutrient simulation further requires daily max-min air temperature. Hourly rainfall data collected by the National Weather Service (NWS) at the nearby Athens Airport were disaggregated to 15-min amounts for use in the model. Daily evaporation and max-min temperature data used in the simulation were also obtained from the NWS and were collected at stations within a 60-mi radius of the watershed.

Two different pesticides, atrazine and paraquat, were selected for study on the watershed. These pesticides were applied and monitored during the EPA-ARS study and reasonable simulation results were obtained with the ARM model. In the 10-yr simulation period of this study, the two pesticides were applied once in the spring of each year at planting time. The application amounts (2.6 lb/ac for atrazine and 2.25 lb/ac for paraquat) and application date were selected based on average values and dates from the 3-yr EPA-ARS study. In the same manner fertilizer (N and P) was applied twice each year (Table II).

Table II. Nitrogen and phosphorus fertilizer applications on the P2 watershed.

Day of Application Each Year	Form Applied	N (lb/ac)	P (lb/ac)
119	sulphate of ammonia	34.0	29.5
162	50% urea 50% ammonia	89.9	0.0

The Arm model simulated hydrology, sediment and pesticides on a 15-min time step. Nutrient washoff was also computed on a 15-min time step although the nutrient transformations in the soil were computed once every 6 hr. When runoff greater than a minimum value (0.10 cfs) occurred, a complete list of runoff constituents was written to a computer file for later analysis. Separate simulation runs were made for atrazine, paraquat and nutrients for each of the three SWCPs and the base conditions.

SIMULATION RESULTS AND DISCUSSION

The continuous information produced by the 10-yr ARM model simulation was analyzed in two ways: (1) mean annual runoff and pollutant losses, and (2) frequency of occurrence of runoff rates, pollutant concentrations, and pollutant flux (*i.e.,* mass removal in mass per minute). The SWCPs are represented only by the parameter changes discussed above. Other effects of SWCPs, such as changes in infiltration characteristics, crop growth, soil fertility, etc. are not considered due to lack of sufficient evidence to indicate appropriate parameter adjustments. Also, the pesticide and nutrient parameters (for example, application rates and time, adsorption coefficients, reaction rates) were the same for all conditions to provide a basis for comparing the effects of specific SWCPs. Thus, increasing pesticide applications to compensate for no-tillage practices was not included in this analysis, but such a practice can be evaluated in a similar manner.

Mean Annual Values

Table III summarizes the mean annual runoff and pollutant losses (in solution and on sediment) for each watershed condition, and provides the percent change for each SWCP from the base conditions. The results show that the total annual runoff decreases with each SWCP, with minor reductions ($<5\%$) for minimum tillage, 10 to 12% for contours, and 30 to 40% reductions for contours and terraces. Since overland flow comprises more than 90% of the total runoff for this small watershed, the reductions for overland flow are similar. Interflow, which travels subsurface for part of its flow path, increases 5 to 6% for both minimum-tillage and contours, and more than 30% for contours and terraces. Thus, the practices that retard and reduce overland flow also increase the subsurface flow. Both interflow and groundwater contributions increase, although groundwater is not present in the runoff from this small watershed.

Sediment loss for each SWCP partially reflects the reductions in surface runoff. Contours and terraces reduce sediment loss from 2.8 ton/ac for base conditions to 1.3 ton/ac, a 55% reduction. The sediment reductions for minimum tillage and contours are comparable, 30% and 37%, respectively, but for different reasons. Contours reduce sediment loss by reducing overland flow whereas minimum tillage reduces sediment loss by minimizing the disruption of the soil surface and protecting the surface with crop residues remaining on the watershed. Under the assumptions of this study, the soil loss reductions are comparable.

Pesticide and nutrient contents of runoff for each SWCP are the result of the changes in runoff and sediment loss discussed above. Although atrazine is detected both in solution and on sediment, the solution component is predominant. Thus, the reductions of atrazine in solution for the SWCPs are in the same order as for total runoff; the negative values increase with minimum tillage, contours, and contours plus terraces. Atrazine loss on sediment is relatively minor and, as expected, the reductions are similar to those for

Table III. Mean annual runoff and pollutant losses simulated with the ARM model.[a]

	Base Conditions	Minimum Tillage	Contours	Contours and Terraces	Percent Change from Base		
					Minimum Tillage	Contours	Contours and Terraces
Total Runoff (in.)	8.03	7.74	7.16	5.21	- 3.6	-10.9	-35.2
Overland Flow (in.)	7.55	7.23	6.66	4.58	- 4.2	-11.9	-39.4
Interflow (in.)	0.48	0.51	0.50	0.63	+ 5.8	+ 4.8	+31.8
Sediment Loss (ton/ac)	2.84	1.99	1.80	1.29	-30.0	-36.8	-54.6
Total Atrazine Loss (lb)	0.163	0.150	0.130	0.063	- 7.7	-20.1	-61.4
In Solution (lb)	0.162	0.150	0.130	0.063	- 7.5	-19.9	-61.3
On Sediment (lb)	0.0013	0.0009	0.0008	0.0004	-30.8	-38.5	-69.2
Paraquat on Sediment (lb)	0.976	0.738	0.647	0.475	-24.4	-33.7	-51.3
Nitrogen (lb/ac)							
Organic	6.30	8.36	4.91	3.84	+32.7	-22.1	-39.0
NH_4 - Solution	1.96	2.02	1.89	1.88	+ 3.0	- 3.9	- 4.2
NH_4 - Adsorbed	0.680	0.508	0.460	0.339	-25.3	-32.4	-50.1
$NO_3 + NO_2$	2.67	2.80	2.89	3.62	+ 4.8	+ 8.4	+35.8
Phosphorus (lb/ac)							
Organic	0.68	0.56	0.52	0.40	-17.6	-23.5	-41.2
PO_4 - Solution	0.376	0.408	0.372	0.367	+ 8.5	- 1.1	- 2.4
PO_4 - Adsorbed	2.21	1.81	1.66	0.45	-18.2	-24.9	-79.7

[a]Obtained from 10-yr simulations (1966-1976) on the P2 watershed in Watkinsville, Georgia.

sediment loss. Total atrazine loss is reduced by 61% for contours and terraces, 20% for contours alone, and 8% for minimum tillage.

Paraquat is a highly ionic compound that is strongly and irreversibly adsorbed to sediment particles. Thus, the values for paraquat loss are shown only for sediment in Table III because paraquat is not transported in solution. For each SWCP, the percent reductions for paraquat and sediment are basically the same.

The nutrient values in Table III are presented by the separate nitrogen and phosphorus forms simulated with the ARM model. With one exception, the sediment-associated nutrients (organic N and P, adsorbed NH_4 and adsorbed PO_4) demonstrate the same relative percent reductions for each SWCP as sediment loss. For minimum tillage, these nutrient forms are reduced by 18 to 25%, 22 to 32% for contours, and 39 to 80% for contours and terraces. The variations reflect the specific reaction and adsorption behavior of the individual components. The only exception to this pattern is the 33% increase in organic N under minimum tillage. This results from the assumption that the crop residues remaining on the watershed deposit organic material that can washoff with the eroded sediment. We assumed 10 lb/ac of organic N was deposited with each fall harvest.

Both soluble NH_4 and soluble PO_4 show relatively minor changes for any of the SWCPs. Both constituents increase slightly under minimum tillage and decrease slightly for contours with and without terraces. Except to show the expected direction of a change, percentage changes less than 10.0 are not likely to be significant within the accuracy of the model. Thus, these SWCPs do not appear to have major impact on the mean annual losses of these constituents.

On the other hand, NO_3 (and NO_2) loss is increased by 36% when terraces are used, with only minor increases under minimum tillage and contours. This change is a result of the increase in interflow when terraces are installed since interflow is the major transporting mechanism for NO_3. The other solution nutrients are not as drastically affected by interflow because they also adsorb onto sediment particles, and thus do not move as readily through the subsurface environment.

Frequency Analysis

Frequency analysis of the simulated information was used to characterize the frequency or probability of occurrence of runoff and pollutant levels under the range of meteorologic and environmental conditions during the 10-yr simulation period, and to determine how these frequencies change with each different SWCP. The frequency curves for runoff and sediment, atrazine, nitrogen, and phosphorus are shown in Figures 6 through 9, respectively.

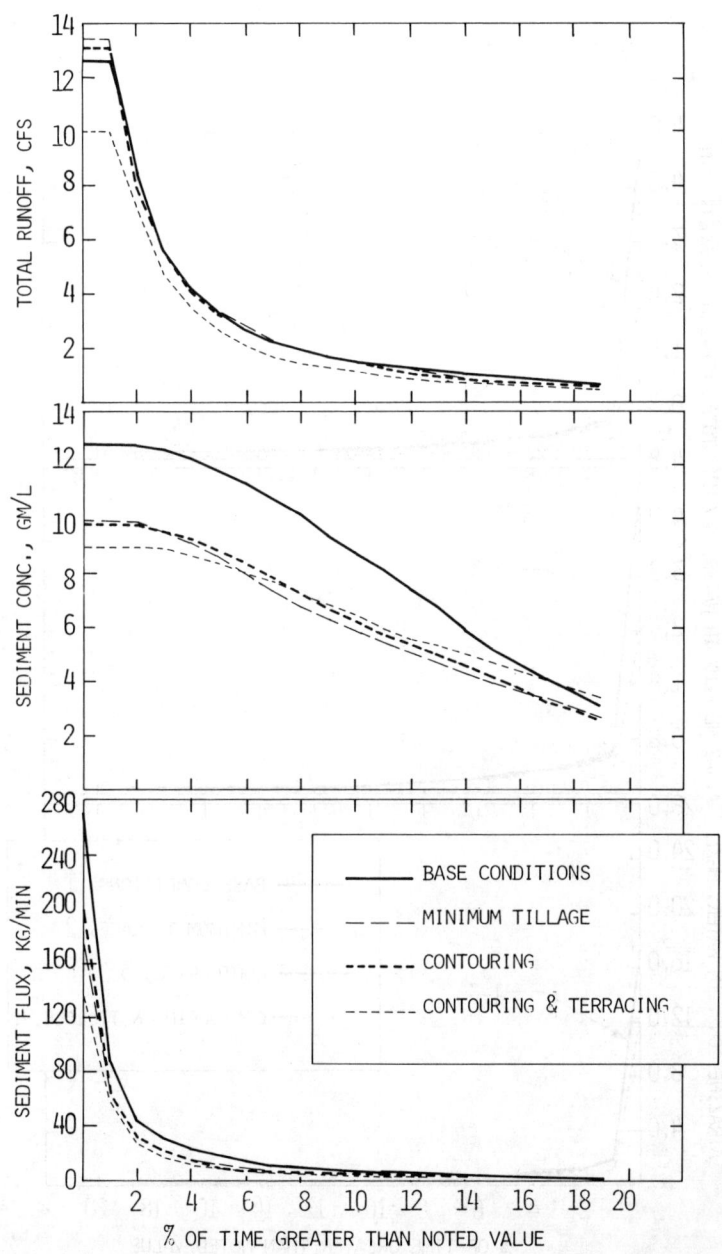

Figure 6. Runoff and sediment frequency analysis (P2 watershed: 1966-1975).

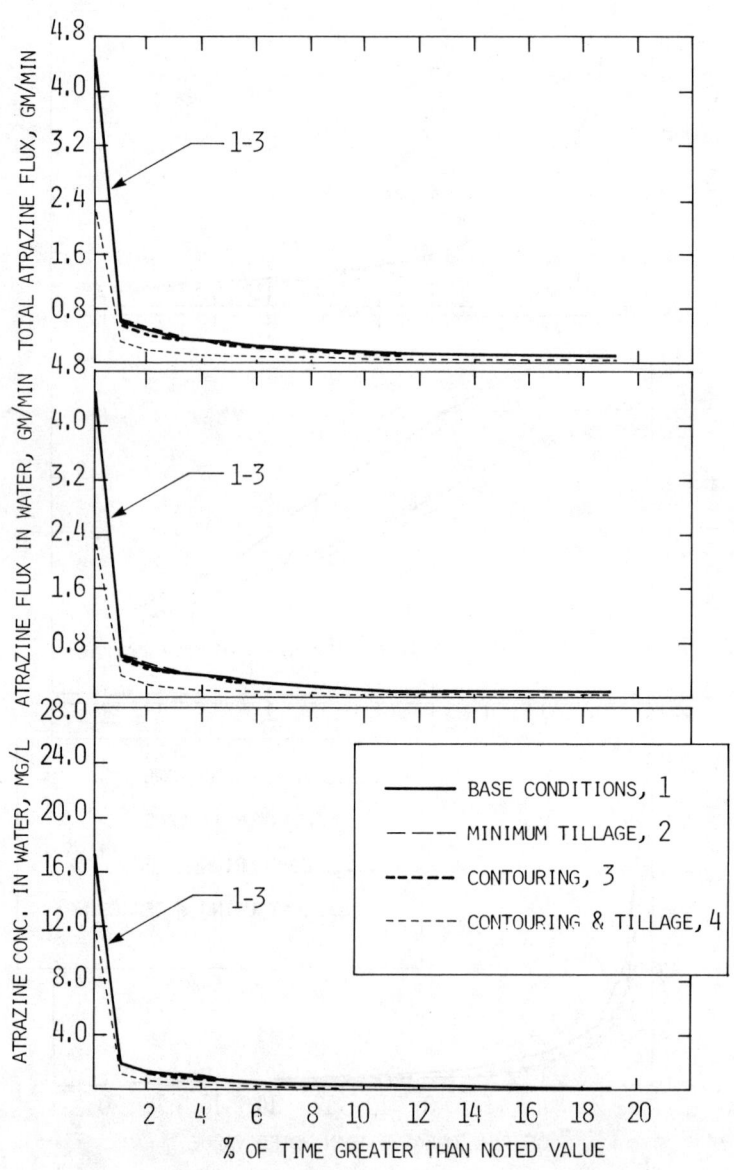

Figure 7. Pesticide frequency analysis (P2 watershed: 1966-1975).

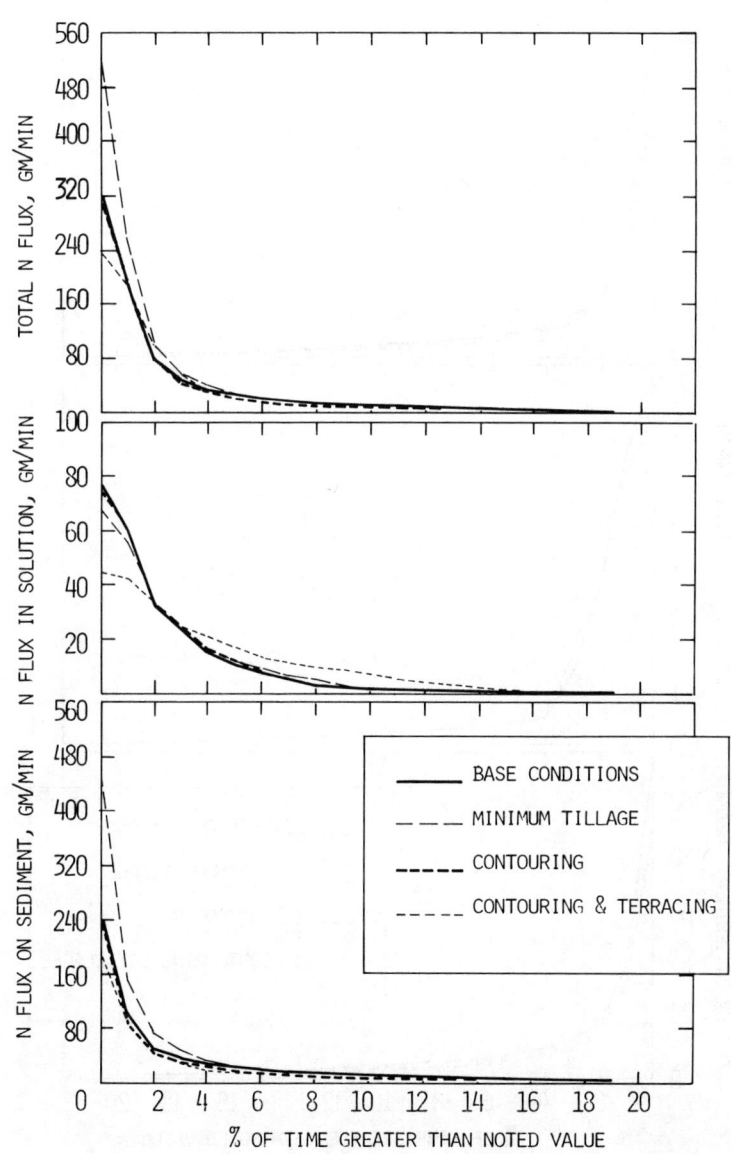

Figure 8. Nitrogen frequency analysis (P2 watershed: 1966-1975).

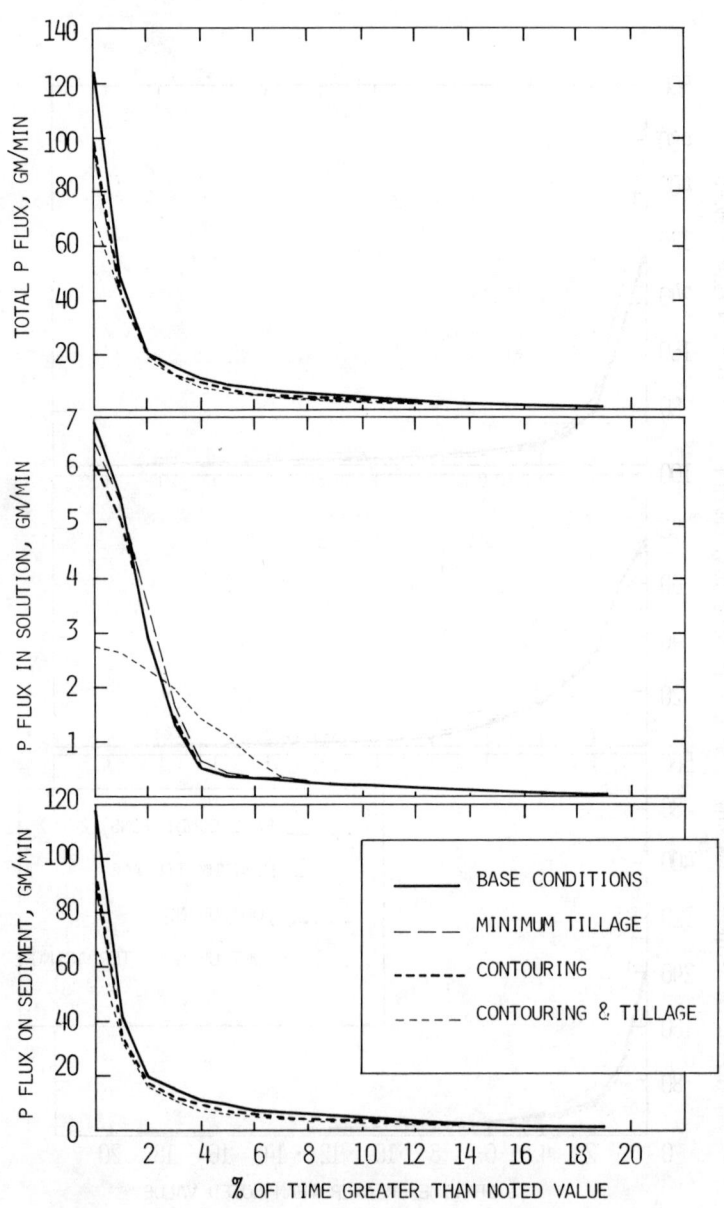

Figure 9. Phosphorus frequency analysis (P2 watershed: 1966-1975).

The curves are presented in terms of the percent of time the particular variable (for example, runoff in cfs) is greater than the ordinate value. Thus, Figure 6 shows that sediment concentrations under terracing and contouring are greater than 9.0 mg/l for 2% of the time (time during which runoff is occurring), whereas base conditions produce sediment concentrations greater than 12.7 gm/l for 2% of the time. Each of the curves can be analyzed in a similar fashion.

The runoff frequency graph (top graph in Figure 6) shows relatively small differences among the curves for base conditions, minimum tillage, and contours, with the combined contours and terraces curve being somewhat lower. The mean annual runoff values in Table III confirm this. The sediment concentration frequency (middle graph in Figure 6) indicates the base condition concentrations are substantially higher than the three SWCPs which produce similar values. However, the sediment flux frequency (bottom graph in Figure 6) more closely demonstrates the relative ranking of the SWCPs as shown by the percent reductions in annual values in Table III. In general, flux or mass removal is more representative of the effects of a particular SWCP because it reflects changes in both concentration and flow.

Figure 7 includes the frequency curves for total atrazine flux, atrazine flux in solution, and atrazine concentration in solution. For 1.0% of the time the atrazine flux is greater than 0.3 gm/min with contours and terraces, and greater than 0.6 gm/min for the other watershed conditions. Since very little atrazine is transported on sediment, there is no significant difference between the flux curves for total atrazine and solution atrazine. The concentration curve in Figure 7 shows that for 1.0% of the time, the atrazine concentration in water is greater 1.0 mg/l with contours and terraces, and greater than about 2.0 mg/l for the other conditions. Unfortunately the curve definition in these figures is not sufficient to clarify the differences between the curves. However, with expanded scales the differences would be more readily apparent.

These frequency curves can be analyzed to determine how often specific runoff volumes, flowrates, pollutant concentrations, or flux rates will occur. To evaluate potential ecologic impact, the frequency curves and toxicity data can be used to estimate how often acute or chronic pesticide levels toxic to specific organisms will exist.

Paraquat flux and concentration frequency curves are not shown here because paraquat exhibits relatively constant concentrations and, except for scale, the flux curves are identical to the sediment flux curves in Figure 6.

Total flux frequency curves for nitrogen and phosphorus and their solution and sediment components are shown in Figures 8 and 9, respectively. Solution nitrogen includes NO_3 (plus NO_2) and soluble NH_4, while sediment nitrogen is comprised of organic nitrogen and adsorbed NH_4. Figure 8 shows that minimum tillage results in the highest total nitrogen flux, its maximum values are even greater than the base conditions. This results from a high sediment nitrogen component, which is mostly organic nitrogen from the crop residues remaining on the watershed. Contours and terraces produce reductions in total and sediment nitrogen flux curves. however, for solution nitrogen the SWCP of combined contours and terraces reduce the high flux rates but increase the rates that occur 98% of the time. Also, contours alone increase the solution nitrogen export above the base conditions for values that occur 2% of the time or less. Decisions on implementing contours and/or terraces should consider the relative impacts of solution and sediment nitrogen.

The phosphorus flux curves in Figure 9 show similar results to the nitrogen flux curves except minimum-tillage does not have as dramatic an impact. Sediment phosphorus includes organic phosphorus and adsorbed PO_4 while solution phosphorus is comprised only of dissolved PO_4. Whereas crop residues were assumed to contribute organic nitrogen at harvesting, no such assumption was made for organic phosphorus. Thus, the sediment phosphorus contributed under minimum tillage would likely be higher if organic phosphorus contributions from crop residues had been included. These preliminary results and assumptions regarding the impact of crop residues are currently being reevaluated. The effects of contours and terraces on solution phosphorus are similar to those noted above for solution nitrogen. Contours produce lower flux rates while terraces and contours produce a flatter frequency curve with the highest flux rates for a portion of the time.

To evaluate the net or overall impact of the alternative practices (and to quantify the differences between the curves), the area beneath the curve for each practice can be calculated and compared. From elementary decision theory this area represents the expected value of the ordinate variable under all conditions; that is, the value of the variable times its probability of occurrence, summed over all possible occurrences. For example, the area beneath the base sediment curve in Figure 6 is the measured units of the y-axis, mg/l; each block (1 x-axis unit*1 y-axis unit) is 0.08 mg/l (4 mg/l*0.02).

The differences in area beneath each curve, or the area between the curves, can be used to evaluate the impact of a particular practice. Table IV lists the area beneath each frequency curve and the percent change for each practice from the base conditions. The percent changes are similar to and generally lower than the corresponding values in Table III. The changes in the expected values in Table IV more accurately represent the overall impact of each watershed alternative under the full range of meteorologic and environmental conditions included in the simulated period. Mean annual values can mask the effects of extreme conditions since two frequency curves can have similar mean values but significantly different expected values. The frequency curves allow investigation of extreme conditions, and expected values incorporate these conditions when alternatives are compared.

Table IV. Frequency analysis of alternative SWCPS using the ARM model.

	Base Conditions	Expected Value[a]			Percent Change from Base		
		Minimum Tillage	Contours	Contours and Terraces	Minimum Tillage	Contours	Contours and Terraces
Total Runoff (cfs)	0.689	0.691	0.661	0.526	+ 0.2	- 4.0	-23.6
Overland Flow (cfs)	0.669	0.668	0.640	0.494	- 0.2	- 4.3	-26.1
Interflow (cfs)	0.028	0.030	0.029	0.038	+ 6.8	+ 4.6	+35.7
Sediment Loss							
Concentration (gm/l)	1.97	1.47	1.49	1.67	-25.1	-24.1	-15.3
Flux (kg/min)	4.74	3.36	3.29	2.59	-29.3	-30.6	-45.5
Total Atrazine Flux (gm/min)	0.0786	0.0782	0.0702	0.0333	- 0.5	-10.7	-57.6
Atrazine Loss in Solution							
Concentration (mg/l)	0.194	0.201	0.190	0.116	+ 3.8	- 1.9	-40.1
Flux (gm/min)	0.0777	0.0775	0.0696	0.0330	- 0.3	-10.4	-57.5
Atrazine Loss on Sediment							
Concentration (mg/l)	0.587	0.613	0.573	0.462	+ 4.5	- 2.3	-21.3
Flux (gm/min)	0.0009	0.0007	0.0006	0.0004	-22.2	-33.3	-55.6
Paraquat Loss on Sediment							
Concentration (mg/l)	50.80	53.50	54.08	55.74	+ 5.3	+ 6.5	+ 9.7
Flux (gm/min)	0.315	0.240	0.231	0.185	-23.8	-26.5	-41.3
Total Nitrogen Flux (gm/min)							
In Solution	7.39	9.62	6.67	6.50	+30.2	- 9.7	-12.0
On Sediment	2.10	2.15	2.17	2.37	+ 2.4	+ 3.3	+12.9
	5.59	7.70	4.68	4.10	+37.7	-16.3	-26.7
Total Phosphorus Flux (gm/min)							
In Solution	2.48	2.10	2.03	1.77	-15.3	-18.1	-28.6
On Sediment	0.173	0.190	0.174	0.154	+ 9.8	+ 0.6	-11.0
	2.36	1.95	1.90	1.62	-17.4	-19.5	-31.4

[a] Area beneath the corresponding frequency curves obtained from 10-yr simulations on the P2 watershed in Watkinsville, Georgia.

CONCLUSIONS

The intent of this paper is to demonstrate the use of the ARM model in analyzing the relative effectiveness of alternative SWCPs for controlling runoff, sediment, and pesticide and nutrient loss. The specific conclusions of the analysis should be considered preliminary pending further analysis and re-evaluation of the assumptions and model limitations. However, the results presented here provide insight into the important characteristics of the various SWCPs studied and how these practices can be represented in a modeling approach. This work has also delineated areas where data collection and future research are needed to improve existing tools and analytical methods of evaluation.

In general, the SWCPs analyzed in this work—minimum tillage, contours, and terraces plus contours—reduce surface runoff, sediment loss, and the associated constituent loads. However, notable exceptions are evident for specific practices. Conditions that reduce and retard overland flow tend to increase subsurface flow components. Interflow increased for each SWCP with a major increase of up to 30% when terraces are present. The implication for pollution control is that soluble pollutant contributions can increase as noted by substantial increases in solution nitrogen (mostly NO_3). Soluble NH_4 and PO_4 did not demonstrate major changes with any SWCP. These nutrients increased slightly with minimum tillage and decreased slightly for contours with and without terraces. Also, atrazine loss, which occurs mostly in solution, decreased with each SWCP, up to a 60% reduction for terraces with contours. Thus, the adsorption characteristics appear to control the amount of reduction of soluble pollutants, while nonadsorbing components such as NO_3 can substantially increase with subsurface flow. The end result in this study was that solution nitrogen loss increased with each SWCP; soluble atrazine (and total atrazine) loss decreased with each SWCP; and solution phosphorus loss increased with minimum tillage, did not change with contours, and decreased with contours and terraces.

Sediment and associated constituents decreased substantially with each SWCP, except minimum tillage. Nutrient contributions from crop residues remaining on the watershed as part of the minimum-tillage practice need to be further evaluated and analyzed.

In this study, terraces with contours is the most effective SWCP for most constituents, followed by contours and minimum tillage in that order. However, the individual practices are defined specifically for this study based on conventional practices in the southeastern U.S. Various combinations of the SWCPs are often used, and the specific names are applied to widely differing practices across the country.

Parameter changes in the ARM model to represent the alternative SWCPs were based on information in the literature and our own judgment and experience in applying the ARM model on other watersheds. When little or no information was available, we tended to be conservative in the parameter adjustments so as not to overstate the impacts of the different practices. Also, limitations within the model and lack of data prevented consideration of other effects such as changes in infiltration characteristics, crop growth, nutrient transformation, etc. Research is needed to define relationships between model parameters and measureable watershed characteristics, and additional data collection is needed to characterize the impacts of various SWCPs as a basis for parameter adjustments.

Although the ARM model is by no means exact in its representation of all soil processes, this work has shown that the model can be used to analyze the relative effects of some alternative SWCPs. Further research is needed to better represent erosion processes, the effects of tillage operations, the transport of soluble substances, pesticide adsorption and degradation mechanisms, and nutrient transformations. However, for most processes the model includes our best current understanding of the important mechanisms. With an awareness of the changes affected by a SWCP and the representation of these changes by the model, the important effects of a SWCP can be investigated.

REFERENCES

1. Donigian, A. S., Jr., and N. H. Crawford. "Modeling Pesticides and Nutrients on Agricultural Lands," EPA 600/3-76-043. Environmental Research Laboratory, Environmental Protection Agency, Athens, GA (1976).
2. Donigian, A. S., Jr., D. C. Beyerlein, H. H. Davis and N. H. Crawford. "Agricultural Runoff Management (ARM) Model Version II: Refinement and Testing," EPA 600/3-77-098. Environmental Research Laboratory, U.S. Environmental Protection Agency, Athens, GA (1977).
3. Crawford, N. H., and A. S. Donigian, Jr. "Pesticide Transport and Runoff Model for Agricultural Lands," EPA 660/2-74-013, Southeast Environmental Research Laboratory, U.S. Environmental Protection Agency. Athens, GA (1973).
4. Crawford, H. N., and R. K. Linsley. "Digital Simulation in Hydrology: Stanford Watershed Model IV," Technical Report No. 39. Department of Civil Engineering, Stanford University, Stanford, CA (1966).
5. Hydrocomp, Inc. "Hydrocomp Simulation Program Operations Manual," 4th ed., Palo Alto, CA (1976).
6. Negev, M. A. "A Sediment Model on a Digital Computer," Technical Report No. 76, Department of Civil Engineering, Stanford University. Stanford, CA (1967).
7. Meyer, L. D., and W. H. Wischmeier. "Mathematical Simulation of the Process of Soil Erosion by Water," Trans. Am. Soc. Agric. Eng. 12(6): 754-758,762 (1969).
8. Onstad, C. A., and G. R. Foster. Erosion Modeling on a Watershed," Trans. Am. Soc. Agric. Eng. 18(2):288-292 (1975).

9. Davidson, J. M., R. S. Mansell and D. R. Baker. "Herbicide Distributions Within a Soil Profile and Their Dependence Upon Adsorption-desorption," Soil Crop Sci. Soc. Florida Proc. (1973).
10. Mehran, M., and K. K. Tanji. "Computer Modeling of Nitrogen Transformations in Soils," *J. Environ. Qual.* 3(4):291-395 (1974).
11. Hagin, J., and A. Amberger. "Contribution of Fertilizers and Manures to the N and P Load of Waters. A Computer Simulation," report submitted to the Deutsche Forschungs Gemeinschaft (1974).
12. Stewart, B. A., D. A. Woolhiser, W. W. Wischmeier, J. H. Caro and M. H. Frere. "Control of Water Pollution from Cropland, Volume I: A Manual for Guideline Development," Agricultural Research Service. U.S. Department of Agriculture, Washington, DC ARS-H-5-1. EPA-600/2-75-026a (1975).
13. Chow, V. T. *Handbook of Applied Hydrology.* (New York: McGraw-Hill Book Co., 1964).
14. Smith, C. N., R. A. Leonard, G. W. Langdale and G. W. Bailey. "Transport of Agricultural Chemicals from Small Upland Piedmont Watersheds," U.S. Environmental Protection Agency, Athens, GA, and U.S. Department of Agriculture, Watkinsville, GA, final report on Agreement No. D6-0381 (1977).
15. Fleming, G., and K. M. Leytham. "The Hydrologic and Sediment Processes in Natural Watershed Areas," Proceedings of the Third Federal Inter-Agency Sedimentation Conference, Denver, CO (March 22-25, 1976).
16. Hydrocomp, Inc. "Sediment Loading and Agricultural Practices in the Kishwaukee and Thorn Creek Watersheds," prepared for the Northeastern Illinois Planning Commission, Hydrocomp, Inc. Palo Alto, CA (1978).

45

INTERACTIVE EFFECTS OF PESTICIDE PROPERTIES AND SELECTED CONSERVATION PRACTICES ON RUNOFF LOSSES: A SIMULATION STUDY

J. D. Dean, L. A. Mulkey

Technology Development and Applications Branch
Environmental Research Laboratory
U.S. Environmental Protection Agency
Athens, Georgia

INTRODUCTION

Impairment of water quality by substances from nonpoint sources is a problem of acute interest in today's environmental studies. The augmentation of pesticide and fertilizer usage in American agriculture over the past few decades[1] has caused an increased concern over the potential damage that these substances may do in ecological systems. This concern is reflected by legislation such as the Federal Water Pollution Control Act Amendments of 1972 that stress the need to properly manage pesticides in agriculture and silviculture and provide stimuli for research into methods to prevent or curtail pollution from such activities.

Even a superficial investigation indicates that the nonpoint source problem does not lend itself to a simple, straightforward solution. The many avenues by which pesticides may move from the application site imply many methods of control. Each of these control techniques is associated with varying effectiveness and varying costs dependent on regional physiography, hydrology, and economic climate. Across-the-board implementation of certain management practices may result in over- or under-design of pollution controls or a less-than-ideal economic strategy in certain areas. The alternative to a universal control approach is selection of best management practices for agriculture and silviculture on a local basis. In fact, the concept of local evaluation is intrinsic to the definition of a best management practice (BMP). Pisano[2] provides this definition:

> The term 'Best Management Practice' refers to a practice that is determined by a state after examination of alternative practices that are considered practicable and most effective in preventing or reducing the amount of pollution generated by a nonpoint source to a level compatible with water quality goals. According to this approach, each state defines its own set of Best Management Practices that will be tailored to meet the specific problems and environmental conditions within that state.

Physical characteristics, economics and hydrology are only three among many factors that confound the selection of BMPs. Basic pesticide properties, for example, may have a profound influence on mechanism and magnitude of pollutant losses from the watershed.[3-6] Other varaibles such as agricultural chemical formulation, application timing, and mode of application can also affect runoff losses. Factors affecting transport of agricultural chemicals from application sites can generally be grouped into five categories as shown in Table I. The categories are not necessarily exclusive and some factors may fit equally well in more than one. The intention is not to rigorously define or classify these variables, but rather to indicate the enormous complexity encountered in selecting optimal practices.

Table I. Some factors affecting magnitude and mode of pesticide losses from agricultural watersheds.

Category		Factor
I	Pesticide Properties	Solubility
		Affinity for sediments
		Persistence
II	Climate	Timing of precipitation
		Magnitude of events
		Temperature
		Radiation
III	Watershed	Slope
		Roughness
		Management practices
IV	Soil	Texture
		pH
		Moisture status
		Hydraulic properties
		Organic content
V	Miscellaneous	Timing of application
		Mode of application
		Formulation
		Crop type

The number of pesticidal variables to be considered is large. For example, a 1973 publication lists more than 500 chemicals then used in Canadian crop protection.[7] Leonard et al.[5] have reported that more than 500 pesticides having some 8,000 formulations are commercially available in the United States. Each has unique properties that may directly affect the mode and magnitude of loss. Moreover, the influence of formulation may result in even greater diversity of losses.

Physiographic variation induces another set of parameters. An Agricultural Research Service-Environmental Protection Agency manual[8] shows the 156 different land resources areas identified by the Soil Conservation Service (SCS). These areas are delineated according to soils, topography, farming practice, land use and climate. Within each area, great diversity still exists among soil type, topography and land use.

Given this rudimentary estimate of variables involved in predicting pesticide runoff losses, a hypothetical example illustrating the extent of the evaluation problem is contrived. Suppose that from the 500+ pesticides available, 20 categories can be recognized with similar properties such as those in category I of Table I. A characteristic compound from each of the 20 groups is chosen to represent the spectrum of the 500. Suppose also that three formulations for each representative compound are to be studied. Of the 156 land resource areas, 100 are assumed to contain potentially arable land. Suppose further that three watersheds can be chosen that have characteristics indicative of the entire resource areas, and that five different conservation practices are to be implemented on each during the course of the investigation. Further suppose that 10 different crop species are to be grown on each of the watersheds, one per experiment. Ten years of concurrent meteorological data might be considered to be sufficient to characterize the effects of climate on each experiment. How many years would be required to assess the effects of all the above variables on pesticide losses from agricultural activities? Even this restricted variable list requires 9.0×10^6 years of data collection. If the three watersheds in the 100 resource areas could be operated simultaneously, it would still require 30,000 years to obtain sufficient data to quantify the effects of the above variables on pesticide loss. Even then, the effects of applying some management practices simultaneously are not taken into account.

Obviously, the resources required for experimental analysis of the effects of all possible management practices are prohibitively large. This is not to suggest that studies of this nature are folly, because much useful information can be derived from them. The fact remains, however, that a detailed experimental study extensive enough to evaluate the effects of even a meager portion of the variables concerned is infeasible. Fortunately, viable alternatives exist for such evaluations. Comprehensive computer models capable of continuously simulating the hydrology of agricultural watersheds, as well as the transport of sediments, pesticides and plant nutrients have become available in recent years.[9-11] The Agricultural Runoff Management Model (ARM) developed by Hydrocomp, Inc., for the U.S. Environmental Protection Agency (EPA) is one such model. It has been used in the study subsequently discussed to demonstrate the utility of models in assessing the interactive impact of conservation practices, pesticidal properties, and climate on the transport of chemicals from agricultural watersheds.

OBSERVED MANAGEMENT IMPACTS

Relatively little research has been done to evaluate the quantitative impact of management practices on pesticide losses from agricultural lands. Perhaps the reason for this is an inability to control many of the factors (especially climatic) that affect transport. A striking characteristic of reported data is inconsistency of results among the different studies. This may be directly attributed to noise caused by uncontrollable variation in the hydrologic cycles that partially govern transport or to incomplete interpretation of the many interactive processes involving the pertinent variables.

Effect of Conservation Practices

The use of conservation practices has been advocated as a means to reduce the total volume of runoff water and sediment losses from watersheds. Because pesticides are lost mainly in runoff water or on sediment (suspended or wash load), it seems reasonable that reduction of overland flow and sediment transport are appropriate for nonpoint source reductions.

Trichell et al.[15] however, reported that runoff losses of 2,4,5-T, picloram, and dicamba were greater from sod plots than from tilled plots under simulated rainfall. Harvey et al.[13] studied the loss of atrazine following conventional tillage, minimum tillage (wheel track method), and no-till. Losses were reportedly greater from the minimum-tillage plots although no "major" differences in losses from the different practices were evident. Smith and associates[14] reported greater losses from no-till than from conventional tillage under natural rainfall conditions. Triplett et al.[12] on the other hand, reported less runoff losses of atrazine and simazine from no-till methods than from tilled watersheds. Asmussen et al.[16] studied runoff losses from plots with grassed waterways, but no comparison was made for the same plots with no grassed waterways.

Perhaps the most comprehensive study recently undertaken to evaluate the impact of management practices on pesticide runoff losses was done at Watkinsville, Georgia, from 1972 through 1975.[17] This project was initiated as a joint effort between the U.S. Department of Agriculture (USDA) and EPA to provide a data base for the development and testing of models for describing pesticide and nutrient transport from agricultural lands. In this 42-month study, seven different pesticides were used for three different crop species. Losses from terraced fields with grassed waterways were compared with losses from those in which no conservation practices were used.

Among the conclusions were that sediment yields were significantly less from watersheds with terraces than from watersheds managed without terraces. Except for one pesticide, paraquat, pesticide yields were not reduced in proportion to sediment yields. The specific impact of terracing and grassed waterways over no conservation practices was very difficult to evaluate because the sizes of the watersheds differed, significant differences in rainfall depth occurred between watersheds, and the duration of the meteorological sequence was too short to evaluate the effect of climate on practice efficiency.

Application Rates

Generally conclusive results exist to define the impact of changing application rates. In all cases reviewed, pesticide runoff increased with application rate. Lutz et al.[18] reported that doubling the application rate of picloram and 2,4,5-T (from 2.24 kg/ha to 4.48 kg/ha) doubled the amount of pesticide available for runoff after 15, 50 and 100 days. Trichell et al.[15] found that losses of 2,4,5-T, picloram, and dicamba were directly related to application rate. They used simulated rainfall (1.3-cm depth) 24 hours after application in their investigation. Barnett et al.[19] also found that doubling application rates resulted in twice the runoff losses of 2,4-D from cultivated fallow land. In two studies[20,21] in Pennsylvania on 14% slopes, a positive correlation was found between application rate and loss of atrazine in runoff.

Application Date Management

Precipitation timing and intensity are instrumental in producing significant pesticide losses. Barnett et al.[19] found a relationship between 2,4-D concentration in runoff water and rainfall intensity and duration. The study was conducted using simulated rainfall on plots 24 hours after application. On the other hand, Triplett et al.[12] found precipitation timing and amount to be insignificant when attempting to predict losses. The difference here is resolved by the concept of pesticide availability. In the former study, transport of the chemical was not in a mass-limiting stage when the event occurred. In the latter study, the pesticides used, namely atrazine and simazine, have relatively short half-lives. When the rainfall events occurred later in the season, the pesticide available for transport was limited and the precipitation inputs had relatively minor effects on total seasonal losses. For pesticides with longer half-lives, losses should be more highly correlated with precipitation inputs. For pesticides with short half-lives, the timing of application is extremely important. Smith et al.[17] found pesticide loss to be correlated with the time between application and the first runoff event. Many researchers have found that the greatest losses occur during the first several events after application, making timing of application a potentially important management factor. Leonard et al.[5] give a rather extensive listing of references containing such results.

Soil antecedent moisture should also be an important factor determining magnitude of losses because soil moisture affects the partitioning of rainfall into overland flow and infiltration water. Rainfall events that occur in combination with wetter soil moisture conditions should potentially produce greater pesticide losses. Davidson et al.[22] found greater losses of fluorometron from wet than from dry soils. Asmussen et al.,[16] however, found that of the 2,4-D lost from plots, about 30% reached the end of a grassed waterway, regardless of the antecedent conditions.

Not only does soil moisture affect the distribution of rainfall into overland flow, infiltrate, etc., but to an extent it affects partitioning of pesticides between the soil and water phases. Leonard et al.[5] point out that water competes effectively for sorption sites preventing extensive adsorption of chemicals to the soil phase and modifying the mobility of the compound in the soil profile. This, in turn, changes the chemical's distribution in the soil profile and its availability for transport in surface runoff or on detached sediments.

Obviously, antecedent moisture conditions can be managed by adjusting the pesticide application date, a consideration that makes timing a doubly important management variable. The need for timely control of target pests, however, may mitigate against timely application to limit runoff losses.

Miscellaneous Practices

Effects of some other practices have also been studied. For example, DDT and toxaphene losses were compared from plots receiving each alone and in combination. Toxaphene applied alone had the least losses, next was loss from DDT applied in combination, then toxaphene in combination. Highest losses occurred from DDT applied alone. Applications were made 12 times during the growing season on cotton. Variations in losses were attributed to differences in formulation.[23]

The effects of crop rotation on losses of DDT, endrin and endosulfan were studied by Epstein and Grant.[24] Losses from a three-crop rotation system of potatoes, oats and sod were less than from continuous potatoes.

The foregoing discussion elucidates several facts:

1. Many of the field pesticide runoff data are inconclusive. Practices implemented in one area give contradictive results when applied in another area. Although some of the paradoxes are explainable, too many have not been adequately resolved.
2. The lack of consistency or universality in the results prevents reasonable extrapolation to other sites.
3. Field data, because of the prohibitive time and costs involved, are necessarily collected on a short-term basis (usually 1 to 2 years) making the effects of hydrologic variation (in time or space) very difficult to discern among effects of other factors.

That field studies alone cannot serve to evaluate the effects of different management strategies is obvious. A methodology is needed whereby the effects of certain variables or combinations of variables can be quantified as to their individual influences on the overall management impact. The need is also apparent for a methodology that considers the stochastic nature of climate together with the deterministic transformation provided by the watershed to evaluate the risk of failure of a given practice or set of practices under a range of conditions.

SIMULATION MODELING ALTERNATIVE

Simulation modeling offers an attractive alternative to the monumental experimental task outlined above. The approach is particularly useful for the study of pesticide runoff because of the numerous compounds for which such information may be desired. Moreover, because new pesticiees have been introduced in the relatively recent past and because new compounds are continually being developed for future use, it is unlikely that long periods of record will ever be available for the more conventional analysis afforded by experimental studies. Indeed, modeling may well be the only alternative for continuing evaluation of pesticide behavior in the environment.

Existing simulation models for estimating pesticide runoff are parametric. That is, the pertinent physical, biological and chemical processes governing pesticide behavior are represented by sets of equations having parameters representing certain of the environmental or pesticide variables. These parameters must be estimated before meaningful results can be expected from the models. One may argue that if parameter estimation or calibration is required before a model is used, then no real advantage is gained because the experimental studies noted earlier are still necessary. For parameters representing variables that tend to be uniquely site-specific, this is indeed the case. Fortunately, many of the pesticide behavioral characteristics can be elucidated by short-term laboratory studies. For example, the sorptive properties of pesticides can be evaluated for given soils using rather well-entrenched techniques. Similarly, persistence data can be obtained in a rather straightforward manner.

The parameter estimation effort now reduces to the more conventional problem of hydrologic and sediment transport calibration. Here field measurements are required, but often historical data exist and the cost incurred for collection of new data is not prohibitive. Once a model has been field-calibrated for hydrology and erosion/sediment transport, the behavior of any pesticide can then be simulated using laboratory-determined parameters for other factors. In many cases, reasonable pesticide parameter values can be inferred from existing data because of the multiplicative use of the basic information. For example, persistence is of interest in pesticide efficacy studies and is usually reported for each existing and proposed compound.

Thus far, the steps involved in simulating behavior of a single pesticide have been suggested. The complicating factor introduced by the large numbers of compounds was noted earlier—the problem remains. Sensitivity analysis performed by the simulation models offers at least a partial solution. If for a "typical" set of watershed management and physiographic conditions, the interaction of the pesticide parameters are evaluated, then subsequent analysis of the model output by "similarity" can be attempted. That is, the properties of a specific pesticide can be compared with similar values in the sensitivity analysis and the results as estimated by the model follow.

The rationale developed thus far can be summarized into a series of steps as follows:

1. Select a parametric pesticide runoff model.
2. Estimate model hydrologic and sediment transport parameters from measured data.
3. Select the type and range of pesticide properties of interest.
4. Perform model sensitivity analysis.
5. Develop model output graphs, nomographs, etc.
6. Evaluate selected impacts of selected compounds by comparison with model output.

The above tasks appear quite simple in concept. In fact, few can presently be accomplished without considerable uncertainty in the results. The questions of model accuracy, data collection errors, computational requirements, and user skills remain important. A considerable effort is ongoing (including experimental watershed studies) to answer these questions for the model used in the study described below. The methodology outlined above is valid. Precision and accuracy of the results remain somewhat ill-defined.

MODEL IMPLEMENTATION FOR SIMULATION STUDY

A simulation study was initiated to illustrate the potential of modeling for evaluating the effects of changing pesticidal, watershed, and hydrologic variables on runoff losses from agricultural watersheds.

Data from watersheds near Watkinsville, Georgia, in the southern Piedmont agricultural region, were used. A complete description of the topography, soils, and other pertinent watershed data may be found in Smith et al.[17] The data were used for parameter estimation in the ARM model. Calibration was accomplished via a procedure described by several other researchers.[10,25-27]

The simulated effects of several variables were studied. The first was pesticide persistence as represented by half-life. The second was pesticide solubility and affinity for sediment. A third effect studied, although not a directly controllable varaible, was that of temporal weather variations on pesticide loss. Finally, the synergistic effects of all three of the foregoing and two different conservation practices were investigated.

Watershed Parameters

A recognized difficulty in assessing effects of management practices on pesticide runoff losses from paired watersheds is the areal difference. For this study, P-03 watershed, which was managed with terraces and grasses waterways, was scaled up to the size of P-01 watershed, which was managed with contouring only. The actual size ratio is 2.70 ha to 1.26 ha, or 2.14:1. Therefore, the area of P-03 was expanded by 2.14 to match the area of P-01. All other parameters remained as per calibration except L, the length of the overland flow plant. A similar scaled-up value for L based on the square root of the area scale was 98.09 m. The maximum recommended value for L is 91.4 m, so the latter value was used. Total runoff has been shown to be fairly insensitive to L (for example, a 40% change in L resulted in a 4% change in total runoff.)[10] Therefore, this perturbation from the actual value of 98.00 (about 7%) should make inconsequential differences in the model output.

Meteorological Sequences

Another difficulty in paired watershed studies is the areal variation in the rainfall that may be indicated by point measurements on each site. To eliminate this bias and ensure equitable effects of meteorology on each watershed, 5-minute rainfall and pan evaporation data measured on watershed P-01 was used on both watersheds in the simulation study.

Pesticide Parameters

Parameters for three different pesticides were used in this study. The adsorption/desorption parameters are listed in Table II. Parameter CMAX is the maximum solubility of the compound in water. DD is the permanent fixed capacity on the soil, K is the coefficient, and N is the exponent in the Freundlich adsorption isotherm. A graph of the isotherm shapes is shown in Figure 1.

Table II. Adsorption-desorption parameters for three mock pesticides.

Parameter	Value		
	I	II	III
CMAX (lb/lb)	0.00001	0.00026	0.00026
DD (lb pesticide/lb soil)	0.0003	0.0	0.0
K	–	10.7	1.8
N	–	1.0	1.6

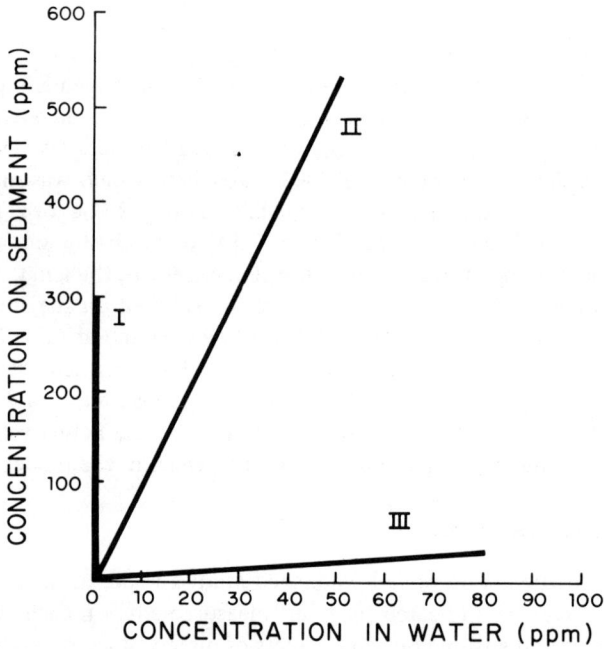

Figure 1. Adsorption isotherms for three mock pesticides.

The selection of these three isotherms was made in an attempt to cover a wide spectrum of sorptive properties. Compound I is essentially totally and irreversibly sorbed on the sediment. Therefore, K and N have no meaning and are omitted from the table for this compound. Compound III has a very weak affinity for the soil, and Compound II is moderately sorbed. The three mock pesticides used bear great resemblance to some marketed pesticides except for the fact that their half-life is variable. Therefore, the names were changed because they do not, as such, represent any one specific chemical.

The half-lives used in this study cover a range in which most pesticides in use today are assumed to fall. Half-life is the amount of time required for half of the pesticide mass to be removed from the watershed assuming first-order kinetics. This property is highly variable based on pesticide chemical makeup and watershed hydrology. Maximum persistence for several classes of compounds are given by Stewart et al.[8] A maximum persistence of 18 months for urea or trianzine herbicides (C/C_0 is the mass percentage remaining on the watershed. On the other hand, organophosphorus insecticides have maximum persistence of about 3 months corresponding to a half-life of 3.5 weeks. Donigian and Davis[27] report some half-lives as low as 2 days. Therefore, the ranges picked for this study (2 days to 24 weeks) seem adequate.

All compounds were applied in equal amounts (7.5 kg/ha).

RESULTS OF SIMULATION STUDY

Table III shows the annual average runoff losses for two conservation practices, three compounds each, as a function of pesticide half-life. Several general observations can be made. For both land treatments and all compounds, runoff losses increase logarithmically with increasing pesticide half-life. This indicates that as half-life increases beyond 6 months, transport-limited, as opposed to pesticide-limited, annual losses begin to occur. The effectiveness of conservation practice decreases with decreasing affinity of the compound for sediments. For strongly adsorbed compounds, the losses were reduced by a factor of two or better, whereas for weakly sorbed compounds, the reduction was minimal and sometimes even negative. Within a given practice, runoff losses decreased with decreasing affinity for sediments. This means that application of compounds that are weakly sorbed (very mobile) will result in a reduction in losses over compounds that travel on the sediment. An exception to the rule here is indicated with terracing and grassed watersheds for Compounds I and II. For short half-lives, the moderately sorbed compound (II) yielded greater losses than the strongly sorbed compound (I). This probably reflects the fact that heavy sediment-producing events did not occur within the pesticide's half-life after application yielding low losses for chemicals moving on the sediment. Significant runoff did occur, however, moving a greater portion of the moderately sorbed compound from the watershed.

Effect of Conservation Practices

The conservation practices studied were contouring and contouring with terraces and grassed waterways. For Compound I, a pesticide moving totally on the sediment, the percentage of loss was decreased from 33.62% of amount applied to 14.74% (3-year average) for contouring and terracing with

Table III. Summary of annual average runoff losses for selected compounds (% of total applied).

weeks	Practice					
	Contour Only			Terraces & Grassed Waterways		
	Compound			Compound		
	I	II	III	I	II	III
2/7	–	0.49	0.17	0.04	0.34	0.11
1	–	2.97	0.51	0.72	2.19	0.38
2	6.62	4.78	0.72	2.35	3.70	0.60
4	14.32	6.51	1.04	5.49	5.19	0.96
6	19.68	7.34	1.29	7.81	5.93	1.22
8	23.27	7.77	1.43	9.40	6.34	1.39
12	27.81	8.31	1.63	11.48	6.81	1.64
16	30.54	8.59	1.77	12.74	7.06	1.79
20	32.35	8.77	1.85	13.59	7.22	1.88
24	33.62	8.89	1.93	14.74	7.34	1.95

grassed waterways, respectively, at the 24-week half-life. For Compound II, a moderately adsorbed pesticide, the conservation practice reduced losses from 8.89 to 7.34%. The reduction effected by conservation practices on pesticide loss of Compound II was practically nil. In fact, losses were slightly greater for the watershed on which terraces and grassed waterways had been implemented. An interesting effect produced by the implementation of conservation practices is shown in Figures 2 and 3. Both figures represent the partitioning of pesticide loss into overland and interflow for a weakly sorbed compound.

The overland flow losses are significantly greater on P-01 than on P-03. When conservation practices are implemented, interflow losses increase with a corresponding decrease in overland flow losses. This agrees with intuition that overland flow losses would be decreased by conservation practices. It is, however, important to note that interflow losses might make up the difference and essentially no benefits would result.

Effect of Half-Life

For all three pesticides, on both watersheds, the percentage of loss increased linearly with the logarithm of the half-life of the pesticide. Figure 4 shows this relationship very clearly. This general trend is to be expected because pesticide availability should be highly correlated with annual losses. It should also be noted from Figures 2 and 3 that half-life has a definite effect

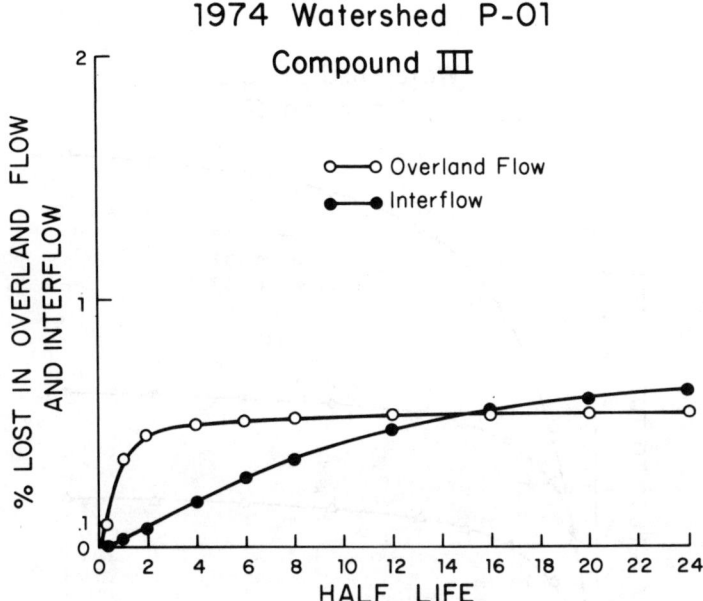

Figure 2. Annual pesticide loss modes from a contoured watershed.

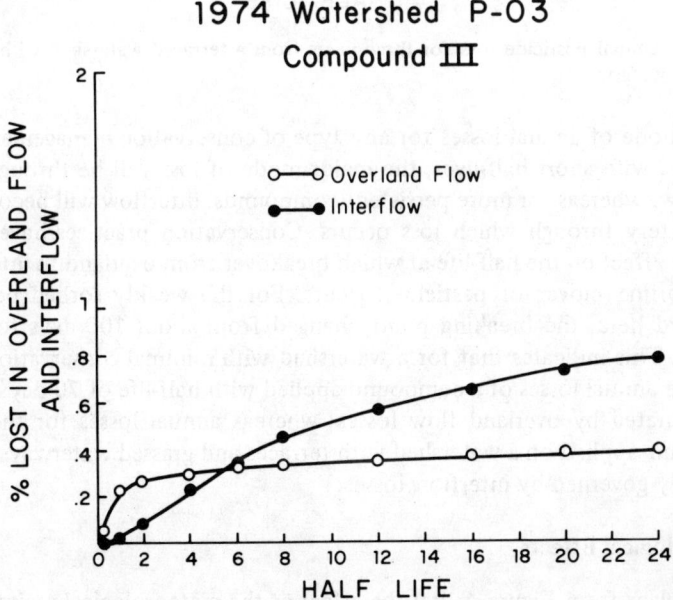

Figure 3. Annual pesticide loss modes from a terraced watershed with grassed waterway.

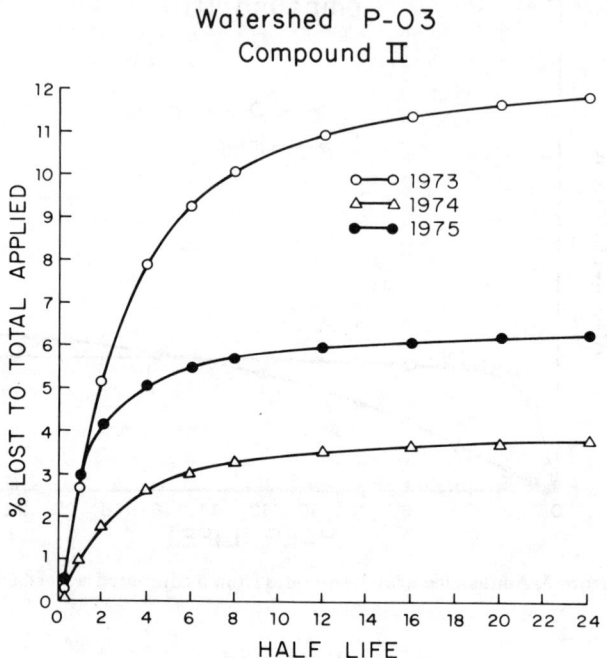

Figure 4. Annual pesticide loss for three years from a terraced watershed with grassed waterway.

on the mode of annual losses for any type of conservation management. For pesticides with short half-lives, the major mode of loss will be through overland flow, whereas for more persistent compounds, interflow will become the major artery through which loss occurs. Conservation practices have a pronounced effect on the half-life at which breakover from overland to interflow, as the prime mover of pesticide, occurs. For the weakly sorbed pesticide illustrated here, the breaking point changed from about 100 days to about 45 days. This indicates that for a watershed with minimal conservation practices, the annual losses of a compound applied with half-life of 70 days would be dominated by overland flow losses; whereas annual losses for the same compound applied on a watershed with terraces and grassed waterways would be heavily governed by interflow losses.

Meteorological Effects

It is clear from Figure 4 that the effect of the meteorological regime has a profound consequence on annual loss. A near 300% difference is noted between the 1973 and 1974 curves, all variables held constant except for application date (which differed by two calendar days) and hydrologic event sequencing. This difference seems to be largely independent of half-life. In 1973, an event occurred 2 days after application, whereas in 1974, a 6-day lag occurred between application and the next event. Also, the magnitude of the first event after application in 1973 was 1.52 cm, but the corresponding event in 1974 produced only 0.23 cm of rainfall. In 1975, a precipitation event occurred 1 day after application with a magnitude of 0.64 cm. Losses of compounds that were weakly sorbed reflected the variability in the climatic influence much more so than more strongly sorbed compounds. Comparison of Figures 4 and 5 very clearly indicates this relationship. Conservation practices had very minor attenuating effects on the loss variability resulting from meteorologic sequence.

Effect of Managing Pesticidal Properties

In Figure 5, all losses for Compound I were associated with the sediment. Figure 3 showed the interaction of the overland flow and interflow modes for a weakly adsorbed compound (III), the sediment associated losses being negligible. Figure 6 shows the modes of annual loss for a moderately sorbed

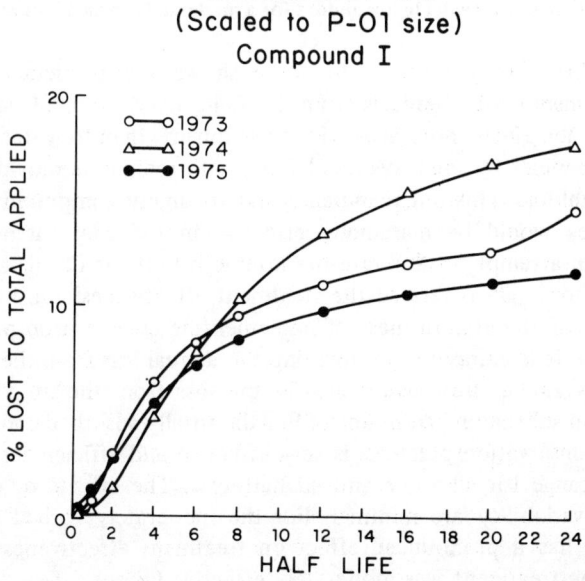

Figure 5. Annual pesticide loss for three years from a terraced watershed with grassed waterway.

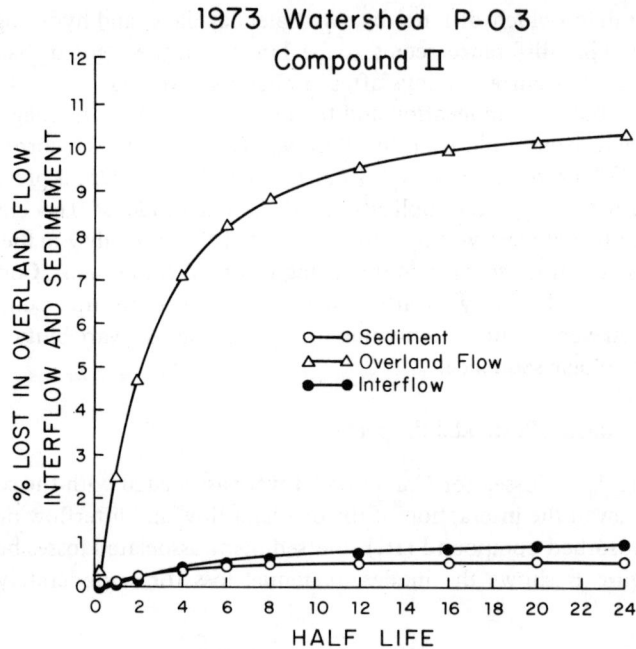

Figure 6. Annual pesticide loss modes for a moderately sorbed compound.

compound (II). Comparison of the three shows how pesticidal properties affect the transport of chemicals from the field. Even though Compound III is more than ten times more attractive to sediments than to water, the prime mover of chemicals by far is overland flow, with sediments moving less than 1% of the total loss. This might indicate that for many compounds conservation practices would be marginally effective in reducing chemical losses. Figure 7 is an attempt to illustrate the interactive effects of all the variables studied. The ordinate is termed the treatment effectiveness ratio (TER) and is a measure of the effectiveness of implementing conservation practices on pesticide loss. It is computed by dividing the annual loss from the watershed with terraces and grassed waterways by the loss from the no-conservation watershed and subtracting from unity. For the totally adsorbed compound (I), the use of conservation practices is very effective and efficiencies are in the 55 to 70% range for all years and all half-lives. The effects of climate on annual loss variability are minimal. For the moderately sorbed compound (II), climate has a pronounced effect on treatment effectiveness. In 1974 and 1975, the treatment was moderately effective (about 30%), but in 1973 the meteorologic sequence was such that effectiveness of terracing with grassed

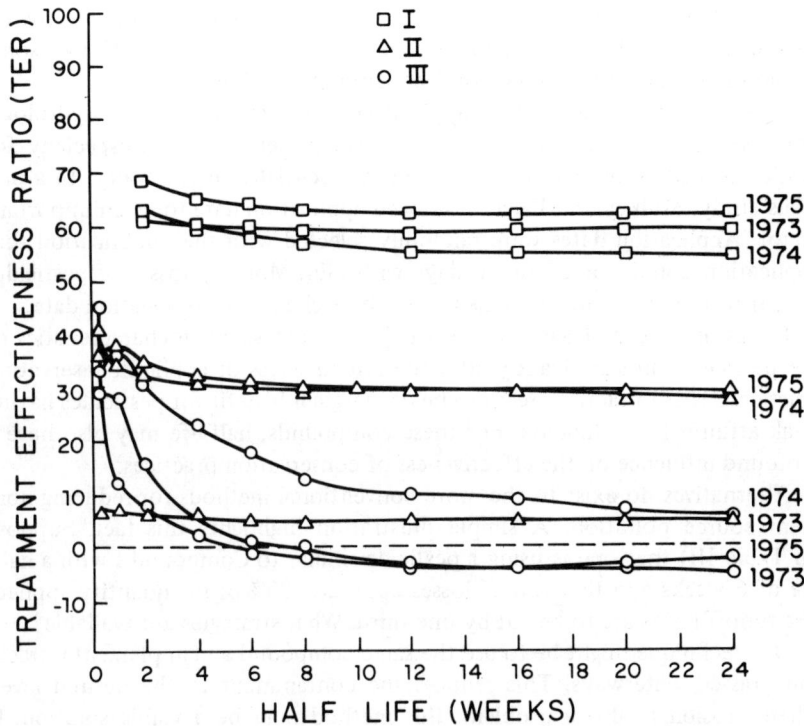

Figure 7. Effectiveness of implementing terraces and grassed waterway for three compounds.

waterway implementation was reduced to about 6%. Half-life had only small influence on effectiveness. For the weakly adsorbed pesticide, Compound III, half-life had signficant effects on the overall loss reduction being moderate for short half-lies and decreasing as half-life lengthened. In combination with certain hydrologic regimes, the TER was even negative for this compound at longer half-lives.

IMPLICATIONS FOR PESTICIDE BMPs

Climate, pesticidal properties, and conservation practices form a complex dynamic continuum that influences the effectiveness of any selected practice or set of practices that might be implemented to control losses of chemicals from agricultural watersheds. As in any decision involving stochastic phenomena, a reasonable estimate should be made of the risk of failure involved with the implementation of such practices. Because of the short period of record, such probabilities have not been determined herein. This study sets some guidelines for a methodology on which to evaluate the effectiveness of certain management practices. The results have shown some possibilities for analysis of data derived from model simulation.

It appears that selection of application date with respect to hydrologic event sequencing may be a promising management variable, especially for pesticides with short half-lives that move essentially in the water. In a previous study, Mulkey and Falco[28] showed application date to be an important factor. Application dates were randomly selected with the qualification that application could not occur on days with rain. Monthly losses of pesticides were shown to be extremely sensitive to these changes in application date.

It has also been shown in this study that the sorptive characteristics of the pesticide can significantly alter the effectiveness of applied conservation practices. These practices seem to be of marginal benefit for pesticides having weak affinity for sediments. For these compounds, half-life may also have a profound influence on the effectiveness of conservation practices.

Alternatives do exist to the more conventional methods for reducing nonpoint source pollution. A simple illustration highlights this fact. Suppose (in Table III) that one is using a pesticide similar to Compound I with a half-life of 6 weeks and that runoff losses approach 20% of the quantity applied. The runoff losses are to be cut by one-third. What strategies are available?

One technique might be to use the same compound and implement terraces and grassed waterways. This reduces the contaminant to the desired level. Where erosion is also a problem, this method may be a viable solution. If this is not the case, however, other less costly alternatives are available. For instance, one might apply a compound with the same sorptive properties having a shorter half-life, which involves no capital outlay. A reduction from 6 to 2 weeks would be sufficient in this case. This, however, may increase the number of spraying operations necessary during the growing season for some purposes. In that case, one might opt for a pesticide with less affinity for sediment (Compound II) with the same half-life. This would also accomplish the desired goal. Another method might be more judicious selection of application date. All of the alternatives to implementation of costly conservation practices involve little, if any, capital investment and produce consistently equitable results.

On the other hand, a system having acceptable losses with Compound II (terraces and grassed waterways) should avoid selection of compounds with longer half-lives or compounds that have greater attraction for sediments.

SUMMARY

It is clear that all the known factors that govern transport and magnitude of pesticide losses should be considered in the selection of BMPs. The use of continuous simulation hydrologic modeling is an alternative to costly and often inconclusive field data collection for this purpose. A methodology is suggested in which field data are used for calibration and parameter estimation, and models are used to evaluate risks of failure for a given management practice or set of practices. This, in turn, gives planners a viable set of options on which to intelligently base decisions for pesticide management.

REFERENCES

1. Caro, J. W. "Pesticides in Agricultural Runoff," in *Control of Water Pollution from Cropland, Volume II: An Overview,* Agricultural Research Service, USDA, Washington, DC, prepared under Interagency Agreement with the U.S. Environmental Protection Agency, Athens, GA, Publication No. EPA-600/2-75-026b (1975).
2. Pisano, M. A. "Nonpoint Source Pollution: A Federal Perspective," *J. Environ. Div. Am. Soc. Civil Eng.* 103(EE3):555-565 (1976).
3. Frere, M. H. "Adsorption and Transport of Agricultural Chemicals in Watersheds," *Trans. Am. Soc. Agric. Eng.* 569-572 (1973).
4. Baily, G. W., R. R. Swank, Jr., and H. P. Nicholson. "Predicting Pesticide Runoff from Agricultural Land: A Conceptual Model," *J. Environ. Qual.* 3(2):95-102 (1974).
5. Leonard, R. A., G. W. Bailey and R. R. Swank, Jr. In *Land Application of Waste Materials,* Soil Conservation Society of America, Ankeny, IA (1976).
6. Pionke, H. B., and G. Chesters. "Pesticide-Sediment-Water Interactions," *J. Environ. Qual.* 2(1):29-45 (1973).
7. Spencer, E. Y. "Guide to the Chemicals Used in Crop Protection," Agriclture Canada, Publication No. 1093 (1973).
8. Stewart, B. A. *et al.* "Control of Water Pollution from Cropland, Volume I: A Manual for Guideline Development," Agricultural Research Service, USDA, Washington, DC, prepared under interagency Agreement with the U.S. Environmental Protection Agency, Athens, GA, Publication No. EPA-600/2-75-026a (1975).
9. Bruce, R. R., L. A. Harper, R. A. Leonard, W. M. Snyder and A. W. Thomas. "A Model for Runoff of Pesticides from Small Upland Watersheds," *J. Environ. Qual.* 4(4):541-548 (1975).
10. Donigian, A. S., and N. H. Crawford. "Modeling Pesticide and Nutrients on Agricultural Lands," Hydrocomp, Inc., Palo Alto, CA, prepared for U.S. Environmental Protection Agency, Athens, GA, Publication No. EPA 600/2-76-043 (1976).
11. Frere, M. H., C. A. Onstad and H. N. Holtan. "ACTMO, An Agricultural Chemical Transport Model," U.S. Department of Agriculture, Hyattsville, MD, Report No. ARS-H-3 (1975).
12. Triplett, G. B., B. J. Conner and W M. Edwards. "Transport of Atrazine and Simazine in Runoff from Conventional and No-Tillage Corn," *J. Environ. Qual.* 7(1):77 (1978).
13. Harvey, R. G., A. E. Peterson, R. L. Higgins and H. W. Pawlson. "Influence of Tillage and Planting Practice on Erosion and Atrazine Runoff," *Weed Sci. Soc. Am. Abstr.* (1976).

14. Smith, G. E., F. D. Whitaker and H. G. Heineman. "Loss of Fertilizers and Pesticides from Claypan Soil," University of Missouri, Columbia, MO, prepared for U.S. Environmental Protection Agency, Athens, GA, Publication No. EPA-660/2-74-068 (1974).
15. Trichell, D W., H. L. Morton and M. G. Merkle. "Loss of Herbicides in Runoff Water," *Weed Sci.* 16:447-449 (1968).
16. Asmussen, L. E., A. W. White, Jr., E. W. Hauser and J. M. Sheridan. "Reduction of 2,4-D Load in Surface Runoff Down a Grassed Waterway," *J. Environ. Qual.* 6(2) (1977).
17. Smith, C. N., R. A. Leonard, G. W. Langdale and G. W. Bailey. "Transport of Agricultural Chemicals from Small Upland Piedmont Watersheds," (in press, 1978).
18. Lutz, J. F., G. E. Byers and T. J. Sheets. "The Persistance and Movement of Picloram and 2,4,5-T in Soils," *J. Environ. Qual.* 2(4):485-488 (1973).
19. Barnett, A. P., E. W. Hauser, A. W. White and J. H. Holladay. "Loss of 2,4-D in Washoff from Cultivated Fallow Land," *Weeds* 15:133-137 (1967).
20. Hall, J. K. "Erosional Losses of s-Triazine Herbicides," *J. Environ. Qual.* 3(2):174-180 (1974).
21. Hall, J. K., M. Pawlus and E. R. Higgins. "Losses of Atrazine in Runoff Water, Soil and Sediment," *J. Environ. Qual.* 1:172-176 (1972).
22. Davidson, J. M., G. H. Brasewitz, D. R. Baker and A. L. Wood. "Use of Soil Parameters for Describing Pesticide Movement Through Soils," Oklahoma State University, Stillwater, OK, prepared for U.S. Environmental Protection Agency, Athens, GA, Publication No. EPA-660/2-75-009 (1975).
23. Bradley, J. R., Jr., T. R. Sheets and M. D. Jackson. "DDT and Toxaphene Movement in Surface Water from Cotton Plots," *J. Environ. Qual.* 1(1):102-105 (1972).
24. Epstein, E., and W. J. Grant. "Chlorinated Insecticides in Runoff Water as Affected by Crop Rotation," *Soil Sci. Soc. Am. Proc.* 32:423-426 (1968).
25. Crawford, N. H., and A. S. Donigian, Jr. "Pesticide Transport and Runoff Model for Agricultural Lands," Hydrocomp, Inc., Palo Alto, CA, prepared for U.S. Environmental Protection Agency, Athens, GA, Publication No. EPA-660/2-74-013 (1973).
26. Donigian, A. S., and N. H. Crawford. "Modeling Nonpoint Pollution from the Land Surface," Hydrocomp, Inc., Palo Alto, CA, prepared for U.S. Environmental Protection Agency, Athens, GA, Publication No. EPA-600/3-76-083 (1976).
27. Donigian, A. S., and H. H. Davis, Jr. "Agricultural Runoff Management (ARM) Model User's Manual: Versions I and II," prepared for U.S. Environmental Protection Agency, Athens, GA (1977).
28. Mulkey, L. A., and J. W. Falco. "Sedimentation and Erosion Control Implications for Water Quality Management," National Symposium on Erosion and Sedimentation by Water, ASAE, Chicago, IL (December 1977).

INDEX

Adirondack Park Agency 97
agricultural chemical transport model (ACTMO) 215
Agriculture Canada 81,91,172
Agricultural Code of Practice, Ontario 163
Agricultural Conservation Program (ACP) 15,342
agricultural practices 7,117
agricultural production 649
Agricultural Runoff Management (ARM) Model 214,509,687, 688,718
Agricultural Stabilization and Conservation Service (ASCS) 13, 67,108,334,342,555
agricultural water pollution control 27
agro-forestry system 33,40,41
alachlor 220
ammonium 231
ammonium-N 251
animal unit 163
animal wastes 6,42
Applied Forestry Research Institute 17
Areawide Waste Management Plan 455
areawide waste treatment management 38
assimilative capacity 322
Association of State and Interstate Water Quality Agencies 14
Association of State Soil Conservation Agencies 14
atrazine 83,84,89,90,709

barnlot 164
barnyard runoff 35,75,180
bed load 299
Best Available Technology (BAT) 53,267
Best Available Technology Economically Achievable (BATEA) 53
Best Management Practices (BMPs) 3,11,17,18,30,33,44,46,57,62, 69,70,93,94,133,264,272,311, 321,370,383,393,419,432,465, 556,563,625,716
Best Practicable Technology (BPT) 53,267
biochemical oxygen demand (BOD) 5,95,458,515,534,546
Black Creek 74,491,499,503
 Demonstration Project 5,63
 Study 65
 Watershed 600
buffer strips 147,586
buffer zone 149,153,155

carrying capacity 287
Cayuga Lake 75
Chattahoochee River 73
chemical oxygen demand (COD) 510
chisel plowing 374,447,620
chloride 234
chlorophyll-a 609
Clean Streams Law of Pennsylvania 341
Clean Water Act of 1977 3,4,13,18, 23,76,77,272,336,384
clear-cut timber harvesting 303

coliform 3,515,520
College of Agriculture and Life
    Sciences (Cornell University)
    17
College of Environmental Science
    and Forestry at Syracuse 17
Comprehensive Coordinated Joint
    Plans (CCJP) 102
Conservation Needs Inventory (CNI)
    62,108,564
conservation practices 715
conservation tillage 124,215,365,
    375
contour
    cropping 474
    farming 124
    plowing 12,111
    stripcropping 71
contouring 402,586,725
contours 697
conventional tillage 324,437,447
Cooperative Extension Service 9,63,
    313,315,334,555,576
copper 83,85,89
Cornell University 17
Corps of Engineers, U.S. Army 468
cost-sharing 6,7,14,76,352,360
Council of State Governments 59,93
County Soil and Water Conservation
    Districts 22
Crop Budget System 387
cropland runoff manual 29
crop management 47
crop production 38,124
crop residues 42
crop rotations 348
Culver Amendment 4,23
Curve Number Runoff Equation 665
Curve Number Runoff Method 697

dairy cattle 170
dairy farm model 659
dairy manure 149
DDT 570,720
Delaware 507
delivery ratio 536
denitrification 139,248,253,256,574
depression storage 441
dieldrin 570

Disaster Loan Fund 336
diversions 74

economic incentives 16
Economics, Statistics and Coopera-
    tives Service 14
endosulfan 83,85,89,90
Environmental Protection Agency
    (EPA) 3,14,25,33,38,59,69,
    96,272,329,383,461,599
environmental quality 38,40,43,119
erosion 94,104,125,145,170,224,
    275,277,295,322,323,324,341,
    359,373,383,393,416,459,475,
    482,485,487,563
    control 59,75,349
eutrophication 73,247,503,609,613
Executive Order 11752 7

Fall Creek 671
Farmers Home Administration
    (FmHA) 14,336
farm production model 649
fecal coliform 638,646
Federal Water Pollution Control Act
    34,38,69,99,263,271
Federal Water Pollution Control Act
    Amendments (1972) 213,393,
    419,599,715
feedlots 25,122,125,164,176,190,
    197,567
    runoff 28,147
fertilizer 6,11,27,30,42,85,90,109,
    133,135,139,220,273,280,360,
    369,379,402,501,563,573
    control 615
fire control practices 121
Fish and Wildlife Service 9
flood control 410
forest ecosystem 286
forest environment 295
Forest and Rangeland Renewable
    Resources Planning Act of 1974
    (RPA) 13
forestry 563,576
Forest Service, U.S. 9,14
forest watersheds 274
Freshwater Wetlands Act 97

INDEX 737

Freundlich isotherm 196
fruit 563,578
fungicides 280

Genesee River Basin Study 94,102
grassed waterways 72,170,585,589, 725
grass seeding and sodding 12
Great Lakes 73,159,168,186,247, 451,465
Great Lakes Basin 5,74,80,159
Great Lakes Water Quality Agreement 159,175

heavy metals 34
herbicides 6,72,220,221,236,280, 369,397
Honey Creek 468,476
Hudson Level B Study 94
Hudson River Basin Study 106
Hydrocomp Simulation Program 689
hydrology 277

Illinois 373,563
Illinois Agronomy Handbook 575
Illinois Environmental Protection Agency (IEPA) 563
indicator bacteria 147
infiltration 147,298,441
insecticides 6,72,280,397
institutional approaches 419,429, 455
Integrated Federal Water Quality Management Program 4,9
Integrated Pest Management Programs 6,9,30,72
interception 298
International Joint Commission (IJC) 74,79,91,175
Iowa 60,359
irrigation 76,125,138
Irrigation Management Services 9
irrigation return flows 34,42
Ithaca, NY 671

Kjeldahl nitrogen 83,84,89

Lake Champlain Level B Study 94, 109
Lake Erie 465,491,503
Lake Erie Wastewater Management Study 468
Lake Michigan 186
Lake Superior 186
land application 145
land conservation practices 19
land grading 74
Land Resource Information System (LRIS) 474
landsat 459
Langmuir isotherm 196
livestock 38,159,161,162,163,166, 170,451
livestock wastes 175,563,566
log extraction 121

manure 75,85,89,90,145,149,159, 162,165,168,176,180,330,501, 593
  disposal areas 38
  management 111
  storage 568
Maryland 60,551
Maumee River 491,499,503
mercury 570
metals 534
methemoglobinemia 573
metribuzin 220
Michigan 455
Michigan Department of Natural Resources 455
Michigan Soil Erosion and Sedimentation Control Act 459
Midwest Research Institute 525
milkhouse 567
milking parlor 567
Mined Land Reclamation Act 97
minimum tillage 12,140,447,451, 474,613,696,718
Missouri River 95
Model Implementation Program (MIP) 8,15
Model State Act for Soil Erosion and Sediment Control 59,74
moldboard plowing 614
monitoring 34

Montana 61
Montana National Streambed and
 Land Preservation Law 63,64
mulches 74

National Association of Conservation
 Districts 14,57,59,74
National Commission on Water
 Quality 34
National Forest Management Act of
 1976 14
National Pollutant Discharge Elimina-
 tion System (NPDES) 332,569
National Rural Clean Water Coordinating Committee 8
New York State 75,581
 Conservation Council 98
 Department of Environmental
  Conservation 17,19,97,581
 Soil and Water Conservation
  Committee 581
nitrate 155,194,226,242,247,520
nitrate-nitrogen 676
nitrification 139,256,574
nitrogen 5,6,48,72,83,89,90,137,
 139,160,176,193,247,273,435,
 501,534,546,573,609,666
nonpoint calculator 525,549
Nonpoint Source(s) 50,62,93,271
 control 3,62,96
 management 17,21,23,24
 pollution 25,33,35,38,44,69,70,
  71,75,108,311,322,340,419,
  599
 problems 19,94
no-till 72,324,346,348,349,355,
 474,482,484,614,718
N-serve 139
nutrient(s) 3,34,73,81,133,134,145,
 153,176,193,273,279,429,491,
 496,509,584,591
 uptake 133

Obers projections 117
Obion-Forked Deer River Basin 12
Occupational Safety and Health Act
 (OSHA) 266
odor control 126

Ontario Ministry of Agriculture and
 Food 91
Ontario Ministry of the Environment
 (OMOE) 80,91
organic-N 251
orthophosphate 83,160
overgrazing 304
overland flow 168
 delivery ratio 165

paraquat 73,220,547,709
pasture 126
pathogens 34,429
PCB 83,85,90
Pennsylvania 61
Pennsylvania Clean Streams Program
 63
pesticide(s) 3,5,6,11,27,34,35,36,
 42,48,73,75,81,104,122,124,
 236,273,280,295,360,369,402,
 460,509,534,563,570,584,693,
 715,716
pet litter 71
phosphate 207,233
phosphorus 5,6,48,72,73,83,84,85,
 88,90,111,137,151,155,159,
 161,175,176,181,193,273,435,
 458,465,468,491,503,510,534,
 546,573,609,666,710
phytoplankton 514,520
plant residue 149
point sources 4
Pollution from Land Use Activities
 Reference Group (PLUARG)
 79,80,91,159,171
potassium 137,176,435
poultry 567
public health 35
Public Laws
 PL-566 409,411
 PL 89-80 101
 PL 92-500 25,38,44,61,69,93,
  94,95,101,127,213,271,329,
  332,336,373,393,404,428,429,
  552,563
 PL 95-192 40
 PL 95-217 34,39,44,54,76,336,
  599
Pure Waters Bond Issue 97

rainfall 151
rainfall energy 443
reduced tillage 124
Resource Conservation and Recovery Act 4
river basin plans 101
road salt 71
runoff 38,145,146,161,168,509, 665
   control 124,125
   retention 170
rural 458
Rural Clean Water Amendment 5
Rural Clean Water Coordinating Committee 6,8,14
Rural Clean Water Cost-Sharing Program 4
Rural Clean Water Program (RCWP) 5,13,26,66,76

Safe Drinking Water Act 4
salinity 34,76,107
Science and Education Administration (SEA) 14,15
Section 201 101
Section 208 3,4,17,23,26,38,44,51, 61,62,67,70,73,93,98,109, 112,127,213,264,265,269, 271,329,404,419,428,552, 563
Section 208 Continuing Planning Process 3,4,9
Section 209 101,109
Section 303 101
sediment 3,5,11,27,28,34,35,42, 73,77,83,84,94,104,134,264, 273,277,289,295,323,396, 408,411,429,449,468,479, 491,509,553,590,662
   delivery ratio 72,325,547,591
   yield 295,298
sedimentation 222,281,289
silage 150,567
silvicultural activities 271
silvicultural management practices 17,281,288
silviculture 117,267,295
Small Business Administration (SBA) 336

Snohomish River Basin 330
Snomet/King County 208 Study 329,336,625
snowmelt 145,151,161,497,504, 613,665
social attitudes 35
socioeconomic impacts 559
sociological aspects 331
soil
   and water conservation 21,140
   erosion 12,27
   loss 27,111,359,362,378,393, 396,397,402,429,451,665
   tests 135
   texture 148
Soil Conservation Districts 39
Soil Conservation Service (SCS) 6, 13,39,58,70,128,315,332,342, 374,383,399,430,459,541,555, 581,665,717
Soil Fertility Testing Programs 6
Soil and Water Conservation Committee 17,22
Soil and Water Conservation Plans 18
Soil and Water Conservation Practices (SWCPs) 51,71,72,74,145,687
Soil and Water Resources Conservation Act of 1977 13,40,384
source management 44
Stanford Watershed Model 689
State Environmental Quality Review Act (SEQR) 97
State Soil and Water Conservation Agencies 39
State Soil and Water Conservation Districts 22,52
State Water Quality Agencies 39
stream bank stabilization 74
Streamside Management Zones (SMZ) 268
stripcropping 12,111,215,345,355
Surface Mining Control and Reclamation Act of 1977 13,14
suspended load 299
suspended solids 83
swine lagoon 200

tax credit 317

temperature 278
terraces 72,349,355,402,449,585,
    589,696
terracing 12,366
Texas 393
Tidal Wetland Protection Act 97
tillage practice 324
timber practices 121
total nitrogen 85,458
total phosphorus 160,117
toxaphene 720
toxic pollutants 4
toxics 94
trace elements 81
transport capacity 146

Universal Soil Loss Equation 62,
    295,324,361,374,384,397,432,
    458,474,487,534,556,607
unit-area loadings 83
urban areas 655
urban runoff 458
U.S. Department of Agriculture
    (USDA) 11,26,39
U.S. Fish and Wildlife Service 54

Virginia 59
volatilization 138,148
voluntary approach 16

Wakarusa River 525
Washington 625
Washington State Department of
    Ecology 332
Water Conservation Law 18
water pollution 3
Water Resources Council 102
water quality 19,28,35,43,79,81,
    90,93,264,272,322,429,507,
    514,563
Water Quality Act of 1965 104
Water Quality Management Program
    23
Water Quality Parameters 83,84
wetlands 27,95
Wild, Scenic and Recreational River
    System 97
windbreaks 12
winter runoff 153

zinc 83,89,90